T0344518

TRANSPORT IN THE ATMOSPHERE-VEGETATION-SOIL CONTINUUM

Traditionally, soil science, atmospheric science, hydrology, plant science and agriculture have been studied largely as separate subjects. These systems are clearly interlinked, however, and in recent years a great deal of interdisciplinary research has been undertaken to understand the interactions better. This textbook was developed from a course that the authors have been teaching for many years on atmosphere-vegetation-soil interactions at one of the leading international research institutes in environmental science and agriculture.

Small-scale processes at the interface of soil and vegetation and in the lower atmosphere may have a profound impact on large-scale processes in the atmosphere and subsurface water. Furthermore, the interaction among soil, vegetation and atmosphere is important for the assessment and monitoring of water resources. This book describes the atmosphere-vegetation-soil continuum from the perspective of several interrelated disciplines, integrated into one textbook. The book begins with the treatment of individual terms in the energy and water balance of Earth's surface, including the role of plants and solutes. A number of these aspects are then combined in the treatment of practical methods to estimate evapotranspiration. This leads to the presentation of a number of integrated applications, showing how the theory of the preceding chapters leads to new insights. The book concludes by presenting integrated hydrological and meteorological models in which the theory of transport processes is applied. The book assumes readers have some familiarity with basic radiation laws, thermodynamics and soil science. However, much of this prerequisite knowledge is also covered briefly in appendices. The text is interspersed with many student exercises and problems, with solutions included.

This textbook is ideal for intermediate to advanced students in meteorology, hydrology, soil science, environmental sciences and biology who are studying the atmosphere-vegetation-soil continuum, as well as researchers and professionals interested in the observation and modelling of atmosphere-vegetation-soil interactions.

ARNOLD F. MOENE is an assistant professor in the Meteorology and Air Quality Group of Wageningen University, the Netherlands. The overarching theme of his research is atmospheric turbulence in relation to Earth's surface. Dr Moene teaches a number of undergraduate and graduate courses related to atmosphere-vegetation-soil interactions (theory and observations) and fluid mechanics. He plays an active role in the organization and development of education in the BSc and MSc programmes on Soil, Water and Atmosphere at Wageningen University. He is a member of the editorial board of the journal *Boundary-Layer Meteorology*, and has authored or co-authored more than 35 peer-reviewed international scientific publications.

JOS C. VAN DAM is an associate professor in the Soil Physics and Land Management Group of Wageningen University, the Netherlands. He is responsible for research and education in the transport of water, solutes, heat and gases in topsoils at the undergraduate and graduate levels. A main focus of his work is physical transport processes and their interaction with vegetation development and micro-meteorology. Dr van Dam is one of the main developers of the widely used ecohydrological model SWAP (Soil Water Atmosphere Plant). He is author or co-author of more than 60 peer-reviewed international scientific publications.

TRANSPORT IN THE ATMOSPHERE-VEGETATION-SOIL CONTINUUM

ARNOLD F. MOENE
Meteorology and Air Quality Group, Wageningen University

JOS C. VAN DAM
Soil Physics and Land Management Group, Wageningen University

CAMBRIDGE
UNIVERSITY PRESS

University Printing House, Cambridge CB2 8BS, United Kingdom

One Liberty Plaza, 20th Floor, New York, NY 10006, USA

477 Williamstown Road, Port Melbourne, VIC 3207, Australia

314-321, 3rd Floor, Plot 3, Splendor Forum, Jasola District Centre, New Delhi - 110025, India

79 Anson Road, #06-04/06, Singapore 079906

Cambridge University Press is part of the University of Cambridge.

It furthers the University's mission by disseminating knowledge in the pursuit of education, learning and research at the highest international levels of excellence.

www.cambridge.org
Information on this title: www.cambridge.org/9780521195683

First published 2014

A catalogue record for this publication is available from the British Library

Library of Congress Cataloging in Publication data
Moene, Arnold F.
Transport in the atmosphere-vegetation-soil continuum / Arnold F. Moene, Wageningen University, Jos. C. van Dam.
pages cm
Includes bibliographical references and index.
ISBN 978-0-521-19568-3 (hardback)
1. Ecohydrology. 2. Micrometeorology. 3. Soil physics. 4. Plant physiology.
I. Dam, J. C. van. II. Title.
QH541.15.E19M64 2013
577.6–dc23 2013027359

ISBN 978-0-521-19568-3 Hardback

Contents

Preface

This book has its roots in courses on Micrometeorology by Henk de Bruin and courses on Soil Physics and Agrohydrology by Reinder Feddes and colleagues at Wageningen University and Research Centre. Most universities teach these subjects in separate courses. In 2007, during a BSc-education reprogramming round at Wageningen University, micrometeorology, soil physics and agrohydrology were brought together in the current course 'Atmosphere-Vegetation-Soil Interactions'. As teachers we had our reservations, but it turned out to work very well.

The interface between atmosphere and land is the location where both domains exchange energy, water and carbon. On the one hand, processes in soil and vegetation influence the development in the atmosphere (e.g., cloud formation). On the other hand, the atmospheric conditions determine to a large extent what happens below the soil surface (e.g., through the extraction of water for transpiration). Many environmental challenges, whether they concern climate change in drought-prone areas, salinization of coastal regions, development and spread of plant pathogens, natural vegetation impoverishment due to deep drainage or low water use efficiency in irrigated agriculture, have their origin in close interactions between atmosphere and land. To understand these processes and solve practical problems, students and professionals should have operational knowledge of transport processes in both domains and be able to understand how the atmosphere affects the land and vice versa.

This book intends to provide a consistent overview of the processes that occur in the continuum that extends from a few metres below the soil surface to roughly a hundred metres above it. It has been a challenge to connect the various disciplines that are active within this continuum: soil physics, ecohydrology, plant physiology and micrometeorology. The result is a unique text that covers all these disciplines on a scientific level that gives students a good preparation for continued education and thesis research. The ample use of up-to-date references to literature provides the student with starting points for further study. Questions and problems are interspersed with the text and answers to all questions are provided.

We gratefully acknowledge the contributions (direct or through inspiration) made by Henk de Bruin, Reinder Feddes and colleagues to the original lecture notes. Furthermore, we thank all the people who were involved in the collection of data that are used as illustrations in the text. Finally, we are grateful to Joel Schröter, Miranda Braam, Bert Holtslag and Reinder Ronda for their numerous comments on the text.

1
The Atmosphere-Vegetation-Soil System

1.1 Introduction

Whereas roughly 70% of Earth's surface is covered by oceans, the remaining 30% of land has a profound influence on processes in the atmosphere (e.g., differential heating, drag, evaporation and resulting cloud formation, composition of the atmosphere). This impact is due to the large variability in the properties (e.g., albedo, roughness, soil type, land cover type, vegetation cover) and states (e.g., soil moisture availability, snow cover) of the land surface. The processes occurring at the land surface are often grouped under the terms biogeophysical and biogeochemical processes (Levis, 2010): they influence the state and composition of the atmosphere both through physical and chemical processes, and biological processes play an important role in both.

Although the interface between Earth and atmosphere is located at the surface, subsurface processes in the soil are of major importance because part of the energy and water exchanged at the surface is extracted from or stored in the soil. Plants play an important role in extracting water from deeper soil layers and providing it to the atmosphere. In return, processes in the soil and plants (e.g., transport of water, solutes, and energy) are strongly influenced by atmospheric processes (e.g., evaporation and precipitation).

The interface between Earth and atmosphere is part of the continuum that ranges from the substrate underlying soils to the top of the atmosphere. The overarching subject of this book is the transport of energy, matter (water, solutes, CO_2), and momentum in the atmosphere-vegetation-soil continuum. In some cases this transport occurs within one of the compartments (e.g., redistribution of solutes in the soil); in other cases exchange over the interface between different compartments takes place (e.g., transpiration of water by plants).

In the context of this book we limit the extent of our subject both in the vertical direction (see Figure 1.1) and in the time scales considered. The lower boundary of the domain is located at that level in the soil where the *yearly* variation in temperature

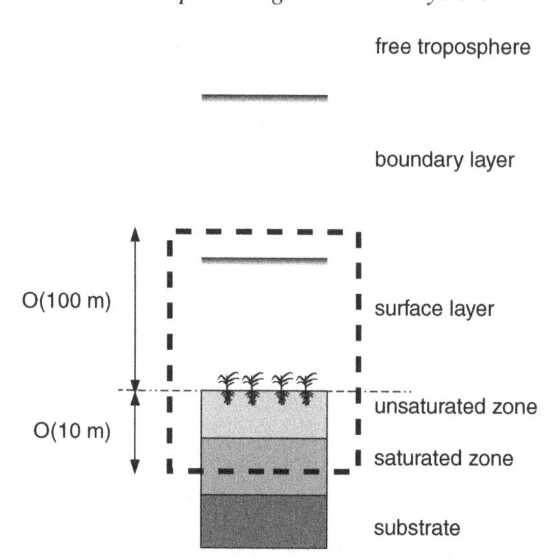

Figure 1.1 Various layers in the soil-vegetation-atmosphere continuum. The rectangular box indicates the vertical extent of the domain covered in this book. The vertical coordinate is roughly logarithmic.

and soil moisture has disappeared, whereas the top is located at the top of the surface layer, which occupies roughly the lower 10% of the atmospheric boundary layer (ABL, the layer where the *diurnal* cycle of surface heating affects the flow). The dynamics of the ABL itself are not part of this book, but are dealt with elsewhere (e.g., Stull, 1988; Vilà-Guerau de Arellano and van Heerwaarden, forthcoming).

The time scales considered range roughly from the diurnal cycle to the yearly cycle, although for the atmosphere turbulent fluctuations on time scales of less than a second are dealt with as well. As an example of variations on time scales from days to a year, Figure 1.2 shows daily rainfall fluxes and simulated daily evapotranspiration and drainage fluxes of a grass on a sandy soil in a Dutch climate and with a deep groundwater table. Although the rainfall fluxes are well distributed over the year, the evapotranspiration and drainage fluxes have a clear yearly pattern. Potential evapotranspiration follows the pattern of solar radiation. In the summer season with a high atmospheric demand, dry periods may cause a large difference between potential and actual transpiration. Drainage occurs mainly in winter periods with a low atmospheric demand and soil at field capacity. Compared to rainfall fluxes, drainage fluxes are much smaller and show a more gradual pattern.

Although the limitation of discussed time scales to a year is arbitrary, it suits the discussion of most processes well. That is not to say that no interesting and relevant processes occur at longer time scales, such as the interannual memory in vegetation cover (e.g., Philippon et al., 2007).

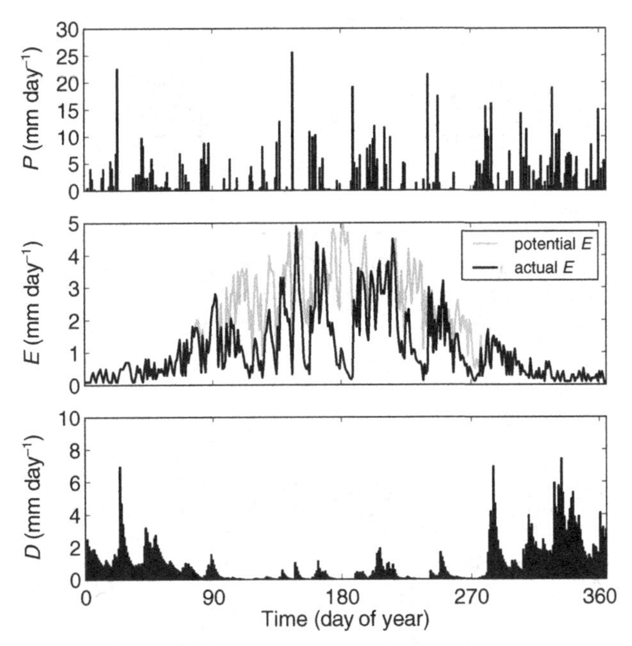

Figure 1.2 Daily rainfall fluxes (top) and simulated daily evapotranspiration (middle) and drainage fluxes (bottom) of grass on a sandy soil in a Dutch climate and with a deep groundwater table. Potential evapotranspiration in the middle panel is the evapotranspiration that would occur if sufficient water would have been available to the plants.

1.2 Conservation of Energy and Mass

To study the processes at Earth's surface in a quantitative way we first need to define that interface clearly. This is done with the help of a control volume that contains the interface. Any difference between the inflow and outflow of a quantity will result in a change in the storage in the control volume. Formally, the conservation equation for an arbitrary quantity can be stated as:

$$\sum F_{\text{in}} - \sum F_{\text{out}} = \Delta S \qquad (1.1)$$

where ΣF_{in} and ΣF_{out} are the summation of all fluxes into and out of the control volume, respectively, and ΔS is the change in storage (see Figure 1.3). To capture all important processes we start with a control volume that contains (part of) the soil column, as well as the part of the atmosphere into which the vegetation protrudes. In the horizontal direction the control volume has an arbitrary size, but horizontal homogeneity is assumed.

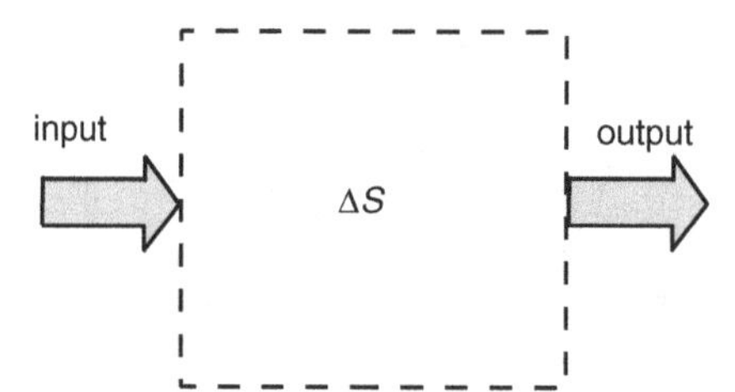

Figure 1.3 Concept of a control volume. Change in storage is due to an imbalance between input and output. Note that inputs and outputs can occur at any face of the control volume, not just the sides.

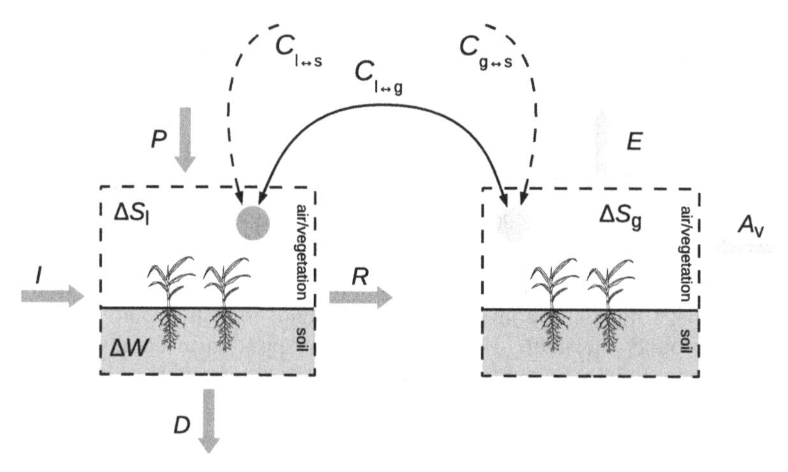

Figure 1.4 Control volume for water: liquid water (left) and water vapour (right). The direction of the arrows holds for typical daytime conditions. Dark grey arrows denote transport of liquid water, whereas light grey arrows are used for transport of water vapour. Arrows between the boxes signify phase changes (e.g., $C_{l\rightarrow g}$ is evaporation): molecules of water do not leave the control volume but only move from one phase to another. The dashed arrows with $C_{l\leftrightarrow s}$ and $C_{g\leftrightarrow s}$ have their other ends located in the control volume for solid water, not drawn here. Other symbols are explained in the text.

1.2.1 Water Balance

Figure 1.4 shows the water balance of the control volume.[1] Because under typical terrestrial conditions water occurs in all three phases (gas, liquid and solid), a distinction has been made between water in the liquid phase and water in the gas phase (ice and snow have been discarded for simplicity, but could be accounted for analogously;

[1] Note that it is implicitly assumed that the control volume has a horizontal extent of one square meter. As a result all fluxes and storage terms, as used in Eqs. (1.2) and (1.3) should be interpreted as fluxes per 1 m^{-2} of ground area. This assumption is not a physical necessity, but simplifies the transition to the other chapters where usually fluxes are interpreted as flux densities (i.e., per unit area).

see Question 1.1). As water can change phase within the volume, all phases have to be accounted for in the total water balance of the control volume. The water balance then reads:

$$P + I - R + A_v - E - D = \Delta W + \Delta S_l + \Delta S_g \quad (1.2)$$

where we distinguish the following transports across the boundaries of the control volume: P is precipitation, I is irrigation (artificial supply of water), R is runoff, D is the drainage rate towards deeper soil layers, A_v is advection of water vapour (which can be positive or negative) and E is the water that leaves the system in gas phase (water vapour; see Section 1.2.3). The different inputs and outputs do not necessarily balance, so that water may be stored in the soil (ΔW, change in soil moisture content), on the soil or on the vegetation in the liquid phase (ΔS_l, e.g., intercepted rain or dew) and in the air (ΔS_g) in the gas phase.

Water molecules can change phase. In Figure 1.4 this is indicated with the phase change terms $C_{l\leftrightarrow g}$ (between liquid and vapour), $C_{l\leftrightarrow s}$ (between liquid and solid) and $C_{g\leftrightarrow s}$ (between vapour and solid). These phase change terms, however, do not occur in Eq. (1.2) because they do not change the number of molecules of water within the control volume but only the phase in which they occur. On the other hand, as energy is released or used when water changes phase, the phase change terms do affect the energy balance, as we see later.

Question 1.1: Figure 1.4 shows the mass balances for water of a control volume extending from the soil into the atmosphere, above the vegetation. Only liquid and gaseous water is dealt with.

a. Sketch the mass balance for solid water.
b. Enumerate all interactions between the mass balances for each of the phases (i.e., which phase changes can occur that exchange water in one phase for another?).

1.2.2 Energy Balance

Whereas in the water balance we have to distinguish between the three phases of water, in the energy balance we distinguish between two forms of energy: sensible heat and latent heat. Sensible heat is the energy contained in a substance that can be extracted by cooling it. On the other hand, latent heat can be extracted only by a phase change. It could be considered as similar to potential energy: a ball on a hill has potential energy that could be extracted when it rolls down the hill. In a similar way, water vapour contains latent heat that would be freed if it condenses to liquid water. In the current context it is customary (but not necessary) to use liquid water as the reference (i.e., ice contains negative latent heat because it would require *input* of heat to bring it to liquid water).

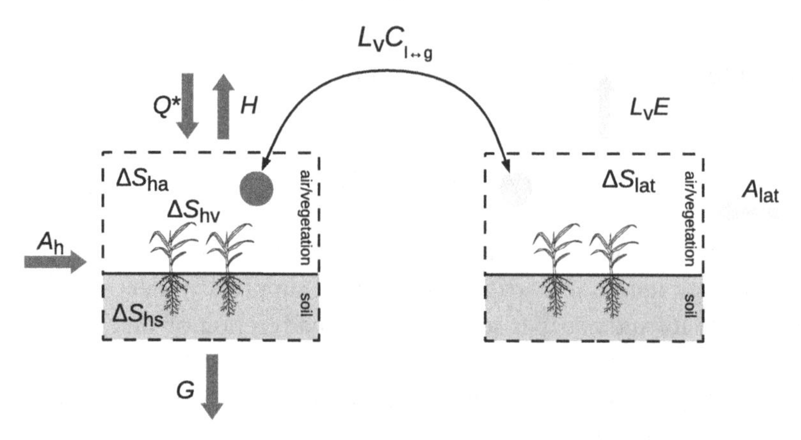

Figure 1.5 Control volume for energy: sensible heat (left, dark grey arrows) and latent heat (right, light grey arrows). Energy is exchanged between sensible and latent heat if water changes phase (here only phase changes between liquid and gas phase are considered: $L_v C_{l\leftrightarrow g}$). Storage of energy in the form of chemical energy in the plants (due to assimilation) as well as some other terms (see text) have been discarded. The direction of the arrows holds for typical daytime conditions.

The control volumes for sensible and latent heat are depicted in Figure 1.5. The complete energy balance equation corresponding to this figure is as follows:

$$Q^* - H - L_v E - G + A_h + A_{lat} = \Delta S_{ha} + \Delta S_{hv} + \Delta S_{hs} + \Delta S_{lat} \tag{1.3}$$

where Q^* is the net radiation (see Chapter 2), H is the sensible heat flux, G is the soil heat flux at the bottom of the control volume and $L_v E$ is the latent heat flux (where L_v is the latent heat of vaporization). If the inputs and outputs do not balance, heat can be stored in the air (ΔS_{ha}), in the vegetation (ΔS_{hv}, i.e., the vegetation becomes warmer) and in the soil (ΔS_{hs}). Finally, A_h and A_{lat} are the net advections of sensible and latent heat.

The latent heat flux plays a special role in the energy balance in the sense that it transports energy through the transport of water vapour. The actual energy consumption related to *evaporation* is contained in the term $L_v C_{l\leftrightarrow g}$, but because this term is internal to the control volume, it does not appear in the energy balance. The energy related to evaporation leaves the control volume as latent heat. In the case that there is a change in the water vapour content in the control volume (i.e., $\Delta S_{lat} \neq 0$) or non-zero advection of water vapour, the transport of latent heat *out of* the control volume ($L_v E$) may be unequal to the energy related to the phase change *within* the volume. Note that the opposite phase change can happen as well: when dew is formed water is transformed from the gas phase to the liquid phase and energy is released.

Some terms have not been accounted for in Eq. (1.3). Part of the solar radiation that hits the vegetation will be used for photosynthesis. This leads to a conversion of radiative energy to the storage of chemical energy in the plant material. This storage term has been discarded (it is of the order of 10–25 W m^{-2} at midday, depending on insolation and the type of vegetation; see Meyers and Hollinger, 2004). Another term in the energy balance, seldom taken into account, is the sensible heat related to the input of precipitation into the control volume. If, for instance, during daytime conditions the temperature of the rain is lower than that of the air that it replaces in the control volume, a net exchange of energy out of the volume occurs. Further, if the cold rainwater percolates into the soil, a significant redistribution of energy within the control volume can occur (see Kollet et al., 2009).

1.2.3 The Link: Evapotranspiration

From Eqs. (1.2) and (1.3) the link between the water balance and energy balance becomes clear: the evaporation appears as a transport of mass in the water balance and as a transport of energy in the energy balance. The total water vapour flux that leaves the system is made up of a number of fluxes *within* the system: soil *evaporation* (E_{soil}), *transpiration* by the plants (T), and *evaporation* of intercepted water (E_{int}). Both transpiration and soil evaporation extract water from the soil subsystem and release it in the air subsystem, whereas in the case of interception the soil is bypassed. The sum of these three terms is called *evapotranspiration* and denoted by E (in the hydrological literature a commonly used symbol is ET).

Evaporation and transpiration occur simultaneously and there is no easy way of distinguishing between the two processes. Apart from water availability in the top soil, the evaporation from a cropped (or more general: vegetated) soil is determined mainly by the fraction of solar radiation reaching the soil surface. This fraction decreases over the growing period as the crop develops and the crop canopy shades more and more of the ground area. When the crop is small, water is lost predominantly by soil evaporation, but once the crop is well developed and completely covers the soil, transpiration becomes the main process. In Figure 1.6 the partitioning of evapotranspiration into soil evaporation and transpiration is plotted in correspondence to leaf area per unit soil surface below it (LAI). At sowing, nearly 100% of E comes from soil evaporation, whereas at full crop cover more than 90% of E comes from transpiration (Allen et al., 1998).

Apart from the direct link between water balance and energy balance through the occurrence of evapotranspiration in both balances, there is also a more indirect link. The two balance equations exactly stand for the two requirements needed for evapotranspiration: water should be available to be evaporated, and energy (through radiation) should be available to actually let the evaporation happen (see also Chapter 7). The availability of water in the soil in turn will largely be determined by the amount of rainfall. These two limiting factors – radiation and rainfall – translate into different behaviour of evapotranspiration in different regions of the world: in regions of abundant rainfall (relative to the evaporative loss) evapotranspiration correlates highly

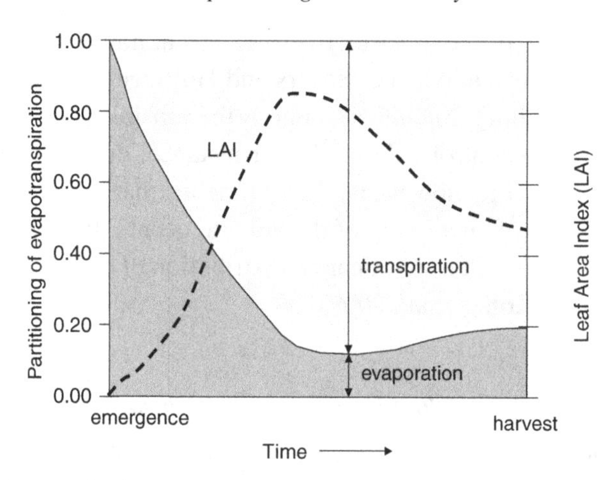

Figure 1.6 The partitioning of evapotranspiration into soil evaporation and transpiration over the growing period of an annual field crop.

with radiation, whereas in more arid regions radiation is not limiting, and evapotranspiration correlates with rainfall (Teuling et al., 2009). Although these correlations sketch the main picture on a seasonal or longer time scale, many regions may show different behaviour on shorter time scales. For example, mid-latitude regions that show a strong correlation between radiation and evapotranspiration may show a stronger dependence on soil moisture and rainfall after a prolonged drought (Teuling et al., 2010 and Chapter 8). On the other hand, semi-arid regions in which the seasonal pattern of evapotranspiration follows the seasonal pattern of rainfall may show a clear correlation of evapotranspiration with radiation *within* the rainy season (e.g., Schüttemeyer et al., 2007).

Question 1.2: Evapotranspiration is a combination of various fluxes: evaporation from intercepted water, soil evaporation and transpiration.

a. What is roughly the impact of each of the terms on the soil moisture content at various depths?
b. At which location is the energy, needed for the phase change from liquid to water vapour, absorbed for each of the fluxes?

1.2.4 Simplified Balances

In many applications the control volumes presented earlier are vertically compressed to become a control surface (see Figure 1.7). Because a surface has no volume, storage terms will disappear. Besides, horizontal advection will vanish as well. Whereas in Eqs. (1.2) and (1.3) all fluxes occurred at the boundaries of the control volume (i.e., at a certain height above the ground or at a certain depth below the surface) now

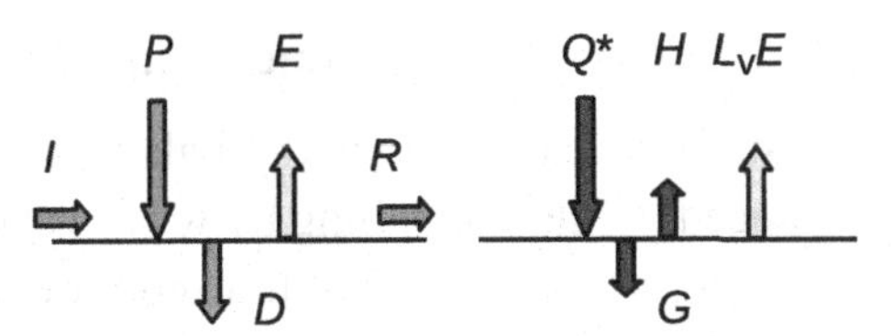

Figure 1.7 Simplified surface water balance (left) and energy balance (right). Very light grey arrows indicate transport of water vapour, light grey arrows show liquid water transport and dark grey arrows indicate energy transport.

all fluxes are supposed to occur *at* the surface: they are called *surface fluxes*. With this redefinition of the fluxes, the surface water balance and surface energy balance become:

$$P + I - R - E - D = 0 \tag{1.4}$$

$$Q^* - H - L_v E - G = 0 \tag{1.5}$$

Although these balance equations are appealing in their simplicity, the compression of the control volume to a surface may lead to problems in the interpretation of observed fluxes. In practice most fluxes will be observed at some height above a canopy, or at some depth below the surface (rather than at the hypothetical surface for which they are supposed to be representative). The omission of storage and advection terms in Eqs. (1.4) and (1.5) may cause a non-closure of the observed water balance or energy balance (i.e., the terms do not add up to zero). To solve this problem one should revert to the full Eqs. (1.2) and (1.3).

Note that the sign convention in Eqs. (1.4) and (1.5) is such that P, I and Q^* are considered as inputs, taken positive when directed towards the surface, whereas the other terms are considered as outputs, being positive when directed away from the surface. This sign convention is often used but arbitrary and other choices are used as well.

A final remark relates to the units of the transports in Eqs. (1.4) and (1.5). Rather than delineating a real surface with a given extent, those equations are usually applied to a unit surface area (e.g., one square meter). This implies that the units of the transport terms in the water balance are either volume per unit area per unit time (volume flux density) or mass per unit area per unit time (mass flux density). The terms in the energy balance have units of energy per unit area per unit time (energy flux density). Often, the terminology is used loosely: the term "density" is dropped and the word "flux" is used where a "flux density" is meant.

To summarize, the various terms in the surface water balance and energy balance are the subject of this book. But the processes are not studied only in isolation, but the interactions between them are at least as important.

1.3 Modes of Transport of Energy and Mass

In the presentation of the surface water and energy balances a number of transport processes have been introduced, without detailing by what means the transport takes place. For the transport of energy, three modes of transport are possible:

- Radiation (transport by propagation of electromagnetic radiation; no matter is needed)
- Conduction (transport of energy through matter, by molecular interactions; matter is needed, but the matter does not move [macroscopically])
- Advection[2] (transport of energy by the movement of energy-containing matter)

In the context of the surface energy balance, all three modes occur: net radiation is radiative transport, the soil heat flux is based on conduction, and the sensible flux is an example of advection (in the sense that turbulent transport involves the motion of energy-containing air).

For the transport of matter (e.g., water or solutes) the two modes of transport are

- Molecular diffusion
- Advection

In the present context, mainly advection is important, but it can have a number of different manifestations. In soil, water flows more or less smoothly, whereas in the atmospheric surface layer transport of matter (e.g., water vapour or CO_2) takes place by turbulence, where air containing the given constituent is moved from one place to another (see Chapter 3). Molecular diffusion plays a role in solute transport in the soil, and in the atmosphere in thin layers adjacent to surfaces (e.g., leaves).

One of the main objectives of this book is to quantify the various fluxes in water balance and energy balance. A generally used method to describe the flux of a quantity is based on an analogy with transport by diffusion on the molecular scale, that is, Fick's law for diffusion of matter and Fourier's law for heat diffusion:

$$F_a = -k_a \frac{\partial C_a}{\partial x} \tag{1.6}$$

where F_a is the flux density of quantity a in the x-direction, C_a is the concentration of a and k_a is the molecular diffusion coefficient for quantity a (k_a has units m^2 s^{-1}). Essential in the case of molecular transport is that k_a is known and depends only on the fluid or solid under consideration and on the state of that fluid or solid (temperature, pressure). For many combinations of transported quantities (e.g., momentum, heat, water vapour) and fluids (e.g., air or water), k_a is known and tabulated.

[2] Here the word 'advection' is used in the sense of large-scale (relative to the molecular scale) motion of matter. Sometimes the term convection is used for this as well, but this may cause confusion with thermal convection, and in some applications convection is the sum of advection and diffusion.

It is very tempting to extend Eq. (1.6) to cases other than molecular transport. In the context of the atmosphere-vegetation-soil system one could think, for example, of the transport of water through the soil and the turbulent transport of heat, water vapour and momentum in the atmospheric surface layer. The shape of such a transport description would be identical to that of Eq. (1.6). But the specification of the diffusion coefficient k is much less straightforward. Coming back to the examples: for water transport in the soil, the diffusivity depends on the porosity of the soil, the shape of the pores and on the water content. In the case of turbulent transport in the atmosphere, the diffusivity depends on properties of the flow, such as the intensity of the turbulence, stratification and the distance to the ground.

Question 1.3. Given Fourier's law for (vertical) heat transport: $H = -\rho c_p \kappa \frac{\partial T}{\partial z}$ where κ is the thermal diffusivity for heat of air (about $2 \cdot 10^{-5}$ m^2 s^{-1}), ρ is the density (about 1.2 kg m^{-3}) and c_p is the specific heat at constant pressure (for dry air, 1004 J kg^{-1} K^{-1}).

a. What are the units of the sensible heat flux H? Check that these are consistent with the units of the right-hand side.
b. What vertical temperature gradient is needed to generate a sensible heat flux of 100 W m^{-2}?

1.4 Setup of the Book

The first five chapters are roughly divided along the lines of the part of the atmosphere-soil-vegetation continuum they discuss. First the two compartments on either side of the atmosphere-soil interface are discussed, followed by the plants that extend in both the soil and the atmosphere:

Atmosphere

- The first part of Chapter 2 covers the interaction of radiation with the atmosphere and Earth's surface, leading to the radiative input to the energy balance as net radiation.
- Chapter 3 deals with the turbulent transport of heat, water vapour and momentum in the atmospheric surface layer. The final aim of that chapter is the description of surface fluxes (in particular of sensible and latent heat) in terms of mean quantities such as vertical gradients or vertical differences of mean temperature and wind speed.

Soil

- The second part of Chapter 2 deals with the transport of heat in the soil (it has been combined with net radiation because net radiation minus soil heat flux provides the energy available for sensible and latent heat flux).
- Chapter 4 presents the basic concepts of water flow in the unsaturated part of the soil: the vadose zone. To address the general flow equation, measurements of soil water pressure head, water content, hydraulic conductivity and root water uptake are treated. Specific attention is paid to infiltration, runoff and capillary rise.

- Chapter 5 focuses on main mechanisms that govern the transport of solutes in soil: convection, diffusion, dispersion, adsorption and decomposition. Salinization of root zones and pesticide leaching to ground- and surface water receive special attention.

Vegetation

- Chapter 6 is devoted to the effect of plants on the transport processes. The transport of water inside the plants, on the one hand, is treated, as this is an important pathway for water from the soil into the atmosphere. The interaction between plants and the atmosphere (rain interception, radiation, wind, heat), on the other hand, is dealt with as well.

Integration

The last three chapters cover various combinations of the material discussed in Chapters 2 to 6.

- Chapter 7 combines the concepts presented in Chapters 2, 3 and 6 (the energy balance, turbulent transport and the effects of plants on evaporation) in a number of ways to develop measurement and modelling techniques for the turbulent fluxes of heat, water vapour and momentum.
- Chapters 8 and 9 take the concepts developed so far to the level of applications.
  - Chapter 8 illustrates how the concepts from the previous examples are combined to study practical situations.
  - Chapter 9 focuses on the way energy and water balances are treated in integrated meteorological and hydrological models.

2

Available Energy: Net Radiation and Soil Heat Flux

2.1 Introduction

In Chapter 1 three different modes of energy transport were identified: radiation, conduction and convection. In the context of the surface energy balance, fluxes of sensible and latent heat are often taken together, as both are transported by turbulent motion. Then net radiation and soil heat flux are summed under the name of *available energy*, in the sense that Q^*-G is the amount of energy available for sensible and latent heat flux. This can be expressed by a rewritten version of the energy balance equation:

$$Q^* - G = H + L_v E \tag{2.1}$$

Once the available energy has been determined, the main question is how this energy is partitioned between H and L_vE.

This chapter first deals with the net radiation flux. Second, the soil heat flux is considered. This includes a brief discussion on surfaces covered by snow, ice and water, as energy transport there takes place by conduction as well. Basic radiation laws are given in Appendix A for reference.

2.2 Net Radiation

Radiative fluxes as they are relevant in the study of the surface energy balance can be split on the basis of their origin (and hence wavelength):

- Shortwave radiation is radiation originating from the Sun (either direct, or after interaction with the atmosphere). It covers the wavelength range of about 0.15–3 μm, corresponding to a surface temperature of the Sun of about 5800 K (the link between the black-body temperature and the peak wavelength of the Planck curve is given by Wien's displacement law; see Figure 2.2a). The spectrum of solar radiation can be subdivided into near-ultraviolet radiation (0.2–0.4 μm), visible light (0.4–0.7 μm), and near-infrared

(0.7–3 μm). The visible wavelength range is also the part of the spectrum that is used by plants for photosynthesis (photosynthetically active radiation [PAR]).

- Longwave radiation originates from either Earth's surface or the atmosphere (with or without clouds). It is characterized by wavelengths in the range 3–100 μm, which corresponds to a black-body temperature of the order of 290 K. This range is also indicated as thermal infrared radiation.

Net radiation is the sum of net shortwave radiation (K^*) and net longwave radiation (L^*):

$$\begin{aligned} Q^* &= K^* + L^* \\ &= K^{\downarrow} - K^{\uparrow} + L^{\downarrow} - L^{\uparrow} \end{aligned} \tag{2.2}$$

where $K^{\downarrow}$ and $K^{\uparrow}$ are the downwelling and upwelling shortwave radiation at the surface, whereas $L^{\downarrow}$ and $L^{\uparrow}$ are the respective longwave radiation fluxes. Another term for $K^{\downarrow}$ is global radiation.

Global radiation originates from the Sun and is modified by the atmosphere (see Section 2.2.2). The upwelling shortwave flux is the reflected portion of $K^{\downarrow}$ and hence depends on the magnitude and characteristics of the global radiation and on the properties of Earth's surface (see Section 2.2.3). Downwelling longwave radiation originates from the atmosphere, both from the clear sky and from clouds (see Section 2.2.4). Upwelling longwave radiation is mainly emitted by Earth's surface (see Section 2.2.5). Finally, Section 2.2.6 summarizes the sum of all terms: the net radiation.

Question 2.1: The distance from the centre of the Sun to Earth is $149.6 \cdot 10^6$ km (equal to 1 astronomical unit [AU]), and the diameter of the Sun is $1.34 \cdot 10^6$ km (Zelik et al., 1992).

a) Verify that the value of the solar constant (see Appendix A) indeed corresponds to a surface temperature of the Sun of roughly 5800 K.

b) Verify that the black-body radiation flux densities at 293 K and 5800 K differ by nearly 7 orders of magnitude.

2.2.1 Interaction between Radiation and the Atmosphere

On its way through the atmosphere, radiation is affected in two ways: by scattering and by absorption.

Scattering

If electromagnetic radiation hits a particle, part of the energy will be scattered in all directions (but not necessarily equally in all directions). This scattered radiation is called diffuse radiation. The case in which scattered light hits another particle and is scattered again is called multiple scattering. The type of scattering depends on the relationship between particle diameter, d_p, and the wavelength of the radiation, λ (see also Figure 2.1):

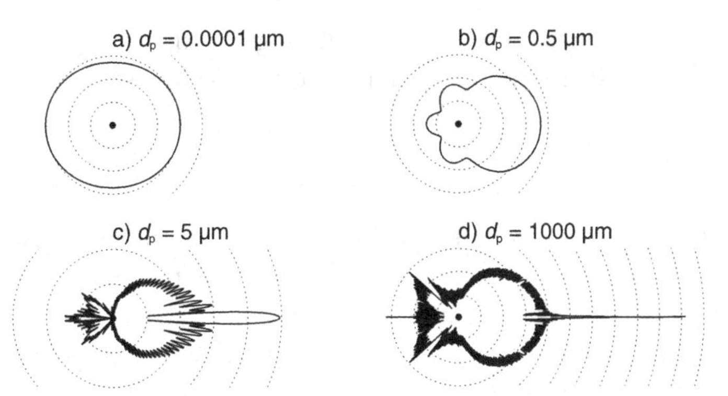

Figure 2.1 Pattern of radiation scattered by spherical particles of various sizes. The logarithm of the normalized intensity is given (with a minimum of 10^{-4} and 10^{-8} for parts **a–c**, and **d**, respectively). Wavelength of the radiation is 0.5 μm, refractive index of the particle is 1.33 and the particle is nonabsorbing. Dashed lines are lines of equal intensity (lines are one order of magnitude apart). Diameters are representative of (**a**) gas molecule (Rayleigh scattering), (**b**) large aerosol (Mie scattering), (**c**) cloud droplets (Mie scattering), and (**d**) cloud droplet (geometric optics).

- $d_p << \lambda$: Rayleigh scattering. The theory of Rayleigh scattering assumes that the particles are spherical, do not influence each other and are smaller than 0.2λ. For the wavelengths considered here, this size requirement implies that Rayleigh scattering is relevant for gas molecules. In the case of Rayleigh scattering, the amount of scattering is equal for the forward and the backward directions (see Figure 2.1a). The extinction for Rayleigh scattering is proportional to λ^{-4}, that is, the shorter the wavelength, the more scattering. One of the consequences of this wavelength dependence is that the sky is blue. Blue light has a short wavelength and hence is scattered more than visible radiation with longer wavelengths. Owing to the direction independence of Rayleigh scattering, the blue light appears to come from the entire hemisphere.
- $d_p \approx \lambda$: Mie scattering. Mie scattering occurs due to interaction of radiation with aerosols. Aerosols are suspensions of small solid and/or liquid particles (e.g., from anthropogenic sources such as industry, biomass burning, or from natural sources as cloud particles, sea spray, desert dust). Although aerosols occur with a large range of diameters, only those with diameters of the order of 0.1–100 μm give rise to Mie scattering in the short wave and long-wave wavelength range. The amount of scattering is only a weak function of wavelength, but the direction in which light is scattered is complex, with various side lobes. Most radiation is scattered in the forward direction, and the proportion of forward scattering increases with increasing particle size. Examples of the angular dependence of scattering by an aerosol and a cloud droplet are given in Figures 2.1b and c.
- $d_p >> \lambda$: Geometric optics. Raindrops, ice crystals, snowflakes and hailstones have a size that is much larger than visible (and infrared) wavelengths. Then scattering is determined by classical optics, such as Snell's law. The scattered (or better, refracted) radiation can have a strong directional dependence. Figure 2.1d shows the scattering pattern for a raindrop. Strong scattering is present in the forward and backward direction. Four distinct peaks (two on each side) are present in the backward direction. Those correspond to the primary and secondary rainbow (at 137° and 130°) respectively (Petty, 2004).

In addition to the relative particle size, scattering also depends on the absorption of radiation by particles: absorption strongly modifies the Mie scattering process (Petty, 2004).

Absorption

Absorption by gases is a spectrally very selective process, in contrast to the scattering of radiation by particles, which is a rather continuous function of wavelength. The most active absorbing gas in the atmosphere is water vapour, which absorbs both in the shortwave and in the longwave part of the spectrum (see Figure 2.2b). Ozone is active at the ultraviolet (UV) side of shortwave radiation. Oxygen plays a minor role in the far-infrared region and in the medium UV part of the spectrum. Finally, methane, nitrous oxide and carbon dioxide play an important role as greenhouse gases (next to water vapour) at near-infrared wavelengths.

The presence of an absorption lines at a given wavelength for a given molecule is a reflection of the fact that the photon with the given wavelength has an energy that corresponds to a transition in the internal energy of that single molecule. But because the interactions between molecules (through collision) influence the energy state of molecules, the exact location, strength and width of the absorption lines also depend on local pressure and temperature (Petty, 2004). Hence, the extinction will be height-dependent for two reasons: the concentration of an absorber may vary with height, and the absorption lines differ with height.

Apart from absorption by gases, radiation may also be absorbed by particles. This absorption depends both on the material of the particle and the wavelength. In the case of liquid water – relevant for the radiative properties of clouds and fog – the absorptivity is low for the wavelength region of visible light, whereas it is high throughout the thermal infrared (Hale and Querry, 1973). The ultimate effect of absorption by particles also depends on the size of the particle and the wavelength of the radiation, as absorption modifies Mie scattering.

2.2.2 Downwelling Shortwave Radiation

For a given location, the downwelling shortwave radiation varies on two predictable time scales: the yearly cycle and the diurnal cycle. This temporal variation in solar radiation is the main driving force in the temporal variation of all terms in the surface energy balance at a given location. On a larger scale the location dependence of the insolation, in particular the variation with latitude, is one of the drivers for the general circulation in the atmosphere. The predictable part of this latitudinal, yearly and diurnal variation of the solar radiation is outside of the scope of this book. The equations describing the variation of solar radiation at the top of the atmosphere can be found in Appendix A.

The final result of all factors that influence the amount of solar radiation at the top of the atmosphere ($K_0^{\downarrow}$), at a given moment and at a given location, can be summarized as:

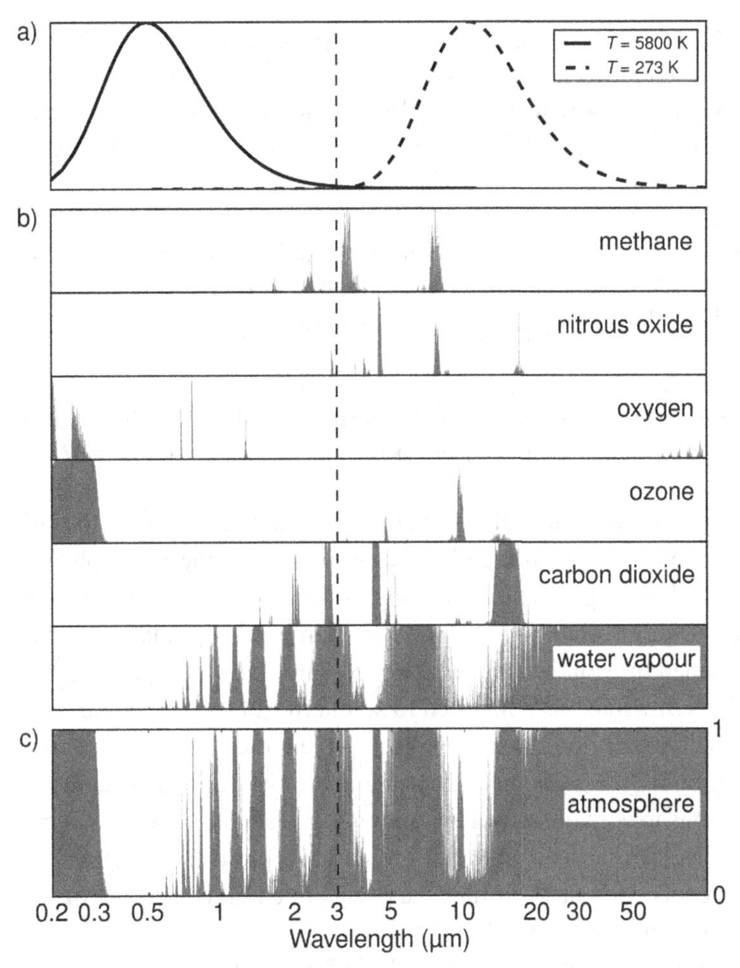

Figure 2.2 Spectra of shortwave radiation and longwave radiation and its absorption in the atmosphere. (**a**) Black-body radiation from objects with temperatures of 5800 K (surface of the Sun) and 293 K (representative of Earth and atmosphere); the radiative flux densities have been normalized by the peak values (which differ by more than 6 orders of magnitude). (**b**) Total absorptivity of the atmosphere (US standard atmosphere) for the most important absorbing species. (**c**) Total absorptivity for all species together. The vertical dashed line separates shortwave and longwave radiation. Spectra determined with the Reference Forward Model (http://www.atm.ox.ac.uk/RFM/, based on GENLN2) (Edwards, 1992), using the HITRAN2008 absorption line database (Rothman et al., 2009) and ozone UV absorption from Brion et al. (1993).

$$K_0^{\downarrow} = \overline{I_0}\left(\frac{\overline{d}_{\mathrm{Sun}}}{d_{\mathrm{Sun}}}\right)^2 \cos(\theta_z) \tag{2.3}$$

where $\overline{I_0}$ is the solar constant (flux density of solar radiation at the mean distance from Sun to Earth), $\overline{d}_{\mathrm{Sun}}$ is the *mean* (over a year) distance between Sun and Earth, d_{Sun} is the *actual* distance between Sun and Earth (depending on the date) and θ_z is the solar zenith angle (angle between solar beam and the normal to Earth's surface), which depends on the location, date and time. In some applications the solar elevation angle (angle between solar beam and horizontal) is used, which is the complementary angle of the solar zenith angle.

Question 2.2: The distance between the Sun and Earth varies through the year and hence the amount of solar radiation that falls on a plane perpendicular to the solar beam. The ratio $\left(\frac{\bar{d}_{Sun}}{d_{Sun}}\right)^2$ can be approximated by $1+0.033\cos[2\pi n_{day}/365]$ where n_{day} is the day number. The solar constant can be taken as 1365 W m^{-2}.

How large (in W m^{-2}) is the variation through the year of the solar radiation arriving at the top of the atmosphere (on a plane perpendicular to the solar beam)?

For a given amount of solar radiation at the top of the atmosphere, the solar radiation at the ground level is determined by the properties of the atmosphere in-between. Cloud cover and the type of clouds are the most important causes of variation. Apart from the day-to-day variability, this variation can also have a latitudinal and seasonal component depending on the local climate (e.g., northern Europe has a low fraction of sunshine hours in winter, as compared to southern Europe; see Figure 2.3).

Besides the presence or absence of clouds, the composition of the atmosphere can vary as well, both on a seasonal time scale and on shorter time scales. This encompasses variations in contents of water vapour (e.g., related to variations in temperature or air mass origin) that lead to variations in absorption (see Figure 2.2b). Furthermore, the aerosols content of air may vary (e.g., due to the presence of sea spray, soot due to biomass-burning, desert dust).

The solar radiation that reaches Earth's surface is affected by the overlying atmosphere in three respects:

1. Total flux density: absorption and backward scattering diminishes the amount of radiation.
2. Directional composition: at the top of the atmosphere all solar radiation comes from the direction of the Sun (a disc with a diameter of about 32 arcminutes); due to scattering, the radiation at Earth's surface comes both from the direction of the Sun, and from the rest of the hemisphere.
3. Spectral composition: radiation is absorbed by atmospheric gases at specific wavelengths and scattering varies with wavelength as well.

In view of point (2), the amount of solar radiation at the surface is therefore decomposed into direct radiation (*S*) and diffuse radiation (*D*) (see Figure 2.4):

$$K^{\downarrow} = S + D = I\cos(\theta_z) + D \tag{2.4}$$

where *I* is the radiative flux density through a plane perpendicular to the solar beam. Because of point (3), the discussion that follows deals with radiation of one wavelength at a time.

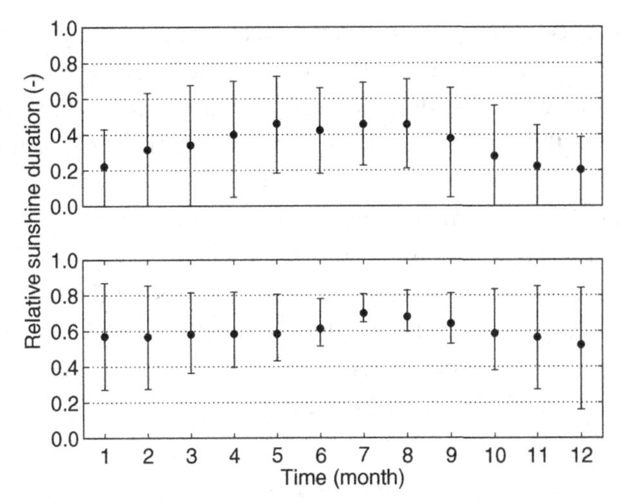

Figure 2.3 Climatology of relative sunshine duration for two European stations: Oslo (top; 59°56′N) and Valencia (bottom; 39°29′N) for the period 1971–2000. Symbols indicate the mean; ranges indicate 25% and 75% percentiles. (Data from Klein Tank et al., 2002)

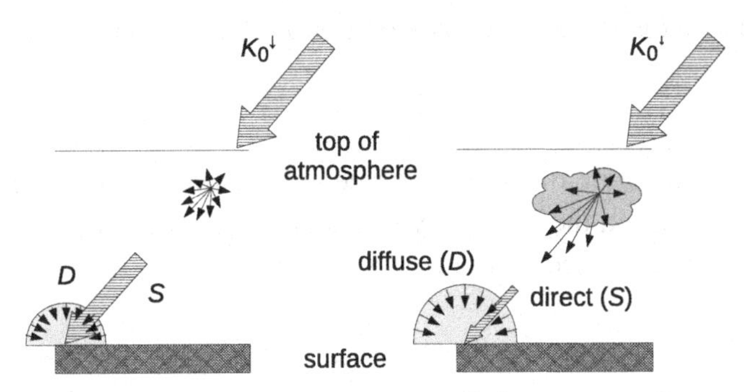

Figure 2.4 Relationship between shortwave radiation at top of atmosphere and direct and diffuse radiation at Earth's Surface for cloudless sky (left) and cloudy sky (right).

Extinction: General

The total effect of the atmosphere on solar radiation will depend on the length of the path of the radiation through the atmosphere. Or more precisely, it will depend on the number of molecules and particles encountered by the radiation. Therefore, the path length is expressed as an optical mass (m_a): the total mass (per unit area) along the path of the radiation:

$$m_a = \int_{\infty}^{0} \rho \, ds \qquad (2.5)$$

where ρ is the density of the air and s is the location along the path (the beam starts at infinity and ends at the surface, $s = 0$). Note that the concept of optical mass is also used for applications where the domain is not the entire atmosphere, but only a finite layer (e.g., between the top of the atmosphere and an elevated location on a mountain). Then the integration in Eq. (2.5) is performed from a height larger than zero and the optical mass will have a smaller magnitude.

The path length through the atmosphere (i.e., where density is non-zero) depends on the solar zenith angle. To characterize the optical mass of the atmosphere, without reference to a certain geometry of the radiation (i.e., a certain solar zenith angle), the vertical optical mass (m_v) is introduced (note the change of direction of integration due to the replacement of s by z, see Figure 2.5):

$$m_v = \int_0^{\infty} \rho \, dz \tag{2.6}$$

Then, if we define the ratio of the true optical mass (m_a) and the optical mass along the local vertical (m_v) as the relative optical mass[1] (m_r), the true optical mass can be decomposed into a component that depends on the atmosphere (m_v), and a component that depends on the direction of the radiation (m_r):

$$m_a = m_v m_r \tag{2.7}$$

If we neglect the curvature of Earth's surface and assume that refraction of light by the atmosphere is absent, then it is easy to see that in the case of slantwise radiation we have $dz = -\cos(\theta_z)\, ds$ (see Figure 2.5), yielding

$$m_r \approx [\cos(\theta_z)]^{-1} \tag{2.8}$$

This approximation is accurate within 1% for solar zenith angles up to 75 degrees (based on the expressions in Kasten and Young, 1989). In principle, deviations from the approximation are different for different atmospheric constituents, owing to differences in their vertical distribution (Iqbal, 1983).

According to Beer's law, the reduction of the radiation along a beam, due to a substance i, can be described as:

$$dI_\lambda = -I_{\lambda 0} k_{\lambda,i} q_i \rho \, ds \tag{2.9}$$

where I_λ is the spectral flux density through a plane perpendicular to the beam, $I_{\lambda 0}$ is the spectral flux density entering the medium, $k_{\lambda,i}$ is the monochromatic extinction

[1] Note that in some literature the relative optical mass as introduced here is called just optical mass. Often, for elevated locations, the height-correction is applied to the relative optical mass through multiplication with (p/p_0), where p_0 is a reference pressure at sea level.

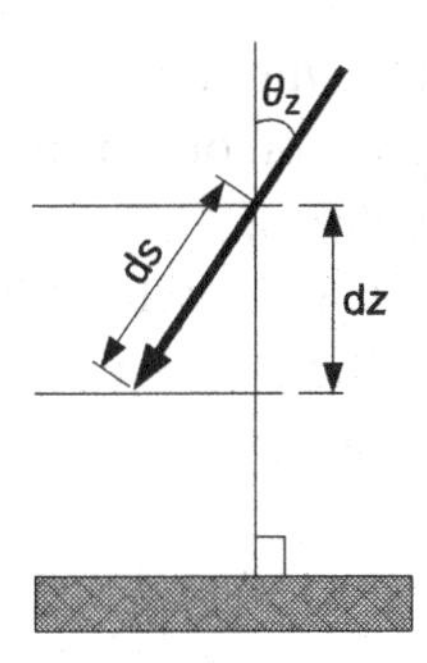

Figure 2.5 Definition of coordinates for radiation travelling through the atmosphere.

coefficient (in $m^2\ kg^{-1}$) for a given substance i and q_i is the specific concentration (mass fraction) of substance i in the air. With the relationship between ds and dz (see earlier and Figure 2.5) and integrating over z we obtain:

$$I_\lambda = I_{\lambda 0} \cdot \exp(-m_r \int_0^\infty k_{\lambda,i} q_i \rho dz) = I_{\lambda 0} \tau_{\lambda\theta,i} \tag{2.10}$$

where $\tau_{\lambda\theta,i}$ is the monochromatic transmissivity for substance i along a path with zenith angle θ_z (dimensionless). The integral in Eq. (2.10) is often referred to as the optical thickness or normal optical depth $\delta_{\lambda,i}$ (Wallace and Hobbs, 2006):

$$\delta_{\lambda,i} \equiv \int_0^\infty k_{\lambda,i} q_i \rho \, \mathrm{d}z \tag{2.11}$$

The transmissivity $\tau_{\lambda\theta,i}$ will have a value between zero and one: it is the fraction of radiation that is transmitted.

In general, the extinction coefficient depends on height: in the case of extinction by absorption the exact spectral location, strength and width of the absorption lines depend on local pressure and temperature and hence on height. But if we would assume the extinction coefficient to be constant and allow only the concentration q_i to vary with height, the optical thickness can be written as $k_{\lambda,i} \int_0^\infty q_i \rho \, \mathrm{d}z$, where the integral represents the total amount of the substance under consideration (e.g., total ozone column, or amount of precipitable water). If we would assume both the extinction coefficient $k_{\lambda,i}$ and the mass fraction q_i to be constant with height, the optical thickness could be replaced by $k_\lambda \cdot q_i \cdot m_v$. This expression can also be used for variable concentrations if q_i is interpreted as a weighted (with ρ_i) average concentration.

The monochromatic transmissivity $\tau_{\lambda\theta,i}$ can be decomposed into a vertical component and an angle-dependent part as follows:

$$\tau_{\lambda\theta,i} = e^{-\delta_{\lambda,i} m_r} = \left(e^{-\delta_{\lambda,i}}\right)^{m_r} = \left(\tau_{\lambda v,i}\right)^{m_r} \tag{2.12}$$

where $\tau_{\lambda v,i}$ is the monochromatic transmissivity for a vertical beam. This shows that $\tau_{\lambda\theta,i}$ depends on the state and composition of the atmosphere (through $\tau_{\lambda v,i}$) as well as on the direction of the beam (through m_r).

The radiation entering the atmosphere will be attenuated by various processes, each with a different extinction coefficient. Because all these processes are working independently, we have (for n processes):

$$\begin{aligned}\tau_{\lambda\theta} &= \tau_{\lambda\theta,1} \cdot \tau_{\lambda\theta,2} \cdot \ldots \cdot \tau_{\lambda\theta,n} \\ &= e^{-(\delta_{\lambda,1}+\delta_{\lambda,2}+\ldots+\delta_{\lambda,n})m_r}\end{aligned} \qquad (2.13)$$

where $\tau_{\lambda\theta,i}$ and $\delta_{\lambda,i}$ are the transmissivity and optical thickness due to process i, respectively. Those processes encompass various types of scattering and absorption. Note that in Eq. (2.13) we have assumed the relative optical masses to be equal for all processes.

Question 2.3: Consider radiation of a certain wavelength λ. For that wavelength, for a substance i, the atmosphere has a transmissivity for a *vertical* beam ($\tau_{\lambda v,i}$) of 0.8. The vertical air mass m_v is 10 000 kg m^{-2} and the specific concentration of substance i is 0.003 kg kg^{-1}.

a) Verify that the value for the vertical air mass has a realistic order of magnitude.
b) What is the value of the extinction coefficient for substance i ?
c) If the solar zenith angle is 40 degrees, what is the transmissivity $\tau_{\lambda\theta,I}$?
d) If the amount of radiation at the top of the atmosphere at wavelength λ is 1.5 W m^{-2} µm^{-1}, what is the amount of radiation arriving at Earth's surface for the situation given under (b)?

Extinction: Scattering and Absorption

Extinction of short wave radiation is due both to scattering and to absorption. Owing to the direction independence of Rayleigh scattering half of the scattered radiation is scattered backward, thus reducing the light intensity. Mie scattering and geometric optics scattering occur mainly in the forward direction, but there is also backward scattering in some specific directions (see Figure 2.1). Hence the net effect of all types of scattering is that a part of the radiation is removed from the beam, either backward or sideways.

Absorption of radiation in the shortwave part of the spectrum takes place in two distinct wavelength regions. At the near-infrared part the most important absorbing gas is water vapour (see Figure 2.2b), which absorbs in a large number of bands. Oxygen, carbon dioxide and methane play a smaller role, all at near-infrared wavelengths. On the other hand, absorption in the UV part of the spectrum is dominated by ozone and oxygen. Ozone is active in a broad band of ultraviolet wavelengths down to 0.2 µm, whereas oxygen absorbs – apart from the Herzberg band around 0.27 µm – at wavelengths below 0.2 µm. Absorption in the UV part of the spectrum is a special case because the energy of the photons is so large that the molecules (ozone and oxygen) are broken up in atoms that reorganize into other types of molecules.

To summarize, the extinction of direct beam radiation can be treated as the sum of a number of processes. In terms of transmissivities:

$$\tau_{\lambda\theta} = \left(\tau_{\lambda\theta,\text{Ray}} \cdot \tau_{\lambda\theta,\text{Mie}} \cdot \tau_{\lambda\theta,\text{geo}}\right) \cdot \left(\tau_{\lambda\theta,\text{gas}} \cdot \tau_{\lambda\theta,\text{O}} \cdot \tau_{\lambda\theta,\text{w}}\right) \tag{2.14}$$

where the first group of transmissivities is related to scattering (Rayleigh and Mie scattering, and geometric optics theory) and the second group to absorption (ozone, water vapour and other, uniformly mixed, gases). It depends on the variability of the concentration of atmospheric constituents whether the extinction due to the respective processes is variable from time to time, or between locations. Whereas $\tau_{\lambda\theta,\text{Ray}}$, and $\tau_{\lambda\theta,\text{gas}}$ will be relatively constant, other transmissivities may be highly variable due to variations in aerosols and clouds ($\tau_{\lambda\theta,\text{Mie}}$), ozone concentrations ($\tau_{\lambda\theta,\text{O}}$) and atmospheric water vapour content ($\tau_{\lambda\theta,\text{w}}$).

The direct beam radiation that enters the atmosphere (I_0) but does not reach the surface ($I_0 - I$) has to go somewhere. There are three fates for this energy:

- Radiation that is scattered in a direction away from Earth's surface leaves the atmosphere as reflected radiation (especially important for cloudy situations).
- Radiation that is scattered in a direction towards the surface reaches the surface somewhere as diffuse radiation.
- Radiation that is absorbed heats up the air where it is absorbed. This in turn will result in extra thermal emission. Note that the air will re-emit the absorbed radiation at a very different wavelength than at which it was absorbed: absorption takes place at wavelength in the shortwave region (left-hand-side of Figure 2.2c), whereas emission occurs in the longwave region, corresponding to the temperature of the air and only at wavelengths where emission lines exist (right-hand side of Figure 2.2c).

Impact of the Atmosphere on Radiation Reaching the Surface

The clearest impact of absorption and scattering on the amount of radiation reaching the surface can be seen when cloudiness changes. Figure 2.6a shows observations of global radiation ($K^{\downarrow}$) for two consecutive, but contrasting days. The input of solar radiation at the top of the atmosphere was nearly identical (only a one-day difference) but the presence of clouds on May 22 leads to a reduction of total incoming radiation of about 75%. One remarkable feature is the high peak around 15 Coordinated Universal Time (UTC) at May 22 which exceeds the value of $K^{\downarrow}$ on the clear day. This is probably due to a combination of direct insolation (sunlight peeking through the clouds) and reflection of solar radiation on the sides of cumulus clouds.

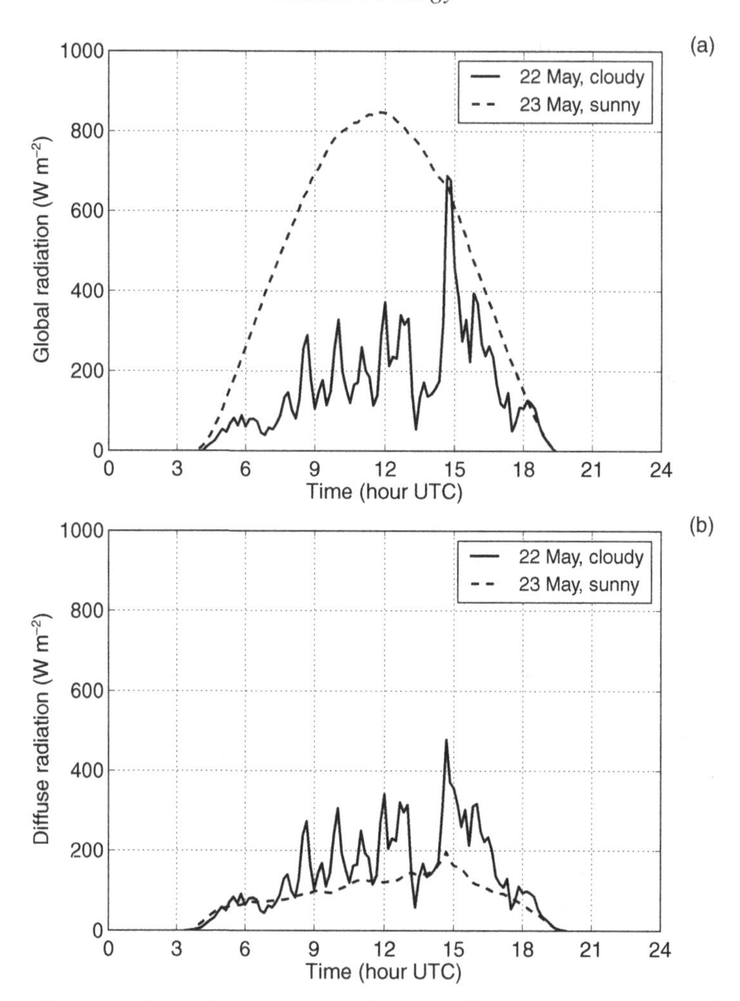

Figure 2.6 Observations of shortwave radiation at Haarweg Meteorological Station for a clear day (May 23, 2007) and a cloudy day (May 22, 2007). (**a**) Global radiation. (**b**) Diffuse radiation.

Not only the *amount* of radiation reaching Earth's surface is influenced by atmospheric composition, but the directional composition as well: how much radiation comes from which part of the hemisphere. Figure 2.6b shows the diffuse radiation (radiation not coming from the direction of the solar disc) on the same days as shown in Figure 2.6a. On the cloud-free day roughly 80% of the radiation comes from the direction of the Sun (i.e., 20% is diffuse radiation), whereas on the cloudy day this is close to zero (all radiation is diffuse).

Figure 2.7 shows the spectral composition of the shortwave radiation at Earth's surface for a clear day. In the total radiation at the surface, the absorptions by ozone (short wavelengths) and water vapour (longer wavelengths) are clearly visible. Furthermore, the spectral peak of the diffuse radiation has shifted to lower wavelengths

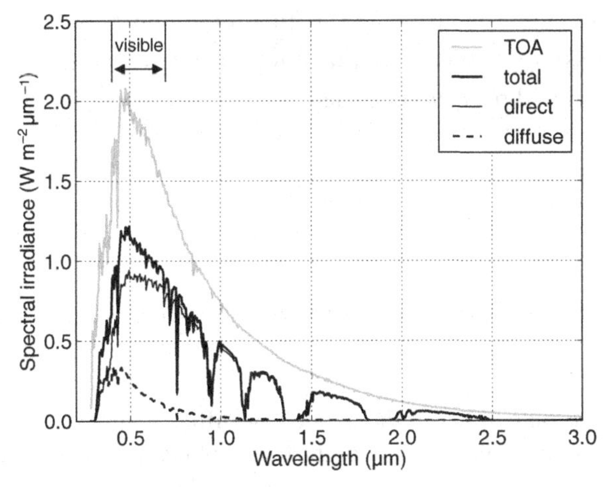

Figure 2.7 Spectral composition of shortwave radiation on May 23, at 52°N, 5°E: at top of atmosphere (TAO), and at ground level: global radiation (total), direct and diffuse radiation. The TAO spectrum is due to Gueymard (2004); the ground level data are based on model calculations with SMARTS version 2.9 (Gueymard, 2001), applied to a US standard atmosphere.

(blue) relative to the radiation at the top of the atmosphere, whereas the direct radiation has shifted in the direction of higher wavelengths (yellow and red) as a result of the wavelength dependence of Rayleigh scattering. These shifts result in the blue colour of the hemisphere (diffuse radiation) and the yellowish colour (at sunset and sunrise reddish) of the Sun (direct radiation).

Global Radiation

In the preceding analysis we have taken into account the directional and spectral composition of the radiation reaching the surface. But in the context of the surface energy balance we are interested in the total amount of solar radiation that reaches the surface (global radiation): $K^{\downarrow}$. Thus, to obtain $K^{\downarrow}$ all radiation reaching the surface has to be integrated over the spectral range of solar radiation and over all directions in vertical (θ_z) and horizontal (φ, azimuth angle).

In the context of global radiation, two other transmissivities are useful to define. First, the broadband beam transmissivity $\tau_{b\theta}$ (the broadband equivalent of spectral beam transmissivity; see Eq. (2.14)) is defined as:

$$\tau_{b\theta} \equiv \frac{I}{I_0}, \tag{2.15}$$

where I_0 is the radiation flux density at the top of the atmosphere, through a plane perpendicular to the beam. This transmissivity is useful in those cases in which a distinction is made between direct and diffuse radiation (see Eq. (2.4)). Second, a broadband transmissivity that does not take into account the direction of the beam is frequently used, denoted by τ_b and defined as:

$$\tau_b = \frac{K^{\downarrow}}{K_0^{\downarrow}}. \tag{2.16}$$

This transmissivity is used when no distinction is made between direct and diffuse radiation.

Question 2.4: The ratio between the amount of diffuse radiation (D) and the global radiation ($K^{\downarrow}$) is an important indicator of the nature of the radiation that reaches Earth's surface. What is (approximately) the value of $\frac{D}{K^{\downarrow}}$ (see also Figure 2.6):

a) On a sunny day without clouds
b) On an overcast day

Question 2.5: See Figure 2.6. On May 22 and 23, at 12 UTC, the solar zenith angle is about 32 degrees. The ratio $\left(\frac{\overline{d}_{Sun}}{d_{Sun}}\right)^2$ is about 0.974 for these dates and the solar constant can be taken as 1365 W m^{-2}.

a) Estimate I at 12 UTC for May 22 and May 23 from Figure 2.6.
b) Estimate the broadband beam transmissivity $\tau_{b\theta}$ at 12 UTC for both days.
c) Estimate the broadband transmissivity τ_b at 12 UTC for both days.
d) Is the difference in broadband beam transmissivity between both days due mainly to differences in absorption, or differences in scattering?

Question 2.6: In this section a range of transmissivities has been introduced. Collect them and note down the following for each of them: symbol, meaning (what does a given value mean in physical reality?) and definition (mathematical relationship to various radiation flux densities).

For τ_b various empirical models exist. An empirical model often used to estimate the daily mean solar radiation ($\overline{K^{\downarrow}}^{24}$) from sunshine duration data is:

$$\frac{\overline{K^{\downarrow}}^{24}}{\overline{K_0^{\downarrow}}^{24}} = \tau_{b,24} = a + b\frac{n}{N_d} \tag{2.17}$$

where n is the hours of bright sunshine, N_d is the day length (in hours) and a and b are empirical constants. The overbar denotes temporal averaging, in this case of a period of 24 hours. Typical values for the Netherlands are $a = 0.2$ and $b = 0.55$ (DeBruin and

Stricker, 2000). Allen et al. (1998) recommend 0.25 and 0.5 for global applications. The coefficient ($a + b$) can be interpreted as the transmissivity of the cloudless sky, whereas a is the transmissivity of the clouds. Other values for the constants can be found in Iqbal (1983). The strength of expression (2.17) lies in the fact that it contains the two dominant variables that determine daily mean global radiation: the yearly cycle (through $\overline{K^{\downarrow}}^{24}$) and cloudiness (through n/N_d). However, it assumes a fixed transmissivity of the cloudless atmosphere.

Question 2.7: Refer to the graphs of relative sunshine duration for Valencia and Oslo in Figure 2.3. The following additional data are given:

	Daylength (hour)		K_0 (W m^{-2})	
	January	June	January	June
Valencia	9.5	14.7	120	356
Oslo	6.6	17.8	29.6	358

a) Explain why Oslo (located much more Northerly) has a higher daily mean radiation at the top of the atmosphere in June than Valencia (figure not needed).
b) Estimate the daily average number of sunshine hours for Valencia and Oslo, for January and June. Explain why the relative contrast in daily sunshine hours between January and June is larger for Oslo than for Valencia.
c) Estimate the daily mean global radiation in Valencia and Oslo for January and June.

A parameter that is often used to characterize the transmissivity of the cloudless atmosphere is the Linke turbidity factor T_L. This factor uses the transmissivity of a clean, dry atmosphere as a reference (as this reference atmosphere is dry, it is a hypothetical atmosphere). In such a reference atmosphere attenuation takes place only due to Rayleigh scattering and absorption by gases other than water vapour (mainly CO_2, O_2 and ozone). The transmissivity of the real atmosphere is expressed in the number of clean, dry atmospheres needed to attain the same transmissivity:

$$\tau_{b\theta} = \left[\left(\tau_{bv,cda} \right)^{m_r} \right]^{T_L} = e^{-\delta_{b,cda} m_r T_L} \tag{2.18}$$

where $\delta_{b,cda}$ is the broadband optical thickness of the clean, dry atmosphere and $\tau_{bv,cda}$ is the vertical broadband beam transmissivity of the clean, dry atmosphere. These values depend on the exact definition of the reference atmosphere (see Ineichen and Perez, 2002), but a typical value for $\tau_{bv,cda}$ is 0.9. As the Linke turbidity factor is a broadband quantity, it will depend on the spectral composition of the radiation arriving at the surface. As the latter depends on the relative optical mass (some wavelengths become depleted if the path length becomes longer), so does the Linke turbidity. Usually, values are reported that are valid for a relative optical mass of 2, denoted by $T_{L,2}$. Typical values for the Linke turbidity factor range from 2 (clear, cold air), through 4 (moist, warm air) to 8–10 (polluted air) (Scharmer and Greif, 2000). Various empirical models for T_L exist, which usually link the turbidity to the amount of water vapour, ozone and aerosols (e.g., Jacovides, 1997; Gueymard, 1998).

The transmissivity that can be derived based on the Linke turbidity is a beam transmissivity (Eq. (2.18)): it can be used to obtain the direct radiation. To obtain the global radiation (direct plus diffuse) using the Linke turbidity, a model is needed (see Ineichen, 2006, for an overview). One example is the model of Ineichen and Perez (2002):

$$\frac{K^{\downarrow}}{K_0^{\downarrow}} = \tau_b = 0.868 e^{-0.0387 m_r T_{L,2}} \tag{2.19}$$

where $T_{L,2}$ is the Linke turbidity at relative optical mass 2.

Question 2.8: See Figure 2.6. On May 22 and 23, at 12 UTC, the solar zenith angle is about 32 degrees. The ratio $\left(\frac{\overline{d}_{Sun}}{d_{Sun}}\right)^2$ is about 0.974 for these dates and the solar constant can be taken as 1365 W m^{-2}.

Estimate $T_{L,2}$ at 12 UTC for both days. Are these reasonable values?

2.2.3 Reflected Shortwave Radiation

Because terrestrial surfaces, under natural conditions, are never so hot that they can emit significant amounts of shortwave radiation, the sole source of upwelling shortwave radiation is reflected solar radiation. Thus the specification of the upwelling shortwave radiation boils down to the specification of the reflectivity of the surface. But, similar to the case of atmospheric extinction, the reflection both has spectral and directional dependencies. The spectral dependence is easily illustrated by the fact that some natural surfaces are green (relatively high reflectivity at a wavelength around 0.53 μm) and others are red (high reflectivity around 0.68 μm). Thus the reflectivity of surfaces is wavelength dependent.

Regarding the directional dependence of reflected radiation three cases can be distinguished (for the geometry, see Figure 2.8):

- Specular reflection: The incoming and reflected light make the same angle with respect to the surface normal ($\theta_{out} = \theta_{in}$) and the direction of reflection is opposite to the direction of the incident beam ($\varphi_{out} = \varphi_{in}$+180°). Specular reflection appears if the irregularities of the surface are small as compared to the wavelength (e.g., a lake under low-wind conditions). The reflectivity for specular reflection strongly depends on the zenith angle (see, e.g., Hecht, 1987): high reflectivity at large zenith angles (low incidence angles) and low reflectivity at normal incidence.

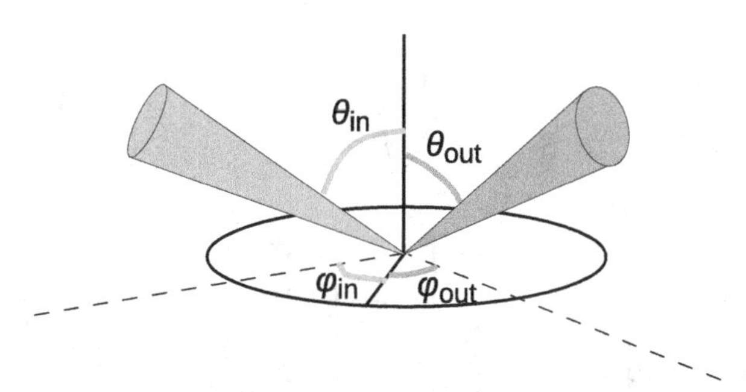

Figure 2.8 Geometry of incoming ('in') and reflected ('out') radiation at a surface. The combination of four angles defines one point in the bidirectional reflectance distribution function (BRDF).

- Lambertian reflection: This occurs when light falls on a Lambertian surface that reflects radiation equally in all directions (diffuse). Hence direct radiation is converted to diffuse radiation.
- General diffuse reflection: Light is reflected diffusely, but not equally in all directions. The total amount of reflected radiation may also depend on the azimuthal direction of the incident radiation. Most natural surfaces fall in this last category.

These three types of reflection can be applied to the surface of individual materials (e.g., leaf tissue or water) or to entire surfaces (e.g., a forest or a lake).

One of the few natural *materials* that exhibits specular reflection is water. Other relatively smooth materials may exhibit specular reflection to some extent, especially at small incidence angles (e.g., ice and leaves). Rougher materials (e.g., flat soil and snow surfaces) usually show some form of general diffuse reflection.

When it comes to natural *surfaces*, smooth water surfaces (windless conditions) exhibit specular reflection and observations (see Table 2.1) show the related strong dependence of reflectivity on solar zenith angle. When waves are formed on the water surface, the specular character of the reflection is – partly – lost. More generally: for most types of surface cover not only the material of which the surface consists matters, but the structure of the surface matters as well (e.g., vegetation, urban areas, soil with micro relief). Three effects may result in a dependence of the reflectivity on the zenith angle of the incident radiation (see Figure 2.9 for an illustration using a simplified structured surface consisting of rectangular objects, for simplicity referred to as "canopy"):

- Radiation that hits the objects will be partly absorbed and partly reflected. The reflected parts can be directed out of the canopy (thus being counted as reflected radiation) or into the canopy. The latter radiation will hit other parts of the canopy and will in turn be partly absorbed and partly reflected (multiple reflection will occur). At small zenith angles (Figure 2.9a), the direct radiation will penetrate deep into the canopy, so that multiple reflection and the loss of radiation due to absorption occur over a large part of the canopy depth. Only a small part of the radiation that

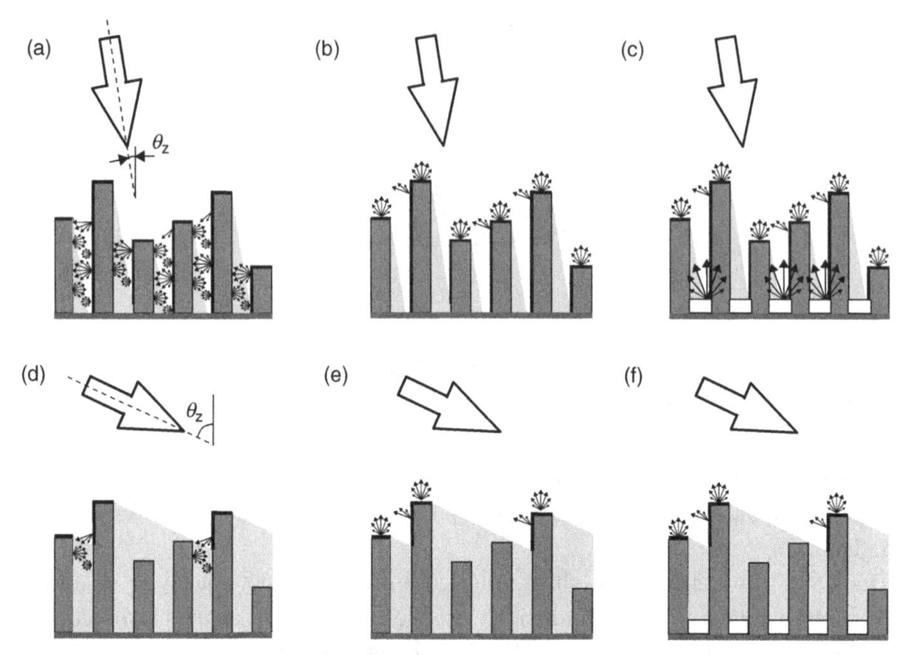

Figure 2.9 Effect of the zenith angle θ_z of incoming radiation on the reflectivity of a structured surface: small zenith angle (top row) and high zenith angle (bottom row). Left (**a** and **d**): effect of multiple internal reflection. Middle (**b** and **e**): effect of objects casting shadows on other objects. Right (**c** and **f**): effect of direct radiation penetrating (or not) to the ground which has a different reflectivity. Black lines on objects indicate direct illumination (only direct radiation is considered). Reflection is assumed to be Lambertian.

penetrated into the canopy will eventually leave the canopy again: the largest part of the radiation is trapped. At large zenith angles (Figure 2.9d) multiple reflection is limited to the upper part of the canopy only, and radiation can more easily escape again. Thus multiple internal reflections cause the reflectivity of a canopy to be lower at smaller zenith angle (high solar elevation).

- Objects cast shadows on other objects. If the zenith angle is small (Figure 2.9b) only few shadows occur: the upward facing parts, as well as a large part of the side walls of the objects are directly lit and hence will be the source of high levels of reflected radiation. On the other hand, if the solar zenith angle is large (Figure 2.9e), a large fraction of the surface of the objects is shadowed by other objects. Hence only a small part of the surface reflects radiation. Thus shadowing causes the reflectivity of a canopy to be higher at higher radiation zenith angle (the opposite of the effect of multiple reflections).
- The surface between the objects will have a different reflectivity than the objects themselves. Hence, the reflectivity of the canopy as a whole depends on the degree to which the radiation can penetrate down to the ground. At small zenith angles the radiation will have a higher probability to reach the surface than at large zenith angles (Figure 2.9c). It will depend on the reflectivity of the underlying surface whether the penetration of radi-

ation to the surface leads to a lower albedo of the canopy as a whole (e.g., dark soil) or to a higher reflectivity (e.g., snow; see Gryning et al., 2001).

Another aspect of the directional dependence of reflectivity is its dependence on the azimuth angle of the incoming radiation. An example where the directional dependence is related to the azimuth angle (φ_{in}) is a row crop. If the radiation enters the crop parallel to the rows, the radiation is reflected both by the crop and by the soil in-between (which may differ in reflectivity). If, on the other hand, the radiation falls on the crop perpendicular to the rows, the reflection comes only from the crop (unless the row spacing is very large, or the zenith angle of the incident radiation is very small).

Apart from the dependence of reflection on the direction of the *incident* radiation, the *reflected* radiation (for a given direction of incidence) may also have a directional dependence (rather than being diffuse, equal in all directions). Here again, a water surface is the clearest example: at low solar altitude the water acts as a mirror, but the observer sees most of the reflected light only if she looks in the direction towards the Sun, and at a zenith angle equal to that of the Sun. Another example is a vegetated surface. If the observer has the Sun at his back, he looks at the illuminated part of the plants, whereas when he is facing the Sun he looks at the shadow side (which will yield a lower amount of reflected radiation).

Question 2.9: Consider a row crop with dark bare soil in-between. The fraction of vegetation cover is 0.5 (i.e., 50% of the soil is covered by vegetation).

a) If the Sun shines in a direction parallel to the rows, what is the albedo of this composite surface (see Table 2.1 for representative values for simple surfaces).
b) As (a), but now a situation where the Sun shines in a direction perpendicular to the rows (and where the direct radiation does not reach the soil). Neglect diffuse radiation.
c) How large is the relative difference in net short wave radiation K^* between the situation mentioned under (a), and (b) (take it relative to K^* of question a).

Bidirectional Reflectance Distribution Function

This complicated combination of the dependence of the reflectivity (r) of a surface on the directions of incident radiation, reflected radiation (and wavelength) is summarized in the bidirectional reflectance distribution function (BRDF) of a surface:

$$r = r(\lambda, \theta_{in}, \varphi_{in}, \theta_{out}, \varphi_{out}) \tag{2.20}$$

The combination of four angles (zenith and azimuth angle of incoming radiation, and zenith and azimuth angle of the observer; see Figure 2.8) defines one point in the BRDF. This implies that the determination of the BRDF of a natural surface is a very laborious job, as radiative fluxes have to be measured under many different incidence angles and many different observing angles.

In the context of the surface energy balance, we are interested in the total amount of radiation reflected from a surface. Therefore, the BRDF can be integrated over all values of θ_{out} and φ_{out}:

$$r(\lambda,\theta_{in},\varphi_{in}) = \int_0^{\pi/2} \int_0^{2\pi} r(\lambda,\theta_{in},\varphi_{in},\theta_{out},\varphi_{out})\, d\varphi_{out}\, d\theta_{out} \tag{2.21}$$

But we cannot simplify further. Hence, the reflected radiation becomes:

$$K^{\uparrow}(\lambda,\theta_{in},\varphi_{in}) = K^{\downarrow}(\lambda,\theta_{in},\varphi_{in})\; r(\lambda,\theta_{in},\varphi_{in}) \tag{2.22}$$

From this equation it is clear that the reflected radiation depends not only on wavelength (of the incoming radiation, and the spectral properties of the surface), but also on the directional composition of the incoming radiation.

Although in the context of the surface energy balance, the direction of the upwelling radiation is irrelevant (leading to the simplified Eq. (2.22)), in the context of remote sensing of the surface reflectance the full BRDF has to be taken into account (see Figure 2.8). In that application, the sensor on board of the satellite views the radiation reflected by Earth's surface into a particular direction (θ_{out} and φ_{out}). This direction depends on the location of the satellite at that particular time. For a cloud-free situation, most of the radiation is direct radiation, coming from a direction θ_{in} and φ_{in}. Finally, remote sensing sensors usually are observing in narrow wavelength bands. Thus, such a remote sensing observation of the surface reflectance in fact is only a small part of the entire BRDF.

Albedo

In many applications, we are not interested in (or have no information on) the spectral and directional composition of the incoming radiation, and therefore we define a broadband (all wavelengths in the shortwave range), hemispheric (all incident angles) albedo:

$$r \equiv \frac{K^{\uparrow}}{K^{\downarrow}} = \frac{\int_0^{\pi/2}\int_0^{2\pi}\int_{\lambda_1}^{\lambda_2} K^{\downarrow}(\lambda,\theta_{in},\varphi_{in})\; r(\lambda,\theta_{in},\varphi_{in})\, d\lambda\, d\varphi_{in}\, d\theta_{in}}{\int_0^{\pi/2}\int_0^{2\pi}\int_{\lambda_1}^{\lambda_2} K^{\downarrow}(\lambda,\theta_{in},\varphi_{in})\, d\lambda\, d\varphi_{in}\, d\theta_{in}} \tag{2.23}$$

The main purpose of showing this ratio of two threefold integrals is to clarify that the broadband hemispheric albedo (which is generally referred to as simply 'albedo') depends not only on the properties of the surface (as we would like to), but also on the characteristics of the incoming radiation. This is due to the fact that the two occurrences of $K^{\downarrow}$ (within the integral) in the numerator and denominator of Eq. (2.23) do

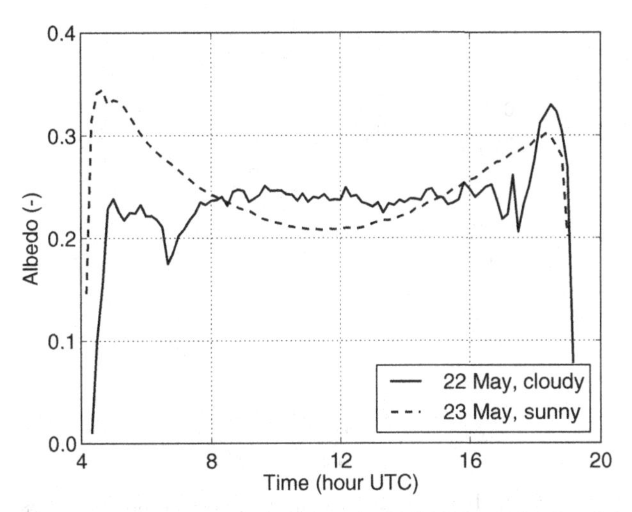

Figure 2.10 Albedo for two consecutive days in 2007: May 22 (cloudy) and May 23 (clear). Observations from Haarweg meteorological station (grass).

not cancel. A simple example to demonstrate this is to look at the time dependence of the albedo on a single day. During one day, the Sun traverses a range of zenith angles and azimuth angles. Figure 2.10 gives an example of the diurnal variation of the albedo for a grass surface in the Netherlands for a clear day and an overcast day. For the clear day (May 23) there is an obvious diurnal cycle with higher albedos in early morning and late afternoon (at large values of the solar zenith angle, or low solar altitudes): in that case the radiation does not penetrate into the grass, but instead is reflected by the top of the leaves. At midday the radiation penetrates into the grass and is trapped due to multiple reflections. In contrast, the albedo on the overcast day is rather constant in time, with a small reduction in the morning. In the late afternoon the clouds have cleared and the albedo is similar to that of the clear day at that time of day.

Finally, as higher canopies have more layers in which radiation can be trapped (see Figure 2.9), the albedo roughly decreases with vegetation height (see Figure 2.11, where aerodynamic roughness has been used as a proxy for vegetation height). This relationship is useful when comparing different vegetation types. However, when one considers the development of vegetation height and albedo for one specific vegetation type (e.g., during a growing season), the relationship may be different, or even reversed (for maize see, e.g., Jacobs and van Pul, 1990).

Typical values of the albedo for different surface types can be found in Table 2.1.

Question 2.10: Consider again the row crop of question Question 2.9. If the sky is completely overcast, and all incoming solar radiation is diffuse radiation, how large will be the difference in albedo between the situation in which the Sun shines in a direction parallel to the rows versus the situation in which the Sun shines in a direction perpendicular to the rows?

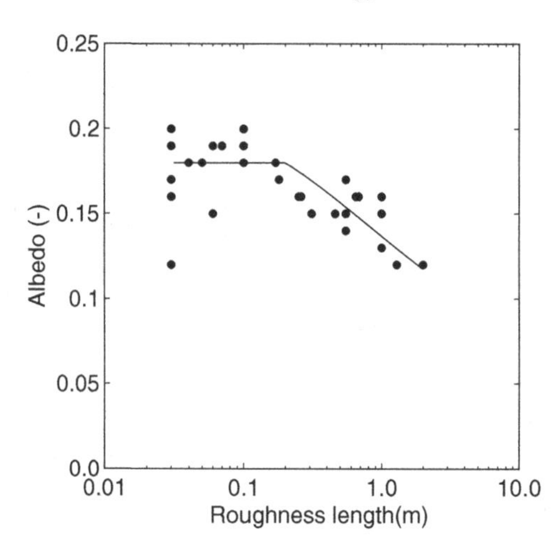

Figure 2.11 Relationship between albedo and aerodynamic roughness (see Chapter 3) for vegetated surfaces from a global database. (Data from Hagemann, 2002)

Table 2.1 Typical values for shortwave albedo (hemispheric, broadband) and longwave emissivity for different surface types

Surface type	Remark	R	ε_s
Ocean	High sun	0.05	0.95
	Low sun	0.1–0.5	0.95
Forest	Tropical rain forest	0.07–0.15	0.98
	Coniferous	0.1–0.19	0.98
	Deciduous	0.14–0.2	0.96
Crops		0.15–0.25	0.96
Grasses		0.15–0.30	0.96
Soils	Dark, wet	0.1	
	Wet sandy	0.1–0.25	0.98
	Dry sandy	0.2–0.4	0.95
	Wet clay	0.1–0.2	0.97
	Dry clay	0.2–0.35	0.95
Snow	Fresh	0.65–0.95	0.95
	Old	0.45–0.65	0.9
Urban areas		0.10–0.27	0.85–0.96

From Garratt (1992) and after Oke (1987).

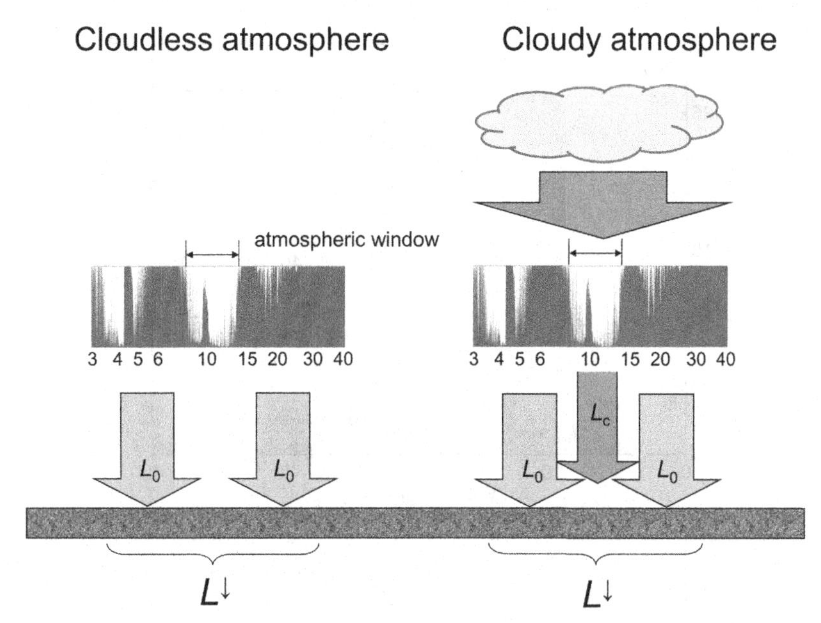

Figure 2.12 Origin of longwave radiation at Earth's surface for a cloudless (left) and cloudy (right) atmosphere. The absorption spectra from Figure 2.2c are reproduced for the wavelength range 3–40 μm.

Question 2.11: Use Eq. (2.23) to explain why the albedo is nearly constant in time on an overcast day (see Figure 2.10)?

2.2.4 Downwelling Longwave Radiation

Because outer space, beyond the atmosphere, is too cold and too much of a vacuum to emit any radiation in the wavelength range that we consider as longwave radiation, the only source of downwelling longwave radiation at Earth's surface is the atmosphere.

The atmosphere contains gases that have strong *absorption* bands in the wavelength region of longwave radiation, notably CO_2, O_3, N_2O, CO, O_2, CH_4 and water vapour (see Figure 2.2). This implies that these gases also *emit* longwave radiation at the wavelength of these absorption bands, in all directions. The amount of longwave radiation received by Earth's surface is hence dependent on the vertical distribution of temperature (which determines the amount of black-body radiation), and the concentrations of the aforementioned absorbing gases (also indicated as greenhouse gases or [GHGs]). Because the temperature of the atmosphere has only a small diurnal cycle (relative to the absolute temperature), the incoming longwave radiation is present day and night and only shows a limited diurnal variation.

The combination of the emission bands of the various gases leads to a broad range of wavelengths with a high absorptivity (Figure 2.2c), with the exception of the

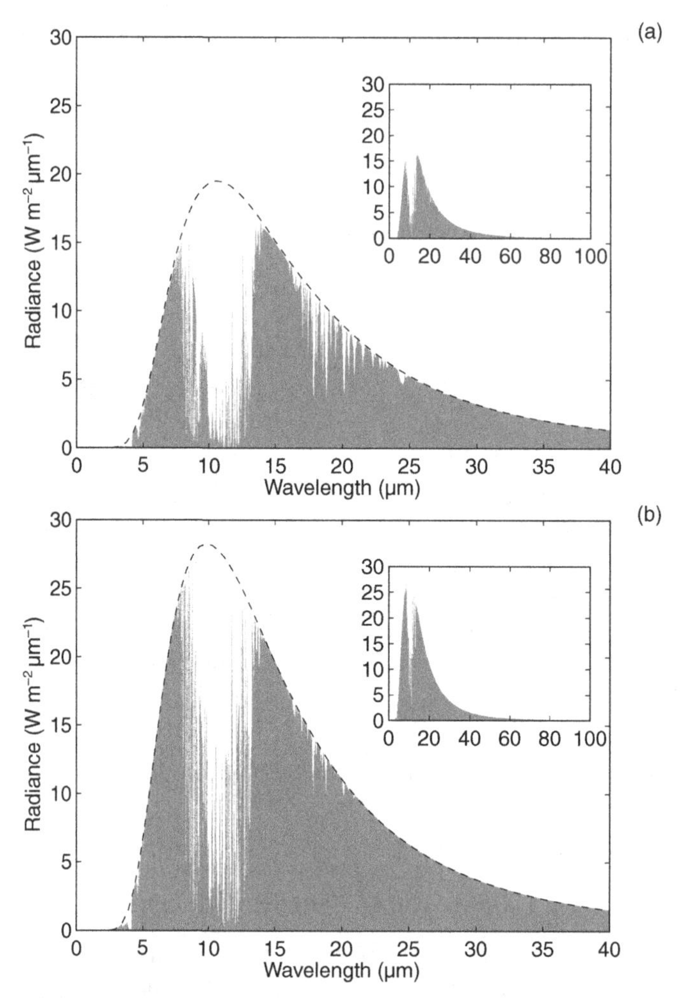

Figure 2.13 Downwelling longwave radiation for a typical mid-latitude winter atmosphere (**a**) and summer atmosphere (**b**). The dashed line is the Planck curve for 273 and 294 K, respectively. The inset shows the full spectral range of longwave radiation. Data computed with the Reference Forward Model (http://www.atm.ox.ac.uk/RFM/, based on GENLN2) (Edwards, 1992), using the HITRAN2008 absorption line database (Rothman et al., 2009).

range between approximately 8 and 13 μm, the so-called 'atmospheric window'. The absence of absorption lines in the atmospheric window implies that the atmosphere (without clouds) neither emits nor absorbs radiation in this wavelength range. Figure 2.12a depicts downward emission of longwave radiation from a cloudless atmosphere. For that case, incoming longwave radiation consists only of radiation in the wavelength ranges 3–8 and 13–100 μm, denoted by L_0.

Figure 2.13 illustrates the spectral composition of downwelling longwave radiation for mid-latitude conditions during winter and summer. The first thing to note is that the magnitude of the downwelling longwave radiation is smaller during winter than during summer. This is due to the lower temperature of the atmosphere. The envelope of the radiation is a Planck curve for a temperature that is representative of the lower 100 m of the atmosphere. Thus, it appears that most of the 'clear-sky' longwave radiation received by the surface originates from the first hundred meters of the atmosphere. This is due to the fact that longwave radiation emitted at higher altitudes in the atmosphere is completely absorbed by the layers below and hence the downwelling longwave radiation does not contain radiation emitted in the higher atmosphere. The atmospheric window is clearly visible in the downwelling radiation: hardly any radiation at wavelengths within the atmospheric window is emitted by the atmosphere.

The spectrum for the winter conditions shows more dents where the radiation is lower than the Planck curve. This is due to the lower water vapour concentration during winter: not all water vapour emission lines are fully used.

In contrast to the clear-sky emission that occurs in distinct molecular absorption lines, clouds emit (and absorb) radiation like a black body in the longwave wavelength region. Because, usually, the cloud base is higher than a few hundred meters, all radiation emitted by clouds outside the atmospheric window is absorbed by the atmosphere between the clouds and the surface. Hence only the radiation by the clouds inside the atmospheric window (denoted by L_c) will reach Earth's surface, that is, that part of the radiation emitted by clouds that is *not* absorbed by the atmosphere between the clouds and the surface. This effect is illustrated in Figure 2.12b. Note, that as the emissivity (and hence the absorptivity) of clouds is close to unity, clouds do not reflect longwave radiation. Hence, longwave radiation originating from Earth's surface (see next section) is absorbed by clouds: the downward longwave radiation originating from clouds is *not* due to the reflection of upwelling longwave radiation from the surface.

Figure 2.14a shows observations of downwelling longwave radiation for a clear and a cloudy day. For most of the day, $L^{\downarrow}$ is 50–100 W m^{-2} higher on the cloudy day as compared to the clear day, owing to extra radiation that is emitted by the clouds, and passes through the atmospheric window (if $L_0^{\downarrow}$ would be identical for both days, this 50–100 W m^{-2} would represent $L_c^{\downarrow}$). Also note that the diurnal cycle of $L^{\downarrow}$ is rather limited.

Complex models have been designed to describe $L_0^{\downarrow}$ and $L_c^{\downarrow}$ in detail as a function of vertical profiles of temperature and atmospheric composition. These models can either make a line-by-line calculation (using the absorption spectra as depicted in Figure 2.2b and yielding spectral fluxes as shown in Figure 2.13) or the absorption spectra can be summarized as absorption bands, hence speeding up the calculation. It is outside the scope of this book to treat these complex models.

A commonly used empirical model for incoming longwave radiation is based on the observation that incoming longwave radiation comes mainly from the lowest few

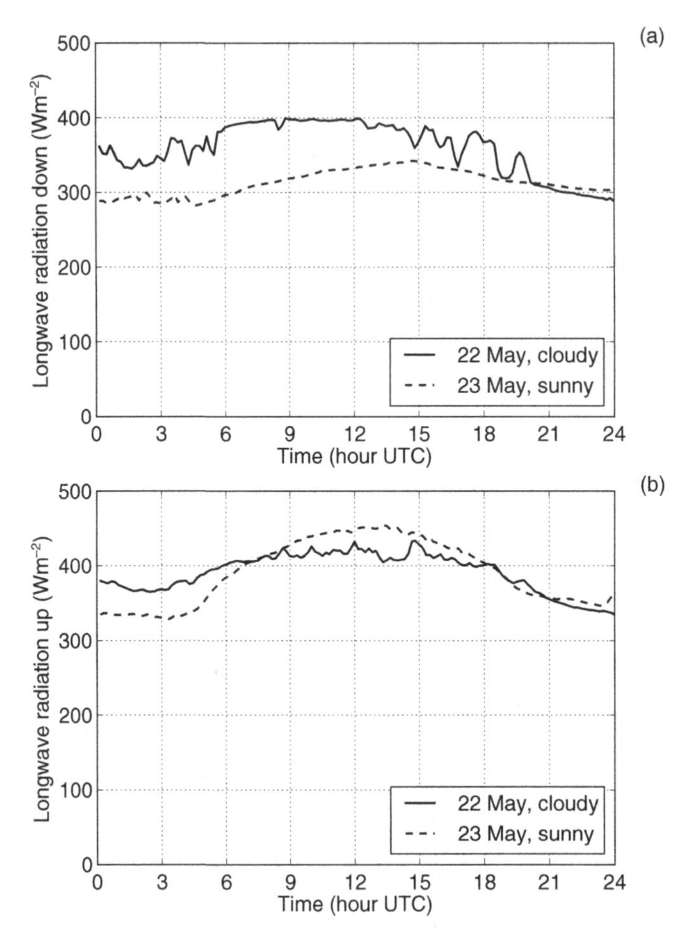

Figure 2.14 Observations of longwave radiation at Haarweg Meteorological station for a clear day May 23, 2007) and a cloudy day (May 22, 2007). (**a**) Downwelling longwave radiation. (**b**) upwelling longwave radiation.

hundred meters (of which the temperature correlates with the temperature at standard level). Furthermore, the atmosphere is assumed to be a grey body, with apparent emissivity ε_a:

$$L^{\downarrow} = \varepsilon_a \sigma T_a^4 \tag{2.24}$$

where σ is the Stefan–Boltzmann constant (= 5.67 10^{-8} W m^{-2} K^{-4}). For cloudless conditions, water vapour is the most variable, radiatively active, constituent. This observation is the basis for many empirical models for ε_a (see, e.g., Crawford and Duchon, 1999). One often used model for ε_a for clear conditions ($\varepsilon_{a,clear}$) is the approximation of Brunt (1932):

$$\varepsilon_a = \varepsilon_{a,clear} = c_1 + c_2\sqrt{e_a} \tag{2.25}$$

where e_a is the water vapour pressure in hPa, and c_1 and c_2 are empirical constants with standard values of 0.52 and 0.065 $hPa^{1/2}$ (but having ranges of 0.34–0.71 and 0.023–0.110, depending on location and season (Jiménez et al., 1987)). In combination with Eq. (2.24) the expression for $\varepsilon_{a,clear}$ yields a model for the clear-sky longwave radiation L_0.

For conditions with clouds, the total incoming longwave radiation can be parameterized as an interpolation between the clear sky emissivity and the emissivity of clouds (equal to one):

$$\varepsilon_a = f_{cloud} + (1 - f_{cloud})\varepsilon_{a,clear} \tag{2.26}$$

where f_{cloud} is the cloud fraction (Crawford and Duchon, 1999). It should be noted that it depends on the height of the cloud base to which extent the air temperature at screen level (as used in Eq. (2.24)) is actually representative for the cloud temperature.

Question 2.12: Ozone and oxygen in the atmosphere absorb radiation mainly in the shortwave region (see Figure 2.2). At what wavelengths is the absorbed energy emitted again?

Question 2.13: Of the total radiation emitted by clouds, only the part inside the atmospheric window reaches the ground. Consider Figure 2.14a.

a) If we assume that around 9 UTC the temperature of the lower atmosphere is identical on May 22 and May23, then how large is the contribution of the clouds to the longwave incoming radiation?
b) At a temperature representative of the temperature of the lower atmosphere, a black body emits roughly 35% of its radiation in the wavelength range of the atmospheric window. What is the temperature of clouds on May 22?

Question 2.14: Consider the observations of downwelling longwave radiation in Figure 2.14a. On May 22 and 23, the air temperatures at screen level at 12 UTC were 19.0 °C and 19.5 °C. At those moments the relative humidity was 88% and 49%, respectively.

a) Compute the atmospheric emissivity for both days at 12 UTC.
b) Estimate the atmospheric emissivity for both days based on Eqs. (2.25) and (2.26).
c) Evaluate the error due to the use of the empirical emissivity of answer (b) in both downwelling longwave radiation and the net longwave radiation (upwelling longwave radiation is given in Figure 2.14b).

2.2.5 Emitted (and Reflected) Longwave Radiation

In good approximation Earth's surface behaves as a grey-body in the longwave part of the spectrum. Consequently, the longwave radiation emitted by the surface can by approximated as

$$L_e^{\uparrow} \approx \varepsilon_s \sigma T_s^4 \tag{2.27}$$

in which ε_s is the broadband emissivity of the surface in the longwave wavelength region and T_s is the surface temperature (in K).

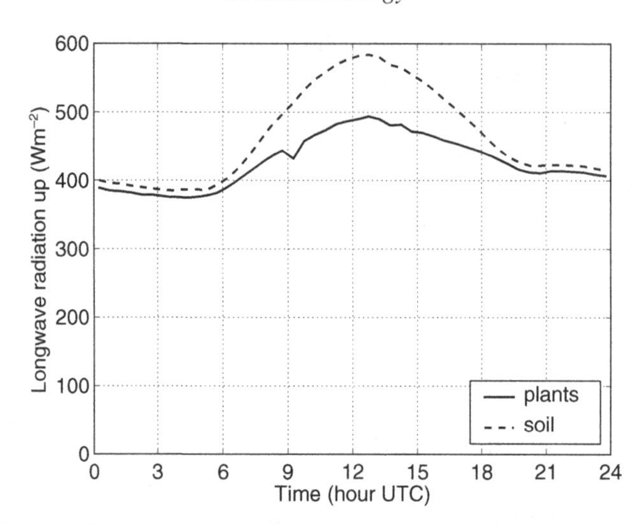

Figure 2.15 Upwelling longwave radiation as observed at a vineyard site in Central Spain (Tomelloso), June 12, 1991. Radiation has been measured separately above the vine plants, and the bare soil in-between.

Typical values for the surface emissivity can be found in Table 2.1. For most surface types ε_s is between 0.9 and 0.99. This implies that there is *some* reflection of longwave radiation.[2] Hence, the total upwelling longwave radiation is the sum of the emitted radiation, $L_e^{\uparrow}$ (see earlier) and the reflected radiation:

$$L^{\uparrow} = L_e^{\uparrow} + (1 - \varepsilon_s)L^{\downarrow} \tag{2.28}$$

Figure 2.14b shows observations of $L^{\uparrow}$ for two days. For the cloud-free day, the diurnal cycle of $L^{\uparrow}$ is larger than for the overcast day. During daytime the higher value of $L^{\uparrow}$ is due to the larger insolation that results in a higher surface temperature. During night time the surface temperature under cloud-free conditions drops lower than for the night with clouds, mainly because the incoming longwave radiation is smaller (see Figure 2.14a).

But the surface temperature is not only coupled to variations in the total incoming radiation (shortwave and/or longwave). Another important factor that determines the surface temperature is the partitioning of the available energy between evapotranspiration and sensible heat flux. This is illustrated in Figure 2.15 showing observations made over a sparse vegetation (mixture of plants and bare soil) in Spain. The upwelling longwave radiation has been measured separately above the plants (which actively transpire) and the dry bare soil. It is clear that the plants emit much less longwave radiation than the bare soil because the surface temperature of the plants is

[2] Per Kirchhoff's law, spectral absorptivity (α_λ) equals spectral emissivity (ε_λ). For an opaque surface, the reflectivity, r_λ then equals $1 - \alpha_\lambda = 1 - \varepsilon_\lambda$. If we extend this to broadband values, integrated over the longwave wavelength region, r_{long} becomes $(1 - \varepsilon_s)$.

lower. This in turn is due to the fact the plants use a large part of the available energy for evapotranspiration, rather than to heat up their leaves and stems (see Chapter 6). The partitioning of available energy between evapotranspiration and sensible heat flux, as well as the coupling between that partitioning and the emitted longwave radiation, is one of the important subjects of the forthcoming chapters.

Question 2.15: Determine the surface temperature at 12 UTC on May 22 and May 23, using the data in Figure 2.14a and b.
a) Assuming a surface emissivity of 0.96 (see Table 2.1)
b) Assuming a surface emissivity of 1.0
c) Assuming a surface emissivity of 0.96 but neglecting the reflection of downwelling longwave radiation

Question 2.16: Determine the difference in surface temperature between the bare soil and the vine plants at 12 UTC for the data in Figure 2.15. Use a reasonable assumption on the surface emissivity. Incoming longwave radiation at 12 UTC is 410 W m^{-2}.

2.2.6 Net Radiation: Sum of Components

Now that all components of the net radiation at Earth's surface have been introduced, they can be combined to yield the net radiation. First we examine the net shortwave and net longwave radiation, K^* and L^* respectively. Figure 2.16a shows that K^* closely follows the diurnal course of the incoming solar radiation (the values at night are not identical to zero owing to finite accuracy of the sensors; see 2.2.7). As compared to K^* and to the individual components $L^{\downarrow}$ and $L^{\uparrow}$, L^* is rather small: ranging between -100 and 0 W m^{-2} (see Figure 2.16b). On the cloud-free day it has a distinct diurnal cycle. On the cloudy day, however, L^* is close to zero. This is due to the fact that with the inclusion of clouds, the atmosphere acts as a nearly black body. Given the fact that the surface also has an emissivity close to 1, L^* is determined mainly by the difference in temperature between the surface and the lower atmosphere. On overcast days this difference will be small. Figure 2.16c shows that the diurnal cycle of net radiation is dominated by that of K^*. However, at night L^* is the only determining factor.

2.2.7 Measurement of Net Radiation

The measurement of net radiation is in principle straightforward. Sensors (radiometers) exist that can measure either the short wave radiation flux density, or the longwave radiation flux density. Combination of two instruments of each (one facing upward, the other facing downward) yields the four components of the net radiation. This is the preferred way of measuring net radiation. Simpler instruments exist that combine all four sensors in one, and yield the net radiation directly.

Shortwave radiation can be measured with a pyranometer, whereas the device used to measure longwave radiation is called a pyrgeometer (see Figure 2.17). The measuring principle of most radiometers is that radiation is absorbed at the top of a sensor. The

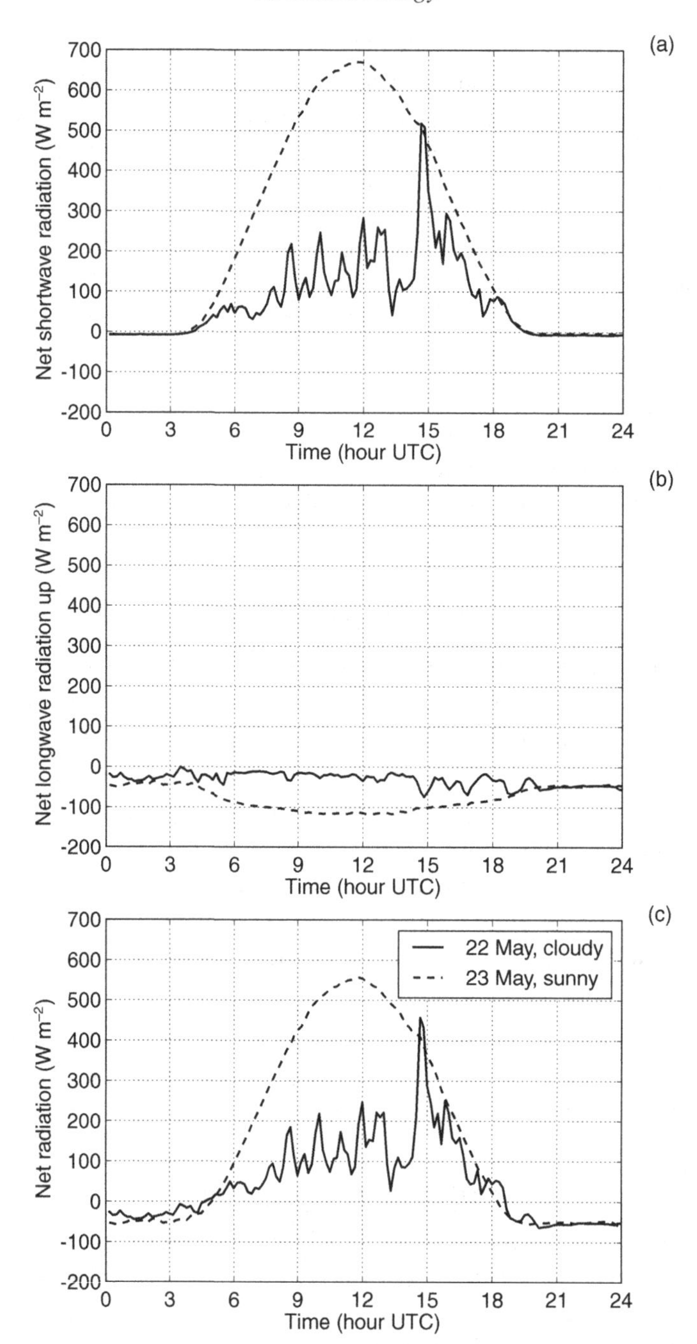

Figure 2.16 Observations of net radiation (and its components) at Haarweg Meteorological station for a clear day (May 23, 2007) and a cloudy day (May 22, 2007). (**a**) Net shortwave radiation. (**b**) Net longwave radiation. (**c**) Net radiation.

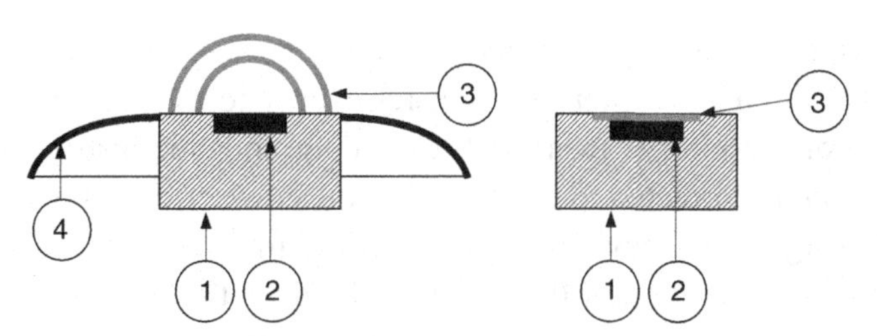

Figure 2.17 Sketch of a pyranometer (left) and a pyrgeometer (right). 1, housing; 2, sensor, 3, transparent dome (left) or longwave transparent filter (right); 4, radiation shield.

temperature difference between the heated top and the cooler bottom then is a measure of the heat flux through the sensor, which is under steady conditions equal to the radiation input at the top. The distinction between a pyranometer and a pyrgeometer is made by the cover (a dome or a flat plate) that protects the radiation absorbing surface from the atmosphere. For a pyranometer this dome is usually made of glass or quartz, limiting the spectral response to 0.29–2.8 or 0.29–4.0 μm, respectively (Gueymard and Myers, 2008). The filter of a pyrgeometer blocks shortwave radiation and transmits longwave radiation (although there may be some absorption). Net-radiometers have a dome that is transparent for both longwave and shortwave radiation.

Whereas a pyranometer absorbs only radiation that is external to the instrument (except for the thermal offset; see later), the radiation received by a pyrgeometer sensor is a balance between the radiation transmitted through the filter, the radiation emitted by the filter (if it is not fully transparent for longwave radiation) and the emission of the sensor. Hence, to derive the incident longwave radiation, not only this net effect (balance between incoming and emitted radiation) needs to be measured, but the sensor's emission as well. This entails – at least – the additional determination of the temperature of the instrument's housing.

For pyranometers two instrument-related error sources can be identified. First, a non-perfect cosine response implies that the absorbing surface does not act as a Lambertian surface. This implies that the sensitivity is not equal for radiation from all directions. Second, there may be thermal errors: the dome is generally cooler than the heated absorbing surface (especially, but not exclusively, during night). Through radiative and convective exchange this cools the absorbing surface, leading to a negative offset (which is particularly visible at night when the instrument should give a zero flux). The thermal offset can in part be suppressed by ventilating the instrument in order that the temperature of the dome is closer to that of the housing. An additional advantage of ventilation is that dew formation is suppressed. Other error sources related to installation and use are insufficient cleaning of the domes and incorrect horizontal alignment of the instrument (Kohsiek et al., 2007).

The main instrument-related error for pyrgeometers is that the filter – which blocks shortwave radiation – will heat up considerably under sunny conditions. If the filter is not fully transparent for longwave radiation it will also have a non-zero emissivity. This implies that the heated filter will transfer energy to the sensor by radiation. If the temperature of the filter is measured (as is done in some instrument types) this radiation input can be corrected for. But in addition, convection within the instrument may transport energy. This may amount to 10–20 W m^{-2} in bright sunshine (Kohsiek et al. 2007). Again, ventilation of the sensor will reduce these errors by reducing the thermal contrast between different parts of the instrument.

The aforementioned error sources for pyranometers and pyrgeometers are equally valid for net-radiometers. An additional problem with net-radiometers is that the sensitivity of the sensor should be equal for longwave radiation and shortwave radiation (i.e., 100 W m^{-2} of longwave radiation should produce the same voltage as 100 W m^{-2} of shortwave radiation). For many instruments these sensitivities are far from equal, leading to errors during daytime in particular, when the relative importance of longwave and shortwave radiation varies considerably (Halldin and Lindroth, 1992).

A final remark concerns the scale and spatial homogeneity of the observed radiation. Upward pointing sensors collect radiation from the entire hemisphere and the measured radiation will not depend much on the exact location of the sensor in a field. The only important consideration is that no obstacles should be located in the field of view of the sensor (unless one is explicitly interested in the effect of those obstacles, e.g. to study the radiation inside a canopy; see Chapter 6). On the other hand, a downward pointing sensor receives radiation from a limited area only: the footprint from which roughly 50% of the flux originates is a circle with radius equal to one time the instrument height. For a 90% recovery, the circle has a radius of three times the instrument height (Schmid, 1997). When measurements are made over a surface with heterogeneous radiative characteristics, the exact location of the sensor will directly influence the observed net radiation.

2.3 Soil Heat Flux

Although invisible to the eye, the heat transport that occurs below the soil surface can be both important and hard to determine. The heat transport can also be modified strongly by the presence of vegetation cover or snow, and when freezing of soil moisture occurs. In this section we first discuss the basics of heat transport in the soil, with special emphasis on the specific properties of soils (as opposed to simple solids or fluids). Second, we look at an idealized case where the temperature at the soil surface varies as a sine (e.g., diurnal or yearly cycle). Then simplified models are treated for bare soil conditions and vegetated surfaces. Finally, snow cover and frost penetration into the soil are dealt with. Note that in this section the vertical coordinate z_d is used, which is taken positive *downward* (i.e., it indicates depth rather than height).

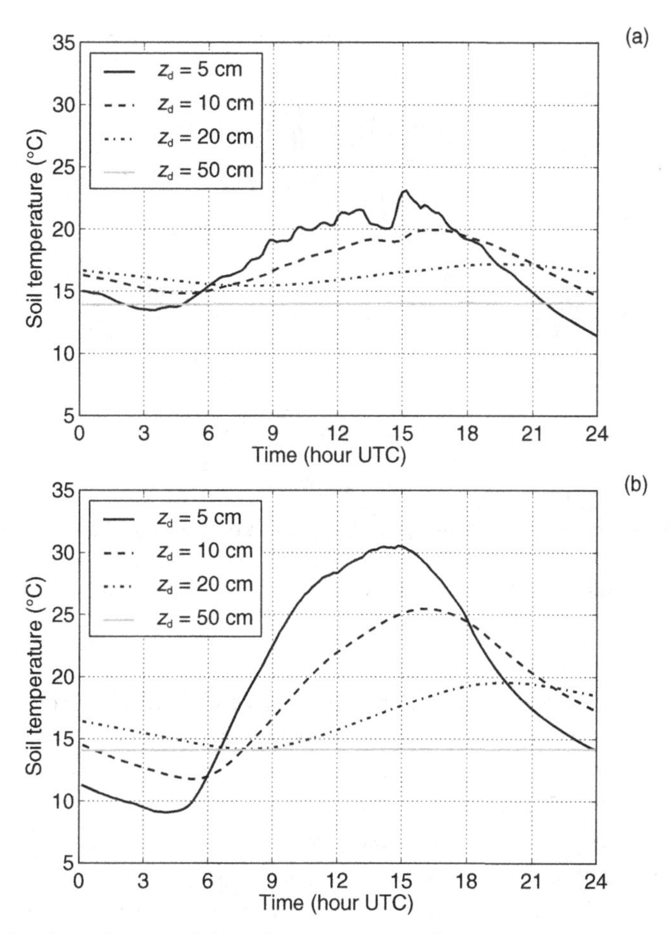

Figure 2.18 Diurnal cycle of the soil temperature of a bare soil plot at four depths at Haarweg Meteorological station on a cloudy day (May 22, 2007, left) and on a sunny day (May 23, 2007, right). Note that the lowest level is actually below grass, but the difference will be insignificant.

2.3.1 Bare Soil

To get an impression of some important features of the soil temperature, the diurnal cycle of the soil temperature at four depths is shown in Figure 2.18 for two consecutive days: a cloudy and a sunny day. Temperatures have been measured at 5 cm, 10 cm, 20 cm and 50 cm below the (bare) soil surface. First the four curves within one plot are compared. A few important features are visible:

- The shape of the temperature curve for May 23 is similar to a sine-wave, although the curve is not exactly symmetric (steep increase in the morning, slow decrease in the afternoon).
- The amplitude of the diurnal variation decreases with depth. On May 22 the amplitudes at the four depths are 4 K, 2.5 K, 1 K and 0 K, whereas on the sunny day the amplitudes are 10 K, 7 K, 3 K and 0 K, respectively.

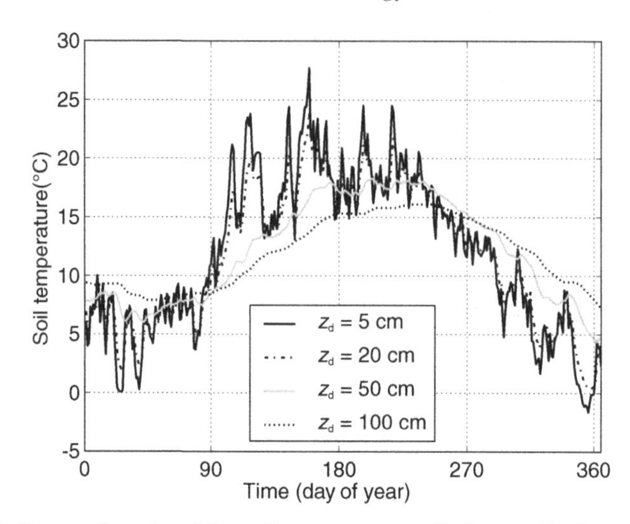

Figure 2.19 Seasonal cycle of the soil temperature of a bare soil plot at four depths (different levels than in Figure 2.18) at Haarweg Meteorological station in the year 2007. Note that the lowest two levels are actually below grass, but the difference will be insignificant.

- The time at which the maximum temperature is reached shifts to later times when going further below the surface. On May 23 the maximum temperature at 5 cm depth occurs at 14 UTC (2 hours after local noon; see Figure 2.6), whereas the maximum at 20 cm depth occurs around 19 UTC, i.e., 5 hours later.
- On the cloudy day, the effect of broken clouds is visible in the observations at 5 cm depth, but the temperatures at greater depth are smooth.
- At 50 cm depth there is no diurnal cycle but a very slight linear increase is visible. The data show an increase of 0.15 K in one day.

Apart from the diurnal cycle, there is also a yearly cycle in soil temperatures. The day-to-day variation of the daily mean soil temperature is shown for the same location in Figure 2.19. Similar features occur as observed for the diurnal cycle:

- The amplitude of the yearly cycle decreases with depth, but is still clearly visible at 50 and 100 cm depth.
- The peak of the temperature shifts to later dates with increasing depth (at 100 cm depth, the maximum temperature is reached in September).
- The short-term (day-to-day) variations are clearly visible at 5 and 20 cm depth, but hardly affect the temperatures at 50 and 100 cm depth. Only the longer cold and warm spells in the end of the year penetrate down to 100 cm.

Although the diurnal and yearly cycles appear to behave very similarly, there is one important difference: the depth at which the cycle is no longer visible (this is where it has been damped). For the diurnal cycle this is somewhere between 20 and 50 cm deep, whereas for the yearly cycle this level will lie well below 100 cm.

The explanation and physical description of the features presented in Figure 2.18 and Figure 2.19 are given in Section 2.3.4, but first the theory of heat transport in soils needs to be treated.

2.3.2 Heat Transport in Soils

This section deals with the transfer of heat in a homogeneous soil, that is, the soil physical properties do not vary in space. Heat transport in the soil mainly takes place by conduction, that is, it is a function of the local temperature gradient and a thermal conductivity λ_s (in W m^{-1} K^{-1}). Hence, the soil heat flux density G is given by:

$$G = -\lambda_s \frac{\partial T}{\partial z_d} \tag{2.29}$$

The soil heat flux in turn may change with depth. This implies that heat is stored in the soil or extracted from the soil: if more heat enters a soil layer at the top than leaves the layer at the bottom, the layer has to heat up. This is expressed by:

$$\rho_s c_s \frac{\partial T}{\partial t} = -\frac{\partial G}{\partial z_d} \tag{2.30}$$

where ρ_s is the density of the soil and c_s is the specific heat capacity (in J kg^{-1} K^{-1}): a change of temperature in time is due to the divergence of the flux with depth.

Combination of Eqs. (2.29) and (2.30) leads to the following diffusion equation:

$$\frac{\partial T}{\partial t} = \frac{\lambda_s}{\rho_s c_s} \frac{\partial^2 T}{\partial z_d^2} = \kappa_s \frac{\partial^2 T}{\partial z_d^2} \tag{2.31}$$

where κ_s is the thermal diffusivity of the soil (in m^2 s^{-1}). Equation (2.31) describes how the temperature in the soil changes in time depending on the shape of the temperature profile (recall that the second derivative of the temperature profile is the curvature (non-linearity) of the profile). With the use of the definition of the thermal diffusivity ($\kappa_s = \lambda_s / (\rho_s c_s)$) given earlier, Eq. (2.29) can then be written in a form that is more familiar in atmospheric applications:

$$G = -\rho_s c_s \kappa_s \frac{\partial T}{\partial z_d} \tag{2.32}$$

(see Chapters 1 and 3 to compare). In the soil the use of volumetric quantities is usually more convenient. Therefore, the product $\rho_s c_s$ is often replaced by the volumetric heat capacity C_s.

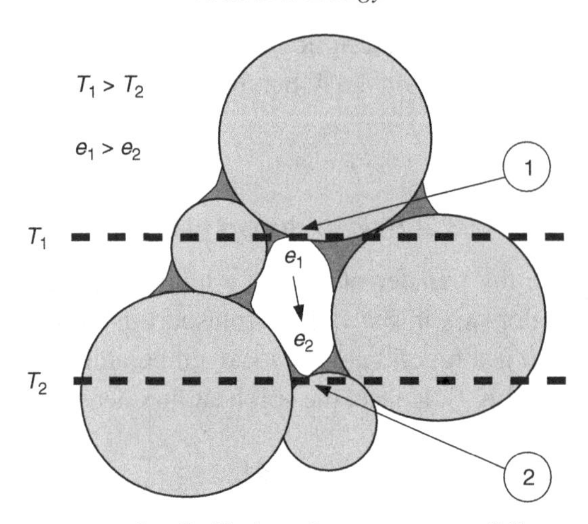

Figure 2.20 Heat transport by distillation: the temperature difference $T_2 - T_1$ causes a difference in vapour pressure $e_2 - e_1$, which in turn induces moisture transport. Water evaporates at location 1 and condensates at location 2, hence transporting latent heat.

Note that we deal solely with heat transport by conduction. However, if soil moisture movement occurs in a direction in which a temperature gradient exists, this transport of liquid water will entail a transport of heat as well. This can be particularly relevant in the case of infiltration of rain (see Kollet et al., 2009). Another mode of heat transport is through the movement of water vapour: the transport of latent heat (see Figure 2.20). The air entrapped in soil pores is generally saturated with water vapour (except for very dry soils). Because the saturated vapour pressure depends on temperature, a gradient in temperature will also entail a gradient in vapour pressure. The latter will induce the transport of water vapour down the vapour pressure gradient (which has the same direction as the temperature gradient). At the low-temperature end of the soil pore the water vapour will condensate, releasing its latent heat (distillation). Van Wijk (1963) shows how this transport can be incorporated by increasing the thermal conductivity of the air phase (it may increase by a factor of 4). Additional modes of heat transport in soils are discussed in Farouki (1986).

Question 2.17: Note down the name and physical interpretation of the following soil thermal properties: λ_s, κ_s, c_s and C_s.

Question 2.18: The increase of the soil temperature (e.g., Figure 2.18b) is an expression of the storage of heat in the soil. Assume that the temperature curve for a depth of 5 cm is representative of the entire layer of 0–10 cm depth.

a) For Figure 2.18b, how much heat is stored in the upper 10 cm in the period between 4 and 15 UTC (assume a volumetric heat capacity of 3.0 10^6 J K^{-1} m^{-3}).
b) What is – over the same time period – the mean difference between the soil heat flux at the top of the soil ($z_d = 0$) and at 10 cm depth?

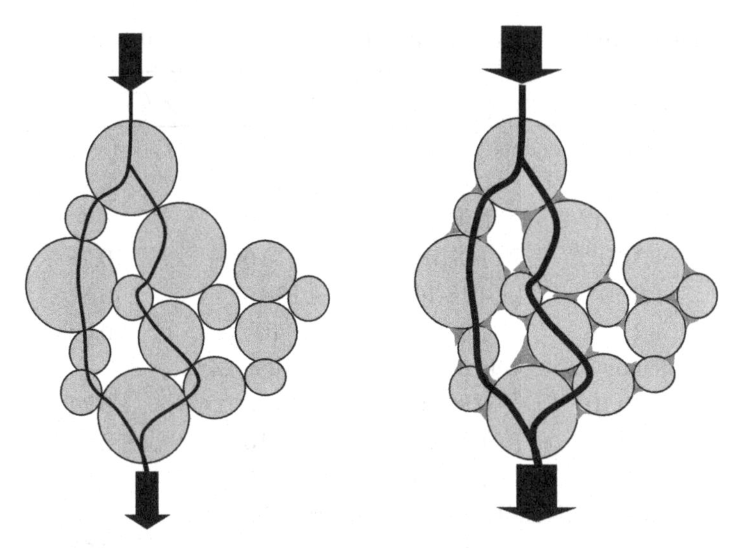

Figure 2.21 Conduction of heat through the soil matrix: a dry soil (left) and a soil with some moisture (right).

2.3.3 Thermal Properties of Soils

The transport of heat critically depends on the thermal properties of the soil. For some of the properties (density and heat capacities) a soil is simply a mixture of three phases: solid particles, water and air, whereas for the conductivity (and hence the diffusivity) the structure of the soil, as well as the water content, is important. Note that the phase 'solid particles' will generally be made up of a variety of materials (quartz, clay minerals, organic material).

We indicate the properties of the three phases with a subscript 'p' (particles), 'w' (water) and 'a' (air), and the respective volumetric fractions of the phases by f_p, θ and f_a. Then density, volumetric heat capacity and specific heat capacity of the soil are given by:

$$\begin{aligned} \rho_s &= f_p\rho_p + \theta\rho_w + f_a\rho_a \\ C_s &= f_pC_p + \theta C_w + f_aC_a \\ c_s &= C_s/\rho_s \end{aligned} \tag{2.33}$$

Thus, both the density and the volumetric heat capacity of the soil depend linearly on the soil moisture content (see Figure 2.22). Table 2.2 lists the thermal properties of the soil constituent used in (2.33), as well as those of a number of typical soils.

For the conductivity λ_s the situation is more complex (see Figure 2.21). The conductivity of the material of the particles is rather large (e.g., for quartz λ_s = 8.8 W m^{-1} K^{-1}; see Table 2.2) whereas air is nearly an insulator (λ_a = 0.025 W m^{-1} K^{-1}). As a result, the transport of heat in a dry soil is limited by the surface area of the points of contact between individual soil particles. The addition of a little water then

Table 2.2 Typical values for thermal properties of various soil materials and soils

Material	ρ (kg m^{-3})	c (J kg^{-1} K^{-1})	C (J m^{-3} K^{-1})	λ (W m^{-1} K^{-1})	κ (m^2 s^{-1})
	Soil components[a]				
Quartz	$2.66\cdot10^3$	$0.80\cdot10^3$	$2.13\cdot10^6$	8.8	$4.2\cdot10^{-6}$
Clay mineral	$2.65\cdot10^3$	$0.90\cdot10^3$	$2.39\cdot10^6$	2.9	$1.2\cdot10^{-6}$
Organic Matter	$1.3\cdot10^3$	$1.9\cdot10^3$	$2.47\cdot10^6$	0.25	$0.10\cdot10^{-6}$
Water	$1.0\cdot10^3$	$4.18\cdot10^3$	$4.18\cdot10^6$	0.57	$0.14\cdot10^{-6}$
Still air	1.2	$1.004\cdot10^3$	$1.2\cdot10^3$	0.025	$21\cdot10^{-6}$
	Sandy soil loosely packed (pore fraction 0.4)[a,b]				
Dry	$1.60\cdot10^3$	$0.80\cdot10^3$	$1.28\cdot10^6$	0.24	$0.19\cdot10^{-6}$
$\theta = 0.2$	$1.80\cdot10^3$	$1.18\cdot10^3$	$2.12\cdot10^6$	2.1	$0.99\cdot10^{-6}$
$\theta = 0.4$	$2.00\cdot10^3$	$1.48\cdot10^3$	$2.96\cdot10^6$	2.5	$0.85\cdot10^{-6}$
	Sandy soil tightly packed (pore fraction 0.33)[a,b]				
Dry	$1.78\cdot10^3$	$0.80\cdot10^3$	$1.42\cdot10^6$	0.29	$0.20\cdot10^{-6}$
$\theta = 0.15$	$1.93\cdot10^3$	$1.06\cdot10^3$	$2.05\cdot10^6$	2.5	$1.2\cdot10^{-6}$
$\theta = 0.33$	$2.11\cdot10^3$	$1.33\cdot10^3$	$2.81\cdot10^6$	2.9	$1.0\cdot10^{-6}$
	Clay soil (pore fraction 0.4)[a]				
Dry	$1.59\cdot10^3$	$0.90\cdot10^3$	$1.43\cdot10^6$	0.15	$0.10\cdot10^{-6}$
$\theta = 0.2$	$1.79\cdot10^3$	$1.27\cdot10^3$	$2.27\cdot10^6$	0.9	$0.4\cdot10^{-6}$
$\theta = 0.4$	$1.99\cdot10^3$	$1.56\cdot10^3$	$3.10\cdot10^6$	1.4	$0.45\cdot10^{-6}$
	Peat soil (pore fraction 0.9)[a,c]				
Dry	$0.13\cdot10^3$	$1.90\cdot10^3$	$0.25\cdot10^6$	0.04	$0.16\cdot10^{-6}$
$\theta = 0.45$	$0.58\cdot10^3$	$3.67\cdot103$	$2.13\cdot10^6$	0.27	$0.13\cdot10^{-6}$
$\theta = 0.90$	$1.03\cdot10^3$	$3.89\cdot10^3$	$4.01\cdot10^6$	0.50	$0.12\cdot10^{-6}$
	Other materials[d]				
Rock	$2.7\cdot10^3$	$0.75\cdot10^3$	$2.03\cdot10^6$	2.9	$1.4\cdot10^{-6}$
Ice	$0.9\cdot10^3$	$2.09\cdot10^3$	$1.88\cdot10^6$	2.5	$1.3\cdot10^{-6}$
Fresh snow	$0.2\cdot10^3$	$2.09\cdot10^3$	$0.42\cdot10^6$	0.1	$0.3\cdot10^{-6}$
Old snow	$0.8\cdot10^3$	$2.09\cdot10^3$	$1.67\cdot10^6$	1.7	$1.0\cdot10^{-6}$

Note that these properties are temperature dependent. Values given here are representative for temperatures in the range 10–20 °C (different sources use different reference temperatures). Data from: [a]De Vries (1963) (see Clauser and Huenges, 1995, for a more extensive review of conductivities of soil minerals and rocks); [b]Smits et al. (2010); [c]O'Donnell et al. (2009); [d]Lee (1978).

makes a large difference. The water will be concentrated in the narrowest parts of the pores (see Chapter 4), that is, close to the contact points between the soil particles. The addition of only a little water will increase the area of contact greatly. Although the conductivity of water is smaller than that of the material of the particles, it is much higher than that of air. As a result the total pathway for heat transport increases.

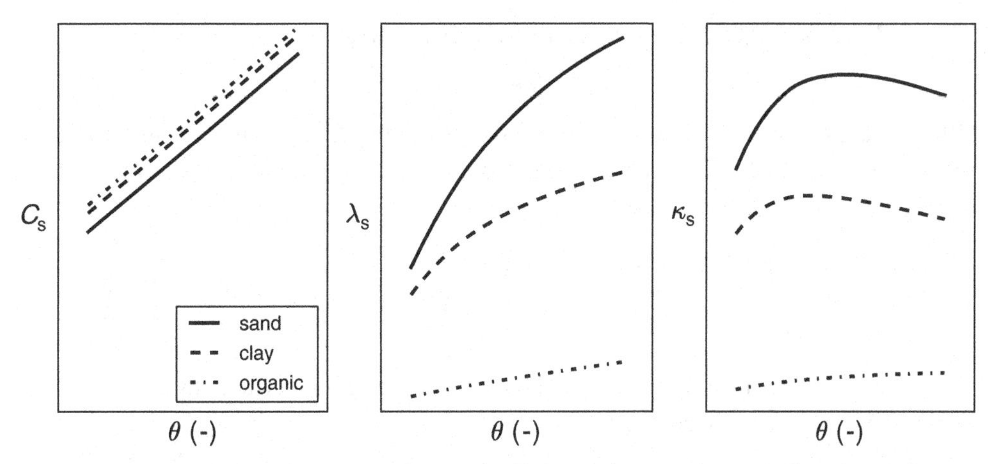

Figure 2.22 Sketch of the dependence of soil thermal properties on soil moisture for three soil types (sandy soil, clay soil, peat): volumetric heat capacity (left), thermal conductivity (middle), and thermal diffusivity (right). (Based on the model of De Vries, 1963)

This effect is most pronounced for dry soils: the addition of a little water increases the thermal conductivity considerably. When the water content is increased further a stage will come where the extra water will not have a large effect: the fact that the conductivity of water is much lower than that of the soil material causes that little extra pathway is added with the addition of extra water.

Various empirical models for the soil moisture dependence of the thermal conductivity exist (see Farouki, 1986, for a review). Peters-Lidard et al. (1998) show that in a land-surface model (such as those discussed in Section 9.2) errors in the estimated soil thermal conductivity not only affect the soil heat flux but also impact on the partitioning of energy between sensible and latent heat flux. In Section 9.1.7 one example of a model for the thermal conductivity as a function of soil composition is discussed.

To see the effect of water content on the diffusivity κ_s we have to take both the conductivity and the volumetric heat capacity into account. Because the increase of λ_s with water content levels off, whereas the increase of C_s with soil moisture is linear, the diffusivity (recall that $\kappa_s = \lambda_s/C_s$) first increases with soil moisture, but at higher soil moisture contents it decreases again. This effect is sketched in Figure 2.22.

Question 2.19: Given a soil with a porosity of 40%, where the matrix (60%) consists of 20% quartz, 50% clay and 30% organic material. The pores are filled with 75% water and 25% air (i.e., soil water content is 30%).

a) Which of ρ_s, c_s, C_s, λ_s or κ_s can be calculated?

b) If possible calculate the thermal soil properties (using Table 2.2).

2.3.4 Semi-infinite Homogeneous Soil with Sine-Wave at the Surface

Soil Temperature

Although the diurnal and yearly cycles of soil temperatures shown in Figure 2.18 and Figure 2.19 are not perfect sines, the analysis of soil temperatures assuming a sinusoidal behaviour is still very useful because:

- To first order the diurnal and yearly cycle are sinusoidal.
- Any signal can be considered as a the sum of sines and cosines with a range of periods (Fourier series).

We consider a semi-infinite homogeneous soil where at the surface, that is, at $z_d = 0$ the temperature is prescribed as:

$$T(0,t) = \overline{T} + A(0)\sin(\omega t) \tag{2.34}$$

in which $A(0)$ is the amplitude of the temperature wave and ω is the frequency of the wave ($\omega = \frac{2\pi}{P}$ with P the wave period). If the diurnal cycle is considered $P = 86\,400$ s (1 day), whereas for the annual cycle $P = 365.25 \cdot 86\,400$ s. With the surface boundary condition given by Eq. (2.34), a lower boundary condition $T(\infty,t) = \overline{T}$ and the assumption that the initial temperature field does not play a role, the solution of Eq. (2.31) is (Carslaw and Jaeger, 1959):

$$T(z_d,t) = \overline{T} + A(z_d)\sin\left(\omega t - \frac{z_d}{D}\right) \tag{2.35}$$

where the amplitude at depth z_d is given by:

$$A(z_d) = A(0)e^{-z_d/D} \tag{2.36}$$

and the penetration depth (or damping depth or e-folding depth) is given by:

$$D \equiv \sqrt{\frac{2\kappa_s}{\omega}} \tag{2.37}$$

The physical interpretation of Eq. (2.35) is as follows:

1. The mean temperature at depth z is identical to that at the surface.
2. The amplitude of the temperature variation decreases with depth.
3. There is a phase shift between the surface temperature and the temperature at depth z_d of z_d/D radians. This corresponds to a time shift of $\frac{z_d}{D}\frac{P}{2\pi}$.

Features (2) and (3) correspond to what we have seen in the observations in Figures 2.18 and 2.19. Feature (1) holds only for the yearly cycle shown in Figure 2.19.

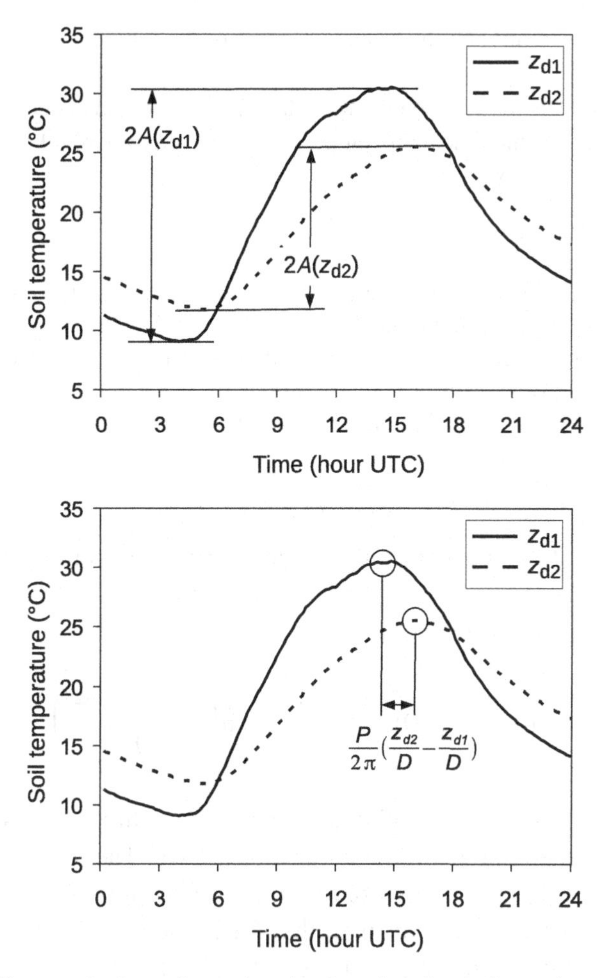

Figure 2.23 Two methods to determine the damping depth from observed soil temperatures at two depths. Top: Ratio of amplitude at two depths yields damping depth through $A(z_{d2})/A(z_{d1}) = e^{-(z_{d2}-z_{d1})/D}$. Bottom: Phase shift between temperature waves at two depths gives damping depth through $t_2 - t_1 = (P/2\pi)(z_{d2}-z_{d1})/D$.

In the diurnal cycle shown in Figure 2.18 the yearly cycle is superimposed, and as a result the temperature at larger depth is not in balance with the mean temperature at the surface.

The model of the soil temperature given by the combination of Eqs. (2.34) and (2.35) can be used to estimate the thermal diffusivity from soil temperature observations at two depths by two methods (see also Figure 2.23):

- If the amplitudes of the temperature wave at two depths (z_{d1} and z_{d2}) are compared, D can be derived from the ratio $A(z_{d2})/A(z_{d1})$ in combination with Eq. (2.36): $A(z_{d2})/A(z_{d1}) = e^{-(z_{d2}-z_{d1})/D}$. From D the diffusivity can be derived, with a known ω. In practice it is usually easier to determine twice the amplitude by taking the difference between the maximum and the minimum value of the temperature in the given period (see Figure 2.23).

- If the time shift of the temperature wave between two depths is compared (e.g., comparing the time at which the temperature reaches its maximum), D can be derived from $t_2 - t_1 = \frac{P}{2\pi}\left(\frac{z_{d2}}{D} - \frac{z_{d1}}{D}\right)$.

Question 2.20: Given soil temperature observations, related to the diurnal cycle. The maximum soil temperature at the surface occurs at 13:00 local time, whereas the maximum at 20 cm depth occurs at 19:30. Assume that the soil is homogeneous.

a) Calculate the thermal diffusivity of this soil.
b) Calculate the amplitude of the soil temperature at 20 cm, relative to the amplitude at the surface.

Soil Heat Flux

Because the soil heat flux depends only on the thermal conductivity and the temperature gradient (Eq. (2.29)), the model for the temperature profile (Eq. (2.35)) also gives a model for the soil heat flux[3]:

$$G(z_d,t) = A(0)e^{-z_d/D}\sqrt{\frac{\omega}{\kappa_s}}\lambda_s \sin\left(\omega t - \frac{z_d}{D} + \frac{\pi}{4}\right) \tag{2.38}$$

Equation (2.38) shows that the amplitude of the soil heat flux decreases with depth in a similar way as the amplitude of the temperature wave. The result of this change of G with depth is that the temperature of the soil changes (flux divergence). The phase shift, however, is slightly different than that for the temperature. Because $\pi/4$ corresponds to one-eighth of the period of oscillation P, it follows from Eq. (2.38) that, at a given depth, the time of maximum heat flux precedes the time of maximum temperature by 3 hours for the daily cycle and by one-and-a-half months for the annual cycle. This seems counterintuitive. But note that we could have taken Eq. (2.38) as the prescribed boundary condition. Then we would have considered the *soil heat flux* as forcing at the surface rather than the *surface temperature* (the first would be more natural in the context of the surface energy balance). In that case Eq. (2.35) would have been found as the solution for $T(z_d,t)$: the interpretation would have been that the T-waves reach their maximum $\pi/4$ rad later than the G-wave at a given depth, which agrees with our intuition.

The change in amplitude of G, the phase shift with depth, and the phase shift between G and T is illustrated in Figure 2.24 with observations from a bare soil in the Negev desert. One of the striking features is that the soil heat flux at the surfaces,

[3] Recall that $\sin(x) + \cos(x) = \sqrt{2}\sin\left(x + \frac{\pi}{4}\right)$.

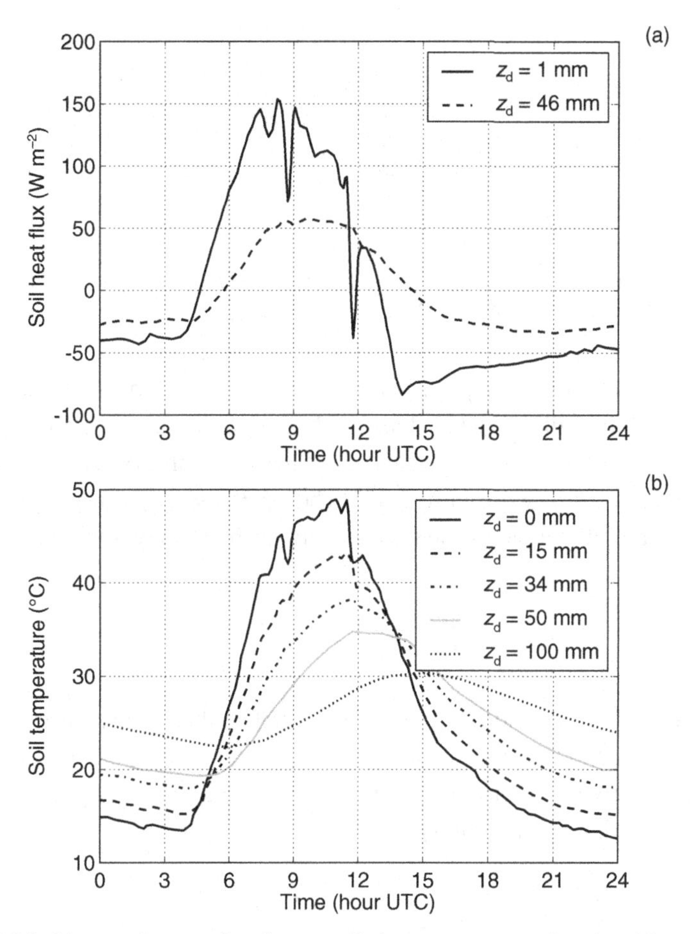

Figure 2.24 Observations related to soil heat transport in the Negev desert (September 30, 1997): soil heat flux just below the surface and at 4.6 cm depth (**a**) and observed soil temperatures at various depths (**b**). Note that local solar time is more than 2 hours ahead of UTC. (Data from Heusinkveld et al., 2004)

as well as the surface temperature show large fluctuations (due to clouds) that are already damped at only a few centimetres below the surface. In the context of sinusoidal temperature variations this can be understood by noting that the damping depth is smaller for oscillations with a frequency higher than the diurnal cycle (as in the case of a cloud shadow). Also note the sharp drop to large negative values of G at sunset due to strong longwave cooling of the hot surface.

2.3.5 Force-Restore Method

Although the theory described in the Section 2.3.2 can be applied directly in operational hydrological and meteorological models, it appears that a complete numerical solution of the governing equations requires too much computational time for some applica-

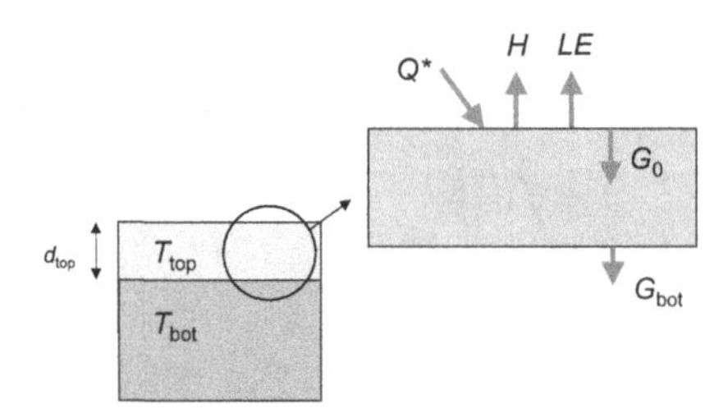

Figure 2.25 Force-restore method: soil column divided in a top soil and infinite reservoir (left); energy balance of the top layer (right).

tions. For that reason often an approximation is used. An example is the so-called force-restore method (Bhumralkar, 1975). This method approximates the soil by two layers: one top layer with temperature T_{top} and a thickness d_{top}, and one infinite layer with a constant temperature T_{bot} (see Figure 2.25). The thickness of the top layer is yet undefined. The time rate of change of the temperature of the top layer is given by:

$$\frac{\partial T_{top}}{\partial t} = \frac{1}{C_s d_{top}} \left(G_0 - G_{bot} \right) \tag{2.39}$$

which is a vertically integrated version of Eq. (2.30). Next we replace G_0 by the sum of the other terms of the energy balance and G_{bot} is taken proportional to the temperature difference between the top layer and the bottom layer and to an integrated conductivity Λ_s (to be determined later):

$$\frac{\partial T_{top}}{\partial t} = \frac{1}{C_s d_{top}} \{ \underbrace{(Q^* - H - LE)}_{\text{force}} - \underbrace{\Lambda_s (T_{top} - T_{bot})}_{\text{restore}} \} \tag{2.40}$$

The first term forces the top layer temperature away from its equilibrium T_{bot}, whereas the second term tends to restore the temperature back to T_{bot}.

Now the layer thickness d_{top} and the conductivity Λ_s need to be determined, subject to the following constraints:

- The temperature T_{top} should have the same amplitude as the surface temperature $T(0,t)$.
- The temperature T_{top} should have the same phase as the surface temperature $T(0, t)$.

These constraints are particularly important to ensure that the sensible heat flux and the upwelling longwave radiation, which both depend on the surface temperature, are correct.

The constraints result in the solution that $\frac{\Lambda_s}{C_s d_{top}} = \omega$ and $d_{top} = \sqrt{\kappa_s / 2\omega}$ (note that d_{top} is proportional to the damping depth, see Eq. (2.37)). Thus both the layer thickness and the proportionality constant in the restore term depend not only on the soil thermal properties, but also on the frequency of the forcing ($\omega = 2\pi/P$).

Hence, a layer thickness that is correct to represent the diurnal cycle, will not be able to reproduce variations with a shorter period (the top layer is too sluggish), or a longer period (the binding to the lower layer is too strong). In weather prediction models this problem is partly circumvented by the introduction of a number of stacked layers, where the thickness of each layer is tailored to take care of variations with a certain period (e.g., minutes, daily, seasonal and yearly, see also Sections 2.3.6 and 9.2.6).

To analyse the role of the turbulent fluxes further, we describe the sensible heat flux in terms of the difference between the surface temperature T_{top} and the air temperature at some height, T_a (see Chapter 3). Furthermore, if we omit the evaporation term, or linearize it in terms of ($T_{top} - T_a$) (see Chapter 7), the sum of the turbulent fluxes can be parameterized as: $H + L_v E = \alpha_{FR}\left(T_{top} - T_a\right)$, where α_{FR} is a combination of a turbulent diffusivity, C_s and d_{top}. This gives: $\frac{\partial T_{top}}{\partial t} = \frac{Q^*}{C_s d_{top}} - \alpha_{FR}\left(T_{top} - T_a\right) - \frac{2\pi}{P}\left(T_{top} - T_{bot}\right)$, which shows that the turbulent fluxes can be interpreted as a restoring term as well (leaving only Q^* as the forcing term).

2.3.6 Vegetated Surfaces

A vegetation cover moves away the active surface (where the interaction with radiation and turbulent fluxes takes place) from the soil to the top of the vegetation. First the vegetation interacts with the atmosphere, and subsequently the energy is transferred by the plant parts and the air between the plants to or from the soil.

From the perspective of the soil heat flux, the effect of this partial decoupling is that the amplitude of the soil heat flux is damped (see Figure 2.26). In the same figure it can also be seen that the short-term variations due to clouds on May 23 are visible only in the bare soil data. From the perspective of the vegetation, the partial decoupling implies that the vegetation layer has its own temperature, which is only loosely coupled to that of the underlying soil. This has important implications, as it is the surface temperature that interacts with the atmosphere through the upwelling longwave radiation and the sensible heat flux (see Chapter 3).

A method to incorporate the effect of the vegetation layer in meteorological models (without dealing with all the details of the transfer of radiation and heat through the vegetation) is the use of an extra vegetation layer, on top of the soil (e.g., Viterbo and Beljaars, 1995). The vegetation layer is considered to have no heat capacity, as vegetation is mainly made up of air (between the plant parts) and the heat capacity of air is rather low. As a result, all energy supplied to the vegetation layer is instantly transferred. The exchange of heat between the vegetation layer (with temperature T_{veg}) and the upper soil layer (with T_{top}) is treated empirically as:

$$G = \Lambda_{veg}\left(T_{veg} - T_{top}\right) \tag{2.41}$$

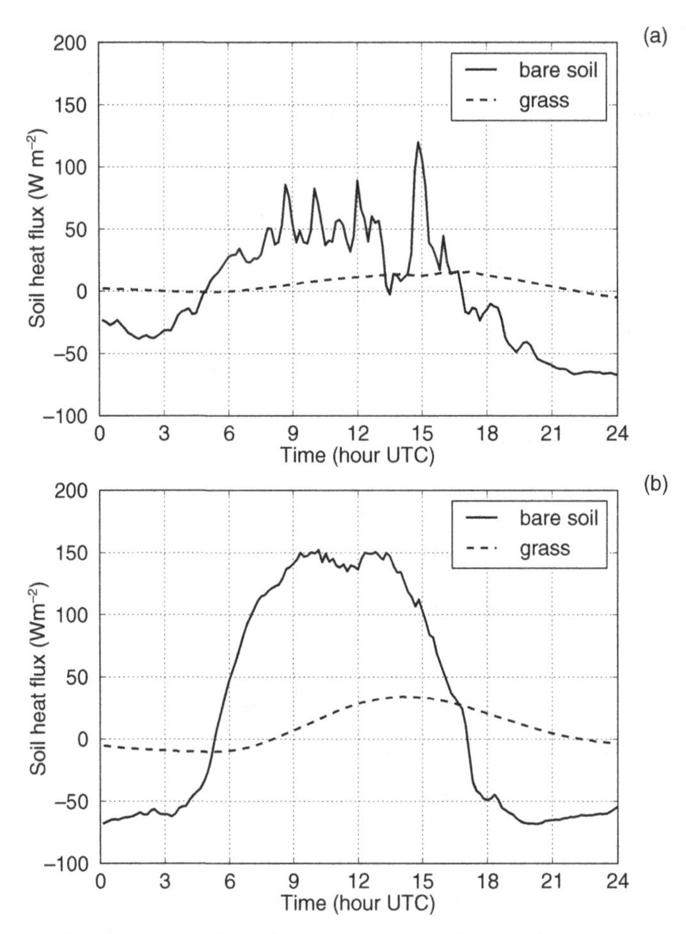

Figure 2.26 Observations of soil heat flux (at 5 cm depth) at de Haarweg Meteorological Station under grass and under bare soil: May 22, 2007 (**a**) and May 23, 2007 (**b**).

where Λ_{veg} is called the skin layer conductivity (compare the empirical description of G in the force-restore method, Eq. (2.40)). A typical daytime value of Λ_{veg} for low vegetation is 5 W m^{-2} K^{-1}, but it does depend on the vegetation type and the fraction of soil covered by vegetation. Furthermore, the value for Λ_{veg} appears to be different between day and night (see ECMWF, 2009).

Another simple model for the soil heat flux, often used in the case that no observations of G are available, is based on the fact that under vegetation the soil heat flux follows a diurnal cycle that is comparable to that of the net radiation (at least during daytime). This leads to the model:

$$G = c_G Q^* \tag{2.42}$$

For grass in the Netherlands it has been found that c_G is about 0.1 (DeBruin and Holtslag, 1982). For taller vegetation such as corn and wheat, c_G is smaller, say, 0.07. If the soil is not completely covered, c_G can vary between 0.1 and 0.5, and the soil heat flux is also no longer in phase with net radiation (Santanello and Friedl, 2003).[4]

If one is interested only in the daily mean fluxes, the soil heat flux will be close to zero: the amount of energy entering the soil at daytime is comparable to the amount leaving the soil at night.

Question 2.21: Given a vegetated surface. The surface soil heat flux is 50 W m^{-2}. If expression (2.41) would be a good approximation of reality, what would be the temperature at the top of the vegetation, if the upper soil layer has a temperature of 20 °C?

2.3.7 Measurement of Soil Heat Flux

The usual way to measure the soil heat flux is to bury a so-called soil heat flux plate at some depth below the surface. This soil heat flux plate has a known thermal conductivity (preferably similar to that of the soil), and by measuring the temperature difference over the plate, the heat flux through the plate can be computed (see Figure 2.27).

If the thermal conductivity of the soil differs from that of the soil heat flux plate the estimated flux will be in error. Heat flux plates with a thermal conductivity that is higher than that of the soil will lead to an overestimation of the flux because the heat will flow preferentially through the heat flux plate. For plates with a conductivity higher than that of the soil the reverse will occur. The following correction for this effect can be used to obtain the real heat flux at depth z_m (Mogensen, 1970):

$$G(z_m) = G_m \left(1 - \alpha \frac{d_m}{\sqrt{A_m}}\left(1 - \frac{\lambda_s}{\lambda_m}\right)\right) \tag{2.43}$$

where G_m is the measured soil heat flux, d_m and A_m are the thickness and area of the plate, λ_m is the conductivity of the plate and α is a shape factor assumed to be 1.70. Though the correction works in the correct direction, it is not always sufficiently large (Sauers et al. 2003). Furthermore, it requires knowledge of λ_s, which can be very variable owing to variations in soil moisture content.

Another source of error is the fact that the plate is buried at a finite depth. Hence, the flux measured will be less than that at the surface (which is the soil heat flux we need in the energy balance equation). Figure 2.24 shows that – certainly for bare soils – the difference in G between the surface and at a commonly used depth of around 5 cm can be very large.

[4] Observations suggest that for sparsely vegetated surfaces the *surface temperature* is quite well in phase with net radiation (rather than the soil heat flux), and hence the soil heat flux should lead the net radiation by 3 hours, according to Eq. (2.38).

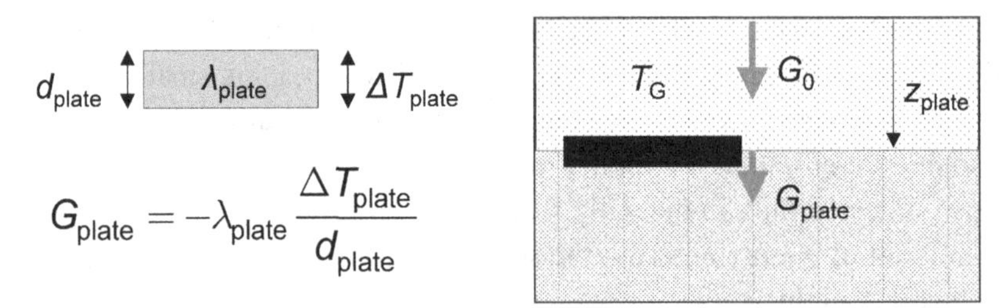

Figure 2.27 Measurement of soil heat flux: soil heat flux plate determines flux from temperature difference (left); soil heat flux plate (black rectangle) is buried at some depth (z_m) so that the heat flux measured by the plate is less than the surface soil heat flux (right).

There are roughly two ways to correct measurements at some depth for this change in G with depth: the *calorimetric* method (e.g., Kimball and Jackson, 1975) and the *harmonic* method (e.g., Horton et al., 1983; Verhoef et al., 1996).

The *calorimetric method* takes into account the heat storage between the surface and the soil heat flux plate (located at a depth z_m). The change in temperature of that layer, T_G, is measured. By integrating Eq. (2.30) with depth, we obtain $\frac{\partial T_G}{\partial t} = \frac{1}{C_s z_m}(G_0 - G_m)$, from which G_0 can be obtained (see Figure 2.27).

In the *harmonic method* temperature observations from at least two depths (z_{d1} and z_{d2}) are needed, in combination with a soil heat flux observation at another depth. The time series at one depth, z_{d1}, is decomposed into a Fourier series, so that not only the sine of the diurnal cycle (with frequency ω) is taken into account but also higher harmonics (with frequencies 2ω, 3ω, etc.):

$$T(z_d, t) = \bar{T} + \sum_{n=1}^{M} \left[A_n(z_d) \sin\left(n\omega t + \phi_n(z_d)\right) \right] \tag{2.44}$$

where $A_n(z_d)$ and $\phi_n(z_d)$ are the amplitude and phase for harmonic n at depth z_d. The depth dependence of the amplitude and phase are $A_n(z_d) = A_n(0)\exp(-\sqrt{n}z_d/D)$ and $\phi_n(z_d) = \phi_n(0) - \sqrt{n}z_d/D$, respectively. The next step is to use the observed soil temperature at the second level z_{d2} to estimate the optimal thermal diffusivity: κ_s is selected such that it produces (with Eq. (2.44) and the expressions for the amplitude A_n and phase ϕ_n) the best approximation for $T(z_{d2}, t)$, in a least-square sense. Then, using the known κ_s, the vertical derivative of Eq. (2.44) is evaluated at the depth of the soil heat flux plate to infer the thermal conductivity λ_s (with Eq. (2.29)). Finally, the vertical derivative of Eq. (2.44) is evaluated at the surface (with the known λ_s) to determine the surface soil heat flux. This entire procedure relies on the assumption that the thermal properties of the soil are uniform. This assumption may be incorrect owing to soil layering, vertical gradients in soil moisture content, and variation in the presence of roots.

Question 2.22: Given a circular soil heat flux plate with the following characteristics: $\lambda_m = 0.8$ W m^{-2} K^{-1}, thickness 5 mm, diameter 10 cm. The flux plate is used in a saturated loosely packed sandy soil.

a) The soil heat flux measured by the sensor is 55 W m^{-2}. What is the real soil heat flux at that location?
b) Given the real soil heat flux calculated in (a), what would be the measured soil heat flux if the flux plate were twice as thick?

Question 2.23: Assume that in Figure 2.18b the soil temperature measured at 5 cm depth is representative of the temperature in the upper 5 cm of the soil (or at least the time rate of change at 5 cm depth is comparable to that in the entire upper 5 cm). The soil heat flux plate (of which the data are shown in Figure 2.26) is installed at a depth of 5 cm.

a) Estimate from Figure 2.18b the instantaneous rate of increase of the soil temperature at 5 cm depth, at 9 UTC.
b) Estimate the heat storage in the layer above the heat flux plate (assume a volumetric heat capacity of the soil of 3.0 10^6 J K^{-1} m^{-3}).
c) Estimate the real surface soil heat flux from the result of question (b), in combination with the observed soil heat flux at 5 cm depth (Figure 2.26).

Question 2.24: Solve this question using the harmonic method. Because the solution requires some iterations, use a spreadsheet program.

From a Fourier analysis of the soil temperature at 5 cm depth the amplitude and phaseshift are determined for the first and second harmonic of the diurnal cycle: $A_1 =$ 6.9 °C, $A_2 = 1.4$ °C, $\phi_1 = -10.10$ hour and $\phi_2 = -9.97$ hour. The daily mean temperature is 20 °C.

At a depth of 10 cm, the observed soil temperature is 18.4 at 10 hours and 22.7 at 16 hours. At 7 cm depth, the soil heat flux observed at 10 hours is 35 W m^{-2}.

a) Determine the damping depth for the diurnal cycle (first harmonic) and the thermal diffusivity.
b) Determine the soil thermal conductivity.
c) Determine the soil heat flux at the surface.

2.3.8 Snow and Ice

The presence of water in the solid phase, either on the soil (snow) or in the soil (ice) has important repercussions for the surface energy balance. First, the thermal properties of snow and ice are different from those of the soil. Furthermore, the presence of solid water implies that an extra phase change (from water to ice, or from snow/ice to water) may occur, implying an extra release or consumption of latent heat.

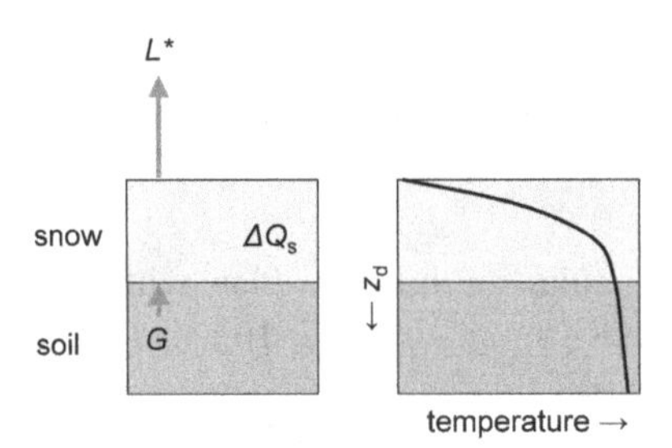

Figure 2.28 Snow pack on top of soil. Left: Energy balance of the snow pack at night with L^* being the only energy flux at the top and ΔQ_s being the change in heat storage. Right: Temperature profile in soil and snow.

Snow Cover

If the soil is covered by snow, this layer of snow could be considered – in terms of heat transport by conduction – as an extra soil layer. However, a number of complications arise owing to the special properties of the snow layer.

- Snow is partly transparent to solar radiation, so that the absorption of solar energy takes place in the entire volume. Hence it is not unambiguous where the 'surface' of the surface energy balance is located.
- Phase changes may occur inside the snow layer owing to melting of the snow (which consumes energy) and refreezing of the snow (where energy is released).
- The snow layer itself is not a fixed porous medium like the soil matrix. The mass balance of a snow pack is a complicated balance between input by snow fall, possibly input by rain (which may or may not freeze in the snow pack), evaporation either from melted snow or directly from the frozen snow (sublimation) and drainage of melt water into the soil. Apart from input and output at the boundaries, internal movement of water (either from melting snow or rain water) transports both water and energy internally.
- The thermal properties of snow are very different from those of the underlying soil (see Table 2.2).

The latter point has a direct influence on the energy balance and the surface temperature, especially at night. This is illustrated in Figure 2.28. At night the only radiative forcing is the net longwave radiation, L^*. The evaporation is usually small. Furthermore, in the case of a night with no or light winds, the sensible heat flux will effectively be suppressed by a strong surface inversion (see Chapter 3). Hence, the effect of the energy extraction by L^* is that heat is extracted from the snow layer so that it cools. Furthermore, the snow layer in turn extracts heat from the ground, leading to an upward soil heat flux. Owing to the very low thermal conductivity of snow (see Table 2.2) a large temperature gradient inside the snow is needed to extract the needed amount of energy. If we assume that the cooling of the snow layer is equal at all depths the profile of the heat flux inside the snow (denoted by G_{snow}) must be linear and can be described as:

$$G_{snow}(z_d) = \frac{d_{snow} - z_d}{d_{snow}} L^* + \frac{z_d}{d_{snow}} G \tag{2.45}$$

where d_{snow} is the depth of the snow layer (in the case that other fluxes at the snow surface, like K^* and H, are not zero, they could be simply included by replacing L^* by the net total flux). Because heat transport is related to the temperature gradient (see Eq. (2.29), rewritten as $\frac{\partial T}{\partial z_d} = -\frac{1}{\lambda_{snow}} G_{snow}$) the above expression for $G_{snow}(z_d)$ can be used to derive, by integration, an expression for the temperature profile in the snow:

$$T_{snow}(z_d) - T_{snow}(0) = -\frac{1}{\lambda_{snow} d_{snow}} \left(\left[z_d \cdot d_{snow} - \frac{1}{2} z_d^2 \right] L^* + \frac{1}{2} z_d^2 G \right) \quad (2.46)$$

where z_d is the depth (positive downward) below the snow surface. If we assume a fresh layer of snow with a thickness d_{snow} of 10 cm, in a typical winter night with $L^*=$ -50 W m^{-2} and no supply of heat from the soil, the temperature difference between the top and the bottom of the snow layer would be 25 K, and the layer would cool by nearly 6 K per hour (rate of change of mean temperature is $L^*/(d_{snow}\ C_{snow})$). This is a dramatic difference with the situation without snow: an equivalent layer of sandy soil would give a temperature difference of 1–2 K between top and bottom, and a cooling of less than 1 K per hour.

In Figure 2.29 an example is given of the variation of temperatures above and under a thin snow pack over the course of one cloud-free day (two half nights). Indeed, the surface temperature, as well as the air temperature just above the snow, drop very quickly as soon as the net radiation becomes negative and less than the supply of heat from the soil. The cooling rate of the snow surface over the period 14–17 UTC is more than 3 K per hour, with a peak cooling at around 16 UTC of more than 4 K per hour. In contrast, the soil temperature at 5 cm depth does not change at all, owing to the insulation by the snow (and the grass). The final temperature difference over the snow pack plus the upper 5 cm of soil is approximately 16 K. The levelling off of the cooling after 18 UTC is consistent with the near-balance between the supply of heat by the soil heat flux and the loss of heat by net radiation, although the magnitude of the cooling suggests that the energy loss was smaller in magnitude than the observed $Q^* - G = -10$ W m^{-2}). The increase of the surface temperature between 23 and 24 UTC is related to an increase of the wind speed to above 2 m s^{-2}, which increases mixing and results in a downward sensible heat flux (the same holds for the period between 0 and 4 UTC).

Apart from the effect of snow on the heat transport by conduction, the occurrence of phase changes in the snow (sublimation or melt) will affect the surface energy balance as well. Sublimation may occur under certain conditions (see Chapter 7): if the air is dry enough and the supply of energy is sufficient (warm air and/or high levels of radiation). For sublimation it is not necessary that the temperature of the snow is at or above the freezing point of water. However, sublimation will be a slow process as the amount of energy involved in sublimation is about 10 times the energy needed for melting only (the latent heat of sublimation is the sum of the latent heats of fusion L_f (~ 0.33 10^6 J kg^{-1}) and vaporization L_v (~2.5 10^6 J kg^{-1})). Hence, a quick decrease in the thickness of a snow

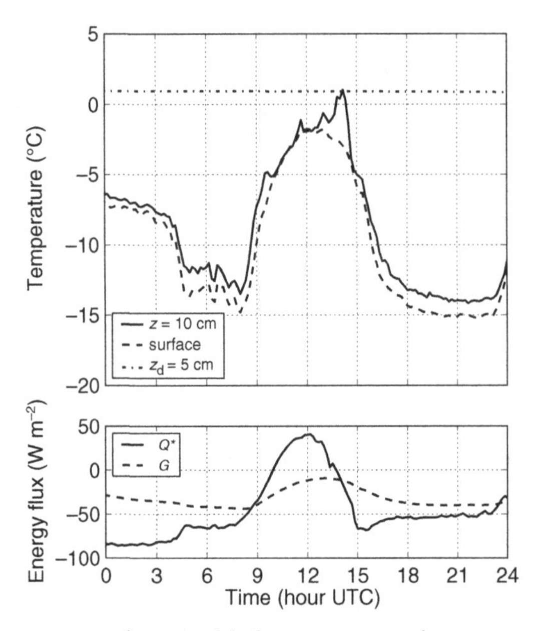

Figure 2.29 Temperatures above and below a snow pack on grass on the Haarweg meteostation, January 6, 2009 (upper panel): air temperature 10 cm above ground level, surface temperature (from emitted longwave radiation, assuming an emissivity of 0.95) and soil temperature at 5 cm below ground level. The snow layer had a thickness of approximately 2 cm and started on January 5. Lower panel shows the net radiation and the soil heat flux at the surface (approximated by the soil heat flux at 2 cm depth, below bare soil).

layer will usually be due to melt. The meltwater subsequently can be removed by infiltration into the soil, surface runoff (in case of saturated or frozen soil) or evaporation.

If the snow cover is continuous the process of snow melt may be slow, because, owing to the high albedo of snow, the net radiation will be low. However, as soon as at some places the (darker) underlying soil, stones or vegetation protrude through the snow, the net radiation will – locally – increase, the supply of heat to the melting process will increase and the decrease of the snow cover will accelerate.

Question 2.25: Given that the latent heat of fusion of water is approximately $0.33\ 10^6$ J kg^{-1}.

a) How much energy is needed to melt completely a layer of old snow of 10 cm thickness (check Table 2.2 for the density of snow)?
b) If the daily mean net energy input into the snow layer (net radiation, supply from the soil, and ignoring the sensible and latent heat flux) is 40 W m^{-2}, then how many days will it take before the snow has completely melted?

Question 2.26: Given the observations in Figure 2.29.

a) Estimate (using Eq. (2.46)) the temperature difference over the snow layer at night around 21 UTC, assuming the snow layer is 2 cm thick. Given the fact that the snow is only one day old, you can assume that this is fresh snow. Is your answer consistent with the data shown in the figure?

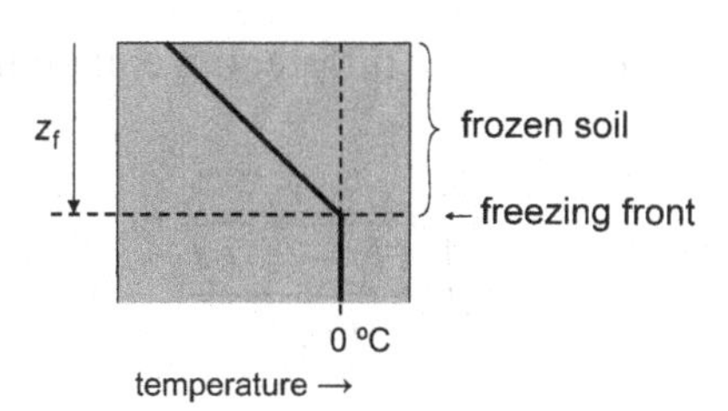

Figure 2.30 Frost penetration into a soil: frozen soil and soil moisture above the freezing front, liquid soil moisture below the freezing front.

b) Because the snow deck thickness may vary from place to place, even within one field, the 2 cm used under (a) is rather uncertain. Therefore, estimate the temperature difference over the snow deck also for layer thicknesses of 1 and 3 cm.

Soil Freezing

Another situation in which solid water and phase changes play a role is the freezing of a soil. If the temperature in the soil is cooled to below the freezing point of water, the soil moisture that is present will freeze. Owing to this phase change latent heat of freezing is released, which will warm the soil again. Freezing takes place at the interface between frozen soil and nonfrozen soil, the freezing front (see Figure 2.30). To cool the soil further, this released latent heat needs to be removed. It can be removed only upward, as the sink of energy is located at the surface. Thus, the downward movement of the freezing front is limited by the ability of the soil to remove the released latent heat.

Let us assume a linear temperature profile above the freezing front (i.e., a constant soil heat flux with depth, see Eq. (2.29)): $T(z_d,t) = T_0 + \frac{z_d}{z_f}(T_f - T_0)$, where T_0 is the temperature at the soil surface, T_f is the temperature at the freezing front (0 °C) and z_f is the depth of the freezing front, which is a function of time. The rate at which latent heat is released when the soil moisture freezes ($Q_{released}$) depends on the soil moisture content θ and the rate at which the freezing front moves downward:

$$Q_{released} = \theta \rho_w L_f \frac{dz_f}{dt} \tag{2.47}$$

where L_f is the latent heat of fusion and ρ_w is the density of liquid water. The amount of heat that can be removed upward is[5]:

$$Q_{removed} = \lambda_s \frac{\partial T}{\partial z_d} = \lambda_s \frac{T_f - T_0}{z_f} \tag{2.48}$$

[5] Regarding the sign, note that in general the heat transport is $Q = -\lambda_s \frac{\partial T}{\partial z_d}$, which would be negative (upward) in this case. Hence $Q_{removed} = -Q$.

Equating Eqs. (2.47) and (2.48) yields a differential equation for z_f (so-called Stefan problem) with the solution:

$$z_f = \sqrt{2\frac{(T_f - T_0)\lambda_s t}{\theta \rho_w L_f}} \tag{2.49}$$

Thus, the penetration of the freezing front has a square root dependence on time, as well as on the temperature at the top of the soil. The square root dependence can be understood as follows. If we assume the surface temperature to be fixed, the temperature contrast between the surface and the freezing front is also fixed. However, as the freezing front progresses downward, the distance over which this temperature contrast occurs becomes larger: the temperature gradient decreases. Because of this decreased temperature gradient less heat that results from freezing can be removed from the soil and hence the freezing front progresses at a slower rate.

This square root dependence can be used in the empirical estimation of frost penetration in soils using the freezing index I_n (see, e.g., Riseborough et al., 2008). I_n is defined as follows:

$$I_n = \sum_{i=1}^{n} (T_i - T_f) \tag{2.50}$$

where T_i is the daily mean air temperature (as an approximation of the surface temperature T_0 in Eq. (2.49)). The summation is started when T_i drops below the freezing point of water (i.e., 0 °C) for the first time. Then the frost penetration is estimated as:

$$z_f = a\sqrt{-I_n} \tag{2.51}$$

where a is an empirical constant that has typical values of 0.03 to 0.06 m $K^{-1/2}$ $day^{-1/2}$. If we compare Eq. (2.49) with Eq. (2.51), we see that the constant a depends on soil type (through λ_s) and soil moisture content θ. Furthermore, the constant a needs to absorb all errors related to the approximation of T_0 by the air temperature.[6] The *mean* effect of this approximation indeed can be taken into account in the value of the empirical constant a. However, day-to-day variations in radiation, wind speed and humidity, as well as the changes in snow cover, will cause a *random* modulation of the relationship between air temperature and surface temperature. Those random fluctuations will decrease the predictive power of Eq. (2.51) on short time scales. Apart from using Eq. (2.51) for the prediction of frost penetration, it can also be used to monitor the removal of frost from the soil, if the summation in Eq. (2.50) is continued after the air temperature has risen above 0 °C again.

[6] The values for a quoted above correspond to a saturated soil with a porosity of 0.4 and thermal conductivities of 0.7 and 2.5 W m^{-1} K^{-1}, respectively (if we neglect the error due to the use of the air temperature).

The method underlying Eq. (2.51) was originally developed for the prediction of the growth of ice on open water (Stefan, 1889). The long-term development of the thickness of ice sheets indeed shows a development similar to Eq. (2.51): a decrease of the growth rate with time due to the increased difficulty to remove the heat released on freezing towards the atmosphere when the ice gets thicker. However, to remove the need for empirical constants (like the a used in Eq. (2.51)) and to allow for day-to-day variations of meteorological conditions, the correct coupling between the atmosphere and the ice surface is essential and hence more elaborate models are needed (see, e.g., DeBruin and Wessels, 1988; Ashton, 2011).

Question 2.27: Given the following observations of daily mean temperatures at the Haarweg meteorological station (in °C, starting on December 15, 2007):

Day	15	16	17	18	19	20	21	22	23
T	1.5	–0.8	–0.4	–1.2	–2.7	–3.8	–4.3	–3.0	1.8
Day	24	25	26	27	28	29	30	31	
T	1.3	2.9	3.8	4.6	6.8	5.8	5.3	3.6	

a) At which date did the maximum frost penetration occur?
b) What was the frost index at that date?
c) Estimate the depth of the frost penetration (assume a typical value of the empirical constant a in Eq. (2.51)).
d) On which date had frost disappeared again from the soil?

2.4 Summary

The energy that is available for transport of heat and water vapour into the atmosphere is equal to the net supply of radiative energy, diminished with the transport of heat into the soil (when storage terms, etc. are omitted).

Radiation exchange at Earth's surface can be decomposed based on the origin (the Sun or the atmosphere or surface) and the direction of the radiation (upward or downward), leading to four composing terms. Downwelling shortwave radiation is highly affected by the geometry of the solar beam relative to the surface. This depends on the geographical position, date and time, leading to seasonal and diurnal variation of the radiation available at the top of the atmosphere. Subsequently, the composition of the atmosphere and the presence of clouds modify this radiation on its way to the surface: it leads to variations in the amount of radiation, the directional dependence (direct versus diffuse) and spectral composition of the radiation. A large part of the downwelling shortwave radiation is absorbed by the surface, but depending on the type of surface (and to a lesser extent the solar angle) a certain fraction is reflected.

Downwelling longwave radiation is emitted both by gases and by liquid (or solid) water in the atmosphere. Emission by gases in the atmosphere closely follows the spectral dependence of a black body, except for the wavelength region between 8 and

13 μm, the so-called 'atmospheric window'. If clouds are present, their main effect is to add radiation within the atmospheric window, thus increasing the downwelling longwave radiation considerably. The longwave radiation emitted by the surface is determined by the surface temperature, and to some extent by the emissivity, although the latter tends to be close to one for most natural surfaces. The temperature of the surface is the result of a complex balance of processes that heat and cool the surface (i.e., all terms of the energy balance together).

The soil below the surface operates as a buffer for heat: during daytime (and during summer) heat is transported into the soil, whereas during night time (and during winter) heat is released from the soil. Heat transport in the soil occurs predominantly through conduction. Hence, the complexity of soil heat transport is related not to the mathematical treatment, but to the specification of the thermal properties. As a soil is a complex mixture of the solid soil matrix, water and air, the thermal properties can vary widely with the composition of the matrix, the porosity and especially the water content.

The soil heat flux is mainly driven by the supply of energy at the surface. Hence the diurnal and yearly cycle also dominate the soil heat flux. The approximation of the temperature at the soil surface by a sinusoidal variation in time provides a powerful framework to study the variation of both temperature and heat flux with time and depth. An important quantity appearing in this framework is the damping depth (or e-folding depth). As this damping depth depends on the frequency of the variation, as well as on the thermal diffusivity of the soil, examination of the extinction of the diurnal or yearly variations of temperature can provide information on soil thermal properties.

The presence of vegetation modifies the dynamics of soil temperatures and soil heat flux as it moves the active surface (where interaction with the atmosphere takes place) away from the soil surface. As canopies are largely made up of air, the general effect of vegetation is to provide an insulating layer on top of the soil.

Snow cover on soils has a similar effect as the thermal conductivity of snow is much lower than that of most soils. Hence, if net cooling occurs at the surface of the snow layer, a large temperature gradient inside the snow is needed to provide the energy lost at the surface. This leads to very low surface temperature. Another effect of solid water (snow or ice) on the surface energy balance is related to phase changes. Melting and sublimation of snow can consume significant amounts of energy. On the other hand, when the soil surface temperature falls below 0 °C, soil moisture will start to freeze, causing a release of latent energy from the soil. As the heat released on freezing needs to be transported towards the soil surface, the penetration of the freezing front slows down as it moves down away from the surface.

3

Turbulent Transport in the Atmospheric Surface Layer

3.1 Introduction

This chapter is concerned with exchange processes between the surface and the atmosphere. According to the surface energy balance, during daytime the net input of energy at Earth's surface ($Q^* - G$) is used to supply heat to the atmosphere and to evaporate water. This heat and water vapour needs to be transported away from the surface. During night time, on the other hand, as we have seen in Chapter 2, the available energy is generally negative and hence the sensible heat is transported downward (water vapour can go either way). The exact partitioning between the sensible and latent heat flux (both during day time and night time) is at this stage not crucial and is dealt with later in Chapter 7.

How does this transport of heat and water vapour from and to the surface occur? If we take heat transport as an example, one option could be to transport the heat by molecular heat diffusion. A typical daytime value for the sensible heat flux could be 100 W m^{-2}, and the thermal diffusivity of air is around $2 \cdot 10^{-5}$ m^2 s^{-1}. Then we can derive from Eq. (1.6) that a vertical temperature gradient of more than 4000 K per meter would be required (note that in this case the transported quantity used in Eq. (1.6) is enthalpy per unit volume: $\rho c_p T$). It is clear that vertical temperature gradients of this magnitude do not occur, so there must be another mode of transport.[1] This is turbulent transport: heat, water vapour (and other gases) as well as momentum are transported by the movement of parcels of air that carry different concentrations of heat, water vapour, etc.

As we will see later, atmospheric turbulence is mostly produced by processes related to Earth's surface: wind shear and surface heating. These production mechanisms have a strong diurnal variation due to the variation in insolation. The part of the atmosphere in which this diurnal cycle of turbulence production (as well as the variation in fluxes of, e.g., water vapour and CO_2) is noticeable is called the atmospheric boundary layer (ABL). The ABL is the turbulent layer between the surface

[1] In fact, very close to the surface such temperature gradients *do* occur, but over a very small distance only.

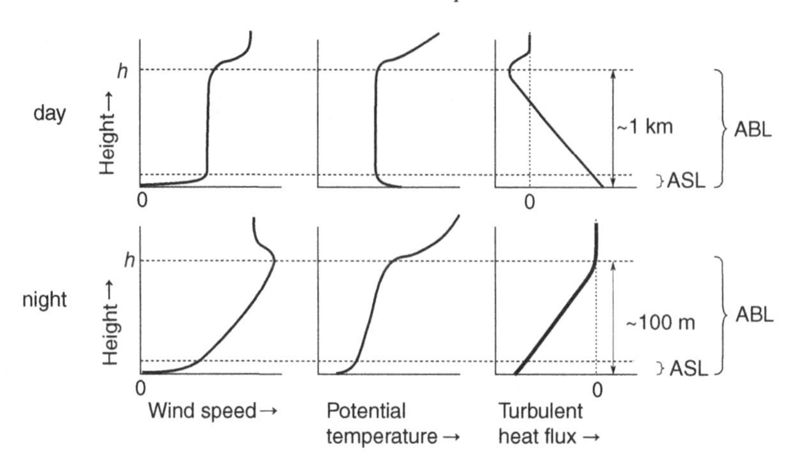

Figure 3.1 Sketch of profiles of mean wind speed, mean potential temperature and the turbulent sensible heat flux in the atmospheric boundary layer (ABL) with depth h. The atmospheric surface layer (ASL) constitutes the lower 10% of the ABL. Note the order of magnitude difference in boundary-layer depth between the day time and night time cases.

and the nonturbulent free troposphere. During the day the ABL is heated from below and convection causes strong turbulent mixing, leading to more or less uniform profiles of, for example, wind and potential temperature (see Figure 3.1 showing idealized profiles and Figure 3.2 showing a vertical cross section of the turbulent fields of temperature, humidity and vertical wind speed). On the other hand, during night time surface cooling stabilizes the ABL, leading to weak turbulence and large gradients. The fact that turbulence is restricted to the ABL can be deduced from the fact that the turbulent fluxes decrease to zero at a certain level near the top of the ABL (see Figure 3.1). During daytime, convection in the ABL can be so strong that penetrating thermals cause an exchange of air between the free troposphere and the ABL (entrainment). This entrainment is visible in Figure 3.1 as a negative sensible heat flux, which is due to the downward transport of warm air from the free troposphere. In Figure 3.2 the entrainment is visible as the inclusion of patches of warm and dry air from the free troposphere into the ABL (e.g., at horizontal location 2 km). Above the ABL, the atmosphere is mostly ignorant of the time of day (e.g., if the temperature at 5 km height is –20 °C at night, it will remain so during daytime, unless large-scale processes such as advection affect the temperature).

Although the entire ABL is linked to processes at Earth's surface, this chapter is restricted to a description of processes close to the surface, roughly in the lower 10% of the ABL, loosely named the atmospheric surface layer (ASL; see Section 3.4.3 for a more thorough discussion). This part of the ABL is characterized by large gradients in temperature and wind speed. Furthermore, the turbulent fluxes do not deviate much from their values at the surface.

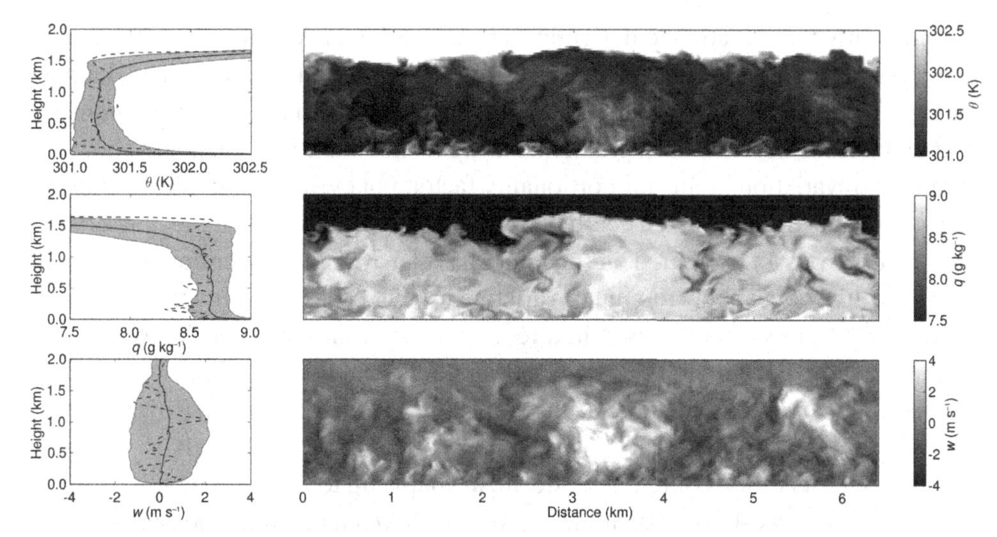

Figure 3.2 Vertical cross section through a convective boundary layer: potential temperature (top), specific humidity (middle) and vertical wind speed (bottom). Profiles at the left show mean profiles (averaged over the cross section shown) and the shading indicates deviations of one standard deviation around the mean. Also shown are the instantaneous profiles at one location (dashed line). Fields originate from a large eddy simulation (LES) with the Dutch Atmospheric LES (DALES; see Heus et al., 2010).

In the forthcoming sections we first explore the characteristics of the turbulent diffusivity that would be needed to relate a turbulent flux to a gradient. Then various ways of characterizing turbulence are dealt with in Section 3.3. Next the transporting properties of turbulence are dealt with in Section 3.4, including the reference technique to measure turbulent fluxes, viz. the eddy-covariance method. The framework of similarity relationships is developed in Section 3.5 and used to derive fluxes from mean turbulent quantities in Section 3.6. In Section 3.7 a summary of this chapter is given, including a concept map. It may be useful, while reading through this chapter, to consult regularly the concept map in Figure 3.22. Note that Appendix B reiterates some basic thermodynamics, gives an overview of various properties of air and lists a range of measures for the amount of water vapour in the air.

3.2 Characteristics of Turbulent Diffusivities

Before dealing with turbulence in more detail, we first examine some characteristics of turbulent transport. As an example we look at the transport of heat. Inspired by Eq. (1.6), we can define a turbulent diffusivity that links the flux of sensible heat to the vertical gradient of the mean temperature:

$$F_{\mathrm{h}} \equiv -\rho c_{\mathrm{p}} K_{\mathrm{h}} \frac{\partial \overline{T}}{\partial z} \tag{3.1}$$

If we could use this equation to determine a turbulent flux, just in terms of the vertical gradient of the transported quantity and a diffusivity, life would be very easy. But note that Eq. (3.1) is really only a *definition* of K_h: we only say that there is a parameter (K_h) that links the flux to a local gradient, but we do not make any statement about the magnitude or variation of this proportionality factor, not even about its sign (although one would hope that a diffusion coefficient is positive). Also note that the temperature in Eq. (3.1) carries an overbar, denoting that this is an average temperature (an instantaneous profile would show too much variation, as can be seen in Figure 3.2).

Equation (3.1) can also be used in a reverse sense: using observations of the flux and the gradient, the turbulent diffusivity can be deduced. To this end, we use data gathered on a sunny day in June, at the Cabauw tower (operated by Royal Netherlands Meteorological Institute [KNMI]) in the centre of the Netherlands.

Although observations of temperature are available up to 200 m height, we restrict ourselves to the lowest 20 m, as in this lower layer we can assume the sensible heat flux to be rather constant with height (within 10%, to be discussed later, Section 3.4.2) so that we can use the *surface* sensible heat flux to represent the flux at a given height. Figure 3.3b shows the temperature profile at two instances: at night time and during mid-day. Temperatures are much lower at night than during the day. The gradient is positive at night and negative at day. Finally, the gradients are larger (in absolute sense) close to the surface than at higher levels. From these temperature profiles we can directly infer the behaviour of the turbulent diffusivity (Figure 3.3d):

- The values of the diffusivities (order of 1 $m^2 s^{-1}$) are much larger than the molecular thermal diffusivity (roughly $2 \cdot 10^{-5}$ $m^2 s^{-1}$).
- The combination of a positive temperature gradient with a negative sensible heat flux gives a positive diffusivity at night. The same result is obtained for daytime with a negative gradient and a positive heat flux.
- From Eq. (3.1) it is clear that the large temperature gradients close to the surface are connected to small values for the diffusivity. One *could* interpret this as: to transport the same amount of energy (we assumed the flux to be constant with height) a smaller diffusivity is needed when the gradient is larger. But nature has a different causality chain: because the diffusivity is smaller close to the surface, a larger gradient is needed to transport the same amount of energy.
- The diffusivities are much higher during daytime than during night time (by one to two orders of magnitude).

To conclude, Figure 3.3c shows the entire diurnal cycle of the diffusivities at three heights. The variation between day and night and the variation with height are clearly visible here as well. There are some undefined points around sunrise and sunset, which are due to the fact that when gradients and fluxes become small, they may change sign at different moments, yielding negative values for the diffusivity.

Essentially, the rest of this chapter is devoted to the variation of the turbulent diffusivities with height and time and how we can understand and describe that variation.

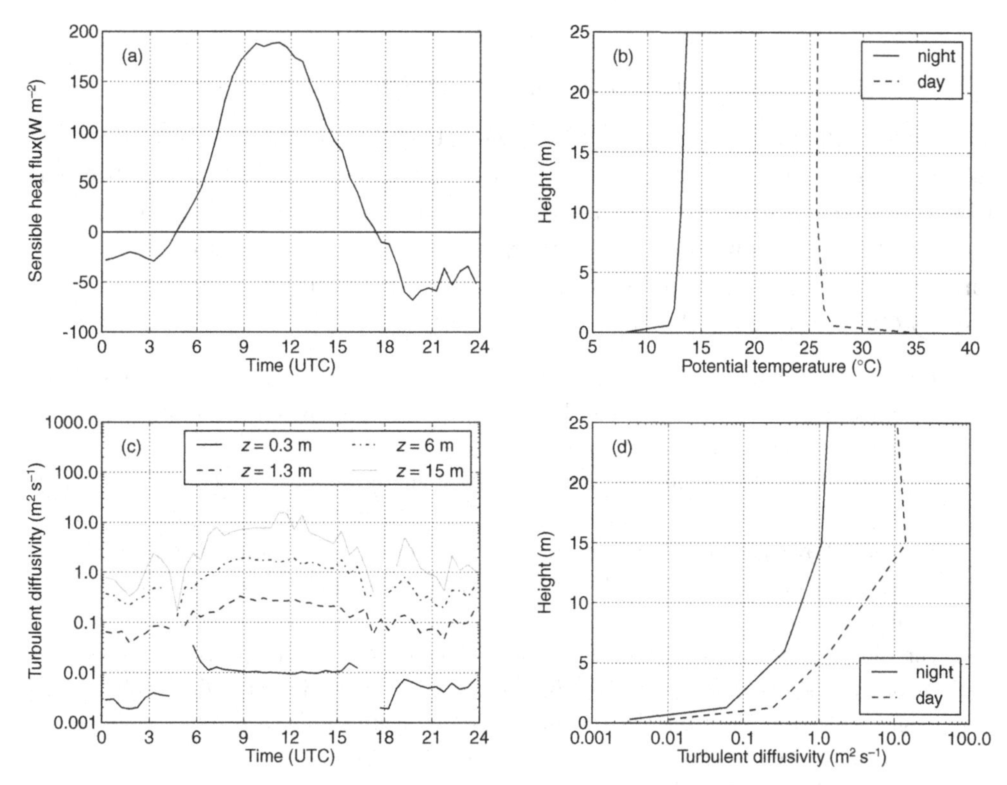

Figure 3.3 Temperature and turbulent diffusivity for heat as derived from observations at Cabauw (The Netherlands). (**a**) Diurnal variation of surface sensible heat flux. (**b**) Profile of potential temperature (night: 2:00–2:30, day: 12:30–13:00). (**c**) Time series of K_h at four heights. (**d**) Profiles of K_h during night time and daytime. (Data courtesy of Fred Bosveld, KNMI)

Question 3.1: Figure 3.3d shows the turbulent diffusivity for heat transport for a night time period and a daytime period. It is clear that the diffusivity increases with height. But because K_h has been plotted on a logarithmic axis (to accommodate the large spread in values), the *exact* dependence of K_h on height z cannot be determined.

a) Create a table of values for K_h for a number (say four) of heights, for daytime and night time separately.

b) Deduce from those values whether K_h increases with height in a linear fashion (i.e., $K_h \sim z$), more than linearly (e.g., $K_h \sim z^{1.5}$), or less than linearly (e.g., $K_h \sim z^{0.5}$). Do this for night time and daytime separately. Note that the exact power is not of interest, only if the increase is stronger or weaker than linear. The answer will become relevant again in Section 3.5.5. Hint: determine for each height interval $\frac{\partial K_h}{\partial z}$; from the height dependence of $\frac{\partial K_h}{\partial z}$ (so in fact from $\frac{\partial^2 K_h}{\partial z^2}$) one can determine whether K_h varies more than linearly or less than linearly with height.

3.3 Turbulence

3.3.1 Qualitative Description

Starting with the pioneering work of Reynolds (1895), turbulent flows have been the subject of scientific research ever since (for a review see, e.g., Monin and Yaglom, 1971). From this research a more or less commonly accepted picture has evolved that describes turbulent flows both qualitatively and quantitatively. Based on this picture some general properties of turbulent flows can be summarized (after Tennekes and Lumley (1972); Lesieur (1993)):

- Turbulence occurs in flows where the nonlinear terms in the governing equations[2] dominate over the linear viscous terms. Those nonlinear terms may involve momentum and/or density (or temperature) variations.
- Turbulent flows are irregular or chaotic in space and time: they are not reproducible in detail.
- Turbulent flows are diffusive: heat, momentum, as well as mass are mixed and transported efficiently by turbulent flows. In many practical applications this is a desirable feature of turbulence.
- Turbulence is essentially rotational and three-dimensional, which is a distinction to other chaotic flows (like, e.g., cyclones). Rotating patches of fluid (loosely called eddies) have length scales ranging from the size of the flow domain (in the ASL this would be the height above the ground) down to the order of millimetres.
- Turbulent flows are dissipative: the kinetic energy of the velocity fluctuations, produced at the largest scales, is dissipated at the smallest scales into heat through viscous forces.

As stated before, turbulence is essentially three dimensional (and time dependent). But very often we are not able to capture the variability of a turbulent flow in all those four dimensions (except with very advanced measurement techniques, and in numerical simulations). To obtain a first glimpse of what turbulence looks like, we will discuss the observed time series[3] of vertical wind speed, temperature, humidity and CO_2 concentration as observed above a savannah vegetation in Ghana (see Figure 3.4). The following remarks can be made:

1. The four signals are indeed chaotic. But some structure is apparent as well. Large deviations from the mean are very rare, whereas smaller deviations are more common. Furthermore, larger deviations from the mean last for some time (e.g., around 11.3, 11.4 and 11.6 hours): so scale and magnitude of the fluctuations are related.
2. The signals of the scalars (i.e., temperature, humidity and CO_2) are asymmetric in the sense that there is a base level from which the signal deviates in only one direction. This

[2] Nonlinear terms are those terms where a property of the flow, in particular a velocity component or density (or temperature) is multiplied with another (or the same property). An example of such a quadratic term is the advection term $\left(u\frac{\partial u}{\partial x}\right)$ occurring in the differential equation that describes the change in time of velocity u.

[3] A time series: so only one dimension varies and x, y and z are fixed.

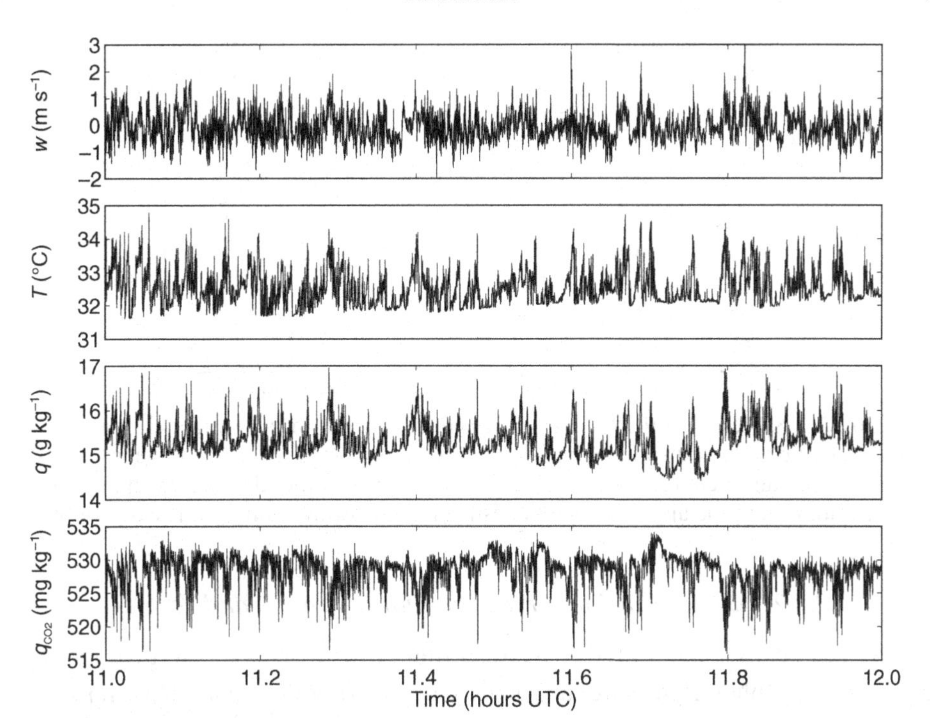

Figure 3.4 One hour of turbulence: instantaneous observations of vertical wind (w), temperature (T), specific humidity (q) and specific CO2 concentration (q_{CO2}). Observations made in Ghana over savannah (October 2001).

base level is related to the mean concentrations in the well-mixed part of the daytime convective ABL (see Figure 3.1 and Graf et al., 2010).

3. The larger (and longer lived) deviations from the mean seem to occur simultaneously for all variables: a positive vertical wind fluctuation coincides with positive fluctuations in temperature and humidity and a negative fluctuation in CO_2: so different quantities are mutually correlated.

The last of the aforementioned remarks is in fact the engine that vertically transports heat, gases (water vapour and CO_2) and horizontal momentum. Figure 3.5 shows a sketch of the engine of vertical turbulent transport. The surface injects a certain amount of heat into the atmosphere (surface sensible heat flux H), and the turbulent motion of air removes that heated air from close to the surface to higher levels, whereas (cooler) air from above replaces the removed air. If this efficient transport mechanism were not present, the air close to the ground would heat up tremendously. Likewise, the surface extracts momentum from the air (slows down the flow) and turbulence replenishes this from higher levels. Turbulent transport is discussed further in Section 3.4.

Question 3.2: Make a sketch similar to Figure 3.5 for the following transports (assign the correct labels and directions to the three arrows):

a) Negative sensible heat flux
b) Positive evaporation

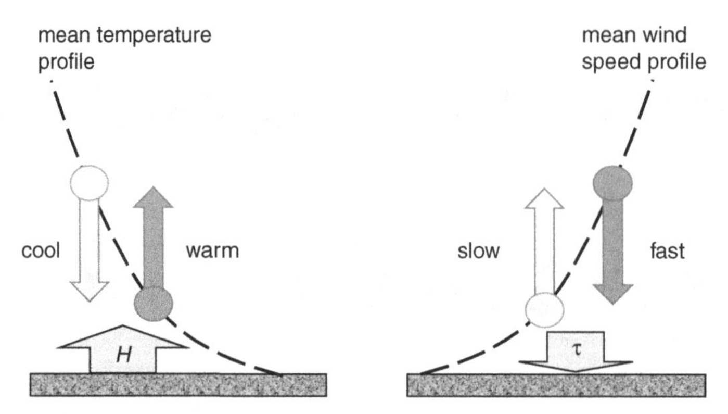

Figure 3.5 Sketch of turbulent transport and its relation to the surface flux and the mean profile of the transported quantity: heat entering the layers close to the ground is removed by turbulent motion of air (left) and surface friction (denoted by τ) removes momentum from the air which is replenished by downward motion of fast air (right).

3.3.2 Intermezzo: Conserved Quantities, Scalars and Vectors

We have seen earlier that turbulent vertical transport involves the vertical motion of air. In the atmosphere, pressure and density decrease with height. This implies that if a parcel of air moves upward it will experience a small decrease in pressure, and consequently it will expand. In the case of an adiabatic process, this expansion will take place at the expense of the internal energy of the parcel; hence its temperature will decrease. But this loss of internal energy can be regained if the parcel is brought back to its original pressure. To eliminate these reversible changes from the analysis, we will use the potential temperature (see Appendix B) from here on.[4] This is one example of a variable that is conserved for adiabatic processes.

Likewise, many indicators for the amount of water vapour (or another gas) change when the pressure (and density) of an air parcel changes (see Appendix B). Only specific humidity q (as well as the mixing ratio) is conserved for adiabatic processes, as q is the ratio of vapour density and air density. Both densities change at the same rate under adiabatic lifting, and hence q does not change.

More generally, a consequence of the change in volume of parcels that move upward or downward is that the description of the contents of that parcel (heat, moisture, momentum) per unit volume is not very useful. Therefore, when vertical motion is involved, we always use specific quantities (i.e., content per unit mass):

- Specific enthalpy: $c_p\theta$ (in J kg^{-1})
- Specific humidity: q (in kg kg^{-1})
- Specific momentum: u (momentum is mass times velocity, hence specific momentum is just velocity: in m s^{-1}).

[4] For the specification of changes in enthalpy due to diabatic (i.e., non-adiabatic) processes the potential temperature can equally well be used as the normal temperature because *changes* in temperature and potential temperature are identical.

An important distinction has to be made between variables that have both a magnitude and a direction (like momentum or equivalently velocity) and variables that have only a magnitude. The first are called *vectors*, whereas the latter are referred to as *scalars* (e.g., temperature, humidity, pressure).

Also note that we often talk about momentum (which is a three-dimensional vector) but actually refer to *horizontal* momentum along the mean flow direction. This quantity is a scalar.

Question 3.3: For an adiabatic process the following holds: $p^{c_p/c_v-1}T^{-c_p/c_v}$ = constant. Thus both temperature and pressure change.

a) Do pressure and temperature change in the same direction (i.e., if one increases, the other increases as well) or in opposite directions in an adiabatic process (values for c_p and c_v can be found in Appendix B)?
b) Use the equation of state for a perfect gas (gas law: $p = \rho RT$) to deduce how the density changes as a function of temperature in an adiabatic process.
c) The same steps as in (a) and (b) can also be used for the partial pressure of water vapour (e): $e^{c_p/c_v-1}T^{-c_p/c_v}$ = constant in combination with $e = \rho_v R_v T$. Hence, deduce the dependence of ρ_v on temperature for an adiabatic process.
d) Show with the results of (b) and (c) that the specific humidity indeed does not change during adiabatic cooling.

3.3.3 Statistical Description of Turbulence

In this section we discuss a number of statistical tools needed in the description of turbulence and turbulent transport.

Reynolds Decomposition

Because turbulent flows are not reproducible in detail, we can treat them only in a statistical sense ("how do things behave on average?"). The first step in this statistical description is the Reynolds decomposition (Reynolds, 1895), which states that a quantity X (might be a wind speed, temperature, etc.) at a given moment and at a given location can be decomposed as:

$$X = \overline{X} + X' \tag{3.2}$$

where $\overline{X}$ is the mean value of X and X' is the deviation from that mean. For $\overline{X}$, in principle only the so-called ensemble mean can be used. The ensemble mean requires that one repeats an experiment (or natural situation) an infinite number of times, under exactly the same conditions. Then the ensemble mean (at a given location and time) is the mean over all those repetitions (thus the mean is space and time dependent). This is – especially for natural systems outside the laboratory – impossible. Therefore, the ensemble mean is generally approximated by a temporal mean (if observations are made at a fixed position) or a spatial mean (if observations are made at different positions).

This approximation is made under the ergodic hypothesis, that is that the ensemble statistics at a given moment are identical to temporal (or spatial) statistics for a given period (or space). Sometimes only weak ergodicity is assumed, restricting the ergodic hypothesis to first and second statistical moments only (Katul et al., 2005).

Thus if we have a time series with N observations X_i, then the estimate of $\overline{X}$ would be $\overline{X} \approx \frac{1}{N}\sum_{i=1}^{N} X_i$. This is a single number, valid for the entire time series of N observations. From the time series X and its mean $\overline{X}$, a new time series can be determined, containing the deviations X' ($X' = X - \overline{X}$). Thus the series X' has N values, like the original time series X. The deviations we are interested in are the turbulent fluctuations, so the period over which averaging should take place should be long enough to remove all turbulent fluctuations from the mean (so that all turbulence signal is contained in the fluctuations), but short enough to prevent nonturbulent fluctuations (such as the diurnal cycle) to influence the deviations X'. The scale that separates the turbulent from the nonturbulent fluctuations is called the (co-) spectral gap. However, it is often not as sharply defined as the word 'gap' suggests (Baker, 2010)). Typical values for the time scale of the (co-)spectral gap in the ASL are 10–30 minutes (Voronovich and Kiely, 2007; see also Section 3.4.2).

A number of computational rules apply for Reynolds averaged quantities (strictly valid only when ensemble means are used):

$$\begin{aligned}
&\overline{X+Y} = \overline{X} + \overline{Y} && \\
&\overline{aX} = a\overline{X} && \text{if } a \text{ is constant} \\
&\overline{a} = a && \text{if } a \text{ is constant} \\
&\overline{\left(\frac{\partial X}{\partial x_i}\right)} = \frac{\partial \overline{X}}{\partial x_i} && \text{with } x_i \text{ is a space or time coordinate} \\
&\overline{(X')} = 0 &&
\end{aligned} \tag{3.3}$$

Statistics of a Single Variable

For a single variable (e.g., the time series of temperature shown in Figure 3.4) the two statistical quantities of interest in the framework of this book are the mean and the variance, $\overline{X}$ and $\overline{X'X'}$, respectively. One could think that, because the variance involves averaging of fluctuations, this should be zero (following the last rule in Eq. (3.3)). But because a squared quantity ($X'X'$, always positive) is averaged, the result is always positive. Besides the variance of X also the standard deviation (σ_X) is often used, which is the square root of the variance. The advantage of the standard deviation is that it has the same unit as the quantity under consideration.

Table 3.1 Statistics of the times series shown in Figure 3.3

	Mean		Variance		Standard deviation	
w	$-7.8 \cdot 10^{-2}$	m s^{-1}	0.27	(m s^{-1})^2	0.52	m s^{-1}
T	32.5	K	0.27	(K)^2	0.52	K
Q	15.3	g kg^{-1}	0.14	(g kg^{-1})^2	0.38	g kg^{-1}
q_{CO_2}	528	mg kg^{-1}	6.0	(mg kg^{-1})^2	2.5	mg kg^{-1}

If we now return to the example time series in Figure 3.4 the four signals can be characterized with the statistics of a single variable. The values are given in Table 3.1. The meaning of the mean value is immediately clear: it indicates the mean level around which the turbulent signal fluctuates. This is a value that we could have estimated by eye from the graphs in Figure 3.4. Note that for the vertical wind speed the mean is close to zero. This is logical, because close to the ground we cannot have mean vertical motion for a long time, as the flow is blocked by the solid surface.

Next we look at the standard deviations (last column in table Table 3.1). In statistics the standard deviation is a measure of the width of statistical distribution. In the analysis of turbulence this can be interpreted as a measure of the magnitude of the fluctuations of the signal around the mean. In Figure 3.4 we cannot easily identify the standard deviation, but bear in mind that for a normal distribution (the data shown are *not* normally distributed) 96% of all samples is located in the range $(\overline{X} - 2\,\sigma_X)$ to $(\overline{X} + 2\,\sigma_X)$. So the difference between the lowest value and the highest value in the time series is roughly four times the standard deviation. Using this rule of thumb, values are obtained that fit rather well with the values in Table 3.1. Finally, the variance is simply the square of the standard deviation. For vertical wind there is a useful interpretation of the variance: $\frac{1}{2}\overline{w'w'}$ is the kinetic energy contained in the vertical wind speed fluctuations (see Section 3.3.5).

Statistics of Two Variables

Considering statistics of two variables, the quantity of interest is the covariance. For two variables X and Y, this is written as $\overline{X'Y'}$. In contrast to the variance, the covariance can be either positive or negative. Furthermore, the covariance can be zero, even if the fluctuations in X or Y are not zero. This happens when X and Y are not correlated.

As an example we look at two covariances here. First we take the vertical wind speed w and potential temperature θ as the two variables. Then the covariance $\overline{w'\theta'}$ is a measure of the amount of heat ($\rho c_p \theta$) transported upward. This can be understood if we zoom in to a small time section of Figure 3.4 (see Figure 3.6). There are a number of periods where the fluctuation in the vertical wind speed is positive (around

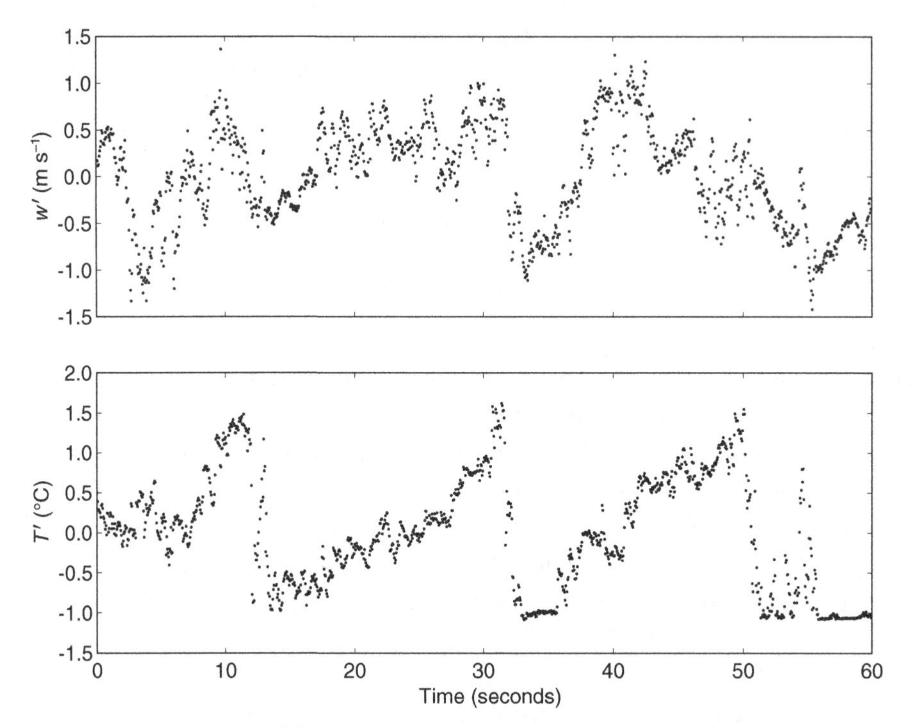

Figure 3.6 Time series of fluctuations of vertical wind speed (top) and temperature (bottom). A sub-sample of 1 minute from Figure 3.4. Each dot represents one observation. Sampling rate is 20 Hz. The ramp structure visible in the lower panel is an expression of the gradual heating of the air by the surface heat flux, followed by an abrupt removal of this warmed air by a sweep that brings down colder air from aloft. This pattern is exploited in the surface renewal method to estimate surface flux (Paw U et al., 1995).

10 seconds, from 20 to 30 seconds and around 40 seconds). During those periods of upward motion, the air is also relatively warm (positive θ fluctuation). Thus, *warm* air is transported *upward*. There are also periods with negative vertical motion (around 5, 35 and 55 seconds). Those are roughly accompanied by negative temperature fluctuations. Thus *cool* air is transported *downward*. The net effect of those motions is that heat is transported upward ($\overline{w'\theta'} > 0$): a positive sensible heat flux. But the picture is not ideal: there are periods in Figure 3.6 where the w and θ signals are not well correlated. The degree to which the signals are correlated can be expressed by the correlation coefficient $R_{w\theta}$. For the general combination of two signals X and Y:

$$R_{XY} \equiv \frac{\overline{X'Y'}}{\sigma_X \sigma_Y} \tag{3.4}$$

If the time series of w and θ, as shown in Figure 3.4, would have been perfectly correlated, $R_{w\theta}$ would be equal to 1 (or −1 for perfect anti-correlation). However, the actual correlation coefficient for this time series is only 0.54. This value for the

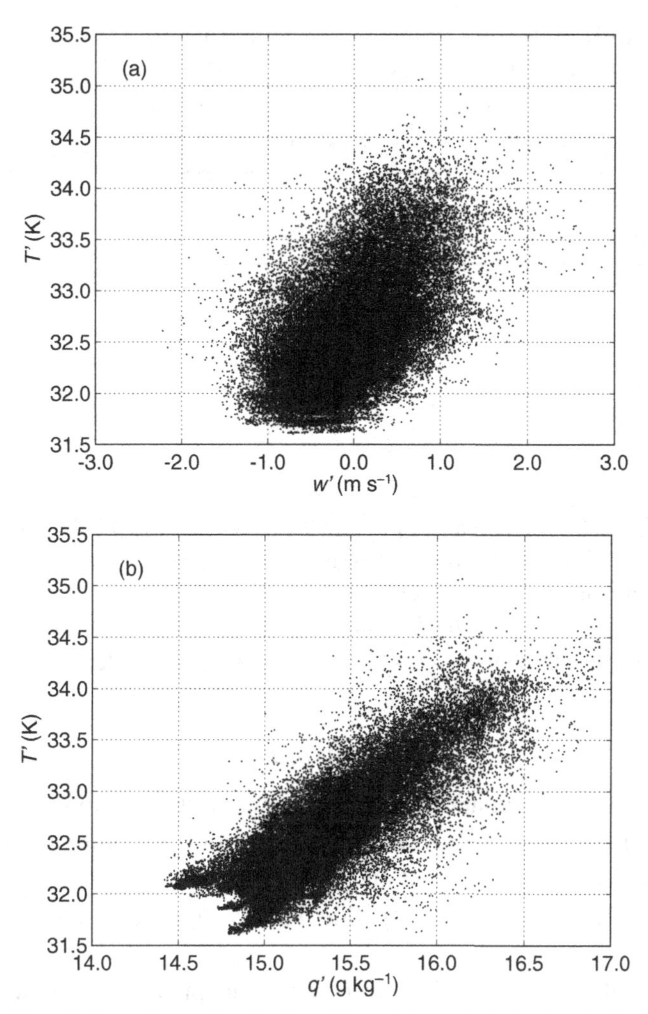

Figure 3.7 Correlation plots of turbulent fluctuations. (**a**) Vertical wind speed and temperature. (**b**) temperature and specific humidity. The data in this figure are the same data as used in Figure 3.4.

correlation coefficient is typical for very convective (unstable) conditions. For neutral conditions $R_{w\theta}$ is much lower, of the order of 0.25 (Moene and Schüttemeyer, 2008; see Section 3.3.5 for the definition of unstable and neutral conditions).

Figure 3.7a shows a scatterplot of all samples of vertical wind speed and temperature. Indeed, there is a positive correlation (higher temperatures go together with higher (positive) vertical wind speed). But it is not a nice one-to-one linear correlation.

The second covariance dealt with here is the covariance between temperature and humidity. Because both heat and humidity are transported by the same mechanism (turbulence), and high temperatures and humidity values are found close to the surface during daytime (and lower values higher up), one would expect temperature and humidity to be well correlated. Indeed, in general they are, with correlation coefficients between 0.8 and 1 (for the present data $R_{\theta q} = 0.82$, see Figure 3.7b).

Question 3.4: Given the following series of quantities X and Y:

X	4	1	5	2	3
Y	10	2	5	7	4

Compute the following quantities:
a) $\overline{X}$ and $\overline{Y}$
b) $\overline{X'}$ and $\overline{Y'}$
c) $\overline{X'X'}$ and $\overline{Y'Y'}$
d) $\overline{X'Y'}$
e) R_{XY}

Question 3.5: In the second panel of Figure 3.4 it can be observed that the temperature seems to have a well-defined lower limit. Explain this lower limit, using the figure of the daytime vertical temperature profile given in Figure 3.3a (Figures 3.3 and 3.4 do not refer to the same situation, but the profile in Figure 3.3 should be at least representative of the profile that occurred during the observations depicted in Figure 3.4).

3.3.4 Buoyancy

Turbulent temperature fluctuations (e.g., as shown in the previous section) give rise to fluctuations of air density. Those density fluctuations, in combination with gravity, in turn cause vertical acceleration of air (buoyancy). First we derive the link between temperature fluctuations and density fluctuations, and subsequently buoyancy is examined.

The equation of state of a perfect gas (such as air) provides the link between pressure, temperature and density: $p = \rho RT$. The presence of water vapour will change the composition of air, and hence the gas constant R (through the molar mass) of the gas mixture. To simplify the analysis, the effect of water vapour on density is moved from the gas constant to the temperature: T is replaced by the virtual temperature T_v and at the same time R (which varies with moisture content) is replaced by the constant R_d, giving the equation of state as: $p = \rho R_d T_v$.

To analyse the equation of state for a turbulent environment, all variables need to be decomposed into a mean and fluctuating component. If it is assumed that $\overline{\rho' T_v'} \ll \overline{\rho}\overline{T}_v$ and $\rho' T_v' \ll \rho' \overline{T}_v$ or $\overline{\rho} T_v'$, then the following expressions can be derived for the mean pressure and the pressure fluctuations:

$$\begin{aligned} \overline{p} &= R_d \overline{\rho}\overline{T}_v \\ p' &= p - \overline{p} = R_d (\rho' \overline{T}_v + \overline{\rho} T_v') \end{aligned} \tag{3.5}$$

Furthermore, given that turbulent pressure fluctuations are of the order of $\rho\overline{u'u'}$ (Wyngaard, 2010) the relative pressure fluctuations (i.e., $\frac{p'}{\overline{p}}$) can be neglected and the following expression for the density fluctuations is obtained:

$$\frac{p'}{\overline{p}} = \frac{\rho'}{\overline{\rho}} + \frac{T_{\rm v}'}{\overline{T}_{\rm v}} \approx 0 \Leftrightarrow \frac{\rho'}{\overline{\rho}} \approx -\frac{T_{\rm v}'}{\overline{T}_{\rm v}} \tag{3.6}$$

Thus, density fluctuations are directly linked to (virtual) temperature fluctuations: a positive temperature deviation leads to a negative density deviation.

The effect of buoyancy on vertical motion (and hence on turbulent fluctuations of vertical wind) can be understood as follows. Assume a parcel of air with volume V and a density ρ that deviates from the density of the surrounding air $\overline{\rho}$ by ρ' (thus $\rho = \overline{\rho} + \rho'$). Two forces act on the parcel:

- The gravitational force acting on the parcel: $F_{\rm g} = -g(\overline{\rho} + \rho')V$
- The upward Archimedes force (weight of the displaced air): $F_{\rm a} = g\overline{\rho}V$

Hence, the net force on the parcel is $F_{\rm net} = -g\rho' V$. Because the acceleration is the force per unit mass, and the mass of the parcel is $(\overline{\rho} + \rho')V$, the acceleration of the parcel is

$$a_{\rm net} = \frac{F_{\rm g,net}}{\overline{\rho} + \rho'} = -\frac{g\rho'}{\overline{\rho} + \rho'} \approx -g\frac{\rho'}{\overline{\rho}} \tag{3.7}$$

With the result of Eq. (3.6) this gives

$$a_{\rm net} = g\frac{T_{\rm v}'}{\overline{T}_{\rm v}} \tag{3.8}$$

Thus if a parcel of air is warmer (or has a higher moisture content) than its surroundings ($T_{\rm v}' > 0$), it will experience a positive (upward) acceleration, whereas a cooler parcel will be accelerated downward. The fact that temperature (and to a lesser extent, moisture) influences the vertical motion in air has led to the term *active scalar* for temperature and humidity. *Passive scalars*, on the other hand, do *not* have a buoyancy effect.

Question 3.6: The virtual temperature can be calculated as $T_{\rm v} \approx T(1 + 0.61q)$ (T is *absolute* temperature in Kelvin; see also Appendix B).

a) Assume we have dry air of a given temperature T (dry air: so the specific humidity equals zero). What would then be the effect of the addition of moisture on the virtual temperature of this air (assume that the temperature of the water vapour is identical to that of the air)?

b) Assume we have two air parcels (at the same temperature and pressure): one *dry* air parcel, the other a parcel *with water vapour*. Which parcel will have the higher density?

3.3.5 Turbulent Kinetic Energy

Turbulence is moving fluid (in our case air). Therefore, the intensity of turbulent motion can best be characterized by considering the fluctuations of the wind speed, in particular by the kinetic energy of those fluctuations. Because turbulence is a three-dimensional phenomenon, we need to take into account motion in all three directions: vertical (w) as well as two horizontal directions (u and v). This results in the following definition of the (specific, i.e., per unit mass) turbulent kinetic energy (TKE, also denoted by $\bar{e}$):

$$\bar{e} \equiv \frac{1}{2}(\overline{u'u'} + \overline{v'v'} + \overline{w'w'}) \tag{3.9}$$

As the TKE is useful to describe the intensity of the turbulent motion, the next step is to find out which processes enhance turbulence (increase TKE) and which processes suppress turbulence. To that end we make a small excursion to more advanced fluid mechanics. From the equations that describe the motion of a viscous fluid in a situation with density stratification[5] for a horizontally homogeneous situation, the following budget equation for TKE can be derived (see, e.g., Wyngaard, 2010):

$$\underbrace{\frac{\partial \bar{e}}{\partial t}}_{\text{I. time change}} = \underbrace{-\overline{u'w'}\frac{\partial \bar{u}}{\partial z}}_{\text{II. shear production}} + \underbrace{\frac{g}{\theta_v}\overline{w'\theta_v'}}_{\text{III. buoyancy}} - \ldots - \ldots - \underbrace{\varepsilon}_{\text{IV. dissipation}} \tag{3.10}$$

where we have omitted (…) two terms that *do* matter, but that mainly redistribute TKE in space. The terms $\overline{u'w'}$ and $\overline{w'\theta_v'}$ are the turbulence fluxes of momentum and virtual heat, respectively (see Section 3.4). The terms shown in Eq. (3.10) have the following interpretation:

I. Change in time of TKE. If it is positive this means that TKE increases, whereas a negative value implies a decrease of TKE.
II. Production of TKE by wind shear. This term is generally positive, since $\overline{u'w'}$ (vertical momentum transport) is generally negative, whereas the mean horizontal wind speed increases with height.

[5] Navier–Stokes equations in combination with the Boussinesq approximation

III. Effect of buoyancy (density fluctuations in a gravity field). This term can either be positive or negative, depending on the sign of virtual heat flux $\overline{w'\theta_v'}$, that is, the covariance of vertical wind and virtual potential temperature. The virtual heat flux can be approximated as[6] (Stull, 1988): $\overline{w'\theta_v'} = \overline{w'\theta'}[1+0.61\bar{q}]+0.61\bar{\theta}\ \overline{w'q'}$. In some applications the entire term III is denoted as the buoyancy flux.

IV. Dissipation of TKE due to molecular friction at the smallest scales. This is always a loss term for TKE.

Terms II and III require some extra attention. The mechanism of shear production (term II) can be understood as follows. The upward displacement ($w'>0$) of an air parcel in a situation with a mean vertical velocity gradient $\left(\frac{\partial \bar{u}}{\partial z} > 0\right)$ produces a deceleration of the air at the level to which the parcel is displaced: $-w'\frac{\partial \bar{u}}{\partial z}$. The displacement per unit time along the direction of the acceleration is u'. Hence, the mean work per unit time is (acceleration times displacement speed) is $-\overline{u'w'}\frac{\partial \bar{u}}{\partial z}$. This work results in the production of turbulent kinetic energy at the expense of the mean kinetic energy of the flow (the terms in Eq. (3.10) can be interpreted as work per mass per time).

Likewise, term III can be analysed. If a parcel experiences a positive acceleration due to a higher temperature (i.e., lower density), and the accompanying displacement per unit time is a positive vertical velocity fluctuation w', positive work is done on the parcel by the Archimedes force, and TKE is produced. If the correlation between w' and θ_v' is negative (downward buoyancy transport), TKE is destroyed.

If we neglect the contribution of moisture to θ_v, $\overline{w'\theta_v'}$ is proportional to the sensible heat flux (see Section 3.4). Thus when the sensible heat flux is positive (upward heat transport), the buoyancy term produces TKE, whereas when the sensible heat flux is negative, TKE is destroyed.[7]

To characterize the role of buoyancy in the production of turbulence, often the ratio of the buoyancy production term and the shear production term is used. This ratio

[6] In terms of heat flux and the Bowen (β) ratio (ratio of sensible heat flux to latent heat flux, see Section 7.1) this becomes: $\overline{w'\theta_v'} = \overline{w'\theta'}\left(1+0.61\bar{q}+0.61\bar{\theta}\,\frac{c_p}{L_v}\beta^{-1}\right)$. Note that, because $0.61\,\bar{\theta}c_p\,/\,L_v$ is of the order of 0.07, the influence of moisture on buoyancy becomes relevant already at Bowen ratios as high as 0.5.

[7] A deviation of the local temperature at a given height from the mean temperature at that height can be interpreted as potential energy. If the stratification is such that the potential temperature decreases with height, this potential energy will be immediately released and converted to TKE. If the potential temperature increases with height potential energy may be converted into TKE and vice versa. Hence the terminology that buoyancy 'destroys' TKE is only partly correct: under stable conditions TKE is converted into potential energy which is partly released back as TKE and partly dissipated due to dissipation of temperature fluctuations. This is the concept of total turbulent energy where turbulent kinetic energy and turbulent potential energy are considered together (TTE, see, e.g., Zilitinkevich et al. 2007).

is called the Richardson number (or flux-Richardson number, to distinguish it from other variants):

$$Ri_{\mathrm{f}} \equiv -\frac{\text{buoyancy production}}{\text{shear production}} = \frac{\frac{g}{\theta_{\mathrm{v}}}\overline{w'\theta_{\mathrm{v}}'}}{\overline{u'w'}\frac{\partial \bar{u}}{\partial z}} \tag{3.11}$$

Note that within the ASL, where turbulent fluxes change little with height (see Section 3.4.3), the only height dependence stems from the wind shear: close to the ground shear is larger than at some larger height, leading to a larger contribution of shear production close to the surface.

A number of situations can be distinguished:

- $Ri_{\mathrm{f}} \approx 0$: There is no buoyancy production/destruction of TKE, only shear production (*neutral* conditions).
- $Ri_{\mathrm{f}} < 0$: TKE is produced both by shear and by buoyancy (*unstable* conditions).
- $Ri_{\mathrm{f}} > 0$: TKE is produced by shear but destroyed by buoyancy (*stable* conditions).
- $|Ri_{\mathrm{f}}|$ is large: The effect of buoyancy dominates over shear production (either very unstable (convective) or very stable, depending on the sign of Ri_{f}).

This analysis already gives a hint about the observations made with respect to the diurnal cycle of the turbulent diffusivity in Figure 3.3. The fact that at night K_{h} was an order of magnitude smaller than during daytime can be understood from the fact that at night time the sensible heat flux is negative, hence $Ri_{\mathrm{f}} > 0$, and TKE is destroyed by buoyancy. Because turbulent motion is needed for efficient transport, the hampering of turbulence by buoyancy will decrease the turbulent diffusivity. The reverse argument holds for daytime conditions with a positive heat flux and enhanced turbulence.

Question 3.7: Suppose we have a flow with the following characteristics:

	u (m s^{-1})	v (m s^{-1})	w (m s^{-1})
Mean	4	2	0
Standard deviation	0.3	0.2	0.2

a) Compute the mean kinetic energy.
b) Compute the turbulent kinetic energy.
c) What are the units of the kinetic energies computed under (a) and (b)?
d) The units given under (c) are not the units of energy. With what quantity should the kinetic energies computed under (a) and (b) be multiplied to obtain a real energy (with the correct units)?

Question 3.8: Verify that the shear production term in Eq. (3.10) is indeed a positive term (and hence a production term).

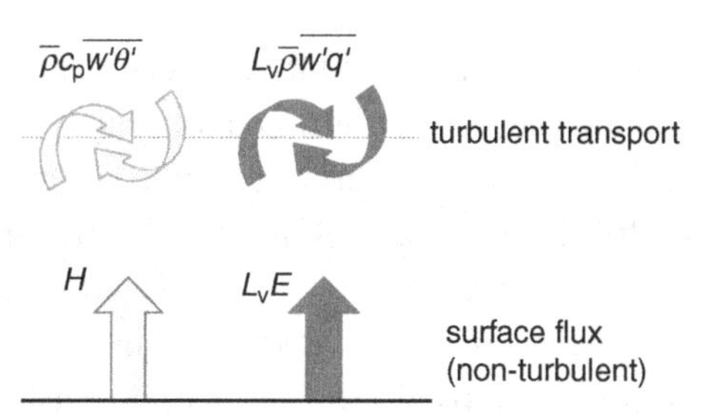

Figure 3.8 Link between surface fluxes (here fluxes of sensible heat and water vapour) and turbulent fluxes. Surface fluxes are relevant for the surface energy and mass balance. Furthermore, they are the source/sink for the turbulent transport.

3.4 Turbulent Transport

3.4.1 Mean Vertical Flux Density

The turbulent transport engine presented in Section 3.3.1 involves the vertical transport of quantities by the vertical motion of air. Thus, at a given location and time, the transport of a quantity with specific concentration X is $w\rho X$ (ρX is the volumetric concentration, or density): a parcel of air with concentration ρX is moving vertically with velocity w. But we are not interested in what happens at a certain moment, but rather in the mean effect of all those parcels moving up and down. Hence we apply the Reynolds decomposition (Section 3.3.3). First we only decompose X:

$$F_X = \overline{w\rho X} = \overline{w\rho(\overline{X} + X')} = \overline{w\rho}\overline{X} + \overline{w\rho X'} \tag{3.12}$$

The term $\overline{w\rho}$ is the total transport of mass. If we neglect the exchange of mass at the surface (in the form of release or uptake of water vapour and CO_2), the mass flux can be assumed to be zero at the surface and in the surface layer (no air is entering or leaving the solid surface). With this step (first only decomposing X) we partly get rid of the mean vertical wind speed $\overline{w}$, which is very hard to measure accurately but can have a considerable impact on the fluxes. Now in Eq. (3.12) only the second term remains, which can be expanded as (applying Reynolds decomposition to w and ρ, and subsequently average):

$$F_X = \overline{w\rho X'} = \overline{\left(\overline{w}\overline{\rho} + \overline{w}\rho' + w'\overline{\rho} + w'\rho'\right)X'} = \overline{w}\overline{\rho' X'} + \overline{\rho}\overline{w' X'} + \overline{w'\rho' X'} \tag{3.13}$$

where we used $\overline{\overline{w}\overline{\rho}X'} \equiv 0$ to omit the first term. It can be shown that the first and the last term on the right-hand side are small compared to the middle one, so that the flux F_X is simply the covariance of the vertical wind speed and the specific concentration X times the mean air density.

Now the question arises as to what is the link between the turbulent flux $F_x = \overline{\rho}\overline{w'X'}$ and the surface fluxes that are the subject of this book. Figure 3.8 shows this link for the sensible heat flux and water vapour flux: heat and water vapour are exchanged between the surface and the atmosphere by nonturbulent transport (H and E). In the atmosphere above the surface the transport takes place by turbulence and is expressed by the covariance of the vertical wind speed and the transported quantity. If turbulent fluxes are considered sufficiently close to the surface, the surface flux and the turbulent flux are very similar because what goes into the atmosphere at the surface has to pass at a few metres above the ground as well (for a more precise discussion, see Section 3.4.3). In the context of this book, this leads to the following link between surface fluxes and turbulent fluxes:

- Transport of heat: The energy flux is $H = \overline{\rho} c_p \overline{w'\theta'}$ (in W m^{-2}), where it should be noted that both the dry air component and the water vapour carry sensible heat: c_p is the specific heat of moist air.[8]
- Transport of water vapour: The mass flux is $E = \overline{\rho}\,\overline{w'q'}$ (in kg s^{-1} m^{-2}); the energy flux is $L_v E = L_v \overline{\rho}\,\overline{w'q'}$ (in W m^{-2}).
- Transport of an arbitrary gas, for example, the mass flux of CO_2: $F_c = \overline{\rho}\,\overline{w'q_c'}$ (in kg s^{-1} m^{-2}).
- Transport of momentum: $\tau = -\overline{\rho}\,\overline{w'u'}$ (in N m^{-2}) (the covariance is usually negative (downward momentum transport), but as the surface stress τ is taken positive, a minus sign is included).

Apart from the energy fluxes (sensible and latent heat flux) and mass fluxes (e.g., water vapour and CO_2), another form of the fluxes is often used in the context of turbulence research: the so-called kinematic flux. This is the mass flux divided by density (comparable to the relationship between dynamic and kinematic viscosity) and in the case of the heat flux the sensible heat flux divided by $\overline{\rho} c_p$. The relationship between the three representations is shown in Table 3.2. It is important to realize that if one of the three representations is known, the others can be determined.

[8] In fact the transported quantity is enthalpy $c_p\theta$ and thus: $H = \overline{\rho}\,\overline{w'(c_p\theta)'}$. As the specific heat c_p depends on specific humidity (see Appendix B), this would lead to extra terms in the Reynolds decomposition: $H = \overline{\rho} c_{pd}\left[\overline{w'\theta'} + 0.84\overline{w'\theta'}\,\overline{q} + 0.84\overline{w'q'}\overline{\theta}\right] = \overline{\rho} c_p \overline{w'\theta'} + \overline{\rho} c_{pd} 0.84\overline{w'q'}\overline{\theta}$. However, the last term in the Reynolds decomposition is erroneous (it suggests that all water vapour transported upward has been heated at the surface from 0 K to the ambient temperature). The error is related to the fact that enthalpy is like potential energy: it can be known only up to an unknown reference value: only differences in enthalpy can be studied, no absolute values. In the context of this book the relevant locations would be the surface and observation height. Assuming the turbulent fluxes to be constant with height (equal to the nonturbulent surface fluxes), the expression for H (at a certain observation level) could be written as: $H = \overline{\rho} c_p \overline{w'\theta'} + \overline{\rho} c_{pd}\, 0.84\, \overline{w'q'}\left(\overline{\theta} - \theta_s\right)$: if the temperature at measurement level is lower than the surface temperature, the water vapour has lost some of its sensible heat (for more details, see van Dijk et al., 2004).

Table 3.2 Relation between the nonturbulent surface fluxes and various definitions of turbulent fluxes, and the context in which they are used. Transport of heat and water vapour are used as examples.

	Heat		Water vapour		Application
	Quantity	Unit	Quantity	Unit	
Surface energy flux	H	W m^{-2}	$L_v E$	W m^{-2}	Energy balance
Turbulence energy flux	$\bar{\rho} c_p \overline{w'\theta'}$	W m^{-2}	$L_v \bar{\rho} \overline{w'q'}$	W m^{-2}	Energy balance
Surface mass flux			E	kg m^{-2} s^{-1}	Water balance
Turbulent mass flux			$\bar{\rho} \overline{w'q'}$	kg m^{-2} s^{-1}	Water balance
Surface kinematic flux	$\frac{H}{\rho c_p}$	K m s^{-1}	$\frac{E}{\rho}$	kg kg^{-1} m s^{-1}	Scaling (Section 3.5)
Turbulent kinematic flux	$\overline{w'\theta'}$	K m s^{-1}	$\overline{w'q'}$	kg kg^{-1} m s^{-1}	Scaling (Section 3.5)

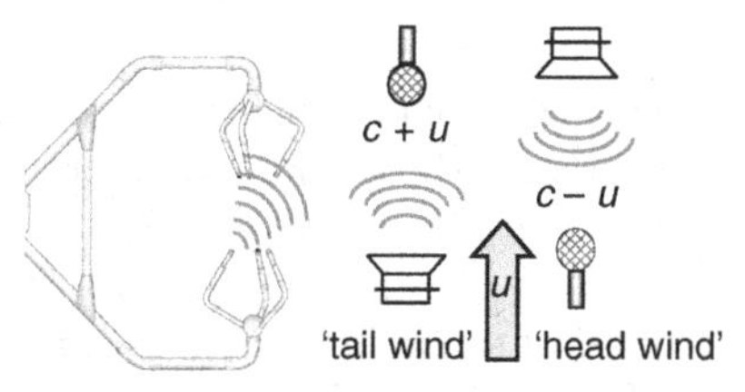

Figure 3.9 Sonic anemometer (in this case a Campbell Sci CSAT, left). Each pair of arms contains a sound source and microphone at both sides. The travel time of the ultrasonic sound pulse depends on the speed of sound (c) and the wind speed (u). With head wind the travel time will be longer than in still air, and with tail wind it will be shorter.

3.4.2 Eddy-Covariance Method

The definition of the turbulent flux as the covariance of the transported quantity and vertical wind speed directly provides a way to determine fluxes from observations. When the fluctuations of, for example, temperature and vertical wind speed are measured simultaneously, one could determine the sensible heat flux from the covariance of the two signals. This is the idea behind the eddy-covariance method in a nutshell. The usual setup is to use a sonic anemometer to measure the wind speed in three orthogonal directions, in combination with one or more gas analysers that measure the concentrations of water vapour and CO_2 on the basis of absorption of electromagnetic by the molecules under consideration. Temperature can either be measured with a fast-response thermocouple, or using the temperature that can be derived from the sonic anemometer (see later).

The sonic anemometer (see Figure 3.9) measures the wind speed by determining the travel time of a (ultrasonic) sound pulse. If the pulse travels in the same direction as the wind is blowing (tail wind) the pulse will travel faster than in still air and hence the travel time will be shorter ($\Delta t = \dfrac{\Delta x}{c+u}$, with Δx the distance between source and receiver of the sound pulse and c the speed of sound). With headwind the situation will be reversed: a longer travel time. If the travel time is measured in both directions simultaneously, both the wind speed and the speed of sound can be determined, as one has two travel times and two unknowns.[9] Because the speed of sound – mainly – depends on temperature, a sonic anemometer can also be used as a fast response thermometer using the following relationship (with T in K):

$$T = \left(\frac{c}{331.3}\right)^2 \frac{273.15}{1+0.51q}$$

To obtain the air temperature the information of the sonic anemometer needs to be combined with information on q from a fast response hygrometer. Note, that if the correction for humidity is not made, the resulting temperature (sometimes called 'sonic temperature') is close to the virtual temperature.

Question 3.9: Consider an idealized sonic anemometer with the path between sound source and receiver pointing vertical. The distance between the sound source and the receiver is Δx =10 cm.

a) First a sound pulse is fired in only one direction, upward. The travel time is observed to be Δt = 0.310 ms. A speed of sound in air is assumed of 330 m s^{-1}. What is the magnitude and direction of the wind speed?
b) On another occasion, two sound pulses are fired, one upward with a travel time of Δt_{up} = 0.295 ms and one downward with a travel time of 0.302 ms. What is the magnitude and direction of the wind speed, and what is the speed of sound?
c) Compute the temperature of the air from the speed of sound determined under (b) (assuming dry air).

A gas analyser uses the absorption of electromagnetic radiation by molecules to measure the concentrations of water vapour and CO_2. Some sensors use infrared radiation (around 4 μm and 2.5 μm for CO_2 and H_2O, respectively; see Figure 2.2), and others ultraviolet radiation (the so-called Lymann-α line around 0.12 μm). The amount of absorption of radiation is related to the number of molecules between source and

[9] The set of equations is: $\Delta t_1 = \dfrac{\Delta x}{c+u}$ and $\Delta t_2 = \dfrac{\Delta x}{c-u}$. With Δt_1 and Δt_2 measured and Δx known (it is a property of the instrument) one can solve for both c and u.

receiver (= concentration or density). Thus gas analysers do not measure the specific concentration q_X, as was used in the definition of the fluxes in the previous section, but the density ρ_X. This problem can be solved in two – equivalent – ways.

The first method is to calculate q_X first from the gas analyser data and temperature, before calculating the flux. In the second method one calculates the flux directly from the measured density:

$$F_X = \overline{w\rho_X} = \overline{(\overline{w}+w')(\overline{\rho_X}+\rho_X)} = \overline{w}\overline{\rho_X} + \overline{w'\rho_X'} \tag{3.14}$$

The term involving the mean vertical wind is *not* equal to zero. It is called the Webb term (Webb et al., 1980). From the fact that the vertical mass flux of dry air with density ρ_d is zero (see Eq. (3.12)), the mean vertical wind speed can be deduced:

$$\overline{w\rho_d} = 0 = \overline{w}\overline{\rho_d} + \overline{w'\rho_d'} \Rightarrow \overline{w} = -\frac{\overline{w'\rho_d'}}{\overline{\rho_d}} \approx \frac{\overline{w'T'}}{\overline{T}}$$

(for the last step, see Eq. (3.6)). Especially for gases with a mean concentration that is high relative to the fluctuations in the concentration (e.g., CO_2) the Webb term is an important contribution to the total flux of Eq. (3.14). Physically this mean vertical velocity (Webb velocity) can be understood as follows (for a situation with positive surface heat flux). Turbulent motions transport relatively cool air downward. At the surface the air is heated by the surface sensible heat flux. As a result the density decreases and the air parcel expands. This expansion of the air close to the surface pushes the entire air column upward.

Gas analysers can either be open path analysers or closed path analysers. In the case of an open path analyser, the absorption measurement takes place at the same location as where the vertical wind speed is measured: the optical path is located close to the measurement path of the sonic anemometer. Currently, open path analysers are available only for H_2O, CO_2 and CH_4. For closed path analysers, air is sampled at the location of the vertical wind speed observations and transported through a tube towards a gas analyser that is located elsewhere. This has the advantage that the gas analyser is not exposed to unfavourable weather conditions such as rain and dew that may disturb the measurements (e.g., Heusinkveld et al., 2008). Furthermore, it allows for the use of gas analysers (e.g., spectrometers) that can detect a large range of gases. Special care needs to be taken to correct for the decorrelation between the vertical wind speed and the gas concentration due to time delays and signal broadening in the transport tube towards the gas analyser (e.g., Moncrieff et al., 1997)

Question 3.10: From the assumption that there is no mean mass flux into or from the surface, the mean vertical wind speed $\overline{w}$ can be computed. Given a situation where the mean temperature is 300 K and the kinematic heat flux $\overline{w'T'}$ is 0.1 K m s^{-1} (thus the sensible heat flux $\overline{\rho}c_p\overline{w'T'}$ is approximately 120 W m^{-2}, assuming a mean density of 1.2 kg m^{-3}). How large is the mean vertical wind speed?

The concept of the eddy-covariance method is simple, but there are a number of caveats:

1. Frequency response: The sensor should have an immediate response to a changing signal. For example, if the air temperature changes by 1 K in 1 second, the thermometer should be fast enough to follow that change (a mercury thermometer will not suffice in this case, and neither will a cup anemometer to measure fast wind speed fluctuations). This is illustrated in Figures 3.10a and b: the sensor of part b) is not able to follow the fast fluctuations of the turbulence and hence it underestimates the magnitude in the temperature variations. If this temperature sensor would be used to measure the sensible heat flux, the flux would be underestimated.
2. Spatial response: The sensors should be able to sense variations at the smallest scales of turbulence that carry significant parts of the flux (i.e., usually down to the scale of mm). Thus a sensor that averages the wind speed over a distance of 1 m will not suffice for measurements a few metres above the ground. As small scale fluctuations also have small time scales, the signal of a sensor that averages over a too long a path will look similar to the signal of a slow sensor (as in Figure 3.10b).
3. Alignment: The sensors should be well aligned with the surface to ensure that the vertical axis (the direction of the w-component of the wind speed) is indeed vertical.
4. Sensor separation: The locations where vertical wind speed fluctuations and concentration of the transported quantity are measured should not be too far separated. If they sample different volumes, part of the correlation will be lost (the larger the distance between the sensors, the smaller the correlation).

Because sensors and measurement setups are usually not as ideal as required, corrections to the computed covariances are needed. Those corrections address issues related to points (1) to (4). On top of that, instrument-related corrections may be needed, because some sensors do not measure exactly the quantity one is interested in (e.g., a gas analyser may be sensitive not only to humidity fluctuations, but also to oxygen fluctuations). If the setup of an eddy-covariance system has been carefully designed, the corrections do not add up to more than 10% of the measured flux (and thus possible errors in the corrections do not have a major influence on the resulting fluxes). The accuracy of observed fluxes (30-minute averages) is 5–10% for the sensible heat flux and 10–15% for the latent heat flux (Mauder et al., 2006).

Apart from the requirements related to the sensors and their installation, there are also important requirements related to the sampling: both sampling frequency and averaging period. As can be seen in Figure 3.4, turbulent signals vary wildly and the correlation between vertical wind speed and the transported quantity is generally not large (Figure 3.7a). The relative statistical error of a covariance (e.g., a flux like $\overline{w'\theta'}$) can be estimated as (after Lenschow et al., 1994):

$$\mathrm{RE}(\overline{w'x'}) = \frac{1}{R_{wx}} \max\left[\mathrm{RE}(\sigma_w), \mathrm{RE}(\sigma_x)\right] \tag{3.15}$$

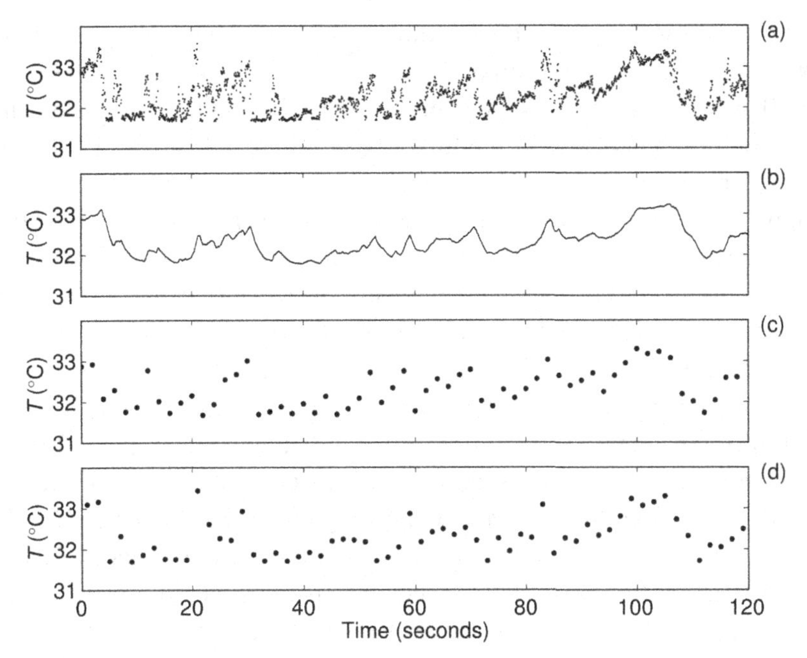

Figure 3.10 Illustration of the requirements for sensor response and number of samples, for vertical wind speed (left) and temperature (right). (**a**) fast sensor, sampled at 20 Hz ($\sigma_T = 0.459$ K). (**b**) slow sensor, sampled at 20 Hz ($\sigma_T = 0.355$ K). (**c**) fast sensor, sampled at 0.5 Hz ($\sigma_T = 0.452$ K). (**d**) fast sensor, sampled at 0.5 Hz, different starting time ($\sigma_T = 0.470$ K). The data are a subset of the data shown in Figure 3.4 (minutes 12 and 13).

where RE denote the relative error and σ_w and σ_x are the standard deviations of the vertical wind speed and the transported quantity, respectively, and R_{wx} is the correlation coefficient between w and x As the correlation between vertical wind speed and, for example, temperature or humidity is not very large in the surface layer (between 0.25 and 0.55), the error in fluxes is larger than the error in the standard deviations. To minimize the error, the error in the estimates of σ_w and σ_x need to be minimized. In general this implies that one tries to maximize the number of samples, which can be done by increasing the sampling rate and by extending the period over which one averages.

The issue of the sampling frequency is illustrated in Figures 3.10c and d. At low sampling frequencies only a small number of samples are gathered and the quantities computed from this limited amount of data will have a relatively large statistical (random) error: the time series of Figure 3.10c overestimates the standard deviation, whereas in part d it is underestimated. On the other hand, the signals in both parts c and d capture the essence of the original signal shown in part d. Increasing the sampling frequency beyond a certain point does not help to minimize the error as the samples become mutually dependent[10] (the extra samples contain little new information).

[10] Sampling rates that are so high that samples are mutually dependent *can* be useful in case one is interested in the temporal or spectral behaviour of the turbulence, rather than fluxes and variances.

More important in this respect is the time span over which averaging takes place. As turbulence varies at many time scales (Figure 3.4), the 2 minutes shown in Figure 3.10 provide only a snapshot, containing just the shorter time scales (the standard deviation of temperature for the full hour of Figure 3.4 is 0.52 K: the 2-minute snapshot underestimates it by 11%). Hence, to quantify fully the variance in the turbulent signal, as well as the covariance, one needs to average over a period that captures all relevant time scales: all scales at which vertical wind and the transported quantity are correlated. For measurements close to the surface (order of 10 m) this usually leads to averaging times of 10–30 minutes (see also Vickers et al., 2009). However, for measurements on high towers (e.g., above tall canopies, or to study the higher parts of the surface layer) averaging periods of 1–2 hours may be needed to capture all flux-carrying scales (Finnigan et al., 2002; Schalkwijk et al., 2010).

Usually, eddy-covariance measurements are made to determine *surface* fluxes. However, the measured flux does not simply originate from a point right below the sensors, but rather from an area mainly upwind of the sensors. Hence, if the area in which measurements are made is not homogeneous in surface conditions, one needs to know what part of the area determines the observed flux (so-called 'footprint') in order to interpret the observations correctly. The size and shape of the footprint depends on (Horst and Weil, 1992):

- Height at which the observations are made: When observations are made higher above the ground, the size of the footprint increases: the instrument can 'look' further.
- Stability: The size of the footprint increases with decreasing mixing: under unstable conditions the footprint is considerably smaller than under stable conditions.
- Roughness of the surface: Turbulent mixing is less intense over smooth surfaces, thus increasing the size of the footprint.

Figure 3.11 shows an impression of the dependence of the shape and size of the flux footprint on stability. The surface area from which 50% of the flux originates is approximately 32, 320 and 1300 times z_m^2 for the three stabilities shown. Thus if the instruments are installed at 10 m above the ground, 50% of the flux under unstable conditions comes from roughly 3200 m^2. The areas for 90% of the flux are roughly 1000, 7000 and 20,000 times z_m^2, respectively.

The eddy-covariance method is widely used to study surface exchange of heat, momentum, water vapour and CO_2 at ecosystem scale. In many locations long-term, multiple-year observations are made. Those cover varying land use types and climate regions. Those long-term observations are often made under the umbrella of regional (mostly continental) networks, which in turn are coordinated in the FLUXNET project (Baldocchi et al., 2001; Baldocchi, 2008). The coordination enables homogenization of processing and archiving so that it becomes feasible to study land–atmosphere exchange processes at multiple sites for multiple years (see also Section 8.4 for an example).

Additional information about the eddy-covariance method can be found in Lee et al. (2004).

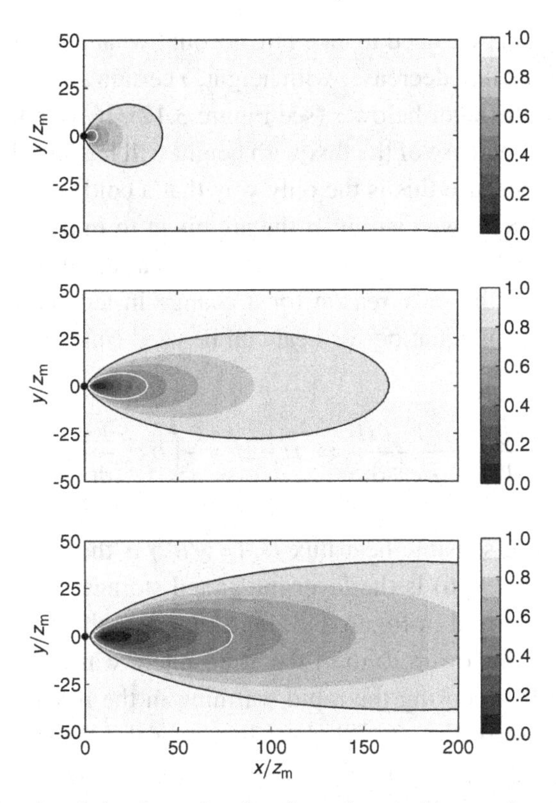

Figure 3.11 Sketch of the footprint for flux observations at location $(x, y, z) = (0, 0, z_m)$ for unstable (top), neutral (middle) and stable (bottom) condition (z_m/L equal to −1, 0 and +0.5; for definition of L see Section 3.5.1). Shades indicate the fraction of the observed flux recovered from the indicated surface. White and black isolines indicate the area from which 50% and 90% of the flux originate, respectively. Figures are based on the model of Kormann and Meixner (2001). Note that horizontal distances are scaled with the observation height z_m.

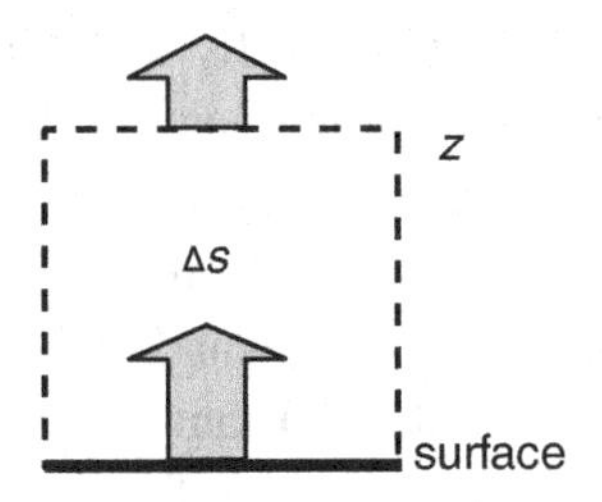

Figure 3.12 A change of the flux with height (flux divergence) leads to change in the storage in the control volume.

3.4.3 The Atmospheric Surface-Layer and the Roughness Sublayer

The turbulent fluxes derived in Section 3.4.1 are defined at an arbitrary height. But if we are interested in exchange processes at Earth's surface, we need to have information about the *surface* fluxes. To determine the link between the surface flux and the flux at a certain height z, we need to take into account what happens between the surface and height z. If the flux decreases with height, a certain amount of the transported quantity is stored in the layer below z (see Figure 3.12). If we take the sensible heat flux as an example, a decrease of the flux with height will lead to a heating of the layer (which is fortunate, because this is the only way that a cold summer night turns into a hot summer day!). Now we can invert the argument to find out to what extent the sensible heat flux varies with height. If we assume that the flux divergence (change of flux with height) is the only reason for a change in temperature, the integrated form of the conservation equation for heat can be used (integrated from the surface to height z):

$$\frac{\partial \bar{\theta}}{\partial t} = -\frac{1}{\bar{\rho} c_{\mathrm{p}}} \frac{\partial H}{\partial z} \Rightarrow H_z - H = -\int_0^z \bar{\rho} c_{\mathrm{p}} \frac{\partial \bar{\theta}}{\partial t} \mathrm{d}z, \tag{3.16}$$

where H is the surface sensible heat flux H_z $(=\overline{w'\theta'})$ is the turbulent flux at z. The right-hand side of Eq. (3.16) is the layer-integrated storage of heat. As an example we look at the day depicted in Figure 3.3. For those data the maximum temperature increase observed in the lower 20 m of the atmosphere was of the order of 6 K per 3 hours or $5.6{\cdot}10^{-4}$ K s^{-1} during the rapid warming in the morning hours. This corresponds to a decrease of the sensible heat flux over this 20 m of about 13 W m^{-2} (assuming $\bar{\rho} c_{\mathrm{p}} = 1200\ \mathrm{J\,K^{-1} m^{-3}}$). Given a total H of 190 W m^{-2} for that time, the relative decrease of sensible heat flux with height was of the order of 7%.

In micrometeorology the layer in which fluxes (not only the sensible heat flux) differ from the *surface* flux by less than 10% is loosely called the atmospheric surface-layer ([ASL] or 'constant flux layer'). The depth of this layer varies during the day and is roughly 10% of the depth of the atmospheric boundary layer (see Figure 3.1). These two occurrences of '10%' are no coincidence. If the sensible heat flux would decrease from the surface value to zero over the depth of the ABL, an allowable 10% flux decrease over the ASL would exactly match a surface layer depth of 10% of the boundary-layer depth. Because of the negative heat flux at the top of the ABL (entrainment flux) this correspondence is only approximate.

Thus at midday, with an ABL depth on the order of 1 km, the ASL has a depth on the order of 100 m. On the other hand, at night and in the morning hours the boundary-layer depth can be 100 m or less and the constant flux layer then is only 10 m deep.

The significance of the concept of a constant flux layer is that in this layer the processes that occur at the *surface* (exchange of heat, water vapour, CO_2, momentum) are the dominant factors to determine the processes occurring in this ASL. Processes in other parts of the atmosphere (in the atmospheric boundary layer or above) are assumed to have a much smaller impact.[11]

[11] It can also be shown (Monin and Yaglom, 1971; Section 6.6) that the Coriolis force has negligible influence in the surface layer.

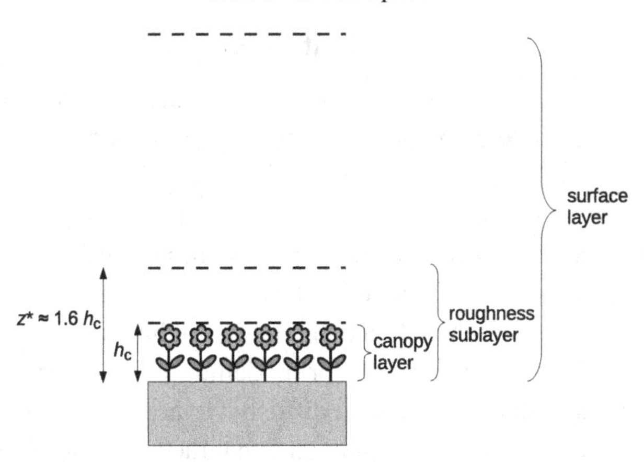

Figure 3.13 Definition of layers within the surface layer (height scale is roughly logarithmic).

Question 3.11: In Eq. (3.16) the change with height of the sensible heat flux is coupled to the change in time of the mean temperature of a layer between the surface and a height z.

a) Derive a similar expression for the change with height of the latent heat flux (take care that the units are correct on both sides of the equality sign).
b) For the same date as depicted in Figure 3.3, the increase of the specific humidity in the lowest 20 m of the atmosphere was approximately 0.25 g kg^{-1} per hour. Compute the change with height of the latent heat flux in between the surface and a height of 20 m (assume reasonable values for the air density and the latent heat of vaporization).
c) Around 8 UTC the latent heat flux was approximately 150 W m^{-2}. How large is the change of the latent heat flux with height as a percentage of the surface latent heat flux?

At the lower boundary of the surface layer Earth's surface is located, covered by trees, grass, stones, ice, buildings, streets, etc. Nearly all natural and man-made surfaces can be considered to be aerodynamically rough. This means that the roughness obstacles (with height h_c, canopy height; see Figure 3.13) are much higher than the thickness of the *viscous sublayer* which is adjacent to every interface between a fluid and a smooth surface (solid or fluid).[12] The fact that the air has to flow around the roughness obstacles implies that between and just above those obstacles the mean turbulent quantities (wind speed, fluxes, etc.) vary horizontally. For example, the wind speed just behind a tree is different from the wind speed between the trees. But even well above the roughness obstacles, the effect of the surface on the profiles of mean quantities (e.g., wind speed and temperature) is visible. The layer in which the spatial variation of the rough surface influences the shape of the profiles is called the *roughness sublayer*. The thickness of this layer is of the order of 1.6 h_c (Wieringa, 1993) but also depends on the structure of the canopy (Graefe, 2004).

[12] The thickness of this viscous sublayer is $\delta \approx 5\nu/u_*$ where ν is the kinematic viscosity. Depending on u^*, δ is of the order of 1 mm (Garratt, 1992). In fact the combination of *surface properties* and the *flow* determine whether a surface is aerodynamically rough.

3.5 Similarity Theory

In the context of this book we are interested in the surface fluxes of, for example, heat and water vapour. In Section 3.4 it was shown that the turbulent fluxes in the surface layer are closely related to those surface fluxes and it was shown how they can be measured. However, in some cases the direct determination of turbulent fluxes is not feasible: for some gases fast response sensors are not available, and in the case of modelling, fluxes need to be modelled in terms of variables that *are* available in a model. Hence we need to find the relationship between the turbulent fluxes and quantities that are accessible for measurement or modelling (e.g., vertical gradients).

In Section 3.1 we saw that the turbulent diffusivity varies with height and meteorological conditions. In Section 3.3 we concluded that the turbulent kinetic energy may play an important role in that variation. But the equation that describes the evolution of the turbulent kinetic energy cannot be solved easily.[13] So, the link between fluxes and gradients (i.e., the derivation of K_h in Eq. (3.1)) cannot be solved from first principles. Therefore, one of the tools often used in fluid mechanics is similarity theory, which assumes that two flows are similar if certain dimensionless characteristics are identical for the two situations.

A real-life example would be a bicycle shop that can sell bicycles in a range of sizes, so the bikes are *not identical*. But if one would take the ratio of the height of the bike and the height of the person who wants to buy it, all bikes would be *similar*, for example, their height may be half the height of the buyer. This ratio can be considered as a similarity law: the dimensionless ratio of bike height and a rider's height is a constant. The bike seller then could use this similarity law by first asking the customer his length, before offering him a bike.

The formal method to determine which are the relevant dimensionless groups in a physical problem, and how they are related, is called dimensional analysis. The details of the method are presented in Appendix C. The main steps in dimensional analysis are important to consider:

1. Find the relevant physical quantities that (may) determine the quantity of interest. For example, one can guess that the temperature at 2 m height depends on the surface sensible heat flux, the wind speed and the height above the surface.
2. Make dimensionless groups out of the quantities selected in step 1.
3. Do an experiment in which all quantities selected in step 1 are measured or obtain existing data.
4. Calculate, from the data obtained in step 3, for each measurement interval the values of the dimensionless groups that were constructed in step 2. If all goes well, the dimensionless groups show a universal relationship (or *similarity relationship*) that can also be used

[13] Differential equations, similar to Eq. (3.10) can be derived that describe the evolution of turbulent fluxes, rather than TKE. One of the omitted terms contains a third-order covariance. We could of course write down an equation for the evolution of that term, but that would include a fourth-order covariance. This shifting of problems is called the 'closure problem': at a certain order, we need to make a model for that higher-order term.

for other, *similar* situations. With such a relationship one needs to measure all variables *but one*: the one that has not been measured can then be calculated from the similarity relationship.

In this section we present one specific set of similarity laws commonly used in surface layer meteorology, viz. Monin–Obukhov similarity theory (MOST; Monin and Obukhov, 1954; see Foken, 2006, for an account of the history of MOST). MOST was developed for use in the surface layer, but above the roughness sublayer (see Section 3.4.3). Although MOST can be applied to many mean turbulent quantities, we focus here on vertical gradients of mean variables (e.g., mean wind speed, mean temperature).

3.5.1 Dimensionless Gradients: Relevant Variables in MOST

The relationship between the flux and the vertical gradient of the transported quantity (Eq. (3.1)) depends on a very large number of factors. To make an analysis possible, MOST is restricted to situations in which the mean turbulent quantities:

- Do not change in time (stationarity).
- Do not change in space horizontally, that is, the only variation is in the vertical direction (horizontal homogeneity).
- Are not influenced by processes occurring outside the surface layer (this appears one of the weak points of MOST, as there is no strict boundary between turbulence in the surface layer and turbulence in the rest of the boundary layer above).

Then, the main ingredients that shape the vertical gradients in the surface layer are:

1. The surface flux of the quantity of interest.[14]
2. The height above the surface (because the domain of the surface layer is bounded at the lower end). In fact, the height above the zero-plane displacement should be used (see Section 3.5.6).
3. The intensity of the turbulent motion. For this we revert to the TKE budget equation (Eq. (3.10)). We assume that mainly the buoyancy production term and the shear production term are relevant.[15] Furthermore, in the shear production term $\frac{\partial \overline{u}}{\partial z}$ itself is shaped by the turbulence, so that only the buoyancy term $\left(\frac{g}{\overline{\theta}_v}\overline{w'\theta_v'}\right)$ and the shear stress ($\overline{u'w'}$) remain as independent variables.

[14] In Monin and Obukhov (1954) local fluxes are used with the restriction that within the surface layer they differ little from the surface fluxes. Here we use surface fluxes, based on the premise that scaling should be based on external variables rather than internal variables In practice, the distinction between local fluxes and surface fluxes is not a major issue, except during stable or transition conditions (see Braam et al., 2012; van de Wiel et al., 2012a).

[15] Dissipation a loss term that is always present and which consumes all net production of TKE.

An important point is that if we do not want to include detailed information on the properties of the underlying surface into the similarity laws, we should restrict the analysis to heights above the roughness sublayer.

To simplify the procedure, a number of scales are defined that have simple dimensions: a velocity scale u_* (friction velocity), a temperature scale θ_*, a moisture scale q_* and a general scale for a scalar x: q_{x*}:

$$
\begin{aligned}
u_* &\equiv \sqrt{\frac{\tau}{\rho}} &&\approx \sqrt{-\overline{u'w'}} \\
\theta_* &\equiv -\frac{H}{\overline{\rho} c_\mathrm{p} u_*} &&\approx -\frac{\overline{w'\theta'}}{u_*} \\
q_* &\equiv -\frac{E}{\overline{\rho} u_*} &&\approx -\frac{\overline{w'q'}}{u_*} \\
q_{x*} &\equiv -\frac{F_x}{\overline{\rho} u_*} &&\approx -\frac{\overline{w'q_x'}}{u_*}
\end{aligned}
\tag{3.17}
$$

where τ is the surface shear stress (momentum transported towards the ground, in N m^{-2}) and F_x is the surface mass flux of x. Note the minus sign in the definitions of the scalar scales. Also note that any kinematic flux can be recovered from the scales in Eq. (3.17) by multiplying with u_* (e.g., $\frac{H}{\overline{\rho} c_\mathrm{p}} = -u_*\theta_*$). Finally, it should be noted that in the scales presented in Eq. (3.17) surface fluxes are used (see footnote 22). But because we restrict ourselves to the layer in which fluxes vary at most by 10% with height (see Section 3.4.3) they may be replaced by the local turbulent fluxes if needed (e.g., $\frac{H}{\overline{\rho} c_\mathrm{p}}$ can be replaced by $\overline{w'\theta'}$ and $\frac{\tau}{\rho}$ by $-\overline{u'w'}$). This is indicated by the ≈ sign.

In MOST, the flux and vertical gradient of the transported quantity are combined into one dimensionless group (remember the ratio of bike height and rider's length). With the scales defined earlier, it is easy to make any vertical gradient dimensionless. For the gradients of $\overline{u}$, $\overline{\theta}$, $\overline{q}$ and $\overline{q}_x$ this gives:

$$
\frac{\partial \overline{u}}{\partial z}\frac{\kappa z}{u_*}, \quad \frac{\partial \overline{\theta}}{\partial z}\frac{\kappa z}{\theta_*}, \quad \frac{\partial \overline{q}}{\partial z}\frac{\kappa z}{q_*} \quad \text{and} \quad \frac{\partial \overline{q}_x}{\partial z}\frac{\kappa z}{q_{x*}}, \tag{3.18}
$$

where the von Karman constant κ is included for historical reasons. The usually employed value for κ is 0.40, but its value (at least in the atmospheric surface layer) is still under debate (Andreas et al., 2006). Note that the dimensionless gradients are positive owing to the inclusion of the minus sign in the scalar scale definitions (e.g., under daytime conditions the vertical temperature gradient is negative, the sensible heat flux is positive, but the temperature scale θ_* is negative).

The ingredients listed under (2) and (3) are combined into another dimensionless quantity, $\frac{z}{L}$ (in the literature sometimes denoted as ζ (zeta)). The definition of $\frac{z}{L}$ is:

$$\frac{z}{L} \equiv -z\kappa \frac{g}{\overline{\theta}_{\mathrm{v}}} \frac{H_{\mathrm{v}}}{\overline{\rho} c_{\mathrm{p}}} \frac{1}{u_*^3} = \frac{z\kappa g \theta_{v*}}{\overline{\theta}_{\mathrm{v}} u_*^2}, \tag{3.19}$$

where $\theta_{\mathrm{v}*}$ is the virtual potential temperature scale. This is comparable to the temperature scale θ_*, but with the surface sensible heat flux H replaced by the surface virtual heat flux H_{v}[16]: $\theta_{\mathrm{v}*} \equiv -\frac{H_{\mathrm{v}}}{\overline{\rho} c_{\mathrm{p}} u_*}$. The inclusion of κ in the Obukhov length is an arbitrary choice made in the past and the inclusion of the minus sign ensures that z/L has the same sign as the Richardson number. L is a length scale, called the Obukhov length. Recall that $\overline{\theta}_{\mathrm{v}}$ is the *absolute* virtual potential temperature, that is, in Kelvin, so that the Obukhov length is not very sensitive to the temperature of the air. We come back to the physical interpretation of $\frac{z}{L}$ in Section 3.5.2.

Question 3.12: Verify that indeed the Obukhov length L has units of length.

Now the central assumption of MOST is that any of the dimensionless gradients given in Eq. (3.18) are a universal function (called ϕ) of $\frac{z}{L}$:

$$\begin{aligned}
\frac{\partial \overline{u}}{\partial z}\frac{\kappa z}{u_*} &= \phi_{\mathrm{m}}\left(\frac{z}{L}\right)\\
\frac{\partial \overline{\theta}}{\partial z}\frac{\kappa z}{\theta_*} &= \phi_{\mathrm{h}}\left(\frac{z}{L}\right)\\
\frac{\partial \overline{q}}{\partial z}\frac{\kappa z}{q_*} &= \phi_{\mathrm{e}}\left(\frac{z}{L}\right)\\
\frac{\partial \overline{q}_x}{\partial z}\frac{\kappa z}{q_{x*}} &= \phi_{\mathrm{x}}\left(\frac{z}{L}\right)
\end{aligned} \tag{3.20}$$

Because the functions $\phi\left(\frac{z}{L}\right)$ are not necessarily the same for each of the quantities, they are identified with a subscript that indicates the variable. The shape of the ϕ-functions cannot be known beforehand and needs to be determined from experiments (step 3 and 4 of dimensional analysis, see page 98).

[16] Based on the approximation of the kinematic virtual heat flux, the *surface* virtual heat flux becomes: $H_{\mathrm{v}} = H\left[1+0.61\overline{q}\right] + 0.61 c_{\mathrm{p}} \overline{\theta}\, E = H\left(1+0.61\overline{q}+0.61\frac{c_{\mathrm{p}}}{L_{\mathrm{v}}}\overline{\theta}\,\beta^{-1}\right)$.

The ϕ-functions are generally called flux–gradient (or flux–profile) relationships because they describe how the flux of a quantity (contained in u_*, θ_*, q_* or q_{x*}) are related to their respective gradients. Physically, the dimensionless gradients can be interpreted as the inverse of transport efficiency as it is the gradient needed to produce a certain flux. If turbulence is intense, only a small gradient is needed to produce a given transport and hence the efficiency is large relative to neutral conditions (small dimensionless gradient and hence a small value for the ϕ-function). This is consistent with the example shown in Figure 3.14, where the unstable part of the function is below the neutral value. The reverse argument holds for the stable side: suppressed turbulence leads to a smaller transport efficiency and hence to a larger dimensionless gradient.

The dimensionless gradients also show that – for a given stability and hence transport efficiency – the magnitude of the vertical gradient of a quantity scales linearly with the surface flux: if the surface flux doubles, the gradient will double as well.

Once the ϕ-functions are known they can be used inversely, for example, to determine the sensible heat from the vertical temperature gradient. This is discussed in Section 3.6.

In the 1960s a number of field experiments were conducted to determine the shape and coefficients of those universal functions (see Dyer, 1974). It happened only then because the instrumentation needed was not available earlier. One of the key experiments was the Kansas experiment (1968). The result for $\phi_{\mathrm{m}}\left(\frac{z}{L}\right)$, from that experiment, is shown in Figure 3.14 as an illustration. The functions used in this book are given in Section 3.5.3 and are shown in Figure 3.17.

In fact, MOST is more general. According to MOST *any* mean turbulence quantity (not only vertical gradients, but also variances, etc.) is a universal function of $\frac{z}{L}$ when the quantity is made dimensionless with a combination of the relevant scale(s) (from the list in Eq. (3.17)), and height z.

Question 3.13: To apply MOST to other mean turbulent quantities, those need to be non-dimensionalized with a combination of relevant scales.

a) How can the standard deviation of temperature (σ_T) be made dimensionless?
b) How can the structure parameter of temperature, C_T^2 (which has units of $\mathrm{K}^2\ \mathrm{m}^{-2/3}$) be made dimensionless?

3.5.2 Physical Interpretation of z/L and Its Relationship to the Richardson Number

The dimensionless group $\frac{z}{L}$ was formed, because we supposed that it would give information on the intensity of turbulence, based on the TKE budget equation. In that sense it should contain the same information as the Richardson number derived

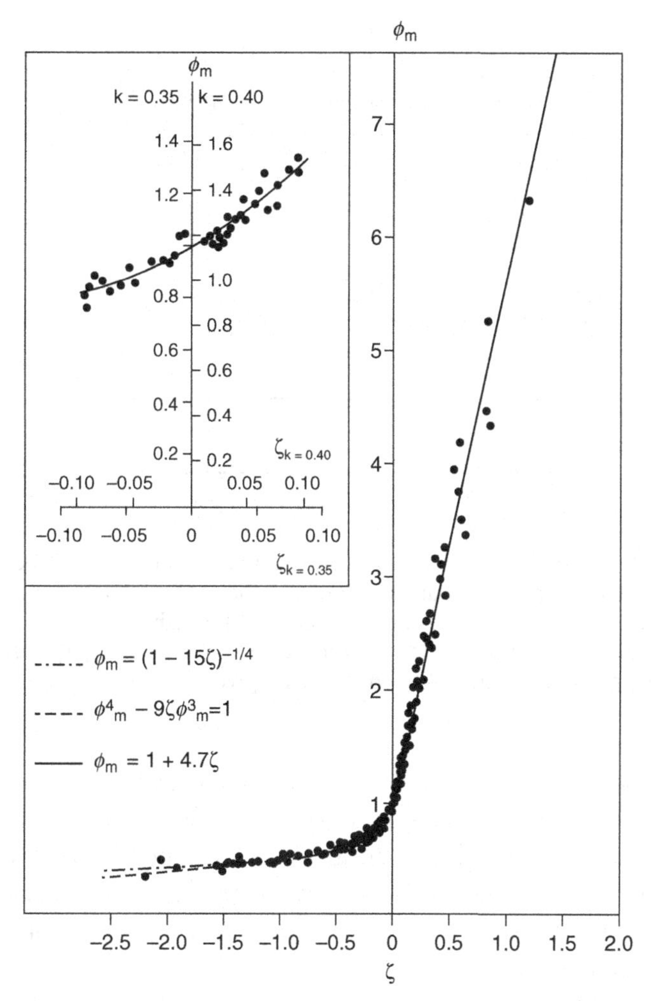

Figure 3.14 Example of a similarity relationship as obtained from an experiment. Here ϕ_m as found in the Kansas 1968 experiment is given (vertical axis), as a function of ζ (or z/L). (From Businger et al., 1971). Note that the functions given in the graph are slightly different than those used here (see Section 3.5.3). (© American Meteorological Society. Reprinted with permission.)

in Section 3.3.5. Indeed, if we approximate the wind shear $\frac{\partial \bar{u}}{\partial z}$ by $\frac{u_*}{\kappa z}$, the Richardson number appears to be equivalent to z/L (see also Section 3.5.6). Thus, the same limiting cases hold for z/L as for Ri_f (see Figure 3.15):

- Positive values of Ri_f and z/L indicate stable conditions (suppression of turbulence by buoyancy).
- Negative values of Ri_f and z/L indicate unstable conditions (enhancement of turbulence by buoyancy).
- Ri_f or z/L equal to zero indicate neutral conditions in which buoyancy does not play a role.

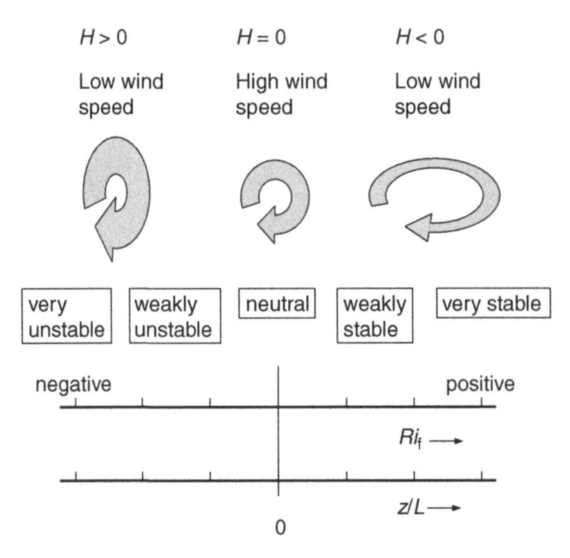

Figure 3.15 Stability parameters (Richardson number and z/L): link between sign and stability. The shape of the arrows indicates the effect of stability on turbulent vertical motion (enhanced for unstable, suppressed for stable). The thickness of the arrows indicates the intensity of the motion. At the top the sign of the sensible heat flux and the typical magnitude of wind speed are given.

With this equivalence of z/L and Ri_f we can also look at the height dependence of z/L. As L depends only on the surface fluxes, the only height dependence stems from z. Hence, below a certain level $|z/L|$ is so small that shear dominates TKE production, whereas high above the surface buoyancy dominates. It can be deduced, from Eq. (3.32), that for unstable conditions the level where shear production and buoyancy production are equal is located at $z \approx 0.6|L|$.

Note that although Ri_f and z/L have qualitatively similar behaviour, this does not mean that they are equal, or even proportional.

Question 3.14: Observations at a given location and time show that the temperature increases with height and that the sensible heat flux is negative.

a) Explain the relationship between the increase in temperature with height and the sign of the sensible heat flux.
b) What is the sign of the Richardson number?
c) What is the sign of the Obukhov length (and hence z/L)?
d) In what way does buoyancy influence the turbulence in this situation?

3.5.3 Similarity Relationships for Gradients

In Section 3.5.1 the similarity relationships for gradients of mean quantities in the surface layer were presented in general form (Eq. (3.20)). The final step in the determination of similarity relationships is the determination of the shape of those relationships from experimental data (steps 3 and 4). This has been an active field of research since the pioneering paper of Businger et al. (1971). Here we present and use only one set of commonly used flux–gradient relationships (the dependence of the dimensionless gradients ϕ on z/L): the so-called Businger–Dyer relationships (Dyer and Hicks, 1970). Although they are widely used, alternative functions are available (reviews in, e.g., Högström, 1996; Wilson, 2001; Foken 2006). The Businger–Dyer flux–gradient relationships are (see also Figure 3.16a):

$$\left.\begin{aligned}\phi_h\left(\frac{z}{L}\right)=\phi_e\left(\frac{z}{L}\right)=\phi_x\left(\frac{z}{L}\right)=\left(1-16\frac{z}{L}\right)^{-1/2}\\ \phi_m\left(\frac{z}{L}\right)=\left(1-16\frac{z}{L}\right)^{-1/4}\end{aligned}\right\}\text{for } \frac{z}{L}\leq 0$$
$$\left.\begin{aligned}\phi_h\left(\frac{z}{L}\right)=\phi_e\left(\frac{z}{L}\right)=\phi_x\left(\frac{z}{L}\right)=1+5\frac{z}{L}\\ \phi_m\left(\frac{z}{L}\right)=1+5\frac{z}{L}\end{aligned}\right\}\text{for } \frac{z}{L}\geq 0 \qquad (3.21)$$

A few things are noteworthy:

- For unstable conditions the ϕ-functions for scalars (temperature, humidity and the general scalar x) are different from that of momentum. For stable conditions however, they are identical. Note that the question whether the flux–gradient relationships (and other similarity relationships) are *really* identical for temperature and other scalars is still an active field or research.
- For neutral conditions ($z/L = 0$) both the unstable and stable formulations tend to a value of one, that is, at $z/L = 0$ both formulations match.
- For unstable conditions, the value of the coefficient that multiplies z/L (that is: 16) can be interpreted as follows. Close to neutral conditions (small $-z/L$) the flux–gradient relationships differ little from a value of one and vary approximately linearly with $-z/L$ (based on a Taylor series expansion, see also Figure 3.16a). On the other hand, for very unstable conditions (large $-z/L$) they vary as $(-16z/L)^{\alpha}$ (with α is equal to $-1/2$ for scalars and $-1/4$ for momentum). The point where the situation is no longer 'close to neutral' but tends to become 'very unstable' is where $-z/L$ is equal to 1/16. At that point the term involving z/L has magnitude one.
- The expressions are well-defined over a limited range of z/L only due to limitations in the range of stabilities available in the data sets. Foken (2006) gives a range of $-2<\frac{z}{L}\leq 0$ and $0\leq\frac{z}{L}<1$ for the expressions given in Eq. (3.21).
- The coefficients 16 and 5 occurring in Eq. (3.21) seem very firm, but they are the outcome of a set of experimental data, and hence are only a best-fit. See Högström (1988) for an extensive overview of alternative expressions. He suggests that variations in

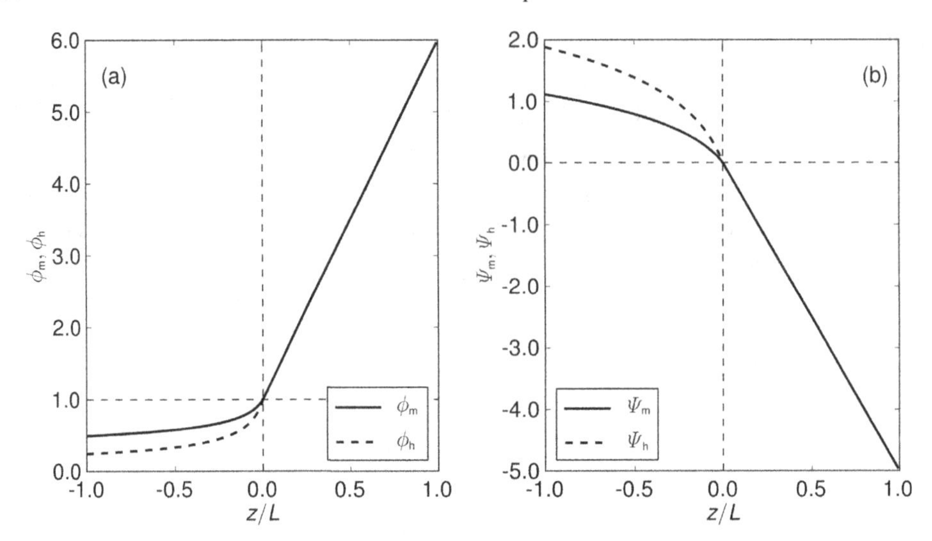

Figure 3.16 Similarity relationships for momentum and heat. (**a**) Flux–gradient relationships according to Eq. (3.21). (**b**) Integrated flux–gradient relationships according to Eq. (3.30).

flux–gradient relationships between different experiments may be due to flow distortion by sensors and comes up with slightly different expressions: $\phi_h = (1-12z/L)^{-1/2}$, $\phi_m = (1-19.3\,z/L)^{-1/4}$ for unstable conditions and, $\phi_h = 1+7.8\,z/L$, $\phi_m = 1+4.8\,z/L$ for stable conditions.

Question 3.15: Various expressions exist for the flux–gradient relationships. One could wonder how much they differ. Determine the relative difference between the Businger–Dyer flux-gradient relationships and the expressions proposed by Högström (1988) (see earlier). Take the Businger–Dyer relationships as a reference. Do this for both heat and momentum, and for the following values of *z/L*: –2, –1, 0, 0.5 and 1.

3.5.4 Gradients and Profiles Under Neutral Conditions

Before dealing with the general case where both shear and buoyancy play a role (Section 3.5.5), we first analyse the consequences of similarity theory for neutral conditions. Because the flux–gradient relationships are identical (and equal to 1) for all quantities, we take only the gradients of horizontal wind speed and potential temperature as examples. Because we are interested in the flux as a function of the gradient, we rewrite the similarity relationship Eq. (3.20) in such a way that the fluxes occur at the left-hand side. To that end we use the fact that $\phi_h = 1$ (neutral conditions) and that $\tau = \rho u_*^2$ and $H = -\rho c_p \theta_* u_*$. Then we obtain:

$$\tau = \rho \kappa u_* z \frac{\partial \overline{u}}{\partial z}, \quad H = -\rho c_p \kappa u_* z \frac{\partial \overline{\theta}}{\partial z} \tag{3.22}$$

If we compare this result to Eq. (3.1) we can conclude that for *neutral* conditions $K_{\rm m} = K_{\rm h} = \kappa u_* z$. Physically this makes sense, because the turbulent diffusivity depends on the intensity of the turbulence (represented by u_* because buoyancy does not play a role under neutral conditions) and the diffusivity increases with height due to the fact that the transporting eddies are larger higher above the surface. These two effects are in accordance with the tendencies observed in Figure 3.3. The diffusivities for other variables will be identical to that of heat, because for neutral conditions the ϕ-functions of all variables are identical.

Often, information on vertical gradients is not available (e.g., in a weather prediction model where variables are known only at discrete levels) or is hard to determine (measurements at a finite vertical distance can only approximate a gradient). Therefore, the integrated version of Eq. (3.20) is also very useful. If we still restrict ourselves to the neutral case ($z/L = 0$), we obtain (for wind and potential temperature):

$$
\begin{aligned}
&\int_{z_{u1}}^{z_{u2}} \frac{\partial \bar{u}}{\partial z}\frac{\kappa z}{u_*}\mathrm{d}z = \int_{z_{u1}}^{z_{u2}} \mathrm{d}z \Leftrightarrow \int_{z_{u1}}^{z_{u2}} \mathrm{d}u = \frac{u_*}{\kappa}\int_{z_{u1}}^{z_{u2}} \frac{1}{z}\mathrm{d}z \Leftrightarrow \\
&\qquad \bar{u}(z_{u2}) - \bar{u}(z_{u1}) = \frac{u_*}{\kappa}\ln\left(\frac{z_{u2}}{z_{u1}}\right) \\
&\int_{z_{\theta1}}^{z_{\theta2}} \frac{\partial \bar{\theta}}{\partial z}\frac{\kappa z}{\theta_*}\mathrm{d}z = \int_{z_{\theta1}}^{z_{\theta2}} \mathrm{d}z \Leftrightarrow \int_{z_{\theta1}}^{z_{\theta2}} \mathrm{d}\theta = \frac{\theta_*}{\kappa}\int_{z_{\theta1}}^{z_{\theta2}} \frac{1}{z}\mathrm{d}z \Leftrightarrow \\
&\qquad \bar{\theta}(z_{\theta2}) - \bar{\theta}(z_{\theta1}) = \frac{\theta_*}{\kappa}\ln\left(\frac{z_{\theta2}}{z_{\theta1}}\right)
\end{aligned}
\tag{3.23}
$$

where z_{u1} and z_{u2} are the heights where the wind speed is known and the temperature is known at $z_{\theta1}$ and $z_{\theta2}$. This logarithmic profile is a cornerstone of surface layer meteorology and represents the shape of wind speed and scalar profiles under neutral conditions.

To show how the flux depends on the observations of wind or temperature at two levels, Eq. (3.23) can be rewritten using the concept of a resistance:

$$
\tau = \rho\frac{\bar{u}(z_{u2}) - \bar{u}(z_{u1})}{r_{\rm am}}, \qquad H = -\rho c_{\rm p}\frac{\bar{\theta}(z_{\theta2}) - \bar{\theta}(z_{\theta1})}{r_{\rm ah}} \tag{3.24}
$$

where $r_{\rm am}$ and $r_{\rm ah}$ are the aerodynamic resistances for momentum and heat transport, respectively (which have units of s m^{-1}). In some applications aerodynamic conductance is used, which is the reciprocal of the resistance (with units m s^{-1}): $g_{\rm a} = 1/r_{\rm a}$.

In Eq. (3.24) the fluxes are proportional to the vertical differences of the transported quantity and inversely proportional to the aerodynamic resistance. This is similar to Ohm's law, where the current is proportional to the potential difference and inversely proportional to the resistance. For neutral conditions the aerodynamic resistances are identical for momentum and scalars (provided the observation levels are the same), and equal to:

$$r_a = \frac{\ln\left(\frac{z_2}{z_1}\right)}{\kappa u_*} \tag{3.25}$$

This expression for the resistance follows directly from Eqs. (3.23) and (3.24) (there is no new physics involved, only mathematics). Note that this resistance depends on the two heights used (just as the diffusivity depends on height). The resistance decreases with increasing friction velocity, and hence with increasing turbulence intensity. Furthermore, the resistance increases when the distance between the two levels increases (which is intuitive).

Question 3.16: Check that under neutral conditions dimensional analysis leads to (3.22).

Question 3.17: Equation (3.25) gives an expression for the aerodynamic resistance under neutral conditions.
a) Given an observed value of the friction velocity u_* of 0.3 m s^{-1}. Compute the resistance between the levels 2.5 m and 8 m (include units in your answer!).
b) Show, using the expression in Eq. (3.25), what the total resistance will be if you have two resistances in series (e.g., one between z_1 and z_2 and another between z_2 and z_3).

Question 3.18: Give the expressions similar to Eq. (3.24) but now for the latent heat flux L_vE) and for the mass flux of a scalar x (which has specific concentration q_x).

Question 3.19: Given the following observations of the mean wind speed at 2 and 4 m height: 2.0 and 2.5 m s^{-1} (assuming neutral conditions, see Eq. (3.23)):
a) Compute the friction velocity from the wind speed observations at 2 and 4 m.
b) Compute the wind speed at 10 m height.

3.5.5 Gradients and Profiles Under Conditions Affected by Buoyancy

Here we repeat the analysis made in Section 3.5.4 (for the wind speed and temperature gradient only), but now for situations in which buoyancy *does* play a role (both unstable and stable conditions). First we rewrite Eq. (3.20) to express the flux in terms of the gradient. This gives:

$$\tau = \rho c_p \frac{\kappa u_* z}{\phi_m\left(\frac{z}{L}\right)} \frac{\partial \bar{u}}{\partial z}, \qquad H = -\rho c_p \frac{\kappa u_* z}{\phi_h\left(\frac{z}{L}\right)} \frac{\partial \bar{\theta}}{\partial z} \tag{3.26}$$

and thus the general expression for the turbulent diffusivities is

$$K_{\mathrm{m}} = \frac{\kappa u_* z}{\phi_{\mathrm{m}}\left(\frac{z}{L}\right)}, \quad K_{\mathrm{h}} = \frac{\kappa u_* z}{\phi_{\mathrm{h}}\left(\frac{z}{L}\right)} \tag{3.27}$$

If we now introduce the actual functional forms of the flux–gradient relationships (from Eq. (3.21)) we obtain:

$$\left.\begin{aligned} K_{\mathrm{m}} &= \kappa u_* z\left(1-16\frac{z}{L}\right)^{1/4} \\ K_{\mathrm{h}} &= \kappa u_* z\left(1-16\frac{z}{L}\right)^{1/2} \end{aligned}\right\} \text{for } \frac{z}{L} \le 0$$

$$\left. K_{\mathrm{m}} = K_{\mathrm{h}} = \frac{\kappa u_* z}{1+5\frac{z}{L}} \right\} \text{for } \frac{z}{L} \ge 0 \tag{3.28}$$

From this we can see that for unstable conditions a decrease of z/L (more negative, hence more unstable) causes K_{m} and K_{h} to increase. On the other hand, for stable conditions an increase of z/L (more stable) causes K_{m} and K_{h} to decrease. Both tendencies are in accordance with what was seen in Figure 3.3.

The results in Eq. (3.28) can be further analysed in terms of the turbulent Prandtl number ($Pr_{\mathrm{t}} \equiv K_{\mathrm{m}}/K_{\mathrm{h}}$), which is an indication of the extent to which momentum and heat are transported in an equivalent way. According to Eq. (3.28) $Pr_{\mathrm{t}} = 1$ for neutral conditions, whereas for unstable conditions $Pr_{\mathrm{t}} < 1$. This relatively efficient transport of heat can be understood from the fact that under unstable conditions vertical wind speed and temperature correlate well as buoyancy is an important cause of vertical motion (R_{wT} roughly doubles from 0.25 to 0.5, going from neutral to very unstable conditions). On the other hand, vertical wind speed and horizontal wind speed show a decrease in correlation. As fluctuations in horizontal wind speed become increasingly dominated by boundary-layer scale motions, the correlation coefficient R_{uw} in the ASL decreases from about 0.6 to 0.15 when going from neutral to unstable (based on Wilson, 2008).Under stable conditions the consequence of Eq. (3.28) is that $Pr_{\mathrm{t}} = 1$ (owing to the fact that the used flux–gradient relationships for momentum and heat are equal). However, there are indications that Pr_{t} is larger than 1 with increasing stability: transport of momentum is hampered less by stability than transport of heat (see for example the flux–gradient relationships of Högström, 1988; Kondo et al., 1978 and Zilitinkevich et al., 2013). There is no full consensus, however, about this dependence of Pr_{t} on stability (Grachev et al., 2007).

Now, along the same lines as in Section 3.5.4, we integrate the gradients in Eq. (3.20) vertically to obtain information on vertical differences:

$$
\begin{aligned}
&\int_{z_{u1}}^{z_{u2}} \frac{\partial \bar{u}}{\partial z}\frac{\kappa z}{u_*}\mathrm{d}z = \int_{z_{u1}}^{z_{u2}} \phi_{\mathrm{m}}\left(\frac{z}{L}\right)\mathrm{d}z \Leftrightarrow \\
&\bar{u}(z_{u2}) - \bar{u}(z_{u1}) = \frac{u_*}{\kappa}\left[\ln\left(\frac{z_{u2}}{z_{u1}}\right) - \Psi_{\mathrm{m}}\left(\frac{z_{u2}}{L}\right) + \Psi_{\mathrm{m}}\left(\frac{z_{u1}}{L}\right)\right] \\
&\int_{z_{\theta 1}}^{z_{\theta 2}} \frac{\partial \bar{\theta}}{\partial z}\frac{\kappa z}{\theta_*}\mathrm{d}z = \int_{z_{\theta 1}}^{z_{\theta 2}} \phi_{\mathrm{h}}\left(\frac{z}{L}\right)\mathrm{d}z \Leftrightarrow \\
&\bar{\theta}(z_{\theta 2}) - \bar{\theta}(z_{\theta 1}) = \frac{\theta_*}{\kappa}\left[\ln\left(\frac{z_{\theta 2}}{z_{\theta 1}}\right) - \Psi_{\mathrm{h}}\left(\frac{z_{\theta 2}}{L}\right) + \Psi_{\mathrm{h}}\left(\frac{z_{\theta 1}}{L}\right)\right]
\end{aligned}
\tag{3.29}
$$

Here Ψ_{m} and Ψ_{h} (psi-functions) are the integrated flux–gradient relationships, defined as:

$$
\Psi_{\mathrm{y}}\left(\frac{z}{L}\right) \equiv \int_0^{z/L}\left[\frac{1-\phi_{\mathrm{y}}(\zeta')}{\zeta'}\right]d\zeta' \quad \text{with subscript y} = \text{m,h,e, or x.}
$$

For each ϕ-function the integral yields a different Ψ-function. Using the expressions for the Businger–Dyer flux–gradient relationships (given in Eq. (3.21)), the following expressions for the Ψ-function can be derived (Paulson, 1970)[17]:

$$
\left.
\begin{aligned}
&\Psi_{\mathrm{m}}\left(\frac{z}{L}\right) = 2\ln\left(\frac{1+x}{2}\right) + \ln\left(\frac{1+x^2}{2}\right) - 2\arctan(x) + \frac{\pi}{2}, \quad \text{with } x = \left(1-16\frac{z}{L}\right)^{1/4} \\
&\Psi_{\mathrm{h}}\left(\frac{z}{L}\right) = 2\ln\left(\frac{1+x^2}{2}\right), \quad \text{with } x = \left(1-16\frac{z}{L}\right)^{1/4}
\end{aligned}
\right\} \text{for } \frac{z}{L} \le 0
$$

$$
\left.\Psi_{\mathrm{m}}\left(\frac{z}{L}\right) = \Psi_{\mathrm{h}}\left(\frac{z}{L}\right) = -5\frac{z}{L}\right\} \text{ for } \frac{z}{L} \ge 0
\tag{3.30}
$$

These functions are depicted in Figure 3.16b. Now, Eq. (3.29) can be rewritten to express the fluxes in terms of vertical differences. The general expression is identical to Eq. (3.24), but the expressions for the resistances are different:

$$
\begin{aligned}
r_{\mathrm{am}} &= \frac{1}{\kappa u_*}\left[\ln\left(\frac{z_{u2}}{z_{u1}}\right) - \Psi_{\mathrm{m}}\left(\frac{z_{u2}}{L}\right) + \Psi_{\mathrm{m}}\left(\frac{z_{u1}}{L}\right)\right] \\
r_{\mathrm{ah}} &= \frac{1}{\kappa u_*}\left[\ln\left(\frac{z_{\theta 2}}{z_{\theta 1}}\right) - \Psi_{\mathrm{h}}\left(\frac{z_{\theta 2}}{L}\right) + \Psi_{\mathrm{h}}\left(\frac{z_{\theta 1}}{L}\right)\right]
\end{aligned}
\tag{3.31}
$$

[17] Wilson (2001) provides alternative expressions for the Ψ-functions, both for the Businger–Dyer flux–gradient relationships used here, as well as for alternative forms of the ϕ-functions.

Again, the expressions for the aerodynamic resistances follow directly from the combination of Eqs. (3.24) and (3.29) and are a mere mathematical consequence.

3.5.6 Similarity Theory: Final Remarks

Other Scalars and Other Turbulence Statistics

In the previous sections we introduced a framework that enables the description of the link between vertical gradients (or vertical differences) and fluxes. The equations given were only for momentum and heat, but given the fact that the flux–gradient relationships presented in Section 3.5.3 are identical for all scalars (heat, humidity and an arbitrary scalar x), the results for heat can also be applied to other scalars.

Furthermore, MOST is applied not only to gradients of mean quantities. Other turbulence statistics used in MOST are, for example, the standard deviation of scalars and velocity components, the structure parameters of temperature and humidity (see, e.g., DeBruin et al., 1993; Li et al., 2012) and the dissipation rate of TKE (e.g., Hartogensis and DeBruin, 2005).

Stability Parameters

In Section 3.3.5 the (flux-) Richardson number was introduced as an indicator of the stability regime in the surface layer and in 3.5.2 it was noted that Ri_f and z/L contain equivalent information, but are not necessarily identical.

Now that we have expressions that link fluxes to gradients, we can investigate the relationship between Ri_f and z/L more precisely. Using the definitions of the flux–gradient relationships, and replacing local fluxes by surface fluxes, Ri_f can be written as:

$$Ri_f = \frac{\frac{g}{\theta_v} u_* \theta_{v*}}{u_*^2 \frac{u_*}{\kappa z} \phi_m\left(\frac{z}{L}\right)} = \frac{\kappa z \frac{g}{\theta_v} \theta_{v*}}{u_*^2 \phi_m\left(\frac{z}{L}\right)} = \frac{z}{L} \frac{1}{\phi_m\left(\frac{z}{L}\right)} \tag{3.32}$$

Thus, for small z/L (where $\phi_m(z/L)$ is close to 1) the flux Richardson number and z/L are identical, but for more stable or unstable conditions this identity does not hold.

Because in some applications (e.g., in atmospheric models), fluxes are not readily available, two other variants of the Richardson number are also often used, viz. the gradient Richardson number Ri_g (where the fluxes are assumed to be proportional to the gradient of the transported quantity) and the bulk Richardson number Ri_b, where the gradients in Ri_g have been replaced by differences:

$$Ri_g \equiv \frac{g}{\overline{\theta}_v} \frac{\frac{\partial \overline{\theta}_v}{\partial z}}{\left(\frac{\partial \overline{u}}{\partial z}\right)^2}, \quad Ri_b \equiv \frac{g}{\overline{\theta}_v} \frac{\Delta \overline{\theta}_v \Delta z}{(\Delta \overline{u})^2} \tag{3.33}$$

The relationship between Ri_f and Ri_g can be derived easily using the flux–gradient relationships:

$$Ri_f = \frac{K_h}{K_m} Ri_g = \frac{\phi_m\left(\frac{z}{L}\right)}{\phi_h\left(\frac{z}{L}\right)} Ri_g \tag{3.34}$$

Thus the relationship between Ri_f and Ri_g does depend on stability (as does the relationship between Ri_f and z/L). If we combine Eqs. (3.32) and (3.34), we find that

$$Ri_g = \frac{\phi_h\left(\frac{z}{L}\right)}{\phi_m\left(\frac{z}{L}\right)} Ri_f = \frac{\phi_h\left(\frac{z}{L}\right)}{\left[\phi_m\left(\frac{z}{L}\right)\right]^2} \frac{z}{L} \tag{3.35}$$

Using the empirical expressions for the flux–gradient relationships (Eqs. (3.21) and (3.34)), we find that:

$$\begin{aligned} \frac{z}{L} &= Ri_g, \qquad &\text{for } \frac{z}{L} < 0 \\ \frac{z}{L} &= \frac{Ri_g}{1-5Ri_g}, \qquad &\text{for } \frac{z}{L} > 0 \end{aligned} \tag{3.36}$$

An important consequence of the stable side of Eq. (3.36) is that the expression for z/L has a singularity at $Ri_g = 0.2$. For that value z/L tends to infinity and the conditions become so stable that all turbulence is suppressed. The value of Ri_g where this happens is called the *critical* Richardson number, Ri_{gc}. The value of 0.2 is a direct consequence of Eq. (3.36), but the exact value of Ri_{gc} is still a subject of debate (see, e.g., Zilitinkevich et al., 2007). For stable stratification a relationship similar to Eq. (3.36) can be derived for the bulk Richardson number (DeBruin, 1982; Launiainen, 1995; Basu et al., 2008):

$$\frac{z_2 - z_1}{L} = \ln\left(\frac{z_2}{z_1}\right)\frac{Ri_b}{1-5Ri_b}, \quad \text{for } \frac{z}{L} > 0 \tag{3.37}$$

Question 3.20: Compute the two production terms (shear and buoyancy production) in the TKE equation (Eq. (3.10)) as well as their ratio (Ri_f) for the following conditions (use the flux–gradient relationship to determine the wind speed gradient, and use a height of 10 m).

a) $H = 140$ W m^{-2}, $L_vE = 250$ W m^{-2}, $u_* = 0.3$ m s^{-1}.
b) $H = -50$ W m^{-2}, $L_vE = 5$ W m^{-2}, $u_* = 0.2$ m s^{-1}.

For the background data assume θ = 295 K, q = 0.010 kg kg^{-1}, ρ = 1.15 kg m^{-3}, c_p = 1015 J kg^{-1} K^{-1}. L_v = 2.45·10^6 J kg^{-1}.

Experimental Determination of Similarity Relationships: Spurious Correlations

The dimensionless groups used to derive the similarity relationships are not fully independent. For instance, ϕ_m and $\frac{z}{L}$ both contain the surface friction, and ϕ_h and $\frac{z}{L}$ both contain surface friction and surface heat flux (see Eqs. (3.19) and (3.20)). In the derivation of similarity relationships from observations this fact is important. The presence of shared variables implies that errors in the observed surface fluxes (both systematic errors and random errors) will result in a simultaneous variation of both dimensionless groups. This may give rise to spurious self-correlation or spurious scatter: correlation or decorrelation between dimensionless groups that is not physical but only due to the sharing of variables.

The direction of this variation – roughly along the similarity relationship or perpendicular to it – depends on whether the shared variable occurs in either the numerator or denominator of both dimensionless groups, or in the numerator of the one and the denominator of the other. If we take the occurrence of u_* in ϕ_m and $\frac{z}{L}$ as an example, we see that it occurs in the denominator of both dimensionless groups. Thus a positive deviation in u_* will result in a negative deviation both in ϕ_m and in the absolute value of $\frac{z}{L}$ (see Figure 3.17). On the unstable side this implies a variation more or less normal to the expected similarity relationship, whereas on the stable side the error in u_* causes a variation along the expected relationship. Hence, measurement errors in u_* will only show up as scatter in $\phi_m\left(\frac{z}{L}\right)$-plots on the unstable side, whereas on the stable side the data will spuriously confirm the expected relationship. The reverse holds for the effect of errors in heat flux measurements on $\phi_h\left(\frac{z}{L}\right)$: those result in scatter on the stable side. See Baas et al. (2006) and Andreas and Hicks (2002) for more details.

The preceding discussion was stated in terms of relative errors. However, under stable conditions the fluxes are small and disturbing factors such as instationarity and intermittency (Klipp and Mahrt, 2004) may cause significant errors in the observed fluxes, hence increasing the relative error. This makes the usefulness of a flux-based similarity theory (like MOST) questionable. Because gradients are usually large during stable conditions (due to the lack of mixing) the use of similarity relationships based on gradients (in terms of Richardson numbers) is sometimes advantageous (Baas et al., 2006).

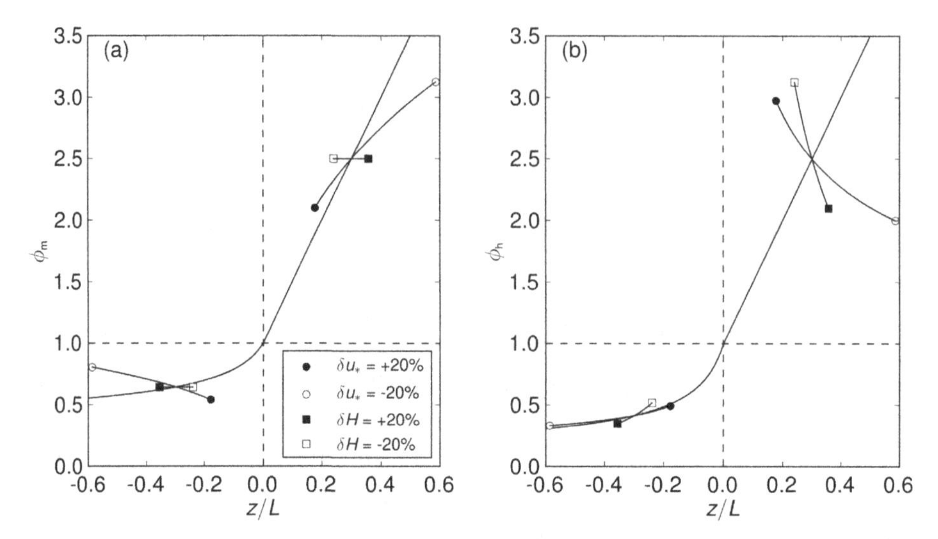

Figure 3.17. Effect of errors in surface fluxes on the derivation of flux–gradient relationships for (**a**) momentum and (**b**) heat. The solid line depicts the reference flux–gradient relationship, whereas the symbols show the effect of an error of 20% in either the surface stress or surface heat flux. If similarity relationship and errors are aligned, the experimentally determined relationship will be partly based on spurious self-correlation.

Zero-Plane Displacement

The similarity relationships have been derived under the assumption that the height above the surface z is a relevant height. However, the effect of roughness obstacles is that they force the flow upward. Hence the height relative to the ground surface underlying the roughness obstacles is not necessarily the most relevant height to describe the height variation of the flow properties. In terms of the plants in Figure 3.13: the flow does not 'know' how long the stems of those plants are, and hence the exact location of the surface on which the plants stand is irrelevant. Therefore, in the relationships derived in this section (and that are used in later sections and chapters) this has to be taken into account by always interpreting the height z as $(Z - d)$ where d is the zero-plane displacement height (or displacement height in short): the height of the surface as the flow experiences it. Then Z is the height above the substrate (Garratt, 1992). For example, Eq. (3.23) should be interpreted as:

$$\overline{u}(Z_{u2}) - \overline{u}(Z_{u1}) = \frac{u_*}{\kappa} \ln\left(\frac{Z_{u2} - d}{Z_{u1} - d}\right) \tag{3.38}$$

Physical interpretations of d are that it is either the height where the drag acts on the canopy elements (Thom, 1971; Jackson, 1981), or it is the equivalent of the displacement thickness (based on mass conservation) as used in fluid mechanics (DeBruin and Moore, 1985).

A first rough guess of the displacement height for a surface with roughness elements of height h_c, is $d = 2/3\ h_c$. More details on the displacement height follow in Section 3.6.2, where it is treated together with the roughness length.

Question 3.21: Given observations of the mean wind speed at 2, 4 and 6 m height: 2.00, 2.64 and 2.97 m s^{-1} (assuming neutral conditions, see Eq. (3.23)). The displacement height for the surface under consideration is 0.5 m.

a) Compute the friction velocity from the wind speed observations at 2 and 4 meters, and from those at 4 and 6 m height. Take into account the displacement height.
b) As under (a), but now ignore the displacement height (set it to zero).
c) For which of the height intervals (2–4 or 4–6 m) is the error in the friction velocity as found under (b) largest? Explain your findings.

Limiting Cases of Stability

For neutral conditions the flux–gradient relationships for stable and unstable conditions coincide: the dimensionless gradients are independent of buoyancy and equal to 1. This is due to the fact that under neutral conditions the virtual heat flux vanishes as a relevant scaling variable in the dimensional analysis. For extremely stable and unstable conditions similar analyses can be made.

In the case of strongly unstable conditions (so-called free-convection) turbulence is produced predominantly by buoyancy and surface shear no longer plays a role. This implies that the friction velocity u_* should vanish as a relevant variable and different scaling variables need to be introduced, based on $\frac{g}{\theta_v}\overline{w'\theta_v'}$ and z only: $u_f = \left(\frac{g}{\theta_v}\overline{w'\theta_v'}z\right)^{1/3}$ and, for example, $T_f = \overline{w'\theta'}/u_f$ (compare to θ_*). The dimensionless gradients made dimensionless with these scales should become constant as there is only one dimensionless group (see Appendix C), that is $\frac{\partial\theta}{\partial z}\frac{\kappa z}{T_f}$. Then if we would rewrite that again in terms of MOST variables (including u_*) one would obtain that $\phi_h \sim \left(-\frac{z}{L}\right)^{-1/3}$. Looking at Eq. (3.28) we see that the Businger–Dyer relationships do not show free-convection scaling in the limit of large z/L. DeBruin (1999) shows that for a limited stability range (down to $z/L = -1$) Eq. (3.28) could be replaced by an expression that *does* exhibit free convection scaling. Whereas mean gradients do not seem to follow free convection scaling, scalar standard deviations and structure parameters do (see, e.g., DeBruin et al., 1993).

For stable conditions both shear and buoyancy play a role. However, if conditions become very stable vertical motion is so much suppressed that the flow no longer experiences the presence of the ground. Hence the height above the ground becomes irrelevant as a length scale, hence the name z-less scaling. Because one still needs a length scale, height needs to be replaced by a length scale that indicates the extent of possible vertical motion. A logical choice would be the depth over which kinetic energy is converted into potential energy in a stratified fluid, a buoyancy length scale (see Van de Wiel et al., 2008). In terms of MOST scaling this leads to an expression of the form $\phi_h \sim \frac{z}{L}$, which is indeed consistent with the expressions given in Eq. (3.28) for large z/L.

Deviations from MOST

The simplicity of MOST is partly due to the strict conditions for its validity: stationarity, horizontal homogeneity and irrelevance of processes in the boundary-layer (above the surface layer). These conditions limit the number of relevant variables. In reality however, violations of one or more of the strict conditions for MOST are the rule rather than the exception: for example, most natural surfaces are heterogeneous and conditions are often non-stationary, as the diurnal cycle is omnipresent. Furthermore, there is no strict separation between turbulence in the surface layer and turbulence in the boundary layer (see, e.g., McNaughton et al., 2007).

One of the consequences of the assumptions underlying MOST is that similarity relationships should be identical for all scalars, and the correlation between scalar fluctuations should be either +1 or −1 (Hill, 1989). This is due to the fact that all mean turbulent quantities (gradients, variances) are determined solely by the vertical contrast over the surface layer, which in turn is related to the surface flux of the quantity under consideration. All scalars have the same source/sink location: the surface.

If not all conditions for MOST are met, decorrelation between scalars may occur. In the case of surface heterogeneity (e.g., dry and wet patches) this is due to the fact that different scalars have different dominating source locations (humidity from the wet patch, temperature from the dry patch, see Moene and Schüttemeyer, 2008). Vertical differences in source location occur if the relative importance of the surface flux and entrainment flux is different for different scalars: differences in entrainment regime (see Moene et al., 2006; Lahou et al., 2010). Other causes for decorrelation of scalars are the active role of temperature and humidity, modulations of the surface layer by the outer layer (and unsteadiness) and advective conditions (see Katul et al., 2008, for an extensive review of literature on this subject).

Conclusion

The main result of Section 3.5 is that vertical fluxes of a certain quantity can be expressed in terms of vertical gradients or vertical differences of the transported quantity. This is an important result for practical applications, which is the subject of the next section.

Question 3.22: In the case of free-convection scaling and z-less scaling MOST dimensionless gradients have specific dependencies on z/L.

a) Show that $\phi_h \sim \left(-\frac{z}{L}\right)^{-1/3}$ implies that scaling is independent of u_*.

b) Show that $\phi_h \sim \frac{z}{L}$ implies that scaling is independent of z.

3.6 Practical Applications of Similarity Relationships

The theoretical (and empirical) framework has been developed. Now it can be used to determine fluxes from observations (or model values). First, the situation is dealt with where values are available at two levels in the surface layer. Second, the lower level will be lowered to the surface, leaving only one observation level in the surface layer.

One remark has to be made here. In the previous sections, the effect of buoyancy has been expressed in terms of the virtual potential temperature (e.g., θ_{v*}, in the definition of z/L). For reasons of simplicity, however, in most of the following sections, except 3.6.1, it is assumed that the moisture contribution to the buoyancy can be neglected. This implies that, with respect to buoyancy, θ_{v*} is equivalent to θ_* and to characterize the production of turbulence only the surface fluxes of momentum and heat are needed.

3.6.1 Fluxes from Observations at Two Levels

The expressions in Eq. (3.29) can be used to derive u_* and θ_* from observations if we would know z/L. However, z/L in turn depends on u_* and θ_{v*} (thus on θ_* and q_*). Hence, we have a system of four equations (equations for the vertical differences $\Delta\overline{u}$, $\Delta\overline{\theta}$ and $\Delta\overline{q}$, and the definition of z/L), and four unknowns (u_*, θ_*, q_* and z/L). This system can be solved iteratively. This iteration would involve the following steps:

1. Compute initial values for u_*, θ_* and q_* based on the observed $\Delta\overline{u}$, $\Delta\overline{\theta}$ and $\Delta\overline{q}$, with $z/L = 0$ (neutral conditions).
2. Compute L.
3. Compute $\Psi_m(z_{u1}/L)$, $\Psi_m(z_{u2}/L)$, $\Psi_h(z_{\theta1}/L)$ and $\Psi_h(z_{\theta2}/L)$ (the latter are also used for humidity).
4. Compute new values for u_*, θ_* and q_* from the observed $\Delta\overline{u}$, $\Delta\overline{\theta}$ and $\Delta\overline{q}$ and the values of the Ψ-functions determined in the previous step.
5. Repeat steps 2 through 4, as long as computed values of u_* and θ_* change significantly from one iteration to the next.

6. After leaving the loop, compute the momentum flux, sensible heat flux and latent heat flux from u_*, θ_* and q_*.

Note that once the above iteration (including the *active* scalar θ or θ_v) has produced values for u_* and θ_*, the fluxes of *passive* scalars can be computed explicitly (without iteration) using expressions similar to Eq. (3.29), but now with a known z/L.

Recall that all heights used in these calculations should be taken relative to the displacement height (i.e., any occurrence of z in fact should be read as $(Z - d)$; see Section 3.5.6).

Question 3.23: Given are the following observations at 2 and 10 meter height: potential temperatures are 281.52 and 280.41 K, and wind speeds are 1.7 and 2.9 m s^{-1}, respectively.

a) Using a spreadsheet program, determine iteratively u_* and θ_* for this situation (simply use θ_* in the definition of the Obukhov length, rather than θ_{v*}).
b) Determine the sensible heat flux (assume ρ =1.15 kg m^{-3} and c_p = 1015 J kg^{-1} K^{-1}).

3.6.2 Fluxes from Observations at a Single Level in the Air and One at the Surface

In many practical applications it is useful or necessary to take the lower level of the equations that describe vertical differences (Eq. (3.29)) at the surface. But this directly poses a problem because taking z_{u1} or z_{t1} equal to zero would imply a division by zero. Furthermore, the surface is located within the roughness sublayer in which the similarity relationships are even not valid. To overcome this problem, the concept of the roughness length z_0 is introduced: the surface value of the variable under consideration is supposed to occur at height z_0 above the surface. Then, if the roughness lengths for momentum and the scalars (see later) are known, as well as the surface values of wind speed and the scalars under consideration, the fluxes can be determined. But the 'if' in the previous sentence is a big 'if'.

Roughness Length: Concept

For the wind speed profile this concept is straightforward: the wind speed should be zero at the surface,[18] that is, the wind does not slip. Then the roughness length z_0 is determined in such a way that the wind speed profile described by Eq. (3.29) becomes zero at $z = z_0$. The roughness length for momentum is also called the *aerodynamic* roughness length. For the determination of the roughness length for momentum one either needs wind speed measurements at two heights or observations at one height, but then including the momentum flux momentum flux. In addition, information on stability in the form of z/L is needed. For neutral conditions (z/L = 0) this procedure is simplified because of the absence of buoyancy effects, and we elaborate the

[18] This is valid for a solid surface. For flow over water the speed at the surface will be equal to the flow speed of the water.

example of wind speed observations at two heights. For neutral conditions we can use the logarithmic profile (see also Eq. (3.23)):

$$\bar{u}(z_2) - \bar{u}(z_1) = \frac{u_*}{\kappa} \ln\left(\frac{z_2}{z_1}\right) \tag{3.39}$$

When $\bar{u}(z_1)$, $\bar{u}(z_2)$, z_1 and z_2 are known, u_* can be determined for this set of observations. Next, with this u_* the roughness length can be determined from:

$$\bar{u}(z_2) - \bar{u}(z_0) \equiv \bar{u}(z_2) = \frac{u_*}{\kappa} \ln\left(\frac{z_2}{z_0}\right) \Leftrightarrow z_0 = z_2 \exp\left[-\frac{\kappa \bar{u}(z_2)}{u_*}\right] \tag{3.40}$$

A more direct way, without the intermediate calculation of u_*, would be to take the ratio of Eqs. (3.39) and (3.40), and calculate z_0 from that (see Question 3.24).

For scalars the extension of the profiles towards the surface is less straightforward. First one needs a value of the concentration of that scalar at the surface. For temperature this is possible because one can estimate the surface temperature from the emitted longwave radiation (provided that one knows the emissivity of the surface). Then the roughness length for heat, z_{0h}, is found by extrapolating the profile to that level where the temperature is equal to the observed surface temperature.

The roughness length for heat is by no means equal to that of momentum. This is due to the fact that the exchange of momentum (i.e., friction) between the air and the surface takes place mainly by pressure forces (form drag), whereas the heat transport between the surface and the air directly adjacent to it occurs by molecular diffusion only. The latter is far less efficient, leading to a relatively high surface temperature for a given amount of heat transport. Hence the temperature profile has to be extrapolated much further down to find the observed surface temperature. This is illustrated in Figure 3.18. Often the roughness length of heat is considered relative to that of momentum, either as a simple ratio z_0/z_{0h} or as $\kappa B^{-1} \equiv \ln\left(\frac{z_0}{z_{0h}}\right)$.

Although the details of the relationship between the roughness length for momentum and the roughness length for scalars is not yet fully clear, it is a necessary concept to provide a link between the sensible heat flux and the surface temperature:

- In atmospheric models the surface temperature is a variable that both enters into the calculation of the sensible heat flux (see later) and in that of the emitted longwave radiation. Besides, the calculation of the soil heat flux is affected indirectly (see Section 2.3). The use of an incorrect value of the roughness length for heat may yield significant errors in the surface temperature and hence in the predicted surface fluxes (e.g., Beljaars and Holtslag, 1991).
- Surface temperatures observed from satellite-borne sensors are often used to estimate the surface energy balance (including the sensible heat flux). In those calculations the roughness length for heat is a critical parameter.

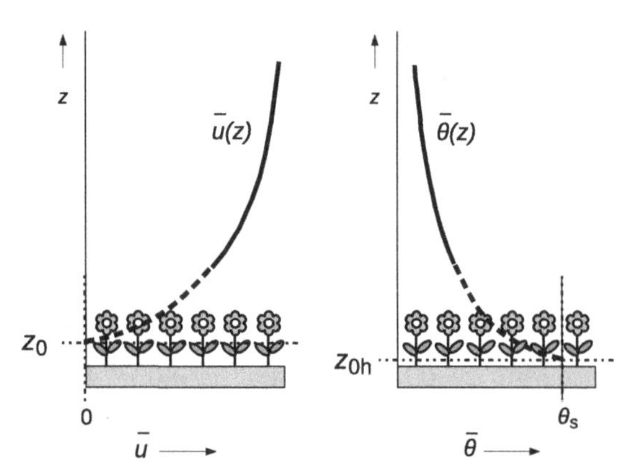

Figure 3.18 Relationship between roughness lengths, surface values and profiles: for momentum (left) and temperature (right). The solid lines are the profiles according to Eq. (3.29). In the region where they are valid the lines are solid. The dashed lines indicate that the profiles are extrapolated downward into the roughness sublayer where the surface layer profiles are not valid.

Question 3.24: Given the observations of Question 3.19, determine the roughness length for momentum for the surface under consideration.

Question 3.25: Given the following wind speed observations: 10 m s^{-1} at 10 m height and 7 m s^{-1} at 2 m. Assume neutral conditions.

a) Compute the friction velocity u_*, and the roughness length z_0.
b) Compute the aerodynamic resistance r_a for $z = 2$ m (i.e., the resistance between the surface and 2 m).

Observations above the same terrain at another moment show the same wind speeds. Now the potential temperature appears to be 20 °C at 10 m height and 21.5 °C at 2 m.

c) Is the surface layer neutrally, stably or unstably stratified?
d) Note for each of the following variables if they will be larger than, smaller than, or equal to the values determined in the questions (a) and (b): u_*, z_0 and r_a (for $z = 2$m). Explain your answers.

Question 3.26: A correct value for the roughness length for heat is important to obtain the correct link between surface temperature and sensible heat flux.

Given are a sensible heat flux of 200 W m^{-2}, an air temperature at 2 m height of 20 °C, a friction velocity of 0.4 m s^{-1} and a roughness length for momentum of 5 cm (ignore the effect of stability). Then, using an expression similar to Eq. (3.40) (but then for temperature: in fact Eq. (3.42) ignoring the stability correction), the surface temperature can be determined, provided that we know the roughness length for heat.

Compute the surface temperature for the following values of the ratio z_0/z_{0h}: 10, 100 and 1000.

Roughness Length and Displacement Height: Values

For a particular surface the roughness length needs to be determined locally (using the method described in the previous subsection). The roughness length for momentum can be considered to be a constant surface property, although for some surfaces (e.g., long grass, water) the roughness may depend on wind speed. At higher wind speeds grass bends and becomes smoother, whereas on water waves will develop that make the surface rougher.

However, local observations are usually not available and therefore one has to revert to tabulated values or simple models. The simplest model is that the z_0 would be proportional to the canopy height h_c, or better, the height of that part of the canopy that extends above the displacement height (Garratt, 1992):

$$z_0 = \gamma_1 \left(h_c - d\right) \tag{3.41}$$

γ_1 is of the order of 0.2 to 0.4. With the rule of thumb that $d = 2/3\ h_c$, this gives z_0/h_c in the range 0.07 to 0.14 (i.e., order 0.1).

The proportionality will depend on the density and distribution of the roughness elements. If there are only few elements per unit area, or if they are very narrow, the flow will hardly be affected by the roughness elements and the constant of proportionality will be smaller than at a medium element density. On the other hand, at high obstacle density (e.g., a dense forest) the flow will no longer enter the region between the roughness elements and will 'skim' the surface, yielding again a lower constant of proportionality (Garratt, 1992). This is illustrated in Figure 3.19.

For the displacement height similar arguments hold with respect to the relationship to obstacle density: at low obstacle density d/h_c will be close to zero since the flow will hardly be lifted. On the other hand for high densities the flow 'skims' the surface and will experience the surface as a rather smooth, lifted surface: d/h_c will tend to one. Typical values for displacement height and aerodynamic roughness length are given in Table 3.3.

Regarding the roughness length for scalars (in particular temperature) a distinction needs to be made with respect to surface type:

- For permeable surface cover (dense packing of small individual elements, e.g., vegetation fully covering the ground) a commonly used value for to $\kappa B^{-1} = 2$, corresponding to z_0/z_{0h} is 7.4 (see, e.g., Garratt and Francey, 1978).
- For bluff body surface cover (nearly impermeable obstacles, or sparse vegetation with patches of bare soil in between) κB^{-1} can be of the order of 4–12 and thus z_0/z_{0h} of the order of 55 to 10^5 (Stewart et al., 1994). For bluff body surfaces the roughness length ratio also depends on wind speed (see, e.g., Malhi, 1996). For sparse vegetation on dry bare soil the large difference between z_0 and z_{0h} can be understood as follows. The main

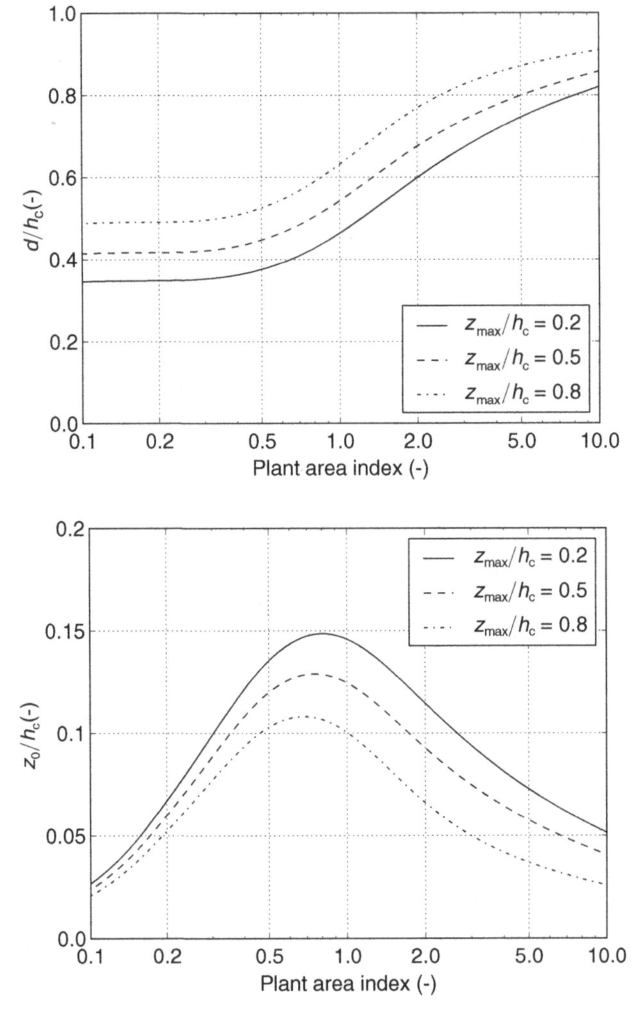

Figure 3.19 Relationship between relative displacement height (d/h_c, top) and roughness length (z_0/h_c, bottom) and the plant area index (one-sided area of plant material per unit ground area). Based on the model of Massman (1997), assuming a triangular vertical distribution of plant material (maximum at z_{max}).

sink of momentum is provided by the vegetation elements that provide a significant form drag and are located well above the soil surface. On the other hand, the main source of heat is located at the soil surface, which is smoother and located below the vegetation elements.

Determination of the Fluxes

For the case that we need to determine fluxes from observations at one height, in combination with surface values, the system of equations of Eq. (3.29) becomes:

Table 3.3 Typical values for aerodynamic roughness length and displacement height for natural surfaces

Surface	Remark	z_0 (m)	d (m)
Water	Still – open	10^{-4}–10^{-3}	—
Ice	Smooth sea ice	10^{-5}	—
Ice	Rough sea ice	10^{-3}–10^{-2}	—
Snow		10^{-4}–10^{-3}	—
Soils		0.002	—
Short grass, moss	h_c : 0.02–0.05 m	0.01	—
Long grass, heather	h_c : 0.2–0.6 m	0.04	0.2
Low mature crops	h_c : 0.3–1 m	0.07	0.5
High mature crops	h_c : 1–2.6 m	0.15	1
Continuous bushland	h_c : 2.3–3 m	0.3	2
Mature pine forest	h_c : 10–27 m	1.2	14
Tropical forest	h_c : 27–31 m	2	30
Deciduous forest	h_c : 10 m	0.8	7
Dense low buildings	Suburb	0.6	3
Regularly built town		1.2	10

After Wieringa (1992). Note that the roughness for water and all vegetations may depend on the wind speed.

$$\bar{u}(z_u) = \frac{u_*}{\kappa}\left[\ln\left(\frac{z_u}{z_0}\right) - \Psi_m\left(\frac{z_u}{L}\right) + \Psi_m\left(\frac{z_0}{L}\right)\right]$$
$$\bar{\theta}(z_\theta) - \theta_s = \frac{\theta_*}{\kappa}\left[\ln\left(\frac{z_\theta}{z_{0h}}\right) - \Psi_h\left(\frac{z_\theta}{L}\right) + \Psi_h\left(\frac{z_{0h}}{L}\right)\right] \quad (3.42)$$

where z_u is the observation height of the wind speed and z_θ is the height of the temperature observation. The Ψ-functions involving z_0 and z_{0h} are generally small and often neglected in practice. Provided that the relevant roughness lengths are known as well as the surface temperature, then an iteration similar to that given in Section 3.6.1 can be used to obtain u_* and θ_*, and hence the turbulent fluxes of momentum and heat.

The determination of fluxes from single level atmospheric data can also be described in terms of resistances (see Eq. (3.24)). Equation (3.24) remains valid, but the definition of the resistance has to be adapted. Apart from the transition from neutral to non-neutral conditions (going from Eq. (3.25) to Eq. (3.31)), now the lower level becomes special. For momentum the lower level is located at z_0 whereas for temperature it will be located at z_{0h}:

$$\tau \equiv \rho u_*^2 = \rho\frac{\bar{u}(z_u)}{r_{am}}, \qquad H = -\rho c_p \frac{\bar{\theta}(z_\theta) - \theta_s}{r_{ah}} \quad (3.43)$$

with:

$$r_{\mathrm{am}} = \frac{1}{\kappa u_*}\left[\ln\left(\frac{z_u}{z_0}\right) - \Psi_{\mathrm{m}}\left(\frac{z_u}{L}\right) + \Psi_{\mathrm{m}}\left(\frac{z_0}{L}\right)\right]$$

$$r_{\mathrm{ah}} = \frac{1}{\kappa u_*}\left[\ln\left(\frac{z_\theta}{z_{0\mathrm{h}}}\right) - \Psi_{\mathrm{h}}\left(\frac{z_\theta}{L}\right) + \Psi_{\mathrm{h}}\left(\frac{z_{0\mathrm{h}}}{L}\right)\right] \qquad (3.44)$$

Note that Eqs. (3.43) and (3.44) are not new, but rather are special cases of (3.24) and (3.31), respectively. Again, an iteration procedure is needed to determine the fluxes, since the resistances depend on stability, which depends on u_* and θ_*. The latter are the quantities we want to solve for.

The difference between the roughness lengths for momentum and scalar (e.g., heat) can also be interpreted in the framework of resistances (see Figure 3.20). The aerodynamic resistance for momentum transport r_{am} is used between the upper level z_θ and z_0. But at z_0 the temperature profile has not yet reached its surface value, that is, $\overline{\theta}(z_0) \neq \theta_s$. The additional step in temperature between z_0 and $z_{0\mathrm{h}}$ is related to an additional resistance. This excess resistance (or boundary-layer resistance, r_{bh}) is due to the molecular exchange of heat directly at the surface. If we consider the total temperature difference between the surface and the observation level z_θ, the resistance to obtain the correct sensible heat flux is the sum of the aerodynamic resistance for momentum and the boundary-layer resistance ($r_{\mathrm{ah}} = r_{\mathrm{am}} + r_{\mathrm{bh}}$).

Question 3.27: Describe the iteration procedure needed to compute the sensible heat flux from single level observations of wind speed and temperature, in combination with assumed values for the roughness lengths for momentum and heat (similar to the iteration procedure given in Section 3.6.1).

In atmospheric modelling the concept of drag coefficients is frequently used to determine fluxes from vertical differences of, for example, wind speed or temperature. Fluxes are then determined using the following expressions:

$$\tau \equiv \rho u_*^2 = \rho C_{\mathrm{dm}}\left[\overline{u}(z_u)\right]^2, \qquad H = -\rho c_p C_{\mathrm{dh}}\overline{u}(z_u)\left[\overline{\theta}(z_\theta) - \theta_s\right] \qquad (3.45)$$

If we compare Eqs. (3.45) and (3.43), we see that the friction velocity implicitly contained in the resistance used in Eq. (3.43) is now replaced by an explicit occurrence of the mean velocity in the drag laws of Eq. (3.45). Using the expressions for the velocity and temperature profiles in Eq. (3.42), the expressions for the drag coefficients become:

$$C_{\mathrm{dm}} = \kappa^2\left[\ln\left(\frac{z_u}{z_0}\right) - \Psi_{\mathrm{m}}\left(\frac{z_u}{L}\right) + \Psi_{\mathrm{m}}\left(\frac{z_0}{L}\right)\right]^{-2}$$

$$C_{\mathrm{dh}} = \kappa^2\left[\ln\left(\frac{z_\theta}{z_{0\mathrm{h}}}\right) - \Psi_{\mathrm{h}}\left(\frac{z_\theta}{L}\right) + \Psi_{\mathrm{h}}\left(\frac{z_{0\mathrm{h}}}{L}\right)\right]^{-1}\left[\ln\left(\frac{z_u}{z_0}\right) - \Psi_{\mathrm{m}}\left(\frac{z_u}{L}\right) + \Psi_{\mathrm{m}}\left(\frac{z_0}{L}\right)\right]^{-1} \qquad (3.46)$$

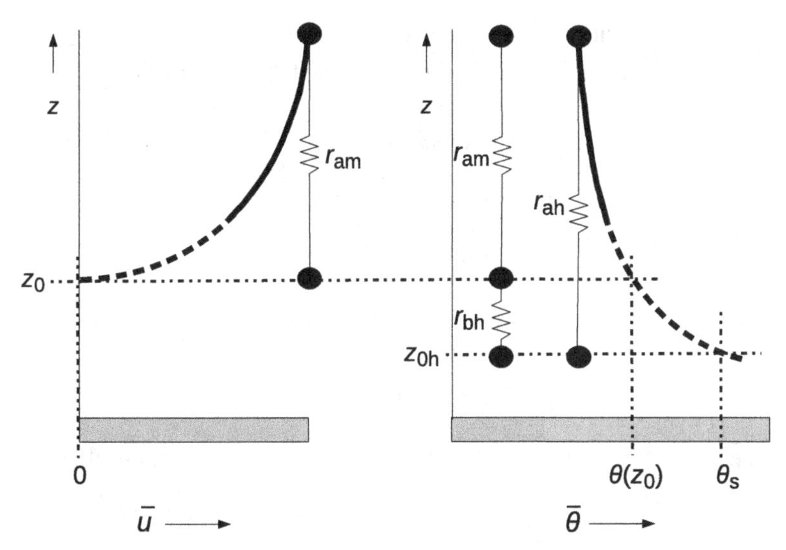

Figure 3.20 Relationship between roughness lengths and aerodynamic resistance for momentum transfer (left) and for heat transfer (right). The aerodynamic resistance for heat can be interpreted as a serial combination of the aerodynamic resistance for momentum and an excess resistance (or boundary-layer resistance) due to molecular diffusion near the surface. Note that the vertical axis is not to scale and the dashed profile is located within the roughness sublayer (after Garratt, 1992).

An advantage of the use of drag laws as opposed to resistance laws is that the drag coefficients do not depend on wind speed. The dependence of the transport on wind speed is explicitly dealt with in the drag law itself. The drag coefficients depend solely on observation heights, roughness lengths and stability.

3.6.3 Analytical Solutions for the Integrated Flux–Gradient Relationships

In the previous sections an iterative procedure was needed to obtain fluxes from observations of wind and temperature at two levels (see Section 3.6.1). However, under the following conditions analytical solutions to the integrated flux–gradient relationships can be found (Itier, 1982; Riou, 1982; DeBruin, 1982):

- The effect of humidity on buoyancy is ignored.
- Wind speed and temperature are measured at the same height (at two levels, denoted as z_1 and z_2).
- The ratio z_2/z_1 is less than about 6.

The existence of an analytical solution implies that no iteration is needed to find the fluxes from vertical temperature and wind speed differences. Although these analytical solutions are not exact (at least not for unstable conditions), they are fairly accurate, especially in view of the accuracy of the *empirical* flux–gradient relationships on which they are based. The solution for unstable conditions is:

$$\frac{H}{\rho c_\mathrm{p}} \approx -\frac{\kappa^2 \Delta\bar{\theta}\Delta\bar{u}}{\left[\ln\left(\frac{z_2}{z_1}\right)\right]^2}(1-16Ri_{\mathrm{b}*})^{\frac{3}{4}} \tag{3.47}$$

$$u_* \approx \frac{\kappa\Delta\bar{u}}{\left[\ln\left(\frac{z_2}{z_1}\right)\right]}(1-16Ri_{\mathrm{b}*})^{\frac{1}{4}} \tag{3.48}$$

where $\Delta\bar{\theta} = \bar{\theta}(z_2) - \bar{\theta}(z_1)$, $Ri_{\mathrm{b}*}$ is an 'effective' bulk-Richardson number, in this case given by:

$$Ri_{\mathrm{b}*} = \sqrt{z_1 z_2}\left[\ln\left(\frac{z_2}{z_1}\right)\right]\frac{g}{T}\frac{\Delta\bar{\theta}}{(\Delta\bar{u})^2} \tag{3.49}$$

For stable conditions, the solution is even exact, due to the particular form (simply linear) of the flux–gradient relationships. Starting from Eq. (3.29) and using the relationship between *z/L* and the bulk Richardson number (Eq. 3.37), the following formulations for the fluxes can be derived (Launiainen, 1995; Basu et al., 2008):

$$\frac{H}{\rho c_\mathrm{p}} = -\frac{\kappa^2 \Delta\bar{\theta}\Delta\bar{u}}{\left[\ln\left(\frac{z_2}{z_1}\right)\right]^2}(1-5Ri_\mathrm{b})^2 \quad \text{if } 0 < Ri_\mathrm{b} < 0.2 \tag{3.50}$$

$$u_* = \frac{\kappa\Delta\bar{u}}{\left[\ln\left(\frac{z_2}{z_1}\right)\right]}(1-5Ri_\mathrm{b}) \quad \text{if } 0 < Ri_\mathrm{b} < 0.2 \tag{3.51}$$

where

$$Ri_\mathrm{b} = (z_2 - z_1)\frac{g}{T}\frac{\Delta\bar{\theta}}{(\Delta\bar{u})^2} \tag{3.52}$$

which is a standard formulation for the bulk Richardson number. It should be noted that these results for the stable case have a singularity at $Ri_\mathrm{b} = 0.2$ (the critical Richardson number). At that stability the fluxes vanish because all turbulence has been suppressed by buoyancy. For values of $Ri_\mathrm{b} > 0.2$ the expressions in Eqs. (3.50) and (3.52) are no longer valid: the fluxes are simply equal to zero.

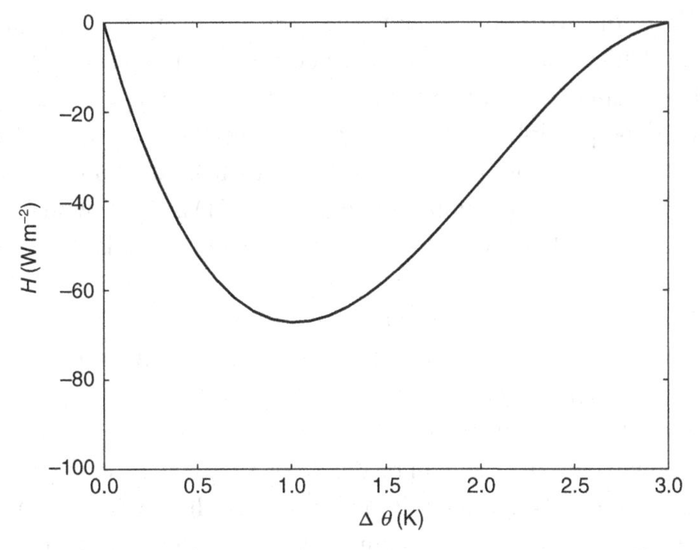

Figure 3.21 Sensible heat flux as a function of vertical temperature difference, given a fixed shear (heights are 10 and 2 m, fixed wind speed difference is 2 m s^{-1}).

For other scalar fluxes (e.g., evapotranspiration or CO_2 transport) expressions similar to Eqs. (3.47) and (3.50) can be used. DeBruin et al. (2000) extended the above framework to situations where temperature and wind speed are not observed at the same height.

3.6.4 Feedback Between Stability and the Sensible Heat Flux for Stable Conditions

The analytic solutions of the flux–gradient relationships (as presented in the previous section) allow for an interesting analysis of the behaviour of the sensible heat flux under stable conditions. The question is: What magnitude of the sensible heat flux is possible at a given vertical temperature gradient?

To answer this question, we assume a fixed vertical wind speed difference ('fixed shear', see van de Wiel et al., 2012a). Then Eqs. (3.50) and (3.52) give a direct solution for the sensible heat flux as a function of the vertical temperature difference. This solution is depicted in Figure 3.21. The striking result is that a given sensible heat flux can be attained with two different vertical temperature differences: a near neutral solution (small temperature difference) and a very stable solution (large temperature difference). The physical interpretation of this result is that in the near neutral case the vertical temperature difference is the limiting factor for the sensible heat flux. On the other hand, for the very stable solution, the turbulence is the limiting factor: turbulence is so much suppressed by buoyancy that despite the large temperature difference, the flux is still small.

This duality in the solution has large practical implications in the modelling of turbulent fluxes under stable conditions (van de Wiel et al., 2007; Basu et al., 2008). As long as the demand for energy, imposed by surface cooling $Q^* - G$, is smaller (in magnitude) than the possible supply of energy by turbulent transport (set by a fixed wind shear) two equilibria are possible: one with a small vertical temperature difference (to the left of the heat flux minimum in Figure 3.21) and one with a large vertical temperature difference. However, if the energy demand exceeds the possible turbulent heat flux, no equilibrium can be reached. Cooling of the surface will proceed (as $Q^* - G - H$ is not equal to zero) and runaway cooling will occur (van de Wiel et al., 2012b). As stability increases (the conditions move towards the right in Figure 3.21), turbulence will be suppressed eventually.

Summarizing: if for a given shear the energy demand is larger than the maximum magnitude of downward flux that can be delivered by turbulence, there is no equilibrium possible and runaway cooling will occur. This cooling will continue until radiative equilibrium is reached ($Q^* = G$) or may halt when the shear increases such that turbulence can sustain a larger (in magnitude) heat flux.

Question 3.28: Make a graph similar to Figure 3.21, but now for the momentum transport (i.e., u_*^2) as a function of the vertical temperature difference, at a given, fixed vertical wind speed difference. Explain the shape of this graph (as compared to that of Figure 3.21).

3.6.5 The Schmidt Paradox

In some applications one is interested in daily averaged fluxes, rather than in instantaneous or hourly averaged fluxes. It would then be tempting to apply the similarity relationships developed in this chapter using 24-hour averaged input data (temperatures, wind speed, etc.) to determine the 24-hour averaged fluxes. For this approach to be valid, it would be required at least that the sign of the 24-hour averaged flux would be consistent with the 24-hour averaged vertical gradient (e.g., if the mean flux is positive, the mean temperature should decrease with height).

In Table 3.4 we investigate this requirement using the data from the same day as those used in Figure 3.3. We see that whereas the mean temperature profile is stably stratified, the mean heat flux is positive: the mean flux goes *against* the mean gradient. This feature has long been recognized and is known as the Schmidt paradox (Schmidt, 1921; see also Lettau, 1979).

To illustrate this further, the 24-hour average turbulent diffusivity has been calculated from the observed sensible heat flux and the observed vertical temperature difference in two different ways:

- The mean diffusivity is calculated from the mean flux and the mean vertical temperature difference. This yields a negative diffusivity, which is inconsistent with the concept that a flux should flow down the gradient.

Table 3.4 The 24-hour average turbulent diffusion coefficient for data shown in Figure 3.3

Quantity	*Value*
$\overline{\theta}^{24}(2\,\text{m})$	20.02°C
$\overline{\theta}^{24}(10\,\text{m})$	20.11°C
$\overline{H}^{24}$	43 W m^{-2}
$\overline{K_h}^{24} = -\dfrac{\overline{H}^{24}/\rho c_p}{\left(\Delta\overline{\theta}^{24}/\Delta z\right)}$ (?)	–2.64 m^2 s^{-2}
$\overline{K_h}^{24} = \overline{K_h(t)}^{24}$	0.96 m^2 s^{-2}

The overbar with 24 to the right signifies a 24-hour mean.

- The mean diffusivity is calculated as the mean of the diffusivities that have been calculated for individual time intervals ($K_h(t)$). This yields a positive diffusivity, as the diffusivities of all individual intervals were positive as well (for all half-hour intervals the flux does flow down the gradient).

The Schmidt paradox can be explained physically as follows. During daytime turbulence is strong owing to the combination of shear production and buoyancy production. Hence the turbulent diffusivity is large and only a small vertical temperature gradient is needed to transport the heat flux imposed by the surface energy balance. In contrast, at night the turbulence is suppressed by buoyancy (stable stratification). Although the flux to be transported is much smaller than during daytime, the required temperature gradient is much larger (and of opposite sign) than the gradient during daytime. To summarize: the mean sensible heat flux is dominated by daytime conditions, whereas the mean temperature gradient is dominated by night time conditions.

Mathematically, the Schmidt paradox is related to the order in which averaging and multiplication are performed. This can be illustrated by decomposing both the diffusion coefficient and the gradient in their 24-hour mean values (denoted by the overbar) and a deviation from that mean (similar to the Reynolds decomposition, here indicated by a double prime):

$$
\begin{aligned}
K_h(t) &= \overline{K_h}^{24} + K_h''(t) \\
\frac{\partial \theta}{\partial z}(t) &= \overline{\frac{\partial \theta}{\partial z}}^{24} + \frac{\partial \theta''}{\partial z}(t)
\end{aligned}
\qquad (3.53)
$$

Then the 24-hour mean flux can be constructed either by first multiplying the diffusion coefficient and gradient for each time interval, and by subsequent averaging, or the order can be reversed:

$$\text{correct: } \overline{H}^{24} = -\rho c_{\mathrm{p}} \overline{\left[\overline{K_{\mathrm{h}}}^{24} + K_{\mathrm{h}}''\right]\left[\overline{\frac{\partial\theta}{\partial z}}^{24} + \frac{\partial\theta''}{\partial z}\right]}^{24} = -\rho c_{\mathrm{p}} \left[\overline{K_{\mathrm{h}}}^{24}\,\overline{\frac{\partial\theta}{\partial z}}^{24} + \overline{K_{\mathrm{h}}''\frac{\partial\theta''}{\partial z}}^{24}\right]$$

$$\text{incorrect: } \overline{H}^{24} = -\rho c_{\mathrm{p}} \overline{\left[\overline{K_{\mathrm{h}}}^{24} + K_{\mathrm{h}}''\right]}^{24}\,\overline{\left[\overline{\frac{\partial\theta}{\partial z}}^{24} + \frac{\partial\theta''}{\partial z}\right]}^{24} = -\rho c_{\mathrm{p}} \left[\overline{K_{\mathrm{h}}}^{24}\,\overline{\frac{\partial\theta}{\partial z}}^{24}\right] \tag{3.54}$$

The difference between the two methods is the covariance term occurring in the correct method, which is missed when the averaging is performed first. Thus, whenever there is a correlation between the variables being multiplied (in this case a negative correlation), the averaging operation should be postponed as long as possible.[19]

Mahrt (1987, 1996) has shown that similar arguments hold for spatial averaging of fluxes and gradients in grid boxes of atmospheric models: if two regions with contrasting sensible heat flux (one positive, one negative) are present within one grid box, the mean vertical temperature gradient in the grid box may be stable, whereas the mean flux would be positive.

To conclude, we return to the calculation of fluxes from mean gradients. It is clear that the use of similarity relationships developed in Section 3.5.5 using 24-hour mean data is invalid. In the example shown here the problem is very obvious because there is a clear discrepancy between the direction of the mean flux and the mean gradient. But using 24-hour mean data is equally dangerous if such a discrepancy is not directly clear. For example, the wind speed will always increase with height, both in the case of 30-minute averaged wind speeds and in the case of 24-hour mean winds. Still the use of the 24-hour mean wind speed to determine the 24-hour mean aerodynamic resistance is questionable: the mean stability does not correctly represent the effect of stability on the mean resistance.

3.7 Summary

Because this chapter involves both complex concepts and a heavy load of equations, a summary in the form of a concept map is appropriate (see Figure 3.22).

We put so much emphasis on turbulence for two reasons. First, turbulence is important because it is the transport mechanism that takes care of the transport of

[19] Note that this rule holds equally for averaging over time periods other than 24 hours (e.g., either shorter or longer time periods).

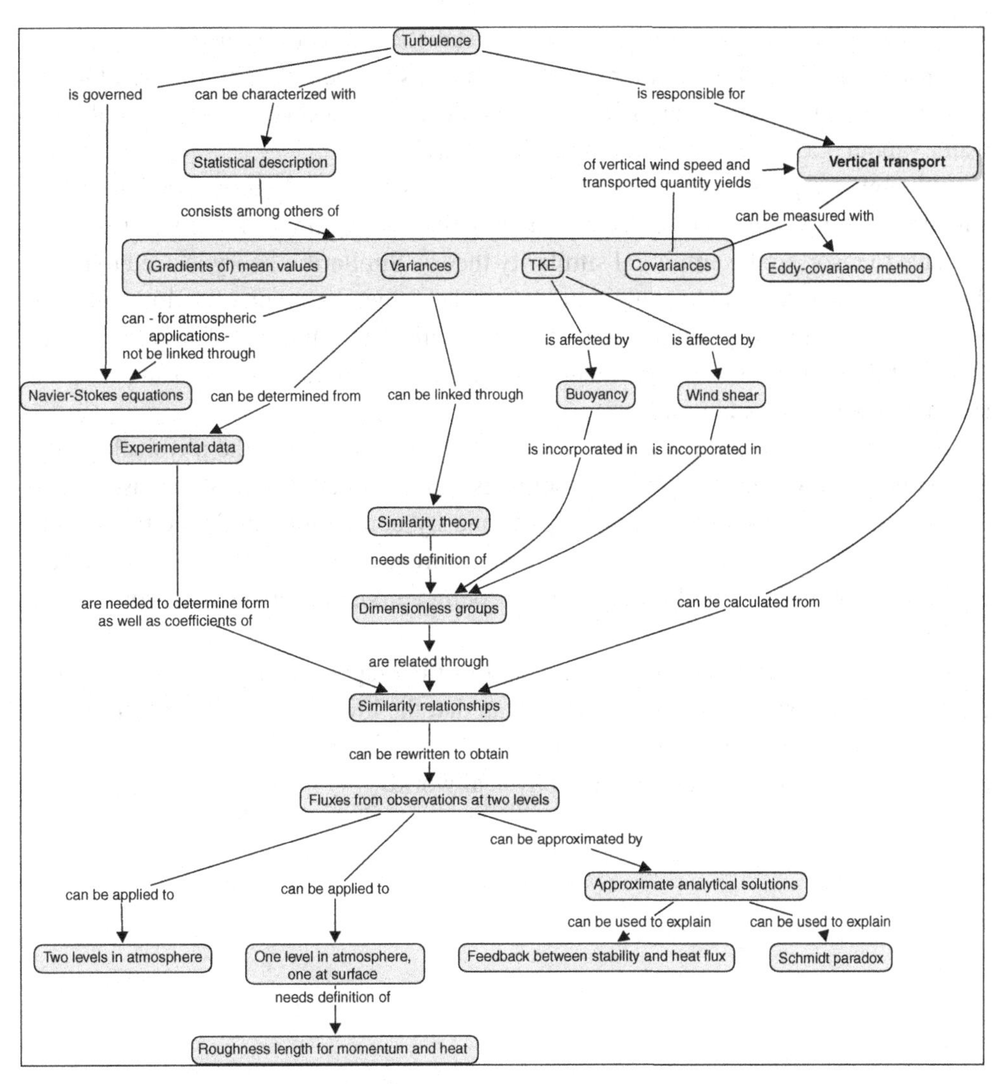

Figure 3.22 Overview of Chapter 3. Everything above 'Similarity theory' relates to the physical reality. Everything below 'Similarity relationships' is a direct – mathematical – consequence of the empirical relationships found by combining similarity theory and experimental data.

heat, momentum and various gases in the atmospheric part of the soil-vegetation-atmosphere continuum. Second, turbulence cannot be described from first principles and therefore we can relate various quantities (e.g., fluxes and mean values) only through empirical relationships.

Turbulence is a specific state of a fluid in which chaotic, though organized movements occur. Because of the chaotic nature, the flow can be described only in terms of statistical quantities. One of those is the turbulent kinetic energy, which is influenced both by wind shear ('mechanical production of turbulence') and buoyancy. In the context of this book the most important statistical quantities are the covariance of vertical wind speed with other quantities such as temperature, humidity, CO_2 and horizontal wind. Those covariances represent the turbulent fluxes (transport) of heat, water vapour, CO_2 and momentum that we are after.

Although the governing equations for the flow of air are known (Navier–Stokes equations), those cannot be solved for any practical atmospheric situation. Therefore, we have to resort to the empirical similarity theory. Similarity theory is used to relate theoretically derived dimensionless groups to each other, based on experimental data. The resulting similarity relationships can be applied in a number of ways to derive turbulent fluxes from vertical gradients and vertical differences of, for example, wind speed, temperature and humidity.

In Figure 3.22, everything above 'Similarity theory' relates to the physical reality. Everything below 'Similarity relationships' is a direct – mathematical – consequence of the empirical similarity relationships found by combining similarity theory and experimental data. If one would have found different similarity relationships (different in form, or different in the values of the coefficients), the derived equations would have been different as well.

The fact that everything in the lower part of the diagram follows *mathematically* from those similarity relations does not mean that the lower part does not bear a relationship to the *physical* reality. On the contrary, the framework of similarity theory enables us to analyse processes in the physical world.

4

Soil Water Flow

4.1 Introduction

Compared to the height of the atmosphere, the depth of the ocean and the thickness of Earth's crust, the permeable soil above the bedrock is an amazingly thin body – typically not much more than a few metres and often less than 1 m. Yet this thin layer of soil is indispensable to sustain terrestrial life. Soil contains a rich mix of mineral particles, organic matter, gases, and soluble compounds. When infused with water, soil constitutes a substrate for the initiation and maintenance of plant and animal life. Precipitation falls intermittently and irregularly, although plants require a continuous supply of water to meet their evaporative demand. The ability of soil to retain soil moisture (and nutrients) is crucial for vegetation to overcome drought periods. Soil determines the fate of rainfall and snowfall reaching the ground surface – whether the water thus received will flow over the land as runoff, causing floods, or percolate downward to the subsurface reservoir called groundwater, which in turn maintains the steady flow of springs and streams. The volume of moisture retained in the soil at any time, though seemingly small, greatly exceeds the volume in all the world's rivers (Hillel, 1998). Without the soil, rain falling over the continents would run off immediately, producing devastating floods, rather than sustaining stream flow. The normally loose and porous condition of the soil allows plant roots to penetrate and develop within it so as to obtain anchorage and nutrition, and to extract stored moisture during dry spells between rains. But the soil is a leaky reservoir, which loses water downward by seepage and upward by evaporation. Managing the top system in water deficit regions so as to ensure the survival of native vegetation as well as to maximize water productivity by crops requires monitoring the water balance and the consequent change of moisture storage (as well as nutrient storage) in the root zone (Hillel, 1998). Soil regulates the amount of evapotranspiration, which is with rainfall the largest component of the hydrological cycle. In weather prediction, climate and environmental research and groundwater recharge, the amount of evapotranspiration plays a key role. Therefore not only a qualitative understanding of the soil water

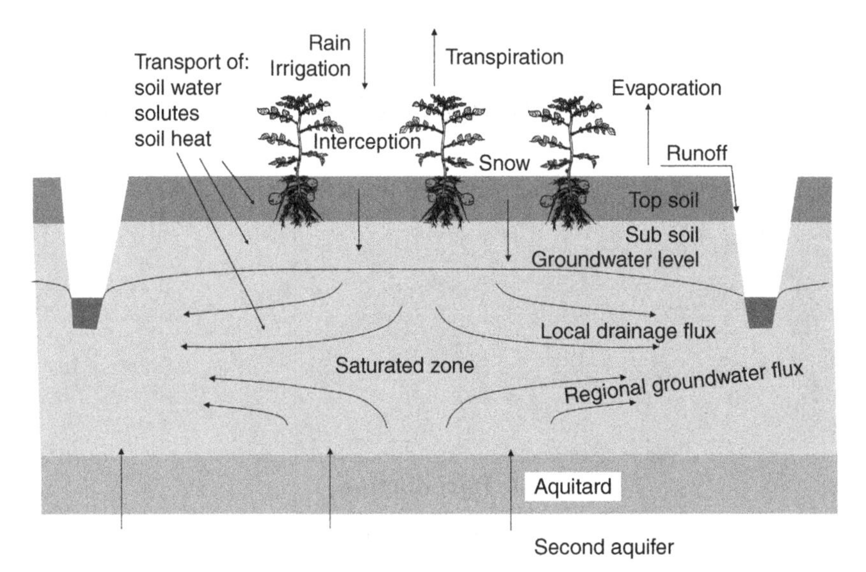

Figure 4.1 Overview of the soil hydrological domain and processes as considered in this chapter.

flow mechanisms is required, but also a precise quantitative knowledge of these processes.

Figure 4.1 shows the domain and processes that we consider in this chapter. The top boundary is located just above the vegetation. The depth of the lower boundary depends on the drainage condition. In the depicted situation with a shallow groundwater level, the lower boundary is situated below the groundwater level in order to include local lateral drainage. The three-dimensional groundwater flow patterns at larger depth belong to the science of hydrogeology and are outside the scope of this book. In case of deep groundwater levels, the domain lower boundary may be located at a few metres below soil surface. The moisture conditions at depths larger than 3 m below the root zone do not affect the soil water flow near the soil surface. The advantage of defining the top and lower boundary of the atmosphere-vegetation-soil continuum in this way is that in this domain the main water flow direction is vertical, which simplifies computation and analysis considerably.

In the region between soil surface and groundwater level, the so-called vadose zone, the volumetric water content shows large rapid fluctuations. This variable water content has a strong impact on other vadose zone processes such as root water and nutrient uptake, biochemical transformations and soil temperatures. Therefore we will consider how we can quantify the water content fluctuations. An important aspect to consider is the soil profile distinct layering with different soil hydraulic properties.

Investigations of soil physical behaviour can be conducted at molecular, pore and macroscopic scales of observation. At the molecular scale, the molecules are the system, and atomic particles like electrons and protons are the system elements. At the

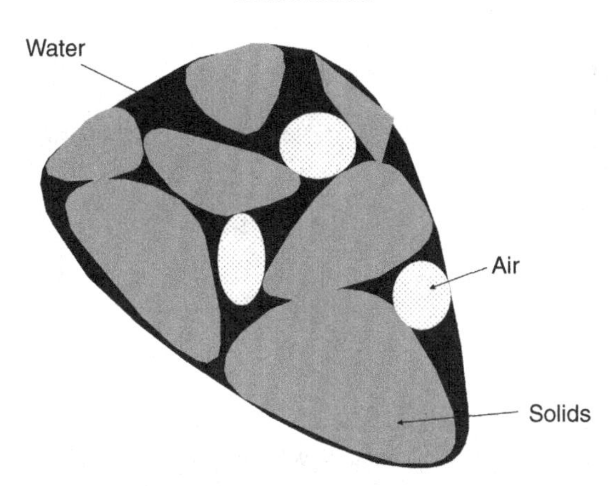

Figure 4.2 The soil system contains three phases.

pore scale, the three phases – solid, liquid and gas – form the soil system (Figure 4.2), and the atoms, molecules and ions are the invisible elements of the system. Because of the exceedingly large number of these elements, it is usually more convenient to choose a volume containing a sufficiently large number of atoms, molecules or ions so that their mean statistical behaviour is relevant (Scott, 2000). A volume enclosing such a continuum molecular mixture is called a *representative elementary volume* (REV). The REV must be large compared to the mean free path of molecules caused by Brownian motion. The concept of REV was developed because of the need to describe or lump the physical properties at a geometrical point. We say that we give to one point in space and time the value of the property of a certain volume surrounding this point. The REV is used to define and sometimes to measure the mean properties of the volume in question. Consequently this concept involves an integration in space. According to De Marsily (1986) the size of the REV is determined by two points:

1. The REV should be sufficiently large to contain a soil volume that allows the definition of a mean global property while ensuring that the effects of the fluctuations from one pore to another are negligible.
2. The REV should be sufficiently small that the parameter variations from one domain to the next may be approximated by continuous functions, in order that we can use differentiation calculus.

Figure 4.3 illustrates how to choose the size of the REV. The size of the REV is generally linked to the existence of a flattening of the curve that connects the physical properties with the spatial dimension. It is an averaging of the soil physical properties within the volume. Obviously, the size of the REV varies widely with soil physical properties, location and time and is somewhat arbitrary (Scott, 2000). The REV concept can be used to integrate from the Navier–Stokes equations of fluid flow at pore scale to the less complicated Darcy's law at the macroscopic scale. In this chapter we start after this integration step and focus on the macroscopic scale with the Darcy

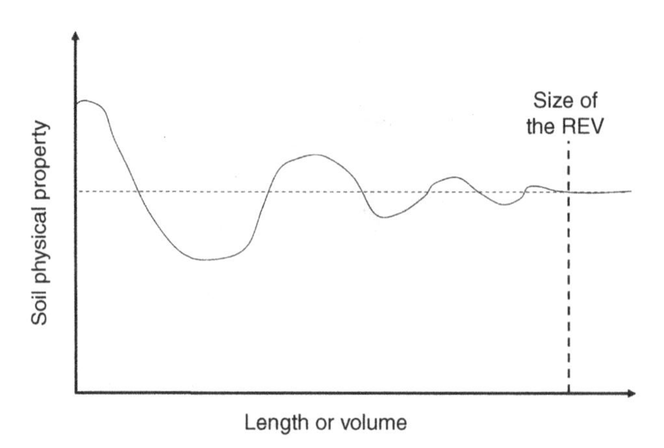

Figure 4.3 Definition of the representative elementary volume (REV).

equation. The soil is considered as a continuum, consisting of a mixture of solid grains, water, solutes and air.

4.2 Field Water Balance

In Chapter 1 we discussed the water balance of soil and an air–vegetation layer. If we omit the air/vegetation layer, we may derive the water storage change ΔW (m) of a soil volume near the soil surface by considering all in- and outflowing water amounts (Figure 4.4):

$$\Delta W = (P + I - E - R - D)\Delta t \tag{4.1}$$

where P denotes precipitation rate (m d^{-1}), I is irrigation rate (m d^{-1}), E is evapotranspiration rate (including evaporation of intercepted water) (m d^{-1}), R is surface runoff (m d^{-1}), D is drainage or deep percolation rate (m d^{-1}), and Δt is the considered time interval (d). All terms are positive except for D and ΔW, which may be either positive or negative. A negative value for the drainage term implies that water is flowing upward into the vadose zone volume (capillary rise).

In field conditions, it is usually possible to measure P, I, and R with adequate precision. Also the profile water content and its changes ΔW can be measured accurately. Evapotranspiration fluxes are more difficult to measure, especially for longer periods. To date no instruments exist to measure percolation fluxes in a soil profile on a routine basis. Also, unless a field has subsurface tile drains, drainage fluxes cannot be measured in the field directly. Consequently, the drainage flux is often determined as the closing term in the water balance. However, we should realize that any error we make in one of the water balance components will affect the accuracy of the closing term. This is especially the case for drainage, which is generally relatively small compared to the evapotranspiration flux. The following question illustrates that relative errors of

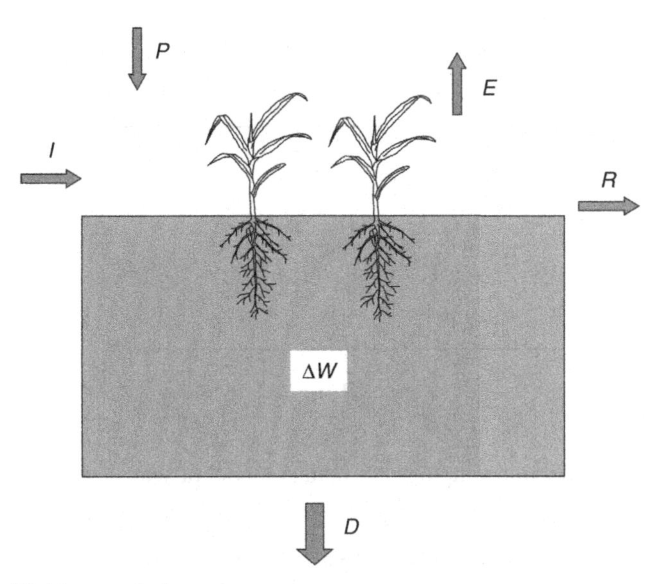

Figure 4.4 Field water balance components.

evapotranspiration in semi-arid regions magnify strongly the relative error of drainage or groundwater recharge.

> **Question 4.1:** In an irrigated field of a semi-arid region the precipitation plus irrigation equals 1100 mm y^{-1} and the actual evapotranspiration amounts 1000 mm y^{-1}. The measurement inaccuracy of the evapotranspiration flux is 10%. How large is the measurement inaccuracy of the average groundwater recharge?

For modelling purposes we may consider the soil as a reservoir with depth D_r, which can be filled by precipitation and gradually releases water to the vegetation and the subsoil (Figure 4.5). Such a model is the Warrilow model (Warrilow, 1986; Kim, 1995), which is used in simple soil routines. All precipitation is assumed to infiltrate, unless the reservoir saturates and surplus precipitation flows away as surface runoff. Reduction of potential evapotranspiration E_p occurs when the soil moisture content drops below a critical value:

$$E(\theta) = \beta_w(\theta) E_p \quad \text{with} \quad \beta_w = \begin{bmatrix} 1 & \text{for} & \theta_c < \theta < \theta_s \\ \dfrac{\theta - \theta_w}{\theta_c - \theta_w} & \text{for} & \theta_w < \theta < \theta_c \\ 0 & \text{for} & \theta < \theta_w \end{bmatrix} \tag{4.2}$$

where E denotes the actual evapotranspiration; θ is the soil reservoir moisture content (m^3 m^{-3}); β_w a reduction coefficient for transpiration (-); and θ_s, θ_c and θ_w the saturated, critical and wilting point moisture content (m^3 m^{-3}), respectively. Percolation

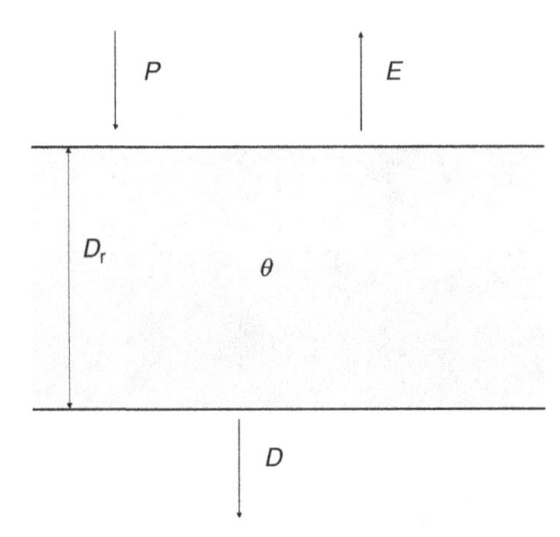

Figure 4.5 Schematization simple water budget model of Warrilow (1986).

is described with free drainage below the soil reservoir, and an exponential hydraulic conductivity function:

$$D = k(\theta) = k_s \left(\frac{\theta - \theta_w}{\theta_s - \theta_w} \right)^n \tag{4.3}$$

where k_s is the saturated hydraulic conductivity (m d^{-1}) and n an empirical constant (-). The water balance provides the change of moisture content with time:

$$\frac{d\theta}{dt} = \frac{1}{D_r} [P - E(\theta) - D(\theta)] \tag{4.4}$$

Question 4.2: Use the Warrilow model to explore soil water dynamics in the root zone with time steps of a day. We have the following input data: $E_p = 5$ mm d^{-1}, $\theta = 0.12$ cm^3 cm^{-3}, $\theta_s = 0.30$ cm^3 cm^{-3}, $\theta_c = 0.15$ cm^3 cm^{-3}, $\theta_w = 0.05$ cm^3 cm^{-3}, $k_s = 10$ cm d^{-1}, exponent $n = 2.0$, rooting depth $D_r = 30$ cm. On the first day rainfall $P = 0$ mm, on the second day rainfall $P = 20$ mm. Calculate the soil water content θ after the first day and after the second day.

Although Warrilow's model may illustrate some mechanisms of soil water uptake by vegetation and soil water redistribution, the model has important limitations:

- Runoff is considered only when a soil becomes saturated; in reality runoff will also occur when the precipitation flux is higher than the maximum infiltration flux into the soil.
- The critical moisture content θ_c depends on the soil type, plant root density and evapotranspiration flux itself; therefore this parameter should be calibrated for a particular situation.
- Free drainage is assumed, which is not valid when shallow groundwater levels occur.

- Capillary rise is not considered.
- The drainage amount is very sensitive to the applied time step.

Therefore in the scientific literature various modifications of this model have been proposed (Kim, 1995). Most of these modifications require extra calibration parameters. In this chapter we explore a level deeper to derive a general framework for soil water flow analysis that uses more constant, physical-based input parameters and is able to address capillary rise, soil layering and runoff.

4.3 Hydraulic Head

As the preceding paragraph shows, we need a more fundamental theory to describe soil water flow near the soil surface in a general way. The concept of *potential energy* forms a solid base for such a theory. Potential energy is the energy a body has by virtue of its position in a force field. For example, a mass possesses greater potential energy in a gravitational field than an identical mass lying below it, because work is required to lift the mass to a higher position. Various forces act on water in a porous material like soil. The gravitational field of Earth pulls the water vertically downward. Force fields that are caused by the attraction of water to solid surfaces pull water in numerous directions. The weight of water and sometimes the additional weight of soil particles that is not compensated by grain pressure also exert downward forces on water lying underneath. Ions dissolved in water have an attractive force for water and resist attempts to move it. An especially important force is associated with the attraction of water molecules for each other and the imbalance of these forces that exists at the air–water interface, the so-called surface water tension (Jury et al., 1991).

The potential energy of water in soil may be defined relative to a reference or standard state, since there is no absolute scale of energy. The *standard state* is customarily defined to be the state of pure (no solutes), free (no external forces other than gravity) water at a reference pressure, reference temperature and reference elevation and is arbitrarily given the value zero (Bolt, 1979). The soil water potential energy is defined as the difference in energy per unit quantity of water compared to the reference state. There are different systems of units in which the total potential and its components may be described, depending on whether the quantity of pure water is expressed as a mass, a volume, or a weight. Table 4.1 summarizes these systems and their units. In soil physics commonly the *energy/weight expression* is used, which results in the very practical dimension length. Consequently, in this chapter we use the energy/weight expression, which is called head instead of potential. In the next sections we consider the hydraulic head of groundwater, soil water and water vapour.

Question 4.3: Consider a water column of 2.0 m height. Express the soil water potential in J kg^{-1}, N m^{-2} and m. Which unit do you prefer?

Table 4.1 Systems of units of soil water potential

Expression	Name	Unit	Dimension
Energy/mass	Chemical potential	$J\ kg^{-1}$	$L^2\ T^{-2}$
Energy/volume	Soil water potential	$N\ m^{-2}$ (=Pa)	$M\ L^{-1}\ T^{-2}$
Energy/weight	Hydraulic head	M	L

4.3.1 Hydraulic Head of Groundwater

Groundwater refers to water below the groundwater level, while soil water refers to water in the vadose zone. Types of energy that may play a role in groundwater are height, pressure, velocity, osmosis and heat. In general in groundwater, the energy amounts due to velocity, osmosis and heat are negligible compared to the energy amounts of height and pressure. For these common situations we may write:

$$H = h + z \tag{4.5}$$

where H is the hydraulic head (m), h is the soil water pressure head (m) and z is the elevation or height (m). For pressure head commonly atmospheric pressure is taken as reference with the value zero. The zero reference for elevation can be taken at any level. In case of hydrostatic equilibrium and atmospheric pressure at the groundwater level, the pressure head is equal to the distance to the groundwater level. The water level in piezometers incorporates both pressure head and elevation. Therefore the water level is equal to the hydraulic head H at the filter (Figure 4.6), which makes piezometers very practical measurement devices.

Question 4.4: In a phreatic aquifer we measure with piezometers the hydraulic heads at two depths. We take $z = 0$ at the soil surface. At piezometer 1 (filter at $z = -100$ cm) the water level occurs at $z = -80$ cm, while at piezometer 2 (filter at $z = -200$ cm) the water level occurs at $z = -90$ cm.

a) Which pressure heads h occur at the filter depths?
b) Which hydraulic heads H occur at the piezometers?
c) Does the groundwater at this location flow upward or downward?
d) Calculate the depth of the groundwater level, assuming a homogeneous soil below the groundwater (hint: in such a situation the gradients dH/dz and dh/dz are constant with depth).

In head diagrams, the elevation, soil water pressure head and hydraulic head are depicted as function of depth. These diagrams are very useful to interpret piezometer and piezometer data and to determine the direction of flow. Figure 4.6 shows the head diagram belonging to the flow situation of Question 4.4. Both elevation and pressure head are linear functions with depth. Therefore also the hydraulic head is a linear function with depth. As H decreases in downward direction, the flow is

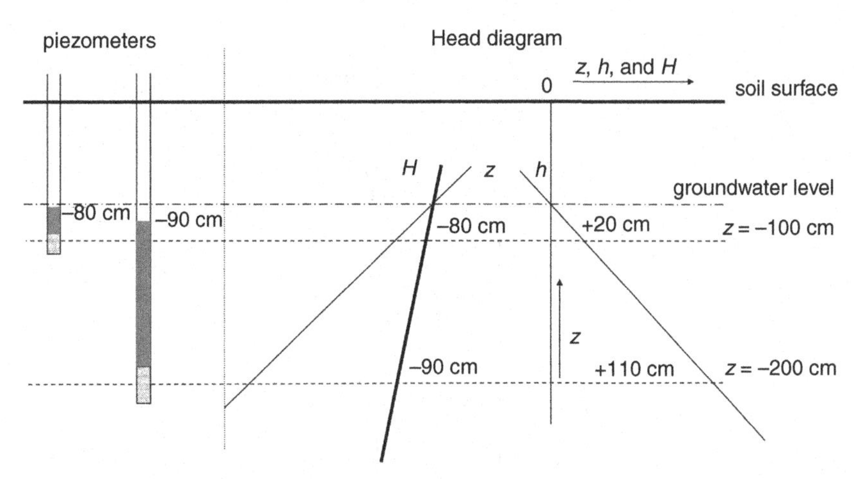

Figure 4.6 Piezometers in shallow groundwater with corresponding Head diagram.

directed downward. Note that in this case the soil water pressure head h increases in the direction of flow. So never use gradients of h to determine the flow direction!

4.3.2 Hydraulic Head of Soil Water

Storage or retention of soil water in the vadose zone is a result of attractive forces between the solid and liquid phases. These '*matrix*' *forces* enable the soil to hold water against forces or processes such as gravity, evaporation, uptake by plant roots, etc. There are three mechanisms for binding of water to the solid matrix (Koorevaar et al., 1983):

- Direct adhesion of water molecules to solid surfaces by London–van der Waals forces
- Capillary binding of water
- Osmotic binding of water in double layers

Water molecules are attracted to solid surfaces by various types of London–van der Waals forces. These are strong, but very short-range forces: they diminish with about the 6th power of the distance. Thus only a very thin water layer is adsorbed in this way around soil particles. Because these adhesive forces are so strong that the water cannot move or be extracted by plant roots, this form of water retention in itself is insignificant for storage, transport and plant growth. However, the adhesive forces, together with the cohesive forces between water molecules, form the basis for capillary binding of soil water. This is the most important mechanism of binding of water in moderate to coarse soils.

Capillary binding might be illustrated by inserting a fine glass tube in a water reservoir (Figure 4.7). Water will rise in the glass tube, until it reaches a maximum height z_c. The forces that act between the water molecules at a water interface are called sur-

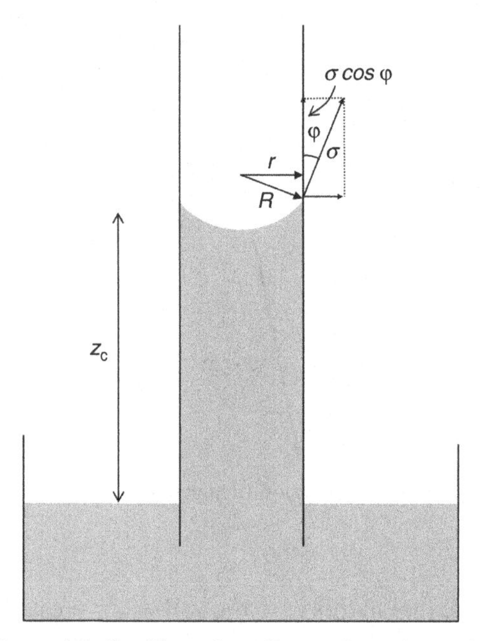

Figure 4.7 Capillary rise of water in a glass tube.

face water tension σ (N m^{-1}). In combination with the adhesive forces (expressed in a so-called wetting angle ϕ), σ determines the maximum height. At equilibrium, the vertical component of the surface water tension is equal to the gravitational force of the lifted water column in the glass tube:

$$2\pi r \sigma \cos\varphi = \pi r^2 z_c \rho g \tag{4.6}$$

were r is the radius of the tube (m), ρ is the water density (kg m^{-3}) and g is the gravitational acceleration (m s^{-2}). This gives for the maximum height z_c:

$$z_c = \frac{2\sigma \cos\varphi}{\rho g r} \tag{4.7}$$

A general figure for surface water tension σ is 0.07 N m^{-1}. In case of clean glass, adhesion is maximum and the wetting angle $\phi = 0^{\circ}$ ($\cos\phi = 1$). In case of clean steel, no adhesion occurs and the wetting angle $\phi = 90^{\circ}$ ($\cos\phi = 0$). Some materials and soils repel water and are called hydrophobic. At these materials wetting angle $\phi > 90^{\circ}$ and $\cos\phi < 0$.

> **Question 4.5:** How much is the capillary rise (mm) of water in a clean glass tube with a radius of 1 mm? And how much for a tube with radius 0.1 mm? Take $\sigma = 0.07$ N m^{-1} and $g = 9.81$ m s^{-2}.

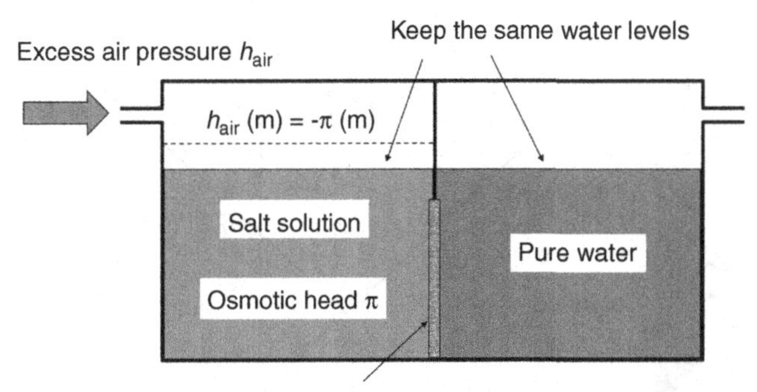

Figure 4.8 Experimental setup to measure the osmotic head of a water solution.

In clay soils the osmotic binding of water in diffuse electric double layers may exceed the capillary binding (Koorevaar et al., 1983). Because water molecules have a dipole moment, ions in the soil water are attracted by the electric field around individual water molecules and tend to cluster around them. The effect of this clustering is to lower the energy state of water. If a membrane permeable to water but impermeable to solutes is used to separate pure water from a solution containing ions, water from the pure side of the membrane will cross over into the solution side. This mass transfer will continue indefinitely, unless stopped by an opposing force.

If the solution is contained in a sealed reservoir such as depicted in Figure 4.8, the pure water entering the volume will expand the salt solution until the increased air pressure balances the ionic attraction of water through the semipermeable membrane. To derive the osmotic head, we might increase the air pressure above the solution until the water levels in both reservoirs are equal (Figure 4.8). In case of common salt solutions, the osmotic head π (cm) can be approximated by (Rhoades et al., 1992):

$$\pi = -400\text{EC} \approx -625\text{TDS} \tag{4.8}$$

with EC the electrical conductivity (dS m^{-1}) and TDS total dissolved solids (mg cm^{-3}). The osmotic head plays an important role in irrigated agriculture, where soil water salinity may hamper root water uptake (Chapter 6).

For soil water movement in the vadose zone, we need to consider only the gravity and pressure head, as water moves freely without a semipermeable membrane (Hillel, 1998). Therefore, similar to groundwater the hydraulic head H (m) is the sum of just two elements, the gravity head and the pressure head:

$$H = z + h \tag{4.9}$$

Unlike the saturated zone, the pressure head will be negative in the unsaturated zone (Figure 4.9). At the groundwater level the pressure head is zero and the hydraulic head is equal to the gravity head or elevation. To measure negative pressure heads,

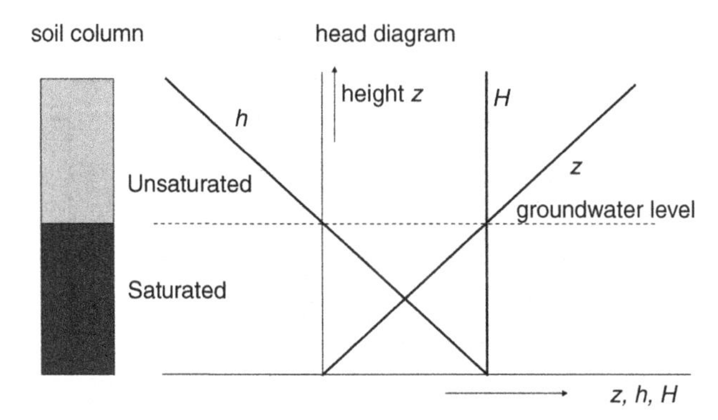

Figure 4.9 Head diagram of variably saturated soil at hydrostatic equilibrium.

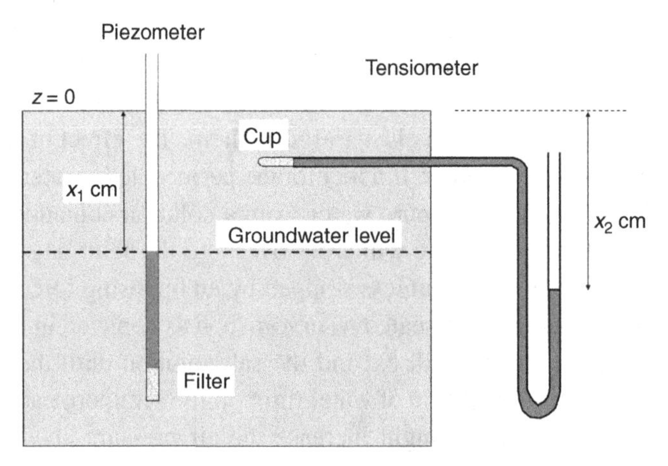

Figure 4.10 Setup to measure the hydraulic head in saturated soil with a piezometer and in unsaturated soil with a tensiometer.

we should use tensiometers, which have a porous cup that is impermeable for air, at least in the occurring pressure head range. Of course the cup should conduct water in order to adjust the water pressure head inside the porous cup to the soil water pressure head around the cup (Figure 4.10).

Question 4.6: Consider the piezometer and tensiometer as depicted in Figure 4.10.

a) How large is the hydraulic head H in case of the piezometer and in case of the tensiometer? Take $z = 0$ at the soil surface.
b) Does the soil water between porous cup and groundwater level flow upward or downward?
c) Can we use a tensiometer to measure the hydraulic head in saturated soils?

In general the pressure head profile will deviate from the 1:1 line. Figure 4.11 shows pressure head profiles above the water table in case of hydrostatic equilibrium (no flow), downward flow (infiltration) and upward flow (capillary rise).

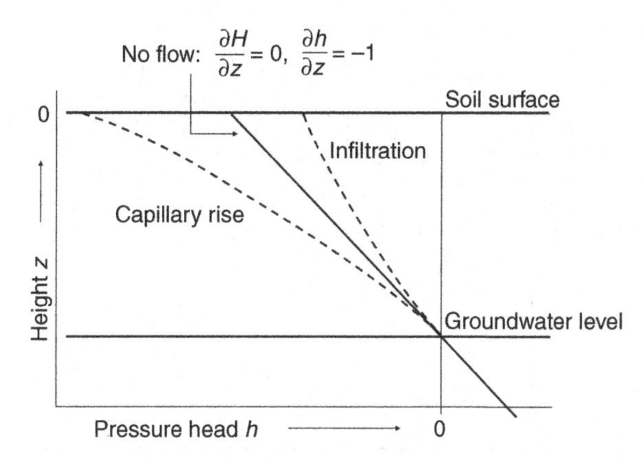

Figure 4.11 Pressure head profiles in case of no flow (hydrostatic equilibrium), infiltration and capillary rise.

4.3.3 Hydraulic Head of Water Vapour

Water is present in the soil not only in the liquid phase, but also in the gas phase as water vapour. At static equilibrium, the total heads in both phases are equal. Water vapour consists of pure water, and thus $\pi = 0$. Also it is not influenced by matrix forces, and thus $h = 0$. Therefore its total head is determined by its vapour pressure e, and by its position in the gravitational field. We may derive the following relation between relative vapour pressure e/e_{sat} (also called relative humidity) and the pressure and osmotic head of soil water with which it is in equilibrium (Koorevaar et al., 1983):

$$\ln\left(\frac{e}{e_{sat}}\right) = 7.5\times10^{-7}\ \text{cm}^{-1}(h+\pi) \tag{4.10}$$

Question 4.7: In the soil plant roots may extract water to about $h = -16\ 000$ cm, also denoted as wilting point. Which relative humidity in the air-filled pores is in equilibrium with this pressure head? Which soil water pressure head corresponds to a relative air humidity of 80%?

4.4 The Soil Water Characteristic

A soil water characteristic or retention curve relates volumetric water content to soil water pressure head. Figure 4.12 shows a retention curve with some typical derived data. Under unsaturated field conditions the soil water pressure head may range over six orders of magnitude: $-10^6 < h < 0$ cm. Because of this large range the pressure head is often depicted on a logarithmic scale, for instance as pF (= log $-h$ cm). Important water contents correspond to field capacity θ_{fc} ($1.7 < pF < 2.3$) and wilting point θ_{wp} ($pF = 4.2$). If at the start of a growing season a soil is at field capacity, the water

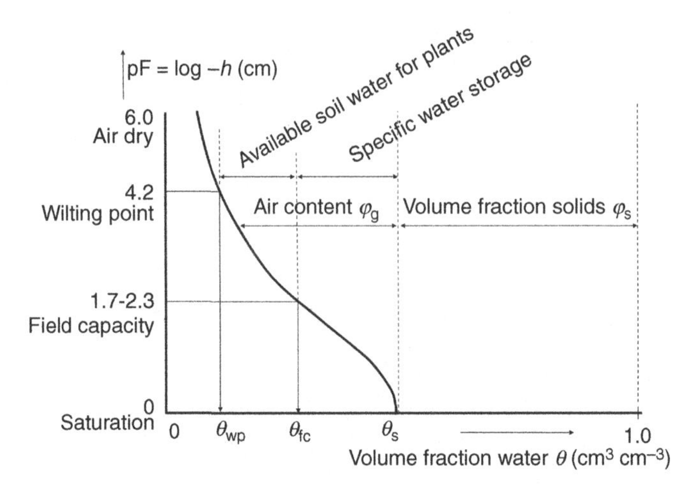

Figure 4.12 Soil water characteristic with derived data.

amount between field capacity and wilting point equals the soil water amount available for plants in dry periods. On the wet side, the water amount between field capacity and saturation equals the rain water amount that can be stored in a soil. In rigid soils the amount of air and water are complementary; therefore the retention function depicts the air content as function of h.

Question 4.8: Consider a loamy soil with the following retention function values: $\theta_s = 0.45$, $\theta_{fc} = 0.34$ and $\theta_w = 0.12$ (cm^3 cm^{-3}).

a) How many millimetres of water can be extracted by roots from the soil in a dry period if the root zone is 40 cm thick and the soil is at field capacity at the start of the period?
b) Suppose the soil is at wilting point conditions and a rain shower of 10 mm occurs. How far does the rain water enter the loamy soil if the soil gets wetted until field capacity?
c) How many cubic metres of water are needed per hectare to bring the top 30 cm of this soil from wilting point to field capacity?
d) How many cubic metres of water can be stored in the top 30 cm if this soil is at field capacity?

Various methods can be used to measure the soil water characteristic, each for a specific pressure head range (Table 4.2). The setup of the sand box apparatus is depicted in Figure 4.13. Undisturbed soil samples are placed on top of a very fine sand or loam layer. This soil layer should stay saturated with water at each applied suction; otherwise air will enter the tubes. Therefore, to reach in the sample $h = -200$ cm, the air entry value of the top layer should be smaller than –200 cm! As this is the smallest air entry value found in natural fine sands and loams, the lowest pressure head in a sandbox is about this value. The sublayer may consist of coarse sand, which allows rapid flushing of air bubbles and quick equilibration with the water level in the

Table 4.2 Laboratory measurement methods of $\theta(h)$

Method	Range (cm)	Reference
Sand box apparatus	$-200 < h < 0$	Klute (1986)
Pressure cell	$-1000 < h < 0$	Kool et al. (1985)
Pressure membrane	$-20\,000 < h < -1000$	Klute (1986)
Vapour equilibration	$h < -20.000$	Koorevaar et al. (1983)

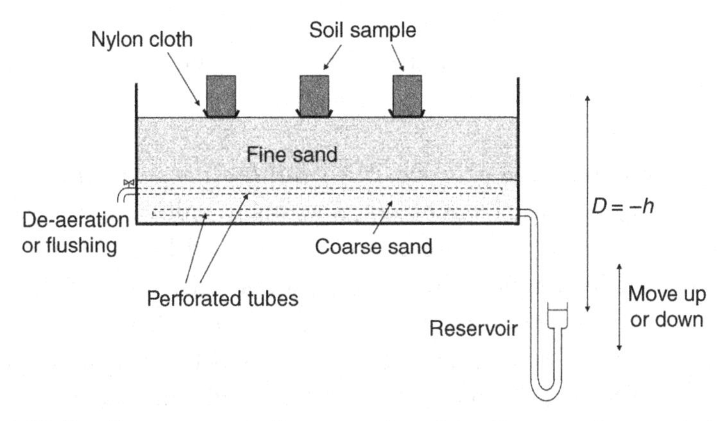

Figure 4.13 Sandbox apparatus for measuring the soil water characteristic at low suctions.

reservoir. A certain pressure head h can be established by placing the reservoir level $D = -h$ cm below the soil samples. After hydrostatic equilibrium is reached, the water content can be measured by gravimetric weighing.

In the range $-20\,000 < h < 0$ cm so-called pressure cells or *pressure membranes* are used. The apparatus consists of an air-tight chamber enclosing a water-saturated, porous ceramic plate connected on its underside to a tube that extends through the chamber to the open air (Figure 4.14). Saturated soil samples are enclosed in rings and placed on the ceramic plate. The chamber is then pressurized, which squeezes water out of the soil pores, through the ceramic, and out of the tube. Similar to the sandbox apparatus, the air entry value of the ceramic plate should be low enough to keep the plate saturated at the applied air pressures. At equilibrium, flow through the tube will cease. At the soil–ceramic plate interface we may write:

$$h = -h_{\text{air}} - z_{\text{tube}} \tag{4.11}$$

When equilibrium is reached, the chamber may be depressurized and the water content of the samples measured. This method may be used up to air gauge pressures of about 20 bars if special fine-pore ceramic plates are used. Because these devices have a very high flow resistance, it may require one week or more before pressure equilibrium has been reached (Jury et al., 1991).

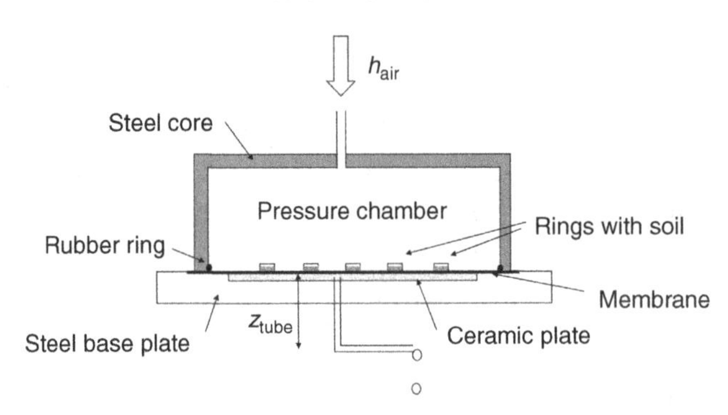

Figure 4.14 Pressure membrane apparatus for measuring the soil water characteristic at high suctions.

In the very dry range ($h < -20\ 000$ cm) *equilibration with salt solutions* is a suitable method. By adding precalibrated amounts of salts, the energy level of a reservoir of pure water may be lowered to any specified level. If this reservoir is brought into contact with a moist soil sample, water will flow from the sample to the reservoir. If the sample and the reservoir are placed adjacent to each other in a closed chamber at constant temperature, water will be exchanged through the vapour phase by evaporation from the soil sample and condensation in the reservoir until equilibrium is reached. Another much applied method to determine water retention in the very dry range is by equilibration with standard laboratory conditions, which correspond to 20 °C and 50% relative air humidity. Using Eq. (4.10), this implies a pressure head $h = -9.2 \times 10^5$ cm in the soil sample after hydrostatic equilibrium has been reached.

Hysteresis of the $\theta(h)$ relation may complicate the measurement and interpretation of the soil water characteristic. The occurrence of various water contents at the same pressure head can be caused by variations of the pore diameter (inkbottle effect), differences in radii of advancing and receding meniscus, entrapped air and swelling/shrinking processes (Hillel, 1998). Gradual desorption of an initially saturated soil sample gives the main drying curve, whereas slow absorption of an initially dry sample results in the main wetting curve (Figure 4.15). In the field partly wetting and drying occurs in numerous cycles, resulting in so-called drying and wetting scanning curves lying between the main drying and the main wetting curve. In practice, often only the main drying curve is used to describe the $\theta(h)$ relation. For instance, a generally applied soil hydraulic database in the Netherlands, known as the Staring series (Wösten et al., 2001), contains only $\theta(h)$ data of the main drying curve. This is due mainly to the time and costs involved in measurement of the complete $\theta(h)$ relationship, including the main wetting, the main drying and the scanning curves, especially in the dry range. Nevertheless, it is obvious that the simulation of infiltration and runoff events with the main drying curve can be misleading in case of significant hysteresis.

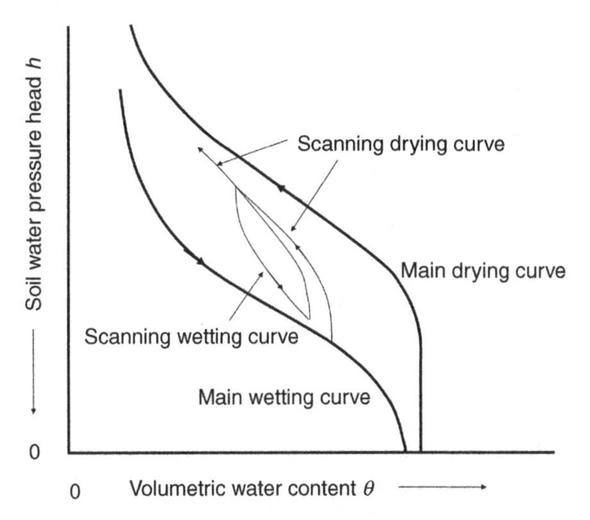

Figure 4.15 Water retention curve with hysteresis, showing the main wetting, main drying and scanning curves.

4.5 Darcy's Law

4.5.1 Saturated Soil

Darcy, a French engineer working for the drinking water supply of Dijon, measured the volume of water Q flowing per unit time through water-saturated packed sand columns of length L (m) and cross section A (m^2) at constant hydraulic head differences $\Delta H = H_1 - H_2$ between the inflow and outflow (Figure 4.16). Darcy derived a linear relation between discharge Q ($m^3\ d^{-1}$) and hydraulic head gradient:

$$Q = Ak_s \frac{H_1 - H_2}{L} \tag{4.12}$$

where k_s is the saturated hydraulic conductivity ($m\ d^{-1}$) which is constant for rigid, saturated soil. Darcy's law may be generalized to apply between any two points of a saturated porous medium provided that the total hydraulic head difference of the water between the two points is known. We will assume that the soil is rigid and saturated and that no solute membranes exist within the water flow paths. Under these restrictions, the total water hydraulic head in saturated soil consists of the sum of the hydrostatic pressure and gravitational potential components. We may eliminate the soil cross section A, and write Darcy's law simply as:

$$q = -k_s \frac{\partial H}{\partial x} = -k_s \frac{\partial (h + z)}{\partial x} \tag{4.13}$$

where q is the soil water flux density ($m\ d^{-1}$) and x is the spatial coordinate (Jury et al., 1991). A sign is needed as the velocity is a vector. The minus sign denotes that

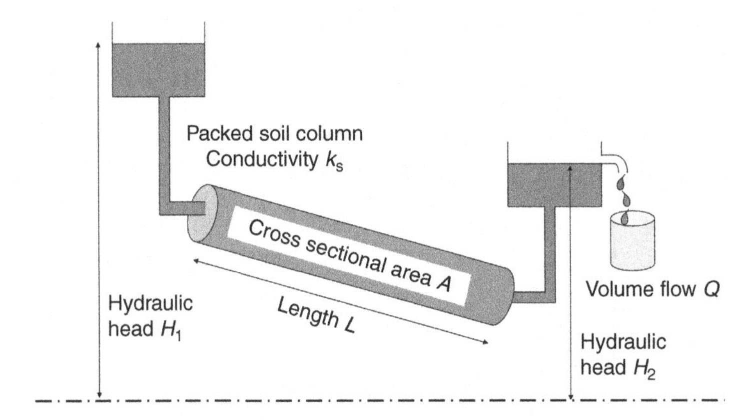

Figure 4.16 Soil column experiment illustrating Darcy's law.

at positive hydraulic head gradients (H increasing with x), water will flow in the negative x-direction.

> **Question 4.9:** A 50-cm-long column containing packed sand with a saturated hydraulic conductivity of 100 cm d^{-1} is placed vertically with the bottom open to the atmosphere. On the top surface of the column 10 cm of water ponds continuously (Figure 4.17). How large is the soil water flux q through the column?

Above question illustrates a fundamental difference between equilibrium problems and flow problems. If the bottom of the column was sealed, then at equilibrium the hydrostatic pressure potential head at $z = 0$ would be 60 cm because the weight of all the water above $z = 0$ is exerted at the bottom. However, when the bottom is open to atmosphere, water will leave the pores at the bottom of the column as soon as any pressure higher than atmospheric pressure (by definition equal to zero) develops. Thus in that case, $h = 0$ at the bottom (Figure 4.17). In other words, in the flow situation the weight of the water in the column is in equilibrium with the viscous resistive forces between water and the porous medium.

Most soil profiles are layered. How can we apply the equation of Darcy to these profiles? Figure 4.18 illustrates steady water flow through a layered saturated soil column containing N layers of thickness L_j and saturated hydraulic conductivity k_j ($j = 1, \ldots N$). We intend to calculate the water flux and hydrostatic pressure distribution given the values of k_j, L_j and ponding water layer. We might replace the heterogeneous profile by a profile with the same height and a uniform, effective hydraulic conductivity, as depicted in Figure 4.18. The total hydraulic head loss between top and bottom of both soil profiles can be written as:

$$\Delta H = q\frac{L_1}{k_1} + q\frac{L_2}{k_2} + ... + q\frac{L_N}{k_N} = q\sum_{j-1}^{N}\frac{L_j}{k_j} = q\frac{\sum_{j=1}^{N} L_j}{k_{\text{eff}}} \tag{4.14}$$

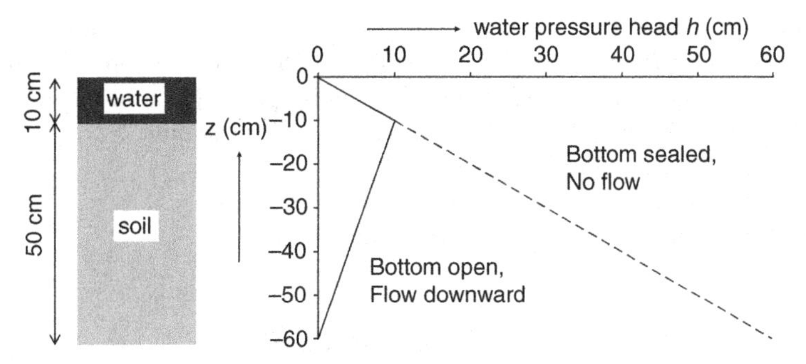

Figure 4.17 Soil water pressure head in sealed and free draining soil column.

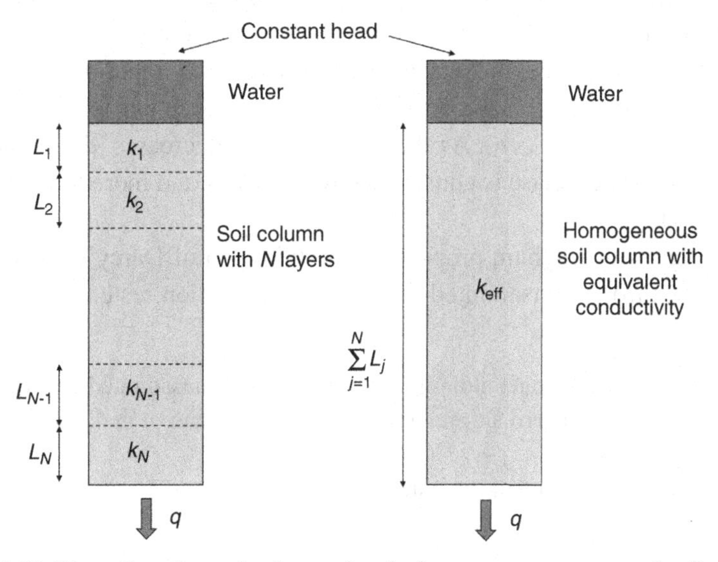

Figure 4.18 Water flow through a layered and a homogeneous saturated soil column, both with the same soil water flux.

The ratio L_j / k_j can also be viewed as a *hydraulic resistance*. In fact, in Eq. (4.14) we are adding the hydraulic resistances of soil layers in series to get the hydraulic resistance of the entire profile. Rewriting Eq. (4.14) results for the effective hydraulic conductivity in:

$$k_{\text{eff}} = \frac{\sum_{j=1}^{N} L_j}{\sum_{j=1}^{N} \frac{L_j}{k_j}} \tag{4.15}$$

With k_{eff} and Eq. (4.14) we can derive the soil water flux q. Subsequently, the pressure drop across any homogeneous layer within the column may be calculated using Darcy's law and the saturated hydraulic conductivity of the involved soil layer.

Question 4.10: A layered vertical column consists of 25 cm of a loam soil with k_s = 5 cm d^{-1} overlain by 75 cm of a sandy soil with k_s = 25 cm d^{-1}. On top of the column a water layer of 10 cm is maintained, and the bottom is open to the atmosphere.

a) How large is the effective hydraulic conductivity of this column?
b) How large is the flux density q through the column?
c) How large is the pressure head at the loam–sand interface?
d) Draw a head diagram of the column, including H, h, and z.

4.5.2 Unsaturated Soil

In unsaturated soils air volumes are present and the water flow channels are smaller than those in saturated soil. The water phase is bounded partially by solid surfaces and partially by an interface with the air phase. In contrast to the positive water pressure found in saturated soils, the water pressure within the liquid phase is caused by water elevation, attraction to solid surfaces, and the surface tension of the air–water interface and is lower than zero. As the water content decreases, the liquid pressure decreases and the water phase is constrained to narrower and more tortuous channels (Jury et al., 1991).

In 1907, Edgar Buckingham proposed a modification of Darcy's law (Eq. (4.13)) to describe flow through unsaturated soil. This modification rested primarily on two assumptions:

1. The driving force for water flow in isothermal, rigid, unsaturated soil containing no solute membranes and zero air pressure potential is the sum of matrix and gravitational potential.
2. The hydraulic conductivity of unsaturated soil is a function of the water content or matrix potential.

In head units, the Buckingham–Darcy flux law may be expressed for vertical flow as:

$$q = -k(\theta)\frac{\partial H}{\partial z} = -k(\theta)\frac{\partial(h+z)}{\partial z} = -k(\theta)\left(\frac{\partial h}{\partial z}+1\right) \tag{4.16}$$

Similar to saturated flow, the flux density q (m d^{-1}) is the water flow per unit cross-sectional area per unit time. Several points should be stressed about Eq. (4.16). First, it is a differential equation that is written across an infinitely small thin layer of soil over which h and $k(\theta)$ are constant. It may not be written across a finite layer of soil unless the water content and matrix head of the layer are uniform. Second, the derivative in Eq. (4.16) is a partial derivative, because in unsaturated soil h may be a function of both z and t. The partial derivative $\partial h/\partial z$ implies that the derivative with respect to z is taken at constant t; it is the instantaneous value of the slope of $h(z)$:

$$\frac{\partial h}{\partial z} \equiv \left(\frac{\partial h}{\partial z}\right)_t = \lim_{\Delta z \to 0} \frac{h(z+\Delta z,t)-h(z,t)}{\Delta z} \tag{4.17}$$

where $(\ )_t$ means that the derivative is evaluated at constant t. Partial derivatives are required for the mathematical description of transient (time-dependent) flow. If the system is at steady state, the partial derivatives reduce to an ordinary derivative because in steady state h depends only on z.

The unsaturated hydraulic conductivity is a strongly nonlinear function of water content or soil water pressure head. Figure 4.19 shows typical curves for a coarse textured (sandy) and a fine-textured (clay) soil. At saturation, the coarse textured soil has a higher conductivity than the fine-textured soil, because it contains large pore spaces, which are filled with water. However, these pores drain at modest suctions, producing a dramatic decrease in hydraulic conductivity in the sandy soil. Eventually, the curves will cross and the sandy soil will actually have a lower hydraulic conductivity than the clayey soil at the same matrix potential, because the latter will retain considerable more water and will contain a larger number of filled pores (Jury et al., 1991).

4.6 Richards' Equation for Water Flow in Variably Saturated Soils

Let's consider the water balance of a small, cubic volume of soil with one-dimensional, vertical water flow (Figure 4.20). The amount of water flowing into the elementary cubic at the bottom, Q_{bottom} (kg d^{-1}), equals:

$$Q_{\text{bottom}} = q\rho\,\mathrm{d}x\,\mathrm{d}y \tag{4.18}$$

where ρ is the water density (kg m^{-3}) and dx and dy are the horizontal cube sides (m).

When the vertical cube side dz (m) approaches zero, the amount of water flowing out of the cubic at the top, Q_{top}, can be calculated with the first derivative only:

$$Q_{\text{top}} = \left(q\rho + \frac{\partial(q\rho)}{\partial z}\mathrm{d}z\right)\mathrm{d}x\,\mathrm{d}y \tag{4.19}$$

The water balance of the cube can than thus be written as:

$$\frac{\partial(\theta\rho)}{\partial z}\mathrm{d}x\,\mathrm{d}y\,\mathrm{d}z = q\rho\,\mathrm{d}x\,\mathrm{d}y - \left(q\rho + \frac{\partial(q\rho)}{\partial z}\mathrm{d}z\right)\mathrm{d}x\,\mathrm{d}y - S\rho\,\mathrm{d}x\,\mathrm{d}y\,\mathrm{d}z \tag{4.20}$$

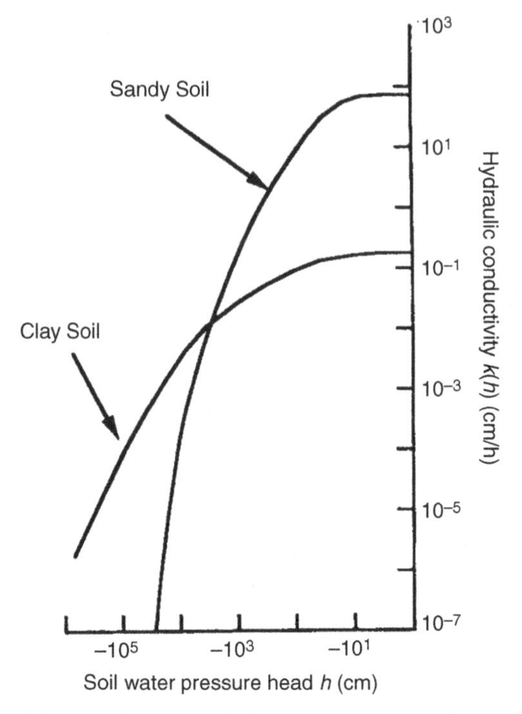

Figure 4.19 Typical hydraulic conductivity curves for a sand and a clay soil.

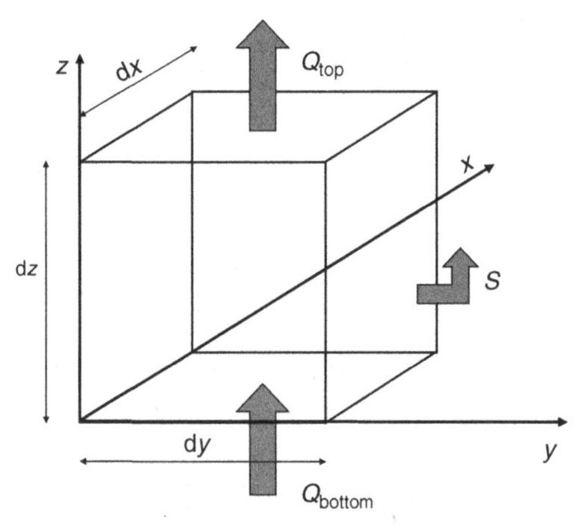

Figure 4.20 Vertical water flow through an elementary cubic soil volume, including root water extraction S.

where S is the water extraction by roots ($m^3\ m^{-3}\ d^{-1}$). If we assume water density ρ to be constant, and divide by (ρ dx dy dz), we arrive at the continuity equation of water in variably saturated soil:

$$\frac{\partial \theta}{\partial t} = -\frac{\partial q}{\partial z} - S \tag{4.21}$$

To derive the general soil water flow equation for variably saturated soils, we combine Eq. (4.16) and (4.21):

$$\frac{\partial \theta}{\partial t} = C(h)\frac{\partial h}{\partial t} = \frac{\partial \left[k(h)\left(\frac{\partial h}{\partial z} + 1 \right) \right]}{\partial z} - S(z) \tag{4.22}$$

where C (= $\partial\theta/\partial h$) is the differential soil moisture capacity (m^{-1}). Equation (4.22) is called *Richards' equation*, and is generally used to solve soil water flow problems in the vadose zone. It is written in pressure head rather than water content, as pressure head is continuous with depth at soil layer transitions. To solve Richards' equation for an arbitrary situation, we should know:

1. The so-called soil hydraulic functions that relate θ, h, and k
2. The actual root water extraction rate S
3. The top and bottom boundary condition
4. The initial soil moisture amounts.

Under strict assumptions some analytical solutions of Richards' equation can be derived. In general the soil hydraulic functions are strongly nonlinear and the field boundary conditions are highly dynamic. In that case numerical solutions are the only feasible way to solve Richards' equation.

4.7 Soil Hydraulic Functions

The soil hydraulic functions relate h with θ (retention function) and k with either θ or h (conductivity function). Although tabular forms of $\theta(h)$ and $k(\theta)$ have been used for many years, currently analytical expressions are preferred for a number of reasons. Analytical expressions are more convenient as model input and a rapid comparison between horizons is possible by comparing parameter sets. Various concepts for modelling hysteresis of the retention function, require analytical soil hydraulic functions. Also scaling, which is used to describe spatial variability of $\theta(h)$ and $k(\theta)$, requires an analytical expression of the soil hydraulic functions. Another reason is that extrapolation of the functions beyond the measured data range is possible. Last but not least, analytical functions allow for calibration and estimation of the soil hydraulic functions by inverse modelling.

Important requirements for analytical expressions of $\theta(h)$ and $k(\theta)$ are that they are flexible in order to describe the wide variability among soils, contain only a few parameters in order to facilitate unique calibration, and that these parameters have some physical meaning, such that they can be related to soil texture, organic matter content and soil bulk density. Van Genuchten (1980) proposed an analytical expression for $\theta(h)$ that met the above requirements and has become widespread among soil scientists:

$$\theta(h) = \theta_r + \frac{\theta_s - \theta_r}{\left(1 + |\alpha h|^n\right)^{\frac{n-1}{n}}} \tag{4.23}$$

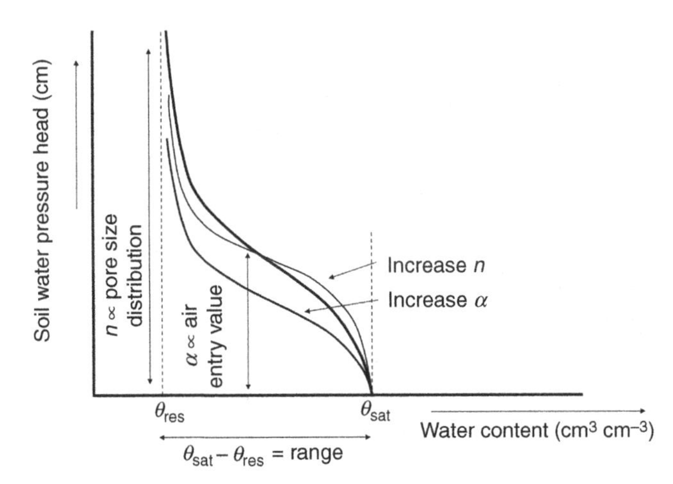

Figure 4.21 Analytical equation for retention function (Eq. (4.23)), showing the effect of the four shape parameters.

where θ_s is the saturated water content ($m^3\ m^{-3}$), θ_r is the residual water content ($m^3\ m^{-3}$), α is an empirical shape parameter (m^{-1}) related to the coarseness of the soil texture and n is an empirical shape parameter (-) related to the width of the particle size distribution. The influence of the four parameters describing $\theta(h)$ is depicted in Figure 4.21.

Earlier Mualem (1976) had introduced a very useful predictive model of the $k(\theta)$ relation based on pore-size hydraulic considerations. In combination with Eq. (4.23), Mualem's predictive model results in the following expression for $k(\theta)$:

$$k(\theta) = k_s S_e^{\lambda} \left[1 - \left(1 - S_e^{\frac{n}{n-1}} \right)^{\frac{n-1}{n}} \right]^2 \tag{4.24}$$

where k_s is the saturated hydraulic conductivity ($m\ d^{-1}$), λ is an exponent (-) related to pore connectivity, and S_e is the relative saturation (-):

$$S_e = \frac{\theta - \theta_r}{\theta_s - \theta_r} \tag{4.25}$$

Figure 4.22 shows the effect of fitting parameters k_s and λ on the conductivity function. Equations (4.23)–(4.25) form the basis of several national and international databanks (e.g., Carsel and Parrish, 1988; Yates et al., 1992; Leij et al., 1996; Wösten et al., 1998; Wösten et al., 2001).

4.8 Infiltration

Infiltration refers to water entry into a soil from one of its edges. Generally, it refers to vertical infiltration, where water moves downward from the soil surface into deeper

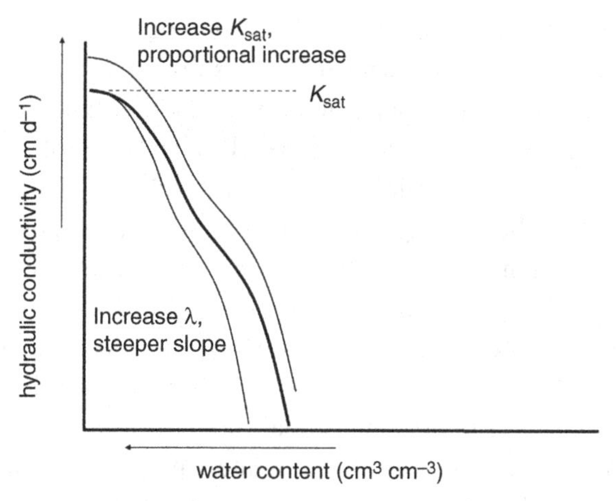

Figure 4.22 Analytical equation for hydraulic conductivity function (Eq.(4.24)), showing the effect of the two extra shape parameters.

soil layers. Vertical infiltration affects directly the runoff amount during high rainfall intensity and therefore receives due attention by hydrologists.

Three main approaches exist to make simple, fast and accurate measurements of infiltration behaviour (Smith, 2002): sprinkler methods, ring infiltrometer methods and permeameter methods. The challenge at sprinklers is to mimic natural rain showers with sufficient accuracy. Difficulties have been experienced in achieving a wide range of application rates while maintaining a drop size distribution and kinetic energy similar to that of natural rainfall. Most sprinkler devices are set up to measure infiltration as the difference between applied rainfall and runoff from an experimental plot. Typically, the plot is bounded and the runoff is routed through a small weir at the downslope side of the plot. In case of ring infiltrometers, single or double rings are inserted into the soil surface, and shallow ponded conditions are created inside the rings. To create one-dimensional vertical flow, either the confining ring must be pushed very deep into the soil or an outer ring with ponded water should be used. A permeameter is distinguished from an infiltrometer by its ability to control the pressure head at the soil surface during infiltration. The major advantage of permeameters is that they are portable and use relatively small volumes of water. This makes them particularly useful for studies on soil spatial variability. Permeameters can also be used to measure infiltration at different suctions to evaluate the effect of macropores. Unlike sprinklers and infiltrometers, the analysis of permeameter experiments is based on three-dimensional infiltration patterns (Smith, 2002).

In general, runoff at a particular location may occur from two types of soil hydraulic limits. In either case the soil surface will be saturated during periods of runoff generation. The first type occurs when the rainfall rate exceeds the rate at which the soil is able to transport water from the soil surface to the subsoil. This type is affected by topography, soil depth, soil hydraulic functions and drainage. The second type occurs when soil is saturated from below. This may occur at relatively low rates of rainfall in humid climates, when downward flow is limited by subsoil layers with low permeability or bedrock, or in seepage situations. The first type might be denoted as 'surface soil control', the second type as 'subsurface soil control'.

Numerical models that solve Richards' soil water flow equation with the proper boundary conditions at the top (e.g., rainfall and potential evapotranspiration rates) and bottom (e.g., relations between flux and pressure head) are able to calculate runoff amounts of both types. An important condition is that the time and space steps of the numerical model near the soil surface are fine enough, as discussed in Section 9.1.2. Although we may meet this condition for one-dimensional models at the plot level, in general multidimensional models based on Richards' equation require too much computation time to simulate runoff in a reliable way at field or larger spatial level. Therefore these models require simplified, semi-empirical methods to approximate runoff amounts. An extensive overview of these methods has been given by Smith (2002). Here we discuss two semi-empirical methods: Horton and Green–Ampt. Both methods refer to the runoff type with surface soil control.

4.8.1 Horton Infiltration Model

Horton (1933, 1939) was one of the pioneers in the study of infiltration in the field. Horton anticipated that the reduction in infiltration rate with time after the initiation of infiltration was controlled largely by factors at the soil surface. These factors included swelling of soil colloids and the closing of small cracks that progressively sealed the soil surface. Compaction of the soil surface by raindrop action was also considered important where it was not prevented by vegetation cover. Horton's field data, similar to those of many other workers, indicated a decreasing infiltration rate for 2 or 3 hours after the initiation of the storm runoff. The infiltration rate eventually approached a constant value that was often somewhat smaller than the saturated hydraulic conductivity of the soil. Air entrapment and incomplete saturation of the soil were assumed to be responsible for this latter finding. Horton used an exponential function to describe the decreasing infiltration rate (Jury et al., 1991):

$$I = I_{\mathrm{f}} + \left(I_0 - I_{\mathrm{f}}\right)\mathrm{e}^{-\beta t} \tag{4.26}$$

where I is the infiltration rate (m d^{-1}), I_0 is the initial infiltration rate (m d^{-1}) at $t = 0$, I_{f} is the final constant infiltration rate (m d^{-1}) that is reached after a long time, and β (d^{-1}) is a soil parameter that describes the rate of decrease of infiltration. The cumulative infiltration I_{cum} (m) follows from integration of Eq. (4.26):

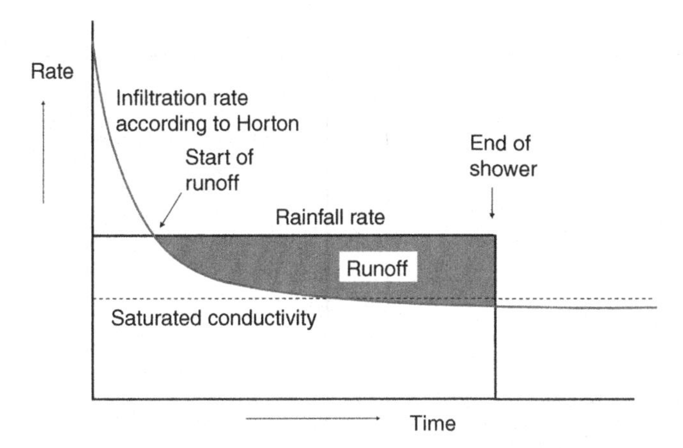

Figure 4.23 Use of the equation of Horton to determine amounts of infiltration and runoff.

$$I_{\text{cum}} = I_{\text{f}}t + \frac{I_0 - I_{\text{f}}}{\beta}\left[1 - e^{-\beta t}\right] \tag{4.27}$$

Figure 4.23 shows how the amount of runoff can be determined with Horton's model. Runoff starts when the infiltration rate is equal to the rainfall rate. Before this time all rainfall did infiltrate. The infiltration in the runoff period follows from Eq. (4.27): take the difference between I_{cum} at the end of the shower and at the start of runoff. The cumulative runoff follows from the difference between total rainfall and total infiltration.

Question 4.11: Runoff occurs during an intensive rain shower with a mean intensity of 25 mm h^{-1} and duration of 2 hours. We use the Horton infiltration equation to calculate the amount of runoff. The Horton soil parameters are: I_0 = 70 mm h^{-1}, I_{f} = 8 mm h^{-1} and β = 1.5 h^{-1}.

a) At which time starts the runoff?
b) How much water has infiltrated at this time?
c) How much rain water infiltrates between the start of the runoff and the end of the shower?
d) How much is the total runoff during this rain shower?

4.8.2 Green–Ampt Infiltration Model

Because infiltration causes the soil to become wetter with time, water at the front edge of the wetting pattern advances under the influence of matrix head gradients as well as gravity. During the early stages of infiltration when the wetting front is near the surface, the matrix head gradients predominate over the gravitational head, and

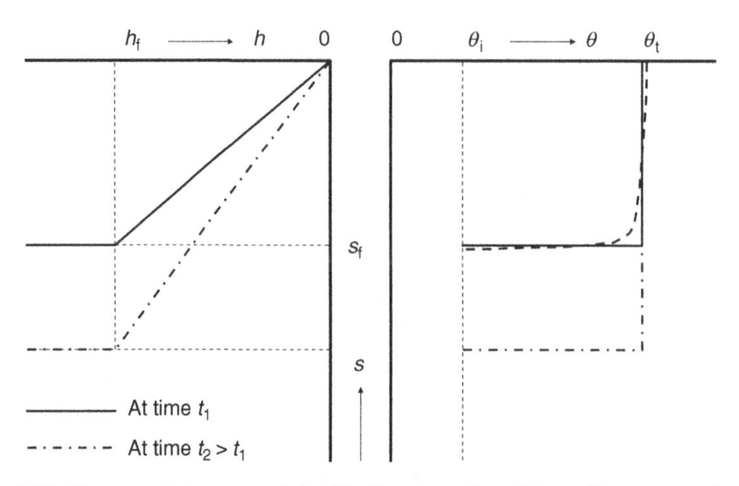

Figure 4.24 Green and Ampt model of infiltration: θ and h profiles at two times.

the infiltration rate will be at a maximum. When the wetting front moves away from the soil surface, the influence of the matrix head will decrease and also the infiltration rate will decrease. The Green–Ampt infiltration model followed this reasoning and adopted the following assumptions (Figure 4.24):

1. Throughout the wetted zone, the volume fraction of water, θ_t, is uniform and constant with time.
2. The change of θ_i to θ_t at the wetting front takes place in a layer of negligible thickness.
3. The pressure head at the wetting front, h_f, has a constant value, independent of the position of the wetting front, s_f.

These assumptions are quite realistic for infiltration into coarse-textured soils with low initial water content, as wetting fronts are generally very sharp under those conditions. From the first assumption it follows that throughout the transmission zone the hydraulic conductivity, k_t, has a constant value. Besides, the flux density is the same everywhere in the transmission zone.

Assuming negligible thickness of the water layer on the soil surface, $h = 0$ at $s = 0$, and using the second and third assumption, Darcy's law for Green–Ampt infiltration can be written as:

$$I = q = -k\frac{\partial H}{\partial s} = -k\left(\frac{\partial h}{\partial s} + \frac{\partial z}{\partial s}\right) = -k_t\left(\frac{h_f}{s_f} - 1\right) \tag{4.28}$$

The cumulative infiltration is:

$$I_{cum} = \int_{s=0}^{\infty} (\theta_t - \theta_i)\,ds = (\theta_t - \theta_i)s_f \tag{4.29}$$

and thus:

$$I = \frac{dI_{cum}}{dt} = (\theta_t - \theta_i)\frac{ds_f}{dt} \tag{4.30}$$

Equations (4.28) and (4.30) yield:

$$k_t\left(1 - \frac{h_f}{s_f}\right) = (\theta_t - \theta_i)\frac{ds_f}{dt} \tag{4.31}$$

When gravity can be neglected (early stage of infiltration or horizontal infiltration), the term $\partial z / \partial s$ in Eq. (4.28) vanishes and Eq. (4.31) becomes:

$$-k_t \frac{h_f}{s_f} = (\theta_t - \theta_i)\frac{ds_f}{dt} \tag{4.32}$$

With rearranging and integrating Eq. (4.32) we obtain:

$$-k_t h_f t = ½(\theta_t - \theta_i)s_f^2 + C \tag{4.33}$$

Because $s_f = 0$ at $t = 0$, the integration constant is zero, and thus Eq. (4.33) becomes:

$$s_f = \left(\frac{-2k_t h_f}{\theta_t - \theta_i}\right)^{½} t^{½} \tag{4.34}$$

Because k_t, θ_t, θ_I and h_f remain constant during the flow process, Eq. (4.34) can be read as $s_f / \sqrt{t}$ = constant, that is, the depth of the wetting front is proportional to the square root of time.

Combination of Eqs. (4.29) and (4.34) leads to (Koorevaar et al., 1983):

$$I_{cum} = \left[-2k_t h_f (\theta_t - \theta_i)\right]^{½} t^{½} = St^{½} \tag{4.35}$$

where the sorptivity S (m $d^{-½}$) is:

$$S = \left[-2k_t h_f (\theta_t - \theta_i)\right]^{½} \tag{4.36}$$

Question 4.12: Does sorptivity S remain constant during the infiltration process?

Question 4.13: Two horizontal columns of the same soil, with initial volume fractions of water θ_1 and θ_2, respectively, are infiltrated with water. Assume θ_t and h_f independent of θ_i. Denote parameters pertaining to sample 1 by subscript 1 and to sample 2 by subscript 2. Using the Green–Ampt model, derive in terms of θ_1, θ_2 and θ_t for both soil columns the ratios of:

a) The distances of the wetting front at time t
b) The cumulative infiltration at time t
c) The times required for the wetting front to reach s
d) The sorptivities
e) The infiltration rates at time t

When gravity cannot be neglected, Eq. (4.31) can be written as:

$$\frac{k_t}{\theta_t - \theta_i} \mathrm{d}t = \frac{s_f}{s_f - h_f} \mathrm{d}s_f \tag{4.37}$$

Integration, using $\frac{s_f}{s_f - h_f} = 1 + \frac{h_f}{s_f - h_f}$, yields:

$$\frac{k_t}{\theta_t - \theta_i} t = s_f + h_f \ln\left(s_f - h_f\right) + C \tag{4.38}$$

The integration constant C can be found from the condition $s_f = 0$ for $t = 0$:

$$C = -h_f \ln\left(-h_f\right) \tag{4.39}$$

The final result is (Koorevaar et al., 1983):

$$t = \frac{\theta_t - \theta_i}{k_t} \left[s_f + h_f \ln\left(\frac{s_f - h_f}{-h_f} \right) \right] \tag{4.40}$$

The physical parameters θ_t, θ_i, k_t and h_f of a given soil must be found experimentally. The values for θ_t and k_t are both near their values at saturation. The pressure head at the wetting front h_f cannot be measured directly, but can be derived from Eq. (4.36) by measuring S, k_t, and $\theta_t - \theta_i$. Values for h_f vary from about –0.05 to 0.8 m for different soils. Once these parameters are known, the time needed for the wetting front to reach a certain depth can be calculated directly with Eq. (4.40). To find s_f for a certain time t is more difficult, because s_f cannot be expressed explicitly as a function of t. Two solution methods can be followed: (1) make a graph of t versus s_f and read for various t from the graph; or (2) apply a numerical technique for root finding. Once s_f is known, the cumulative infiltration and the infiltration rate can be derived from Eqs. (4.28) and (4.29).

Question 4.14: For a fine sandy loam k_t = 1.38 cm d^{-1}, θ_i = 0.1, θ_t = 0.5 and h_f = –40 cm.

a) Calculate s_f, I and I_{cum} for horizontal infiltration in this soil at t = 30 minutes.

b) Do the same for $t = 60$ minutes.
c) Compare the results of the calculations for $t = 30$ minutes and $t = 60$ minutes.

Question 4.15:

a) Find s_f at $t = 60$ minutes for vertical infiltration into the soil of Question 4.14, using Eq. (4.40). Use the following procedure:
   - Calculate t using the value found for s_f in Question 4.14b.
   - Calculate t also or a somewhat larger value of s_f (larger, because gravity is included now).
   - Find a better guess for s_f by linear interpolation.
   - Calculate again t for the value of s_f just found.
   - If necessary repeat these steps.

b) Calculate also I and I_{cum} using the value found for s_f in a.
c) Compare the results with those for horizontal infiltration (Question 4.14b).

4.9 Capillary Rise

Soil water will flow vertically upward if the hydraulic gradient $\partial H / \partial z < 0$, as depicted in Figure 4.11. Upward soil water flow is called capillary rise and occurs in prolonged dry periods. Especially if the groundwater level is within 1 m of the root zone or soil surface and the subsoil has a loamy texture, capillary rise may be considerable. We can calculate the amount of capillary rise straight from the Darcy equation and the soil hydraulic functions. In prolonged dry periods, we may assume soil water flow between groundwater level and root zone bottom to be more or less stationary:

$$q(z) = \text{constant} = -k(h)\left(\frac{\mathrm{d}h}{\mathrm{d}z} + 1\right) \tag{4.41}$$

Partial derivatives have been replaced by normal derivatives because h depends only on z, not on t. We may rewrite Eq. (4.41):

$$\mathrm{d}z = \frac{-k(h)}{q + k(h)}\mathrm{d}h \tag{4.42}$$

If we integrate Eq. (4.42), we get the height Z, which corresponds to a certain h above the groundwater level:

$$Z = \int_0^Z \mathrm{d}z = -\int_0^h \frac{k(h)}{q + k(h)}\mathrm{d}h \tag{4.43}$$

If we know the $k(h)$ relation, we may numerically solve Eq. (4.43) for various values of q. Figure 4.25 shows examples for coarse sand and light loam. If $h = -16\,000$ cm at the bottom of the root zone (wilting point), light loam will still transport 2 mm d^{-1}

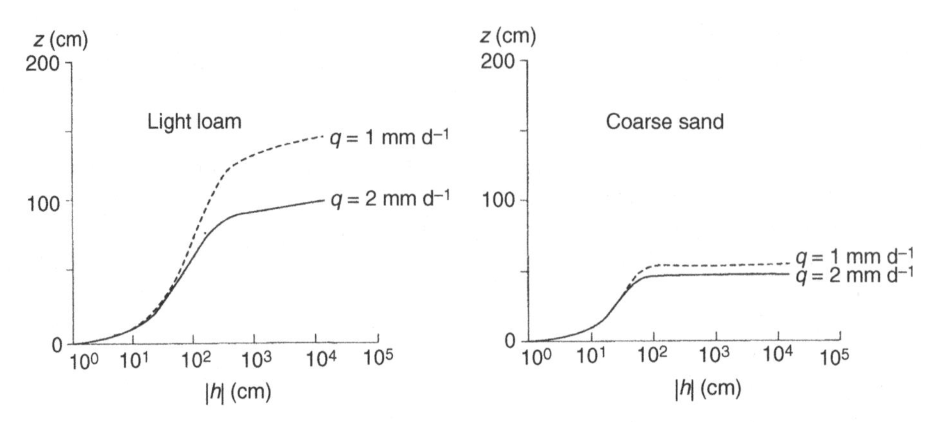

Figure 4.25 Soil water pressure head profiles of light loam and coarse sand at two flux densities.

if the groundwater level is not more than 95 cm below the root zone. The amount of 2 mm d^{-2} is more or less the amount a crop needs to survive dry periods. Smaller fluxes can be transported over larger distances. At light loam soils drought damage seldom occurs as these soils have favourable capillary rise properties and a large water holding capacity. In coarse sand, capillary rise is rather limited: if the groundwater level is more than 65 cm below the root zone, hardly any capillary rise will occur. Also heavy clay (not shown) can transport only limited amounts of water. If the groundwater level is more than 70 cm below the root zone in clay soils, the transport will be insufficient to meet the water demand of crops in dry periods.

Question 4.16: We want to calculate the amount of capillary rise in a loamy soil with Eq. (4.43). The soil hydraulic functions can be described with the Van Genuchten parameters: $\alpha = 1.0$ m^{-1}, $n = 1.40$, $\theta_{res} = 0.00$ m^3 m^{-3}, $\theta_{sat} = 0.45$ m^3 m^{-3}, $K_{sat} = 0.03$ m d^{-1} and $\lambda = -1.0$. Use a modelling tool (e.g., Excel) to make a graph similar to Figure 4.25.

4.10 Measurement of Soil Water Pressure Head

4.10.1 Piezometer

A piezometer is a tube of a few centimetres inner diameter with a permeable filter at a certain soil depth (Figure 4.26). If the filter is below the groundwater table, a piezometer is partially filled with water. The diameter of piezometer tubes is chosen such that large that capillary rise and resistance to water flow are negligible. As a result, any variation in hydraulic head that may arise inside the piezometer is instantaneously equalized. Thus, even if the soil pressure head at the filter is changing rapidly, it can be assumed that the hydraulic head inside the tube is uniform and equal to the hydraulic head of the soil water at the filter. Note that the pressure head inside the tube varies with depth.

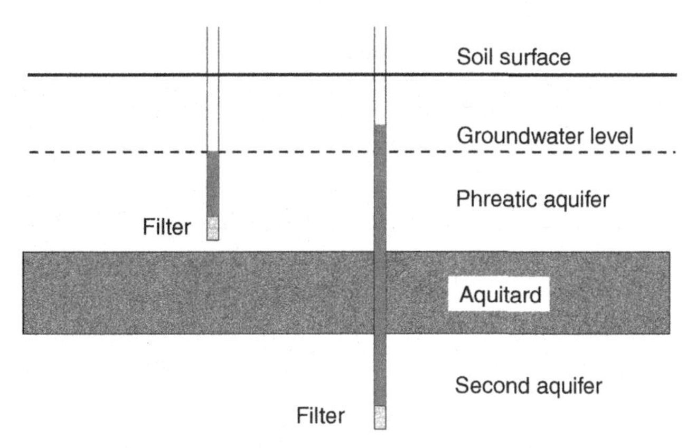

Figure 4.26 Setup of two piezometers to determine the seepage across an aquitard.

Question 4.17: Consider a situation as depicted in Figure 4.26. In the shallow tube the filter is at $z = -170$ cm and we measure a water level at $z = -80$ cm. In the deep tube the filter is at $z = -350$ cm and we measure a water level at $z = -55$ cm. The saturated hydraulic conductivity of the aquitard equals $k_s = 1$ cm d^{-1}, the thickness of the aquitard equals 50 cm.

a) Calculate the hydraulic head of both tubes.
b) Calculate the soil water pressure head at both filters.
c) How large is the upward seepage flux density?

4.10.2 Tensiometer

Piezometers cannot be used to measure *negative pressure heads* in the vadose zone, because any water in the tubes will be adsorbed by the soil. Negative pressure heads are measured with so-called tensiometers. A tensiometer consists of a liquid-filled unglazed porous ceramic cup connected to a pressure measuring device, such as a vacuum gauge, via a liquid-filled tube (Figure 4.27). If the ceramic cup is embedded in soil, the soil solution can flow into or out of the tensiometer through the very small pores in the ceramic cup. Analogously to the situation discussed for piezometers, this flow continues until the pressure head of the water in the cup has become equal to the soil water pressure head around the cup.

The vacuum gauge does not indicate the pressure in the cup when there is a difference in height between the two, such as in Figure 4.27. The liquid in the tube between the cup and the vacuum gauge is at static equilibrium and thus the pressure in this liquid increases linearly with depth. Therefore the pressure head of the liquid in the cup is:

$$h = h_{\text{gauge}} + \Delta z_1 + \Delta z_2 \tag{4.44}$$

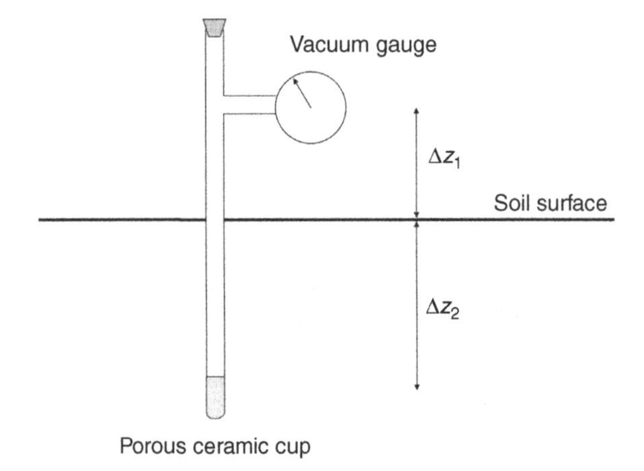

Figure 4.27 Tensiometer with vacuum gauge.

With $z = 0$ at the soil surface and without osmotic head, the hydraulic head equals:

$$H = h_{\text{gauge}} + \Delta z_1 \tag{4.45}$$

Whereas there is no resistance to flow in piezometers, so that they are always instantaneously at equilibrium with the soil water at the lower open end, this is not necessarily true for tensiometers. The porous cup usually presents considerable resistance to flow and the water pressure inside may adjust only slowly to changes in the soil water pressure head at the cup. Also, if the hydraulic head in the soil is not uniform, the tensiometer will indicate only an average of the soil hydraulic head around the cup.

Question 4.18: The cups of tensiometers 1 and 2 are at a depth of 60 and 80 cm, respectively. The gauges are 20 cm above the soil surface. The gauge of tensiometer 1 indicates $h_{\text{gauge}} = -90$ cm.

a) Draw the potential diagram, assuming that the water in the soil is at hydrostatic equilibrium.
b) Calculate the gauge reading of tensiometer 2.

Question 4.19: At another moment the tensiometers of the former question indicate $h_{\text{gauge}} = -90$ cm for tensiometer 1 and $h_{\text{gauge}} = -100$ cm for tensiometer 2.

a) Draw the potential diagram for this new situation, assuming H is linear with z.
b) What is the height of the groundwater table?
c) How can you easily determine the difference in hydraulic potential of the soil water at the two cups?

Instead of vacuum gauges, mercury manometers can be used to measure the pressure head of the liquid in the tensiometer (Figure 4.28). The pressure head in the cup relative to atmospheric pressure can be calculated by starting at the flat air–mercury

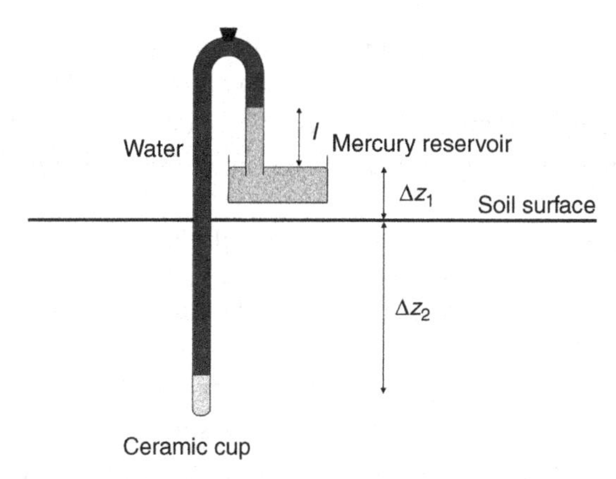

Figure 4.28 Tensiometer with mercury manometer.

interface in the reservoir where the pressure head is zero, and moving through the manometer to the cup:

$$h = -12.6l + \Delta z_1 + \Delta z_2 \tag{4.46}$$

Question 4.20: Two tensiometers are installed in a soil profile and connected to the same mercury reservoir. The cups of tensiometers 1 and 2 are at depths 40 and 80 cm, respectively. The mercury level in the reservoir is at 10 cm above the soil surface. The length of the mercury column in tensiometer 1 is 7.5 cm and that of tensiometer 2 is 9.0 cm.

a) Draw the potential diagram assuming H is linear with z.
b) Calculate the height of the groundwater table.
c) What do you expect of the mercury levels in the manometer tubes when the water in the soil is at static equilibrium?

Nowadays, pressure transducers are frequently used instead of pressure gauges of mercury manometers. A pressure transducer has a sensitive membrane, which converts liquid pressures into voltages. With these devices pressure heads can be measured very accurately and automatically registered. Also pressure transducers require very small displacements of their sensing element to register changing pressures. Therefore tensiometers with pressure transducers require only minute volumes of liquid flowing into or out of the tensiometer cup, which makes them much more sensitive than, for instance, mercury manometers. Although high sensitivity is very desirable in the vadose zone with its small soil water fluxes, one negative effect is that readings are more strongly influenced by changing temperatures, mainly due to thermal expansion and contraction of the liquid and tubing.

The lowest soil water pressure head that can be measured with tensiometers is equal to the vapour pressure of water (–977 cm at 20 °C). In practice, however, already at pressure heads below –900 cm problems arise due to expansion of gas bubbles. The measurement range of tensiometers is limited further by the size of the largest pores in the cup. If the pressure in the cup falls below the air-entry value of the largest pores, air will enter the tensiometer and all the water may be adsorbed by the soil. In practice an air-entry value near –900 cm is chosen. Smaller pores increase the hydraulic resistance of the ceramic cup and thus the reaction time of the tensiometer.

Question 4.21: What is the equivalent diameter of the largest pore in a tensiometer cup that can be used to measure pressure heads of –900 cm? Hint: Use Eq. (4.7) for capillary rise with surface water tension $\sigma = 0.07$ N m^{-1} and wetting angle $\phi = 0°$.

Because air is compressible and has a large thermal expansion coefficient, isolated air bubbles inside tensiometers make them very sluggish in following changes in soil water pressure and make them sensitive to temperature changes. If air is abundant in the system, it can make accurate measurements altogether impossible. The air problems can be nearly eliminated by filling tensiometers with de-aerated water and selecting ceramic cups with pores smaller than 2 μm. However, air may still diffuse through the tubing connecting the cup with the pressure measuring device and through the water in the ceramic cup pores. The former can be eliminated by using impermeable tubing, such as copper, but the latter can be reduced only by using thick ceramic cups of low porosity which increase the reaction time of the tensiometer. Therefore, tensiometers must be flushed periodically with de-aerated water to drive accumulated air out of the system (Koorevaar et al., 1983).

4.11 Measurement of Soil Water Content

4.11.1 Gravimetric and Volumetric Soil Water Content

The quantity of water in soil may be expressed as volumetric water content θ (cm^3 cm^{-3}) or as gravimetric water content w (g g^{-1}). Figure 4.29 defines the volumes and masses of solids, water, air and pores in a soil. The *volumetric* water content is the volume of liquid water per volume soil and is calculated as $\theta = V_w/V_{total}$, where V_w is the water volume and V_{total} is the total soil volume. The *gravimetric* water content is the mass of water per mass of dry soil and equals $w = M_w/M_s$, where M_w is the water mass and M_s the solid mass (note that M_s is used, and not M_{total}). As the density of the solid phase varies in natural soils, volumetric water contents are easier to use than gravimetric water contents. For instance, if we know that a soil has a volumetric water content θ, we may directly calculate the water storage in a soil layer with thickness Δz as the product $\theta \times \Delta z$ cm. Therefore volumetric water contents are commonly used in applied soil physics. The two most applied methods to determine the soil water content are by oven drying and by time domain reflectrometry, which are discussed in the next sections.

Question 4.22: Sometimes gravimetric water contents should be converted to volumetric water contents. Derive from Figure 4.29 that these water contents are related by

$$\theta = \frac{\rho_d}{\rho_w} w$$

where ρ_d is the soil dry bulk density (g cm^{-3}) and ρ_w is the density of water (= 1 g cm^{-3}).

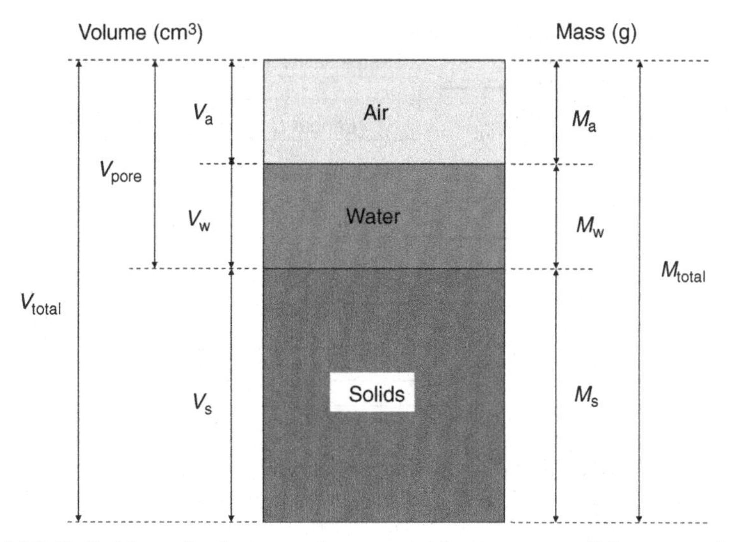

Figure 4.29 Definition of volumes and masses with respect to solids, water, air, pores and total.

4.11.2 Measurement by Oven Drying

This method is very straightforward. One weighs the moist soil sample, dries it to remove the water and weighs the soil sample again. The standard drying method is oven drying at 105 °C for 24 hours. This removes the interparticle water, but keeps the water molecules trapped between clay layers that do not affect flow or extraction by roots (Gardner, 1986). To determine the volumetric water content from the weights, either the dry bulk density ρ_d (g cm^{-3}) or the volume of the soil sample should be known. Important limitations of this gravimetric method are that it is destructive (one can measure only once at the same place) and it cannot be automated.

Question 4.23: A can of moist soil is brought to the laboratory and weighed, dried and reweighed. The following data were recorded:

Mass of can with moist soil 165 g
Mass of can with dry soil 145 g
Mass of empty can 20 g

a) Calculate the gravimetric water content w.
b) Calculate the volumetric water content θ and the dry bulk density ρ_d of this soil, using the volume of the undisturbed soil sample = 100 cm^3.

4.11.3 Measurement by Time Domain Reflectrometry

In the past various nondestructive methods have been proposed based on nuclear radiation, for example, the gamma-ray attenuation method and the neutron attenuation method (Jury et al., 1991). With these methods a huge amount of soil water content

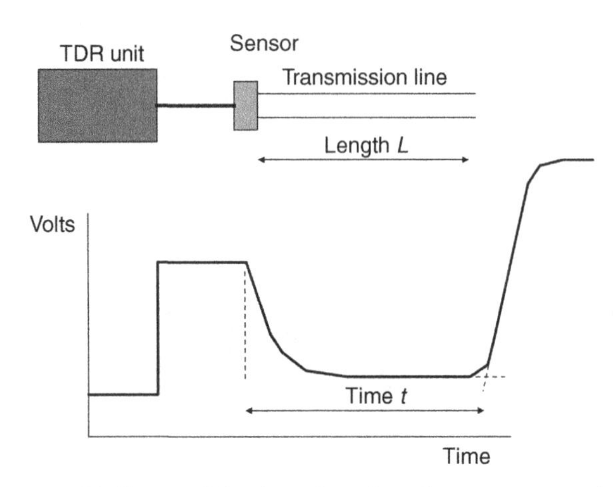

Figure 4.30 Setup of the time domain reflectrometry method.

data have been collected, but these methods have two main disadvantages: the invisible danger of nuclear radiation and the need for site-specific calibration. These disadvantages were eliminated with time domain reflectrometry (TDR). In this method, a so-called TDR unit emits electromagnetic waves along two or three parallel transmission lines that are installed in the soil (Figure 4.30). The reflections of the emitted waves can be visualized with an oscilloscope as function of time or distance. These reflections contain information on the velocity of the electromagnetic wave in the soil. This velocity appears to be a direct function of soil water content, which can be explained as follows.

The dielectric behaviour of a material is physically characterized by its *permittivity*. The relative permittivity, ε, of a material is generally defined as the factor by which the capacitance of a plate capacitor increases when the vacuum or air between the plates is replaced by that material. Thus, per definition, for vacuum and air $\varepsilon = 1$. Relative permittivities are also called dielectric constants, which is somewhat misleading as ε varies with electromagnetic frequency, temperature, and water content.

The permittivity depends in the first place on the polarization in an electrical field. The permanent dipole of water molecules yields the extremely high permittivity $\varepsilon_{water} \approx 81$ (at 18 °C), whereas for most mineral soil components, $\varepsilon_{soil} \approx 5$. Owing to this large difference, the volumetric water content θ of a soil can be determined indirectly by measuring its effective permittivity, ε, if the calibration relationship $\theta(\varepsilon)$ for the particular soil and dielectric measuring equipment is known. According to basic physics, the propagation velocity, v, of an electromagnetic pulse travelling along a wave guide is:

$$v = \frac{c}{\sqrt{\varepsilon}} \tag{4.47}$$

where ε is the effective permittivity of the medium around the wave guide and c is the velocity of light in vacuum (3×10^8 m s^{-1}). Thus, ε and θ can be determined by measuring v of an electromagnetic pulse travelling along a sensor embedded in soil (Dirksen, 1999).

Question 4.24: A common length L of the TDR sensor is 0.10 m. How long is the travel time of an electromagnetic wave from one end of the sensor to the other and return if the sensor is in water? And how long is the travel time if the sensor is in air?

The relation $\theta(\varepsilon)$ was measured by Topp et al. (1980) for a number of soils:

$$\theta = \left(-530 + 292\varepsilon - 5.5\varepsilon^2 + 0.043\varepsilon^3\right) \times 10^{-4} \tag{4.48}$$

Equation (4.48) has given satisfactory results for many soils and contributed greatly to the rapid introduction and development of TDR for automated water content measurements. For very accurate results and for soils with very low bulk density, high clay or organic content and specific mineralogical properties, soil-specific calibration might be necessary (Roth et al., 1992; Dirksen and Dasberg, 1993).

4.12 Measurement of Hydraulic Conductivity

The hydraulic conductivity varies over many orders of magnitude, not only between different soils but also for the same soil as function of its water content. As a result, $k(\theta)$ functions are hard to measure accurately. Many direct and indirect methods have been proposed as listed in Table 4.3 (Klute, 1986; Dirksen, 1999; Hopmans et al., 2002).

There is no single "universal" method that is suitable for all soils and circumstances. Criteria for selecting a suitable method include theoretical basis, control of

Table 4.3 Laboratory measurement methods of $k(\theta)$

Method	Range (cm)	Reference
Suction cell	$-100 < h < 0$	Klute and Dirksen (1986)
Crust method	$-100 < h < 0$	Bouma et al. (1983)
Drip infiltrometer	$-100 < h < 0$	Dirksen (1991)
Evaporation method	$-800 < h < 0$	Wendroth et al. (1993)
Pressure cell	$-1000 < h < 0$	Van Dam et al. (1994)
Sorptivity method	$-1000 < h < 0$	Dirksen (1979)
Hot air method	$-10\,000 < h < -100$	Van Grinsven et al. (1985)
Centrifuge method	$-1000 < h < 0$	Nimmo et al. (1987)
Spray method	$-250 < h < 0$	Dirksen and Matula (1994)
Tension disc infiltrometer	$-100 < h < 0$	Simunek et al. (1998)

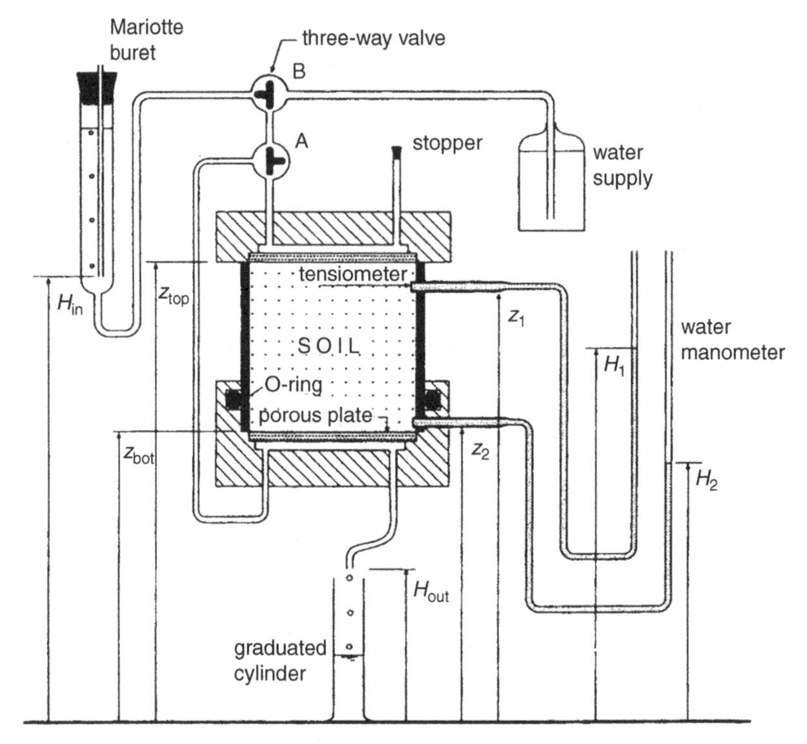

Figure 4.31 Experimental setup to measure hydraulic conductivity at small tensions (Dirksen, 1999).

initial and boundary conditions, accuracy of measurements, error propagation in data analysis, range of application, duration of method, equipment and check on measurements (Dirksen, 1999).

In case of hysteresis in the $\theta(h)$ relation (Figure 4.15), the $k(h)$ relation will also show hysteresis. As the hydraulic conductivity is mainly determined by the water content, $k(\theta)$ relations show only minor hysteresis and are therefore preferred above $k(h)$ relations.

Figure 4.31 shows an experimental setup for measuring hydraulic conductivity at small tensions. Steady hydraulic head gradients are imposed on a sand column by a Mariotte buret at the inlet and a drip point at the outlet. The hydraulic gradient within the vertical soil column is measured with tensiometers connected to water manometers. Unit hydraulic gradient (gravitational flow) resulting in uniform water content can be obtained by gradually adjusting the externally imposed hydraulic heads.

The soil column is clamped between two porous plates. These plates are needed to keep air out of the tubes, which are filled with water at $h < 0$! The air entry value of the porous plates and tensiometers should be small (about –20 cm) to have a highly saturated hydraulic conductivity. The top porous plate is just pressed against the soil cylinder. The small open spaces at the top porous plate and tensiometers do not have to be sealed, because they do not conduct water at $h < 0$. To the contrary, these spaces are needed to provide access for air that needs to enter the sand column if it is to desaturate when pressure heads are lowered and vice versa. If no air can enter or leave the soil column, the water content cannot change with pressure head. The water influx is measured with the Mariot buret and the outflux with a graduated cylinder. The pressure head in the soil column can simply be lowered by increasing the soil column height with respect to the inflow and outflow levels.

Question 4.25: During an experiment with the small tension setup depicted in Figure 4.31 the following data were measured: $H_1 = 30$ cm, $H_2 = 26$ cm, $z_1 = 42$ cm, $z_2 = 38$ cm, influx = outflux = 100 cm^3 d^{-1} and cross section soil sample = 20 cm^2.

a) Have the conditions of steady state and unit hydraulic gradient been achieved in the experiment?
b) Which $k(h)$ data pair results from this experiment?
c) Suppose we increase only the level of the soil column. How does this affect the pressure head inside the soil column and the water flux?

4.13 Measurement of Root Water Uptake

In an unsaturated soil, water flows mainly in the vertical direction z. Root water uptake patterns generally can be derived by applying the water balance equation to a given volume of soil in combination with measurements of soil water pressure head or soil water content. Figure 4.32 shows the calculation procedure in case of tensiometer measurements. Let us apply the method to a homogeneous, vegetated soil, where at various depths during the growing season the soil water pressure head h is measured. The purpose of the tensiometer measurements is to determine the root water extraction at different soil depth. Table 4.4 lists for three depths ($z = -15, -25$ and -35 cm) at two times ($t = 150$ d and $t = 160$ d) the measured h values. The soil water retention curve and hydraulic conductivity function of the soil were measured separately in the laboratory. The relevant section of these functions is listed in Table 4.5.

First calculate the water storage in the layer $-30 < z < -20$ cm at the beginning and end of the time interval. We may assume that the water content at the centre of a depth interval equals the average water content of that depth interval. Therefore, the water amount at $t = 150$ d equals $\theta(h_i^j)\ \Delta z = 0.325 \times 10.0 = 3.25$ cm, and at $t = 160$ d equals $0.175 \times 10.0 = 1.75$ cm.

Next calculate the incoming and outgoing soil water flux at $t = 150$ d and $t = 160$ d. We assume that, at vertical spatial steps of 10 cm, the average unsaturated hydraulic conductivity corresponds to the arithmetic average of k. Therefore at $t = 150$ d and

Table 4.4 Measured soil water pressure heads h (cm)

Height z (cm)	t = 150 days	t = 160 days
–15	–156	–224
–25	–112	–148
–35	–80	–120

Table 4.5 Laboratory data of soil moisture characteristic and hydraulic conductivity function

Pressure head h (cm)	Water content θ (cm^3 cm^{-3})	Hydraulic conductivity k (cm d^{-1})
–80	0.375	0.090
–112	0.325	0.016
–120	0.295	0.010
–148	0.175	0.0045
–156	0.150	0.0019
–224	0.050	0.00022

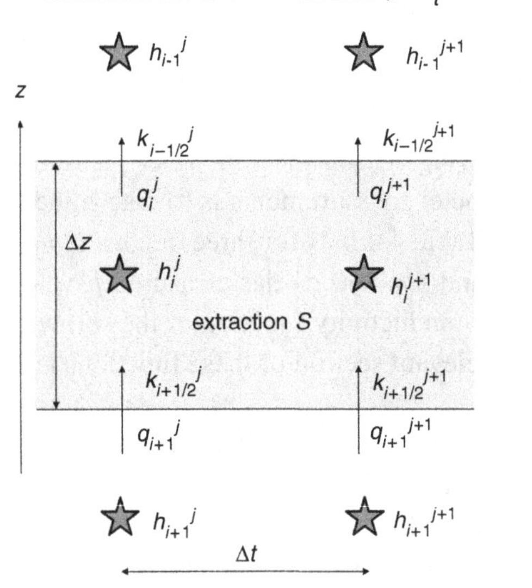

Figure 4.32 Calculation procedure of root water extraction at different depths in the root zone, using tensiometer measurements. The subscript refers to height level, the superscript refers to time level. The soil water fluxes are based on the Darcy equation (4.16), the water balance on the continuity equation (4.21).

$z = -20$ cm, $k_{i-½}^{j} = (0.016+0.0019) / 2 = 0.0090$ cm d^{-1}. Applying Darcy's law, the flux at $t = 150$ d and $z = -20$ cm amounts:

$$\begin{aligned} q_i^j &= -k(-20)\frac{h(-15)-h(-25)}{\Delta z} - k(-20) \\ &= -0.0090\frac{-156-(-112)}{10} - 0.0090 = +0.0306 \text{ cm d}^{-1} \end{aligned} \tag{4.49}$$

Be mindful of a proper use of the signs! The flux $q_i^{\,j}$ has a positive sign, which means that the flow is directed upwards. In the same way we can calculate the other mean values for k and the soil water fluxes $q_{i+1}^{\,j} = +0.0156$ cm d^{-1}, $q_i^{\,j+1} = +0.1166$ cm d^{-1}, and $q_{i+1}^{\,j+1} = +0.0167$ cm d^{-1}. The upward fluxes gradually increase in time.

Finally we can derive the extraction S with the water balance (Figure 4.32):

$$S = \frac{q_{i+1}^{\,j} + q_{i+1}^{\,j+1}}{2} - \frac{q_i^{\,j} + q_i^{\,j+1}}{2} + \left(\theta\left(h_i^{\,j}\right) - \theta\left(h_i^{\,j+1}\right)\right)\frac{\Delta z}{\Delta t} \tag{4.50}$$

which gives:

$$\begin{aligned} S &= \frac{0.1166+0.0167}{2} - \frac{0.0306+0.0156}{2} + \frac{3.25-1.75}{10.0} \\ &= 0.043 + 0.150 = 0.193 \text{ cm d}^{-1} \end{aligned} \tag{4.51}$$

So in this example the root layer at $-30 < z < -20$ cm contributes 1.93 mm d^{-1} to the plant transpiration.

Figure 4.33 shows the measured root water uptake pattern of a red cabbage crop growing on sticky clay in the presence of a 90-cm deep groundwater table (Feddes, 1971). At the top of the profile the magnitude of the root extraction rate is generally small due to a smaller root density and lower water contents. Downward the extraction rate increases to a certain maximum and next decreases to zero at the root zone bottom. As the soil dries in the growing season, the zone of maximum root water uptake moves from shallow to greater depths. Later in the season water uptake from the upper layers becomes relatively less important. Most of the water is absorbed from the zone with higher water contents near the water table (Figure 4.33). As discussed in Section 6.2, the main factors that affect root water uptake are potential transpiration, root density distribution (which may change in time), soil water pressure head profile, soil hydraulic conductivity and plant wilting point.

4.14 Summary

Soil water flow forms the basis for analysis of root water extraction, surface runoff, groundwater recharge and solute transport. Reservoir models as Warrilow may

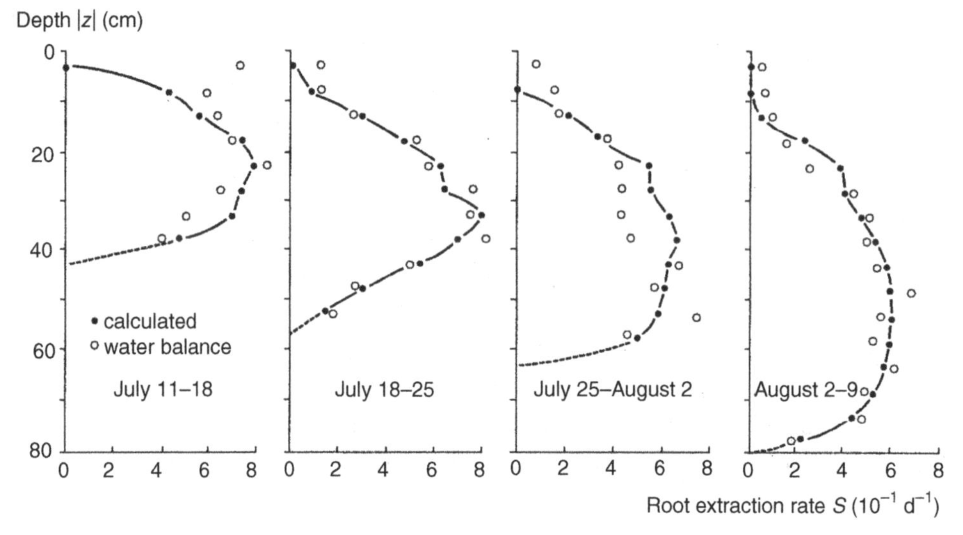

Figure 4.33 Example of measured variations of root water uptake with depth and time of red cabbage grown on a clay soil in the presence of a 90–110 cm groundwater table obtained from water balance studies over 4 consecutive weeks (after Feddes, 1971).

provide a first approximation of the water balance components. The hydraulic head concept offers a more accurate and more general theoretical framework for soil water flow. The main components of the hydraulic head, both in the unsaturated and the saturated zone, are the gravitational and pressure head. Darcy's law relates water flux densities to hydraulic head differences, and applies also to both the unsaturated and saturated zone. Richards' equation combines Darcy's equation and the water balance and forms the basis of most physically based hydrological models. Semi-empirical methods are discussed for infiltration (Horton and Green–Ampt) and capillary rise. The soil hydraulic functions, which are required to solve the Richards' equation, relate soil water pressure head, water content and hydraulic conductivity. Analytical soil hydraulic functions based on Mualem–Van Genuchten, are described. Practical measurement methods are treated for water contents, pressure heads, hydraulic conductivities, soil water retention function and root water uptake.

5

Solute Transport in Soil

5.1 Introduction

At the soil surface, nutrients, pesticides and salts dissolved in water infiltrate the soil. The residence time of these solutes in the vadose zone may have a large effect on soil and groundwater pollution:

- Organic compounds are mainly decomposed in the unsaturated zone, where the main biological activity is concentrated.
- Many plants have no active roots below the groundwater level and therefore extract water and nutrients only from the soil in the unsaturated zone.
- Whereas in the unsaturated zone the transport of solutes is predominantly vertical, in the saturated zone solutes may disperse in any direction, threatening groundwater extractions and surface water systems.

Therefore, to manage soil and water related environmental problems effectively, proper quantification of the transport processes in the unsaturated zone is important (Beltman et al., 1995). For a number of reasons in delta areas relatively much attention is paid to solute transport in soils. In delta areas like the Netherlands the population density is high, the chemical industry is intensive, the agrochemical input in the agriculture is huge, the sedimented soils are very permeable, the groundwater levels are shallow and the groundwater recharge fluxes are large due to the humid climate.

Question 5.1: Why are the aforementioned factors a reason to pay more attention to solute transport in soils?

At field scale level, flow and transport processes can be described in a physical way, as weather conditions, vegetation, soil characteristics, drainage situation and cultivation are well defined for individual fields. For regional analysis, model simulations at field scale level may form the basic unit, which are combined with geographic information systems (Singh, 2005). However, also within a field spatial variability of soil characteristics may cause a large variation of solute fluxes (Biggar and Nielsen,

1976; Van de Pol et al., 1977; Van der Zee and Van Riemsdijk, 1987). Most of this variation is caused by variation of the soil hydraulic functions, preferential flow due to macropores in structured soils or due to unstable wetting fronts in nonstructured soils. In general it is impossible to determine the range and correlations of all relevant physical parameters. A practical approach is to measure for a period of time the solute concentrations in the soil profile and drainage water and apply calibration or inverse modelling to determine 'effective' transport parameters (Groen, 1997). Another approach is the use of Monte–Carlo simulations, where the variation of the transport parameters is derived from stochastic parameter distributions of comparable fields (Boesten and Van der Linden, 1991). Jury (1982) proposed to use transfer functions, which consider the transport processes within a soil column as a black box, and just describe the relation between solutes that enter and leave the soil column. The main limitations of the transfer function approach are that it requires field experiments to calibrate the transport parameters and that extrapolation to other circumstances is risky because of its stochastic rather than physical basis.

Question 5.2: Mention five methods that can be used to analyse the variability of solute fluxes within a field.

This chapter focuses on the vadose zone transport of salts, pesticides and other solutes that can be described with relatively simple kinetics. We consider the processes convection, dispersion, adsorption, root uptake and decomposition. Processes outside the scope of this chapter are (1) volatilization and gas transport, (2) transport of non-mixing or immiscible fluids (e.g., oil and water), (3) chemical equilibria of various solutes (e.g., between Na^+, Ca^{2+} and Mg^{2+}), and (4) chemical and biological chain reactions (e.g., mineralization, nitrification).

5.2 Solute Flux through Soil

The three main solute transport mechanisms in soil water are diffusion, convection and dispersion. *Diffusion* is solute transport that is caused by the solute gradient. Thermal motion of solute molecules within soil water causes a net transport of molecules from high to low concentrations. The solute diffusion flux J_{dif} (kg m^{-2} d^{-1}) is generally described by Fick's first law:

$$J_{\text{dif}} = -\theta D_{\text{dif}} \frac{\partial C_1}{\partial z} \tag{5.1}$$

with θ the volumetric water content (m^3 m^{-3}), D_{dif} the diffusion coefficient (m^2 d^{-1}) and C_1 the solute concentration in soil water (kg m^{-3}). J_{dif} is very sensitive to the actual water content, which affects the effective cross-sectional transport area and the tortuosity of the solute path.

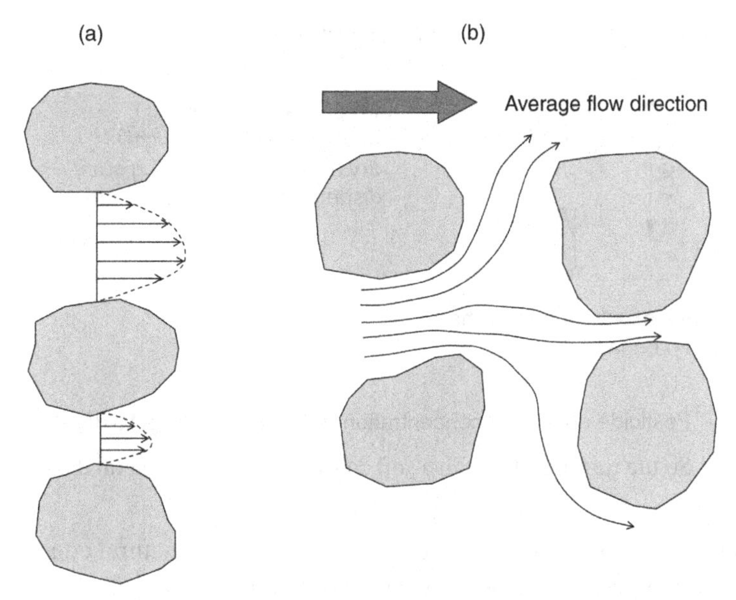

Figure 5.1 (**A**) Flow velocity variation within a pore. (**B**) Flow velocity variation in the flow direction due to the pore network. (After Bear and Verruijt, 1987)

The bulk transport of solutes occurs when solutes are carried along with the moving soil water. This flux is called the *convective* flux, J_{con} (kg m^{-2} d^{-1}), and can be calculated using the average soil water flux q (m d^{-1}):

$$J_{con} = q C_1 \tag{5.2}$$

In the case of water flow, the Darcy flux q, which is averaged over a certain cross section, is usually sufficient. In the case of solute transport, we need to consider the water velocity variation between pores of different size and geometry and even the variation within pores (Figure 5.1).

These differences of water velocities cause some solutes to advance faster than the average solute front, and other solutes to advance slower. The overall effect will be that solute concentration differences are smoothened or dispersed. Solutes seem to flow from high to low concentrations. If the time required for solutes to mix in the lateral direction is small compared to the time required for solutes to move in the flow direction, the *dispersion* flux J_{dis} (kg m^{-2} d^{-1}) appears to be proportional to the solute gradient (Bear, 1972):

$$J_{dis} = -\theta D_{dis} \frac{\partial C_1}{\partial z} \tag{5.3}$$

with D_{dis} the dispersion coefficient (m^2 d^{-1}). Note that the dispersion flux in Eq. (5.3) has a mathematical expression similar to that of the diffusion flux in Eq. (5.1). Under

Steady state water flow through soil column with:

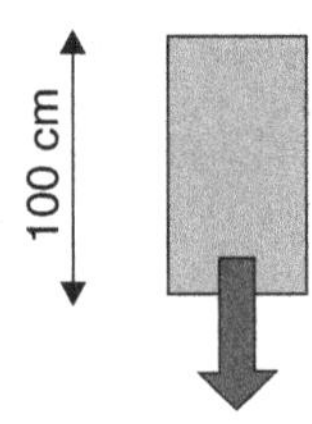

water flux q = −1 cm d^{-1}
water content θ = 0.5 cm^3
dry bulk density ρ_d = 1 g cm^{-3}
dispersion length L_{dis} = 1 cm

linear adsorption isotherm
no decay

Pesticide applied at concentration of 1 mg L^{-1} during 10 days

Figure 5.2 Solute transport through a soil column with some general data.

laminar flow conditions, which is almost always the case in natural conditions, D_{dis} is proportional to the pore water velocity v (= q/θ) (Bolt, 1979):

$$D_{dis} = L_{dis} v \tag{5.4}$$

with L_{dis} the dispersion length (m), which depends on the scale over which the water flux and solute convection are averaged. Typical values of L_{dis} are 0.5–2.0 cm in packed laboratory columns and 5–20 cm in the field, although they can be considerably larger in regional groundwater transport (Jury et al., 1991). Unless water is flowing very slowly through repacked soil, the dispersion flux is much larger than the diffusion flux.

> **Question 5.3:** Assume a soil with the following conditions: q = 2 mm d^{-1}, θ = 0.25 cm^3 cm^{-3}, L_{dis} = 5 cm, and D_{dif} = 0.156 cm^2 d^{-1}. What is the percent of the diffusion flux with respect to the dispersion flux?

The total solute flux J (kg m^{-2} d^{-1}) is the sum of the diffusion, convection and dispersion flux:

$$J = J_{dif} + J_{con} + J_{dis} = qC_l - \theta(D_{dif} + D_{dis})\frac{\partial C_l}{\partial z} \tag{5.5}$$

We here illustrate the transport processes for a soil column. Assume a steady-state water flux with soil physical properties as listed in Figure 5.2. A pesticide with a concentration of 1 mg L^{-1} is applied during a period of 10 days. Figure 5.3 shows the resulting solute concentrations in the case of only convection and in the case of convection plus dispersion after periods of 10 and 50 days. When dispersion is included, the solute profile has a shape similar to the normal Gauss distribution. Some solutes

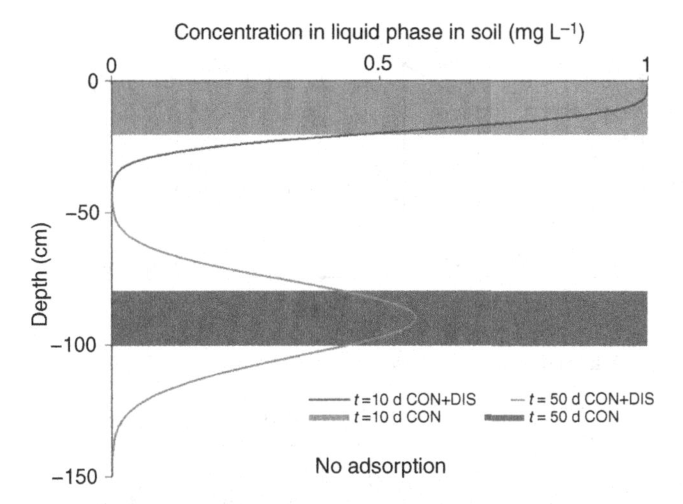

Figure 5.3 Solute concentration profiles after 10 and 50 days in case of convection only and in case of convection plus dispersion respectively (experimental data Figure 5.2).

move faster and other solutes stay behind compared to the average solute velocity equal to $q/\theta = -2$ cm d^{-1}. Whereas near the soil surface the concentrations are close to 1 mg L^{-1}, deeper in the soil dispersion causes a gradual decline of the concentrations. Note that the surface areas of the solute profiles are equal at all times, whether dispersion is included or omitted.

Question 5.4: Why are the surfaces below the solute profiles equal for each case?

5.3 Convection–Dispersion Equation

By considering conservation of mass in an elementary cubic volume (Figure 5.4), we may derive the mass balance or continuity equation for solute transport:

$$\frac{\partial C_{\mathrm{T}}}{\partial t} = -\frac{\partial J}{\partial z} - S_{\mathrm{s}} \tag{5.6}$$

where C_{T} is the total solute concentration in the soil system (kg m^{-3}) and S_s is the solute sink term (kg m^{-3} d^{-1}) accounting for decomposition and uptake by roots.

The solutes may be dissolved in soil water or may be adsorbed to organic matter or clay minerals:

$$C_{\mathrm{T}} = \rho_{\mathrm{b}} C_{\mathrm{a}} + \theta C_{\mathrm{l}} \tag{5.7}$$

where ρ_{b} is the dry soil bulk density (kg m^{-3}) and C_{a} is the solute amount adsorbed (kg kg^{-1}). By combining Eqs. (5.5)–(5.7) and defining the effective diffusion

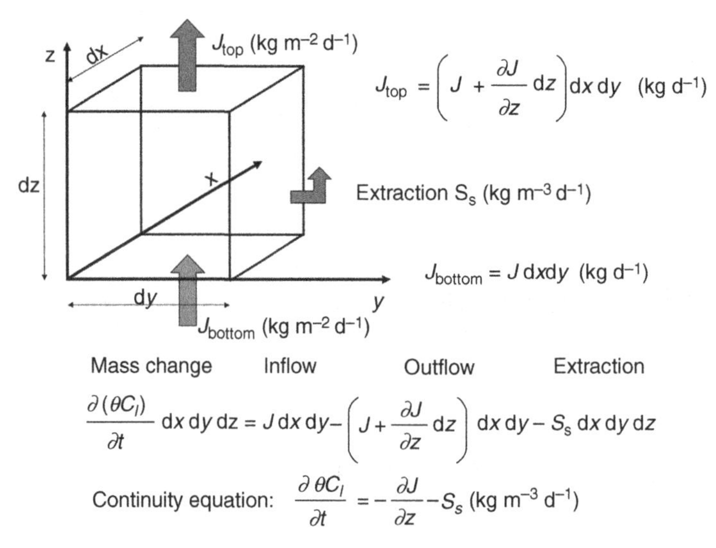

Figure 5.4 Derivation of the continuity or mass balance equation for solute transport in a soil.

coefficient $D_e = D_{dif} + D_{dis}$ (m^2 d^{-1}), we may derive the widely used convection-dispersion equation for solute transport in the unsaturated zone (Biggar and Nielsen, 1967):

$$\frac{\partial}{\partial t}\left(\rho_b C_a + \theta C_l\right) = \frac{\partial}{\partial z}\left(\theta D_e \frac{\partial C_l}{\partial z}\right) - \frac{\partial}{\partial z}\left(q C_l\right) - S_s \tag{5.8}$$

Question 5.5: Check the units of all terms in Eq. (5.8).

A versatile, public domain model which solves Eq. (5.8) analytically is the STANMOD model (Šimůnek et al., 1999). The solute profiles in the figures of this chapter were calculated with this model.

5.4 Transport of Inert, Nonadsorbing Solutes

The most simple soil chemical transport processes are those that involve nonvolatile dissolved solutes that neither react ($S_s = 0$) nor adsorb to soil solids ($C_a = 0$). Examples are chloride or bromide ions flowing through soil that has only negatively charged or neutral minerals. These ions do not react chemically (except for anion exclusion) with the kind of ions normally found in soil solution and are not attracted to clay or organic matter surfaces. Thus they may act as "tracers" of the water flow pathways (Jury et al., 1991).

Imagine experiments similar to Figure 5.2, in which water is flowing uniformly at steady state through a homogeneous soil column of length L that is at constant water content. In this case the general transport equation (Eq. (5.8)) reduces to:

$$\frac{\partial C_1}{\partial t} = D_e \frac{\partial^2 C_1}{\partial z^2} - v \frac{\partial C_1}{\partial z} \tag{5.9}$$

Suppose that at $t = 0$ we add a tracer with concentration C_0 and steady water velocity v to a soil column with $C_1(z) = 0$. The general analytical solution of Eq. (5.9) for such initial and boundary conditions is (Radcliffe and Simunek, 2010):

$$C_1(z,t) =$$

$$C_0 \left[\sqrt{\frac{v^2 t}{\pi D_e}} \exp\left(-\frac{(z - vt)^2}{4 D_e t} \right) + \frac{1}{2} \operatorname{erfc}\left(\frac{vt - z}{\sqrt{4 D_e t}} \right) - \frac{1}{2}\left(1 + \frac{vz}{D_e} + \frac{v^2 t}{D_e} \right) \exp\left(\frac{vz}{D_e} \right) \operatorname{erfc}\left(\frac{-z - vt}{\sqrt{4 D_e t}} \right) \right] \tag{5.10}$$

Question 5.6: Take the transport experiment of Figure 5.2 with $v = -2$ cm d^{-1} and $L_{dis} = 10$ cm. Calculate with Eq. (5.10) the solute concentrations at $z = -100$ cm for $t = 30, 40, 50, 60$ and 80 days. Compare your results to Figure 5.5. (Hint: Use Excel to calculate the various terms of Eq. (5.10). Note that Excel cannot calculate the complementary error function of a negative argument. In that case use $\operatorname{erfc}(-x) = 1 + \operatorname{erf}(x)$).

We might monitor the chloride concentration at the outflow end $z = L$ of the column. The plot of outflow concentration versus time (Figure 5.5) is called an outflow curve, or a "breakthrough" curve (representing the solute breaking through the outflow end). The centre of each of the solute fronts, drawn for different values of the dispersion length L_{dis}, arrives at the outflow end of the column at the same time $T_{res} = L / v = 100 / 2 = 50$ d, called the breakthrough time (Figure 5.5). When dispersion is neglected ($L_{dis} = 0$), all the solutes move at identical velocity v, and the front arrives as one discontinuous jump to the final concentration C_0 at $t = T_{res}$. This model, in which dispersion is neglected, is called the "piston flow" model of solute movement (Jury et al., 1991). The effect of dispersion on the breakthrough curve is to cause some early and late arrival of solutes with respect to the breakthrough time. This deviation is due to dispersion and a small amount of diffusion ahead and behind the front moving at velocity v and becomes more pronounced as L_{dis} and thus D_e becomes larger.

According to the piston flow model, the incoming solute replaces the water initially present in the soil, or, equivalently, pushes this water ahead of the solute front like a piston. Thus one may calculate solute concentration with the piston flow model by estimating how long it will take to replace the water between the point of entry and the final location. For example, we want to calculate the time required to transport nitrate (a mobile ion) from the bottom of the root zone to groundwater $L = 10$ m below. The average water content of the subsoil $\theta = 0.15$

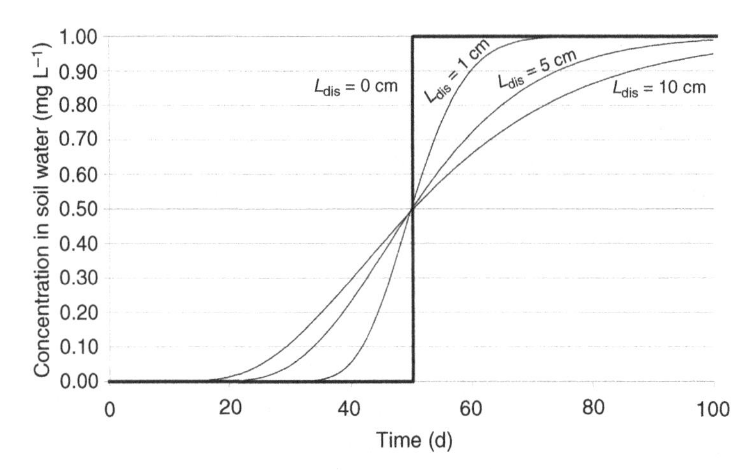

Figure 5.5 Breakthrough curves for the experimental data listed in Figure 5.2 and for different values of dispersion length (L_{dis} = 0, 1, 5 and 10 cm).

and the average annual recharge rate $q = 0.25$ m y^{-1}. The residence time T_{res} can be calculated as:

$$T_{res} = \frac{L}{v} = \frac{\theta L}{q} = \frac{0.15 \times 10}{0.25} = 6\,\text{years} \tag{5.11}$$

Question 5.7: In the Netherlands the recharge of the groundwater amounts to ca. 250 mm y^{-1}. Determine the residence time of inert, nonadsorbing solutes in the unsaturated zone in case of the Veenkampen (peat soil, $\theta = 0.64$, $L = 0.5$ m) and Otterlo (sand soil, $\theta = 0.14$, $L = 20.0$ m).

Often a narrow solute pulse, rather than a front, might be added to a soil. Figure 5.6 shows for the soil column of Figure 5.2 the corresponding outflow curves for a narrow pulse input $C_0 = C(0, t)\,\Delta t$ (kg d m^{-3}) in which Δt is the relatively short application time. The mathematical solution to Eq. (5.9) for a solute pulse added to a clean soil is (Jury and Sposito, 1985):

$$C(z,t) = \frac{-zC_0}{2\sqrt{\pi D_e t^3}} \exp\left(-\frac{(z - vt)^2}{4D_e t}\right) \tag{5.12}$$

A pulse of 1 day was applied with the concentration $C(0, t) = 1000$ μg L^{-1}. As Figure 5.6 shows, the maximum concentration decreases rapidly due to dispersion. Note also that the curves are asymmetric, because as time increases, dispersion causes a larger spreading (Jury et al., 1991).

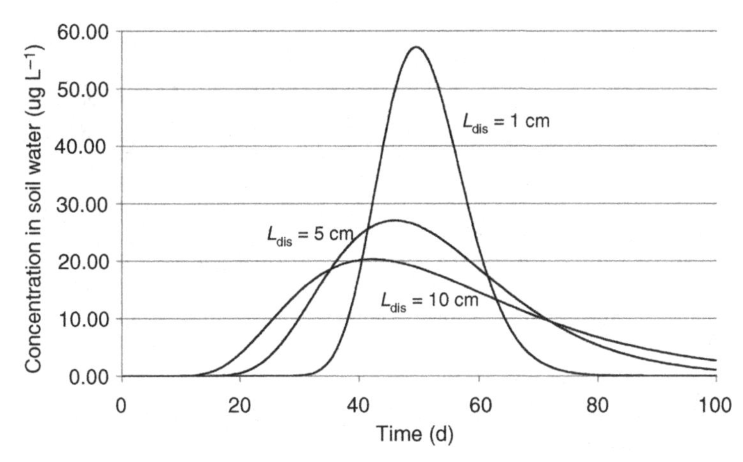

Figure 5.6 Breakthrough curves for a solute pulse at soil surface (C(0, t) = 1 mg L^{-1}, Δt = 1 d), showing the spreading at different values of dispersion length (L_{dis} = 1, 5 and 10 cm).

Question 5.8: In the case of the soil column of Figure 5.2 and the solute pulse C_0 = C(0, t) Δt = 1000 (μg d L^{-1}), calculate the concentration at the bottom of the soil column at t = 40, 50 and 60 d. Check your answer with Figure 5.6.

5.5 Transport of Inert, Adsorbing Chemicals

Certain chemicals, although they do not react chemically or biologically in soil, adsorb to soil solids like clay platelets and organic matter. For these chemicals, the transport equation (Eq. (5.8)) may be written as (assuming homogeneous soil and steady-state water flux):

$$\frac{\rho_{\mathrm{b}}}{\theta}\frac{\partial C_{\mathrm{a}}}{\partial t}+\frac{\partial C_{\mathrm{l}}}{\partial t}=D_{\mathrm{e}}\frac{\partial^2 C_{\mathrm{l}}}{\partial z^2}-v\frac{\partial C_{\mathrm{l}}}{\partial z} \tag{5.13}$$

Equation (5.13) differs from (5.9) only in the left term, which represents the rate of change of adsorbed amount of solutes. To solve Eq. (5.13), we should know the relation between the adsorbed concentration C_{a} and the dissolved concentration C_{l}. This relation at equilibrium is called *an adsorption isotherm*. Figure 5.7 shows different kinds of isotherm shapes with their analytical expression as found for various compounds in soil. The linear adsorption isotherm may be expressed as:

$$C_{\mathrm{a}}=S_{\mathrm{d}}C_{\mathrm{l}} \tag{5.14}$$

where S_{d} is the slope of the isotherm (m^3 kg^{-1}), also known as the distribution coefficient. If we assume that the linear isotherm is valid at all times in soil (i.e., the

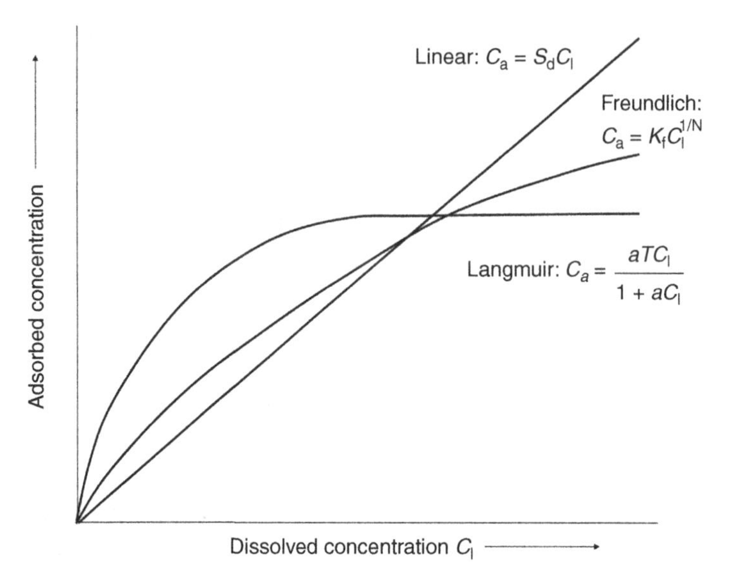

Figure 5.7 Three main adsorption isotherm shapes with their analytical expression.

dissolved and adsorbed phases are instantaneously in equilibrium), then the time derivative of C_a may be written as:

$$\frac{\partial C_a}{\partial t} = S_d \frac{\partial C_l}{\partial t} \tag{5.15}$$

Using Eq. (5.15), Eq. (5.13) can be written as:

$$\left(1 + \frac{\rho_b S_d}{\theta}\right)\frac{\partial C_l}{\partial t} = D_e \frac{\partial^2 C_l}{\partial z^2} - v\frac{\partial C_l}{\partial z} = R\frac{\partial C_l}{\partial t} \tag{5.16}$$

where $R = 1 + \rho_b S_d / \theta$ (-) is defined as the *retardation factor*. The retardation factor equals the total solute amount in a soil volume divided by the dissolved solute amount. Finally, we may divide each side of Eq. (5.16) by *R*, producing:

$$\frac{\partial C_l}{\partial t} = D_R \frac{\partial^2 C_l}{\partial z^2} - v_R \frac{\partial C_l}{\partial z} \tag{5.17}$$

where $D_R = D_e / R$ and $v_R = v / R$ are the retarded dispersion coefficient and solute velocity, respectively.

Let us add adsorption to the column leaching experiment of Figure 5.2. Figure 5.8 shows the effect of adsorption without considering dispersion. In the case of $R = 2$, the solutes move at a velocity $v/2$ cm d^{-1}. Note that the surface of the area below the curve is 50% of the area in case of $R = 1$ (no adsorption) because at $R = 2$ only 50% of the solutes are in the soil water solution. Figure 5.9 includes the effect of dispersion.

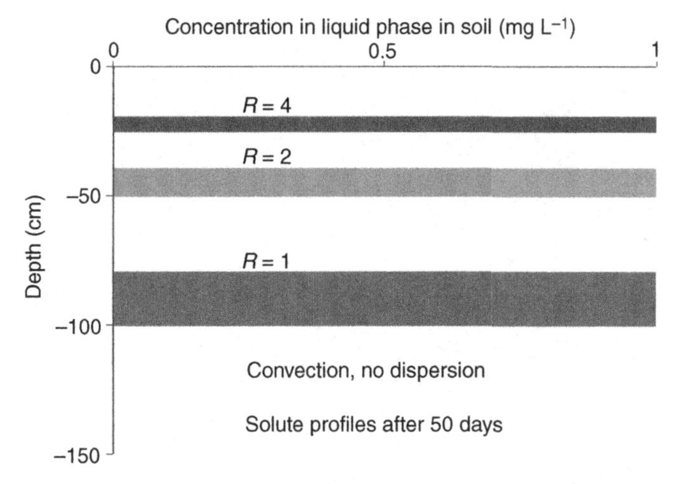

Figure 5.8 Effect of adsorption on the solute concentration profiles after 50 days. No diffusion and dispersion are assumed. The experimental data of Figure 5.2 apply.

Note the large difference in solute profile in case only convective fluxes are considered (piston flow) and in case dispersion and adsorption are included.

Equation (5.17) has the same mathematical form as Eq. (5.9), and therefore it has the same solution as the one for Eq. (5.9), which is illustrated in Figures 5.5 and 5.6. The only difference is that for adsorbing chemicals the solute breakthrough time $T_{\mathrm{res,R}}$ equals:

$$T_{\mathrm{res,R}} = \frac{L}{v_{\mathrm{R}}} = \frac{RL}{v} = RT_{\mathrm{res}} \qquad (5.18)$$

Thus for steady-state conditions, an adsorbing solute would take R times as long to reach a certain soil depth compared to a mobile solute.

Figure 5.10 provides an illustration of the breakthrough curve of our column leaching experiment. Without adsorption the average residence time amounts to 50 d. In the case of adsorption this residence time increases in proportion to the retardation factor. Figure 5.10 also shows that the solute dispersion increases in the case of adsorption. Although the dispersion coefficient in Eq. (5.17) is reduced by a factor R, the travel time is increased, so that the amount of solute spreading as a function of time observed at the outflow end actually appears to be greater than that of a mobile chemical (Jury et al., 1991).

Question 5.9: In the case of a vadose zone 10 m thick with volumetric water content $\theta = 0.15$ and recharge $q = 0.25$ m y^{-1}, the travel time of mobile nitrate amounts to 6 years. For the same hydrologic conditions, calculate the residence time of an adsorbing pesticide with a linear distribution coefficient and $S_{\mathrm{d}} = 2$ cm^3 g^{-1}. The soil has a dry bulk density $\rho_{\mathrm{b}} = 1.5$ g cm^{-3}.

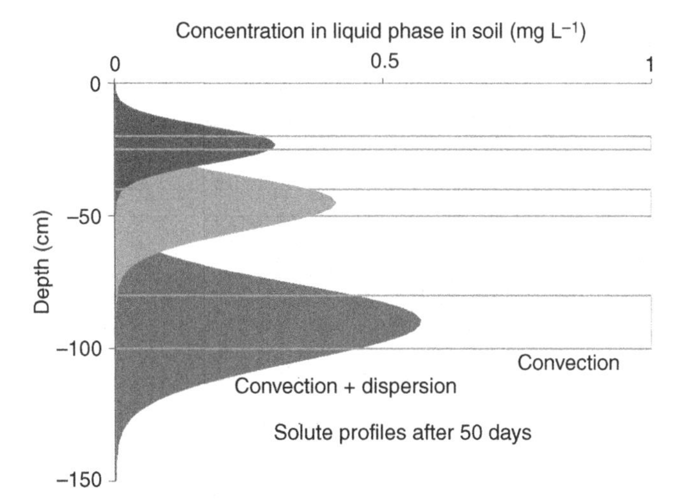

Figure 5.9 The effect of convection, dispersion and adsorption on the solute concentration profiles after 50 days. The experimental data of Figure 5.2 apply.

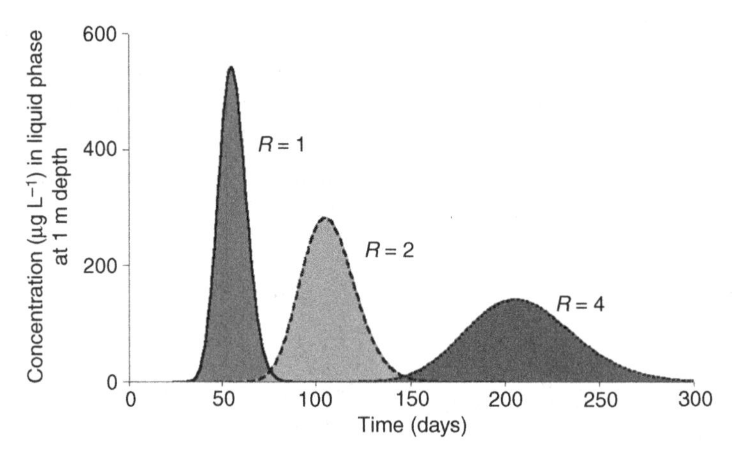

Figure 5.10 Breakthrough curves at different retardation factors. No decomposition occurs. The experimental data of Figure 5.2 apply.

5.6 Reactions of Chemicals in Soil

Many organic chemicals in soil decompose by microbial or chemical reactions. Although this reaction may depend in a complex manner on temperature, pH, microbial population density, carbon content and chemical history, for optimum conditions first-order kinetics in general provide a useful approximation (Hamaker, 1972). A chemical amount $M(t)$ (g) subject to first-order decay loses material at a rate proportional to its mass. This loss rate can be expressed mathematically as:

$$\frac{\mathrm{d}M}{\mathrm{d}t} = -\mu M \tag{5.19}$$

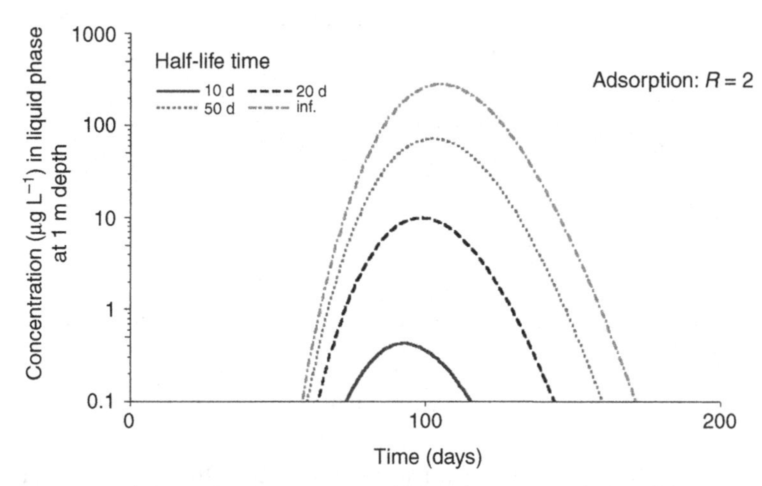

Figure 5.11 Breakthrough curves at retardation factor R = 2 for different rates of decomposition. The experimental data of Figure 5.2 apply.

where μ is the first-order decomposition parameter (d^{-1}). Suppose we have the amount $M = M_0$ at $t = 0$. Integration of Eq. (5.19) results in an exponential decline of M_0 with time:

$$M(t) = M_0 e^{-\mu t} \tag{5.20}$$

This means that 50% of the mass is left at $t = (\ln 2)/\mu$, which is called the half-life T_{50} (d).

Question 5.10: Derive the expression for half-life $T_{50} = (\ln 2)/\mu$ with Eq. (5.20).

A mobile chemical with first-order decay can be simulated with the convection–dispersion equation (Eq. (5.8)), putting $S_s = \mu\, \theta\, C_l$. Therefore, for steady flow conditions (θ = constant) the transport equation may be written as:

$$\frac{\partial C_l}{\partial t} = D_e \frac{\partial^2 C_l}{\partial z^2} - v \frac{\partial C_l}{\partial z} - \mu C_l \tag{5.21}$$

Proceeding with the leaching experiment of Figure 5.2, Figure 5.11 shows the breakthrough curves in case of convection, dispersion, adsorption (R = 2) and different half-lives T_{50}. In case of T_{50} = infinity, no decay occurs and the breakthrough curve is similar to the one in Figure 5.10 with R = 2. At the soil surface we started with solute concentrations equal to 1000 µg L^{-1}. Note how effective the solute concentration decrease is owing to the combined effect of adsorption and decomposition: from 1000 µg L^{-1} to less than 1 µg L^{-1}! Figure 5.12 shows that in case of higher adsorption, the decomposition is much more effective. By increasing the retardation factor from

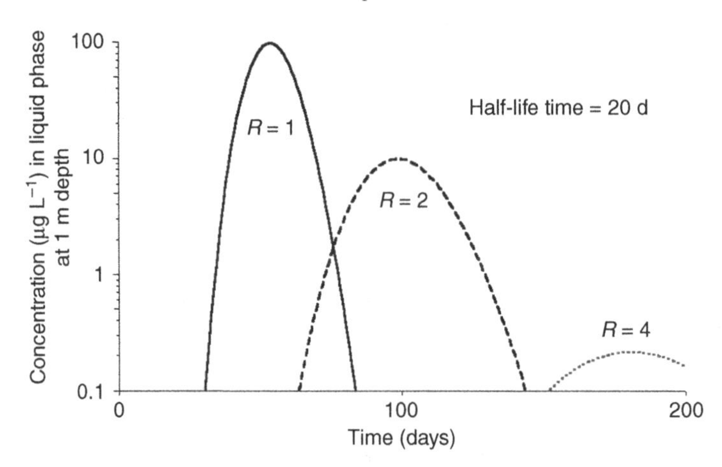

Figure 5.12 Breakthrough curves at half-life T_{50} = 20 d for different amounts of retardation. The experimental data of Figure 5.2 apply.

$R = 1$ to $R = 4$, the solute concentrations at 1 m depth decrease from 90 to 0.23 μg L^{-1}. Because of the very effective immobilization of pesticides when they are adsorbed and simultaneously decompose, screening programs of pesticides focus on high values of both R and μ.

In the case of more reactive solutes we might neglect diffusion and dispersion and use the piston flow model. The chemical moves with a velocity $v = q / \theta$ and reaches a distance $z = L$ in a time $T_{res} = L / v$. If the chemical was added as a front of concentration C_0 at time $t = 0$, according to Eq. (5.20) it will have a concentration $C_0 e^{-\mu T_{res}} = C_0 e^{-\mu L/v}$ when it arrives at $z = L$. If it is added as a pulse of mass M_0, the pulse mass that passes $z = L$ will be $M_0 e^{-\mu L/v}$ (Jury et al., 1991).

Question 5.11: We continue with the soil column of Figure 5.2 and the solute pulse $C_0 = C(0, t)\, \Delta t = 1000$ (µg d cm^{-3}). In the case of piston flow with first-order decay, how much solutes will ultimately leach from the 1 m high soil column in case of no adsorption and T_{50} = 50 days? And at the same half-life with adsorption $R = 4$?

5.7 Salinization of Root Zones

In many regions of the world, rain at agricultural fields has to be supplemented with irrigation water in order to have viable crop production. Water from canals and groundwater contains more salts than rain water and in the long run may salinize the root zone. In this paragraph we quantify the salinity profiles in irrigated soils. We assume (1) no solute reactions, (2) no solute dispersion, and (3) no plant uptake of solute. These assumptions are justified for long-term salinization processes (10–30 years).

The steady-state soil water flow equation at any depth in a root zone with uniform root water uptake S (d^{-1}), can be written as:

$$q(z) = q_0 - Sz \tag{5.22}$$

where q_0 is a mean infiltration flux at the soil surface (m d^{-1}) (negative, as the flux is directed downward) and z is the soil depth, defined as positive upward and zero at soil surface (m). When dispersion is neglected and solutes do not precipitate, dissolve or enter plant roots, the steady-state solute flux should be constant with depth:

$$J(z) = q(z)C_l(z) = \text{constant} = q_0 C_0 \tag{5.23}$$

where C_0 is the average concentration in the infiltration water. Combination of Eqs. (5.22) and (5.23) gives the solute concentration as function of depth:

$$C_l(z) = \frac{q_0 C_0}{q(z)} = \frac{q_0 C_0}{q_0 - Sz} \tag{5.24}$$

We may simplify this expression by defining the *leaching fraction* L_f:

$$L_f = \frac{\text{drainage rate}}{\text{irrigation rate}} = \frac{q_0 + SD_r}{q_0} \tag{5.25}$$

where D_r is the rooting depth. Solving Eq. (5.25) for S and substituting it into Eq. (5.24), gives the concentration profiles in terms of L_f:

$$C_l(z) = \frac{C_0}{1 + (L_f - 1)\dfrac{|z|}{D_r}} \tag{5.26}$$

Figure 5.13 shows a plot of C_l/C_0 versus z/D_r for various values of the leaching fraction. Note the rapid increase of salt concentrations near the bottom of the root zone. The solute concentration at the bottom of the root zone equals:

$$C_{max} = C_l(D_r) = \frac{C_0}{L_f} \tag{5.27}$$

Common values for the leaching fraction are 0.10–0.20. This yields salinity concentrations in the percolation water that are 5–10 times as large as the salinity concentration in the infiltration water.

Question 5.12: Consider an irrigated field with a crop completely covering the soil. The irrigation is applied for such a long time and so often that the soil water fluxes and salinity concentrations in the root zone hardly change in time. Therefore we may consider a steady-state situation. The average irrigation amount equals 6.0 mm/d, the average transpiration amounts 5.4 mm d^{-1}, and the salinity concentration $C_0 = 0.4$ mg cm^{-3}. Consider a uniform root density and root water extraction pattern over the rooting depth.

a) Which salinity concentration do you expect in the centre of the root zone?
b) Which salinity concentration do you expect at the bottom of the root zone?

Consider now a triangular root density profile, with the maximum root density at the soil surface. We may assume that the soil water extraction by roots is proportional to the root density.

c) Which salinity concentration do you expect in the centre of the root zone?
d) Which salinity concentration do you expect at the bottom of the root zone?

To quantify the loss of crop yield due to salinization, scientists have long searched for a proper way of averaging the strongly nonlinear salinity profile (Jury et al., 1991). Raats (1975) developed a very useful approach with plausible, generally valid assumptions. He defined the weighted average salinity $\bar{C}$ of the root zone as the average salinity of the soil water extracted by the plant:

$$\bar{C} = \frac{\int_{-D_r}^{0} S(z) C_l(z) \mathrm{d}z}{\int_{-D_r}^{0} S(z) \mathrm{d}z} \tag{5.28}$$

where the denominator is equal to the plant transpiration T (m d^{-1}). In steady-state conditions, $C_l(z) = C_0 q_0 / q(z)$ and $S = -\mathrm{d}q / \mathrm{d}z$. Thus Eq. (5.28) may be written as:

$$\bar{C} = \frac{-q_0 C_0}{T} \int_{-D_r}^{0} \frac{\mathrm{d}q}{\mathrm{d}z} \frac{\mathrm{d}z}{q} = \frac{-q_0 C_0}{T} \int_{q_0 L_f}^{q_0} \frac{\mathrm{d}q}{q} = \frac{-q_0 C_0}{T} \ln\left(\frac{1}{L_f}\right) \tag{5.29}$$

Finally, using Eq. (5.25) and $S\, D_r = T$, we may write Eq. (5.29) as (Jury et al., 1991):

$$\frac{\bar{C}}{C_0} = \frac{1}{1 - L_f} \ln\left(\frac{1}{L_f}\right) \tag{5.30}$$

Note that the average salinity of Eq. (5.30) is independent of the shape of the water uptake distribution and depends only on the leaching fraction L_f and C_0! This average concentration might be used in combination with the Maas and Hoffman criteria to estimate crop yield loss (Section 6.2.4).

Question 5.13: Which average root water uptake salinity $\bar{C}$ do you calculate for the field in Question 5.12 with the uniform root density? And which average root water uptake salinity for the same field with the triangular root density?

5.8 Pesticide Pollution of Groundwater

To be effective, pesticides should be mobile enough to reach their target organism and persistent enough to eliminate this organism. However, persistency and mobility are unfavourable solute properties from an environmental point of view (Section 5.6).

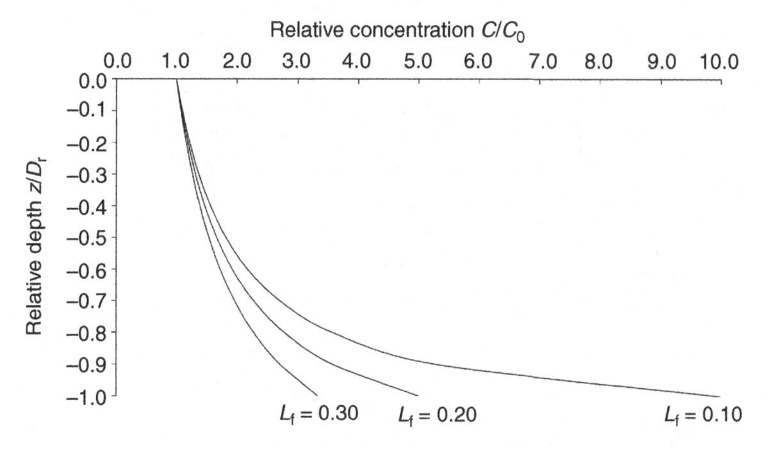

Figure 5.13 Relative salinity concentrations in the root zone for different leaching fractions in case of a uniform root water extraction.

Early compounds, such as DDT and Dieldrin, had a very low mobility in soil and therefore did not have the potential to reach groundwater. However, they were very persistent and thus reached the food chain by exposure through the atmosphere or migration in ground- and surface waters (Jury et al., 1991).

Modern legislation intends to prevent pesticide leaching towards the groundwater. Implicit in this legislation is the idea that pesticides can be screened based on their environmental fate properties at the time of their development to make preliminary decisions about their pollution potential. As an example, Boesten and Van der Linden (1991) calculated pesticide leaching to groundwater at 1 m depth for a sandy soil continuously cropped with maize. They assumed first-order transformation, equilibrium Freundlich adsorption (Figure 5.7) and proportional uptake of water and pesticides by plant roots. Figure 5.14 shows the calculated leaching amount as a function of half-life (T_{50}), adsorption (S_d) and application season (spring or autumn). Pesticide leaching is very sensitive to both T_{50} and S_d: changing these coefficients by a factor of 2, changes the amounts leached typically by about a factor 10! Autumn applications result in much higher leaching of nonsorbing pesticides with short half-lives than spring application (difference about 2 orders of magnitude).

Question 5.14: What are the two main reasons that nonsorbing pesticides with short half-lives leach about 100 times more in autumn compared to spring?

5.9 Residence Time in Groundwater

To analyze solute leaching to surface water systems, we should know the solute residence time in the groundwater system. In the vadose zone, soil water flow is predominantly vertical. Below the groundwater level, water is subject to the

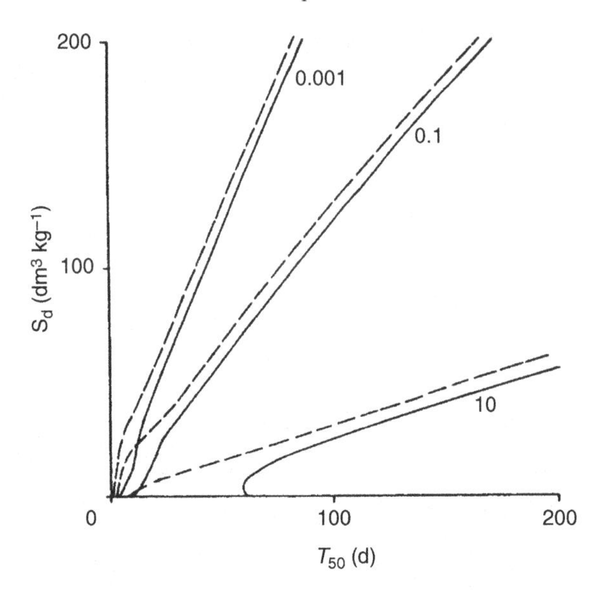

Figure 5.14 Contour lines showing the fraction of pesticide leached below 1 m depth as function of distribution coefficient S_d and half-life T_{50}. Solid lines correspond to application in spring (May 25), dashed lines to application in autumn (1 November) (Boesten and Van der Linden, 1991).

prevailing hydraulic head gradients and may flow in any direction. Sophisticated two- and three-dimensional models have been developed to simulate water flow and solute transport in the saturated zone. Many environmental studies require interaction of the unsaturated and saturated zone into one model. Although a few of those models exist (e.g., MODFLOW, HYDRUS-2D), it is more common either to simplify the groundwater flow and simulate accurately the unsaturated zone, including the interaction with vegetation and atmosphere, or simplify the unsaturated zone and simulate accurately the groundwater flow system.

In this paragraph we consider simplified groundwater flow by making the following assumptions:

a) Steady groundwater flow, for example, the recharge rate multiplied by the total recharge area, is equal to the discharge of groundwater to surface water.
b) Relative thickness changes of the saturated zone are small compared to the aquifer thickness.
c) The thickness of the aquifer and its porosity are constant.
d) Groundwater flow is horizontal, with uniform velocity over depth.

With these assumptions we derive the residence time of solutes in a groundwater aquifer between parallel canals (Figure 5.15). The horizontal, uniform pore water velocity v_x (m d^{-1}) at distance x (m) from the groundwater divide follows from mass conservation:

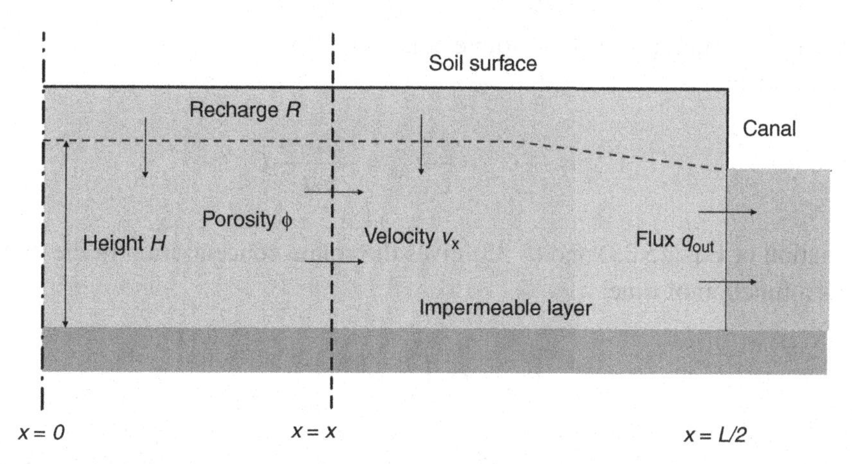

Figure 5.15 Schematization with symbols of stationary flow to parallel canals. The groundwater divide occurs at $z = 0$.

$$v_x = \frac{Rx}{\phi H} \tag{5.31}$$

where R is the groundwater recharge (m d^{-1}), ϕ is the porosity (-) and H is the aquifer thickness (m). By integration we may derive the residence time T_{res} (d), the time needed to flow from $x = x$ to the canal at $x = L/2$:

$$T_{res} = \int_{x=x}^{L/2} \frac{dx}{v_x} = \frac{\phi H}{R} \int_{x=x}^{L/2} \frac{dx}{x} = \frac{\phi H}{R} \left[\ln x\right]_{x=x}^{x=L/2} = \frac{\phi H}{R} \ln\left(\frac{L}{2x}\right) \tag{5.32}$$

Thus the entrance place x for residence time t equals:

$$x = \tfrac{1}{2} L \, e^{-\left(\frac{Rt}{\phi H}\right)} \tag{5.33}$$

We may draw streamlines similar to Figure 5.16. The figure represents the situation at time t, where solutes in the denoted area have reached the canals. As we deal with a streamline pattern at steady-state conditions, we may write for the water fluxes and defined lengths (Figure 5.16):

$$\frac{q_{out} \ell_1}{q_{out} \ell_2} = \frac{R \ell_3}{R \ell_4} \quad \Rightarrow \quad \frac{\ell_1}{\ell_2} = \frac{\ell_3}{\ell_4} \tag{5.34}$$

where q_{out} is the water flux density at $x = L/2$. Suppose initially no solute were present in the groundwater and from $t = 0$ onwards the concentration in the recharge water amounts to C_{in} (kg m^{-3}). Compared to the convective processes in such a groundwater system, we may neglect the diffusive and dispersive transport (Duffy and Lee, 1992). Then the average solute concentration in the drainage water C_{out} (kg m^{-3}) follows from:

$$C_{out} = \frac{\ell_1}{\ell_2} C_{in} = \frac{\ell_3}{\ell_4} C_{in} = \frac{½L - x}{½L} C_{in} \qquad (5.35)$$

Combination of Eqs. (5.33) and (5.35) gives the solute concentration in the drainage water as a function of time:

$$C_{out} = C_{in}\left(1 - e^{\frac{-Rt}{\phi H}}\right) \qquad (5.36)$$

We may extend this equation for the case the initial groundwater concentration amounts C_{orig} (kg m^{-3}):

$$C_{out} = C_{in} + \left(C_{orig} - C_{in}\right) e^{\frac{-Rt}{\phi H}} \qquad (5.37)$$

Question 5.15: Extend Eq. (5.36) for the general case with an initial groundwater concentration, which should result in Eq. (5.37).

Equation (5.37) has a much broader application than the concentration fluctuations in drainage water at parallel canals or drains. Also groundwater concentrations of tube wells, and outlet concentrations of perfectly mixed lakes and reservoirs, can be described with Eq. (5.37) (Van Ommen, 1988).

Question 5.16: Consider an area of intensive agriculture with an unsaturated zone of 4 m thick and a root zone of 1 m thick. The mean water content in the unsaturated zone amounts to 0.20 cm^3 cm^{-3} and the precipitation surplus amounts to 300 mm y^{-1}.

a) How long is the mean residence time of nitrate in the unsaturated zone? The adsorption of nitrate is negligible.

The phreatic aquifer is 5 m thick, has a porosity of 0.25 cm^3 cm^{-3} and has an aquitard at the bottom. We may neglect the vertical groundwater fluxes over the aquitard. Further, the phreatic aquifer is drained with canals with a distance in between of 1 km. Owing to long-term high amounts of fertilizer and manure, the nitrate concentration in the groundwater has risen towards 50 mg L^{-1}.

b) Because of restricted application of fertilizers and manure, the concentration of nitrate in the percolation water from the root zone decreases from 50 mg L^{-1} to 10 mg L^{-1}. How high is the nitrate concentration in the drainage water flowing into the canals 5 years after this decrease in the root zone? You may neglect the nitrate decomposition in the unsaturated zone below the root zone. Take into account the residence time in the unsaturated zone.

c) How long will it take from the decrease in the root zone until the nitrate concentration in the drainage water flowing into the canals has decreased from 50 to 20 mg L^{-1}?

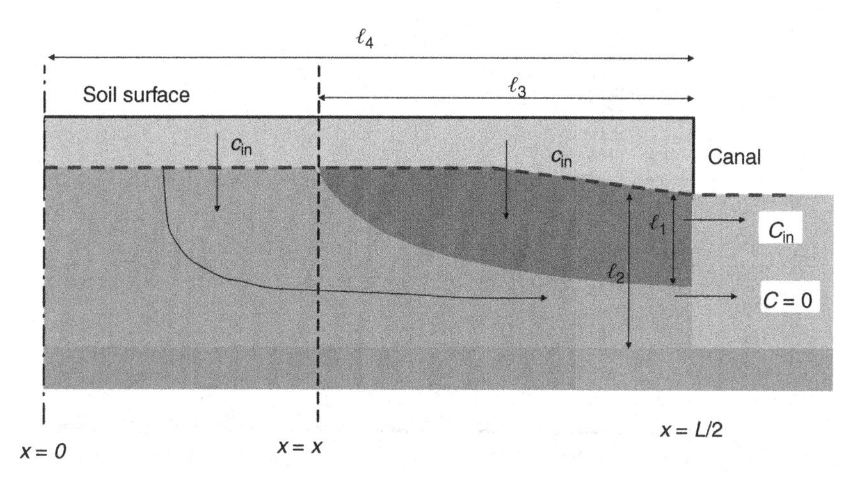

Figure 5.16 Adopted streamline pattern to determine the mean concentration of soil water flowing into the canals.

5.10 Simulation of Solute Transport

Various analytical solutions of solute transport in unsaturated and saturated soil exist, as shown in this chapter. However, most of the analytical solutions require a homogeneous soil, steady-state conditions and linear reaction coefficients. In field soils we deal with heterogeneous soils, rapidly changing conditions and nonlinear reactions. Numerical simulation models of water and solute transport as HYDRUS (Šimůnek, 1998a) and SWAP (Chapter 9) take these field conditions into account. Also, numerical models may solve the general water flow equation (Chapter 4) and solute transport equation (this chapter) simultaneously, which allows to examine all kind of interactions between water flow, solute transport and crop growth. Examples are the effect of salinity on root water uptake and crop growth and the effect of crop growth on evapotranspiration.

An extensive study on pesticide leaching in heterogeneous field soils has been performed by Groen (1997). Field data were collected at experimental, drained fields with soil textures loamy sand and cracked clay, and cultivation of tulips, potatoes and fruits. Four commonly used pesticides were applied, together with the tracer bromide. The laboratory measurements included for each soil layer the hydraulic functions, the adsorption isotherm and the decomposition as function of temperature. The field measurements for model input included rainfall, reference evapotranspiration and drainage rates as function of groundwater level. In addition field measurements were conducted for model calibration and validation. These measurements consisted of periodic water contents and bromide/pesticide concentrations in the soil profile and in the drainage water. After calibration, the SWAP model was used to formulate design and management criteria to decrease pesticide leaching. The amount of leaching turned out to depend mainly on application time, weather conditions, preferential flow and pesticide absorption and decomposition.

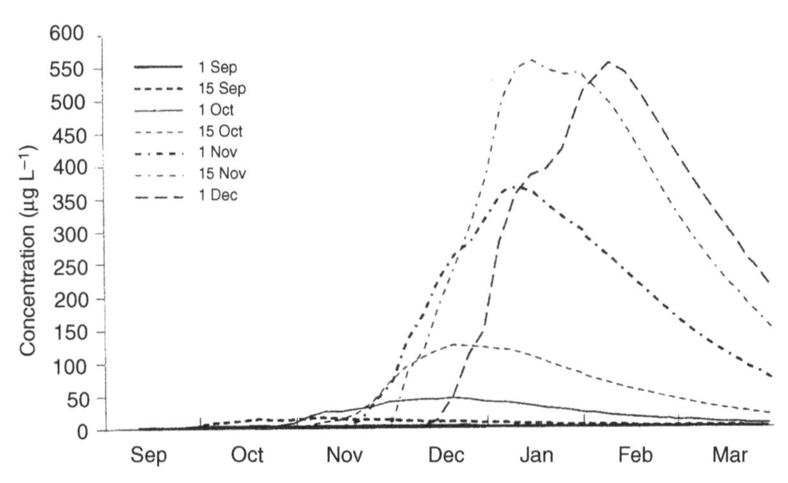

Figure 5.17 Average concentration of 1,3-dichloropropene in the drain pipes over the period 1960–1989 after application of 85 kg ha^{-1} at various dates for a loamy soil (Groen, 1997).

As an example, Figure 5.17 shows the average 1,3-dichloropropene concentrations in the drain pipes over the period 1960–1989 in case of several application dates. The highest concentrations are reached when the pesticide is applied around 15 November. Leaching at earlier applications is less due to less groundwater recharge and higher temperatures; leaching at later application is less as the pesticide reaches the groundwater during springtime when soil water percolation decreases and soil temperature rises. Application at 15 October instead of 15 November results on average in about five times lower concentrations in the drains. The scenario studies showed some effective measures to decrease pesticide leaching: (1) allow pesticide application only during summertime, (2) adopt new drainage criteria with increased drain depth and (3) increase ploughing depth (Groen, 1997).

Question 5.17: Why do you think these three methods are effective to reduce pesticide leaching on clay soils with macropores?

5.11 Summary

Transport of nutrients, pesticides and salts affects directly the environmental quality of soil- and groundwater. Convection is the main transport process in the vadose zone. Diffusion and dispersion smooth out solute concentration differences. Adsorption and decomposition play a key role in the minimization of solute leaching to groundwater. The general convection–dispersion equation is derived and applied to various compounds. Analytical solutions for solute pulses and breakthrough curves are discussed. Salinization of irrigated soils allows the application of steady-state formulations. Special attention is paid to the representative concentration for salinity stress in case of irrigated soils. For pesticide screening, numerical analysis of leaching to groundwater at extreme weather conditions is an important tool. In addition to the solute residence time in the vadose zone, the solute residence time in simple groundwater systems is derived.

6

Vegetation: Transport Processes Inside and Outside of Plants

Plants serve as an intermediary between the atmosphere and the soil: they efficiently transport soil moisture into the air and at the same time ingest atmospheric CO_2 for their growth. This chapter deals with the transport of water inside plants (from the root to the stomata), the link between water uptake and dry matter production and the modification of the near-surface atmosphere by vegetation, including microclimate, dew and rainfall interception.

6.1 Functions of Water in the Plant

Water performs many essential functions within plants. As a chemical agent, water facilitates many chemical reactions, for instance in assimilation and respiration. Water is a solvent and a transporter of salts and assimilates within the plants. Water enables the regulatory system of the plant, as it carries the hormones and substances that are required for plant growth and functioning. Water confers shape and solidity to plant tissues. If the water supply is insufficient, herbaceous plants and plant organs that lack supporting tissue will lose their strength and wilt. The hydrostatic pressure in cells depends on their water content and permits cell enlargement against pressure from outside, which originates either from the tension of the surrounding tissue or from surrounding soil. The large heat capacity of water greatly dampens the daily fluctuations in temperature that a plant leaf may undergo, due to the considerable amount of energy required to raise the temperature of water. Energy is also required to convert liquid water to vapour that transpires from leaves, causing cooling due to evaporation. Without these temperature compensating effects, plants would warm up much more and eventually die from overheating. Owing to these effects, transpiration rates can be estimated from surface temperatures, obtained by infrared thermography using remote sensing from aeroplanes or satellites (Ehlers and Goss, 2003).

In addition, in quantity, water is an important constituent of plants. The composition of roots, stems and leaves of herbaceous plants is 70–95% water. In contrast, water comprises only 50% of ligneous tissues, and dormant seeds contain only

5–15% water. The daily transpiration of plants is large compared to the amount of water contained in plants. For example, consider a mid-season wheat crop with a dry weight of 4 t ha^{-1}. During a dry and windy summer day, the crop may lose 6 mm of transpiration, which has to be extracted from soil moisture. This 6 mm of water corresponds to 60 t ha^{-1}, or 15 times the dry weight of the crop! If we assume that the plant tissues consist on average of 85% water, the wheat crop contained $85/15 \times 4 = 22.67$ t water ha^{-1}. Compared with this store of water in the plant tissues, $60/22.67 = 2.6$ times more water was extracted from the soil and passed on to the atmosphere within one day. The amount of water that is transpired daily by plants is generally 1–10 times more than the water stored in them. Therefore each day plants should extract about the same amount of water from the soil as they lose by transpiration to the atmosphere. Compared with the amount needed for cell division and cell enlargement, the transpiration amount is 10–100 times more, and compared to the needs for photosynthesis it is even 100–1000 times greater (Ehlers and Goss, 2003)!

6.2 Root Water Uptake

6.2.1 Functions of Roots

Roots grow into the soil, anchoring the plant and nourishing the growing shoot by providing water and mineral nutrients. Water is transported from soil to root by mass flow, driven by a difference in hydraulic head between the root surface and the surrounding soil. The hydraulic head gradient generates the convective flow of water towards the root surface. Those plant nutrients that are dissolved in the soil solution are inevitably drawn to the root with the convective flow of water. Roots may not take up some solutes as quickly as they arrive with the water. Consequently the soil in the vicinity of the roots will become enriched with nutrients such as calcium and magnesium (Jungk, 2002).

Besides mass flow, nutrients will be transported to the root surface by the process of diffusion. Transport by diffusion is triggered by nutrient uptake into the plant itself, which lowers the nutrient concentrations at or near the root surface. Diffusion is the more important transport process for nutrients such as phosphorus and potassium, which are present at only small concentrations in the soil solutions and for which the convective flow is insufficient (Jungk and Claassen, 1989).

Roots are not only effective in removing those nutrients from soil solution that are in an available form, but it is significant that they are also able to secrete protons, organic acids and chelating agents. Using these materials they can modify the availability of certain nutrients for themselves. These root exudates stimulate the release of ions from soil minerals, and therefore the bioavailability of macronutrients, such as phosphorus, and some micronutrients that normally are only sparingly soluble (Jungk, 2002).

Most of the material secreted by plant roots is in the form of mucilage, much of it originating in the root cap cells. Root mucilage provides an attractive environment for microorganisms that use components as a substrate for their growth, and in doing so will release other mucilaginous material. The mixture of plant and microbial mucilages is called mucigel. Mucigel has been shown to link clay particles, thereby increasing the cohesion of soil materials and leading to the formation of microaggregates.

The population density of microbes in the *rhizosphere*, the soil that is directly influenced by root activity, may be 10–200 times greater than that in the bulk soil. This increased concentration of soil microorganisms can be either beneficial or detrimental to the plant depending on the species that dominate. Microbial activity may increase availability of nutrients by mineralization of organic forms or by increasing the solubility of mineral forms. For instance, there are phosphorus-solubilizing fungi (*Penicillium balaji*), some of which have been exploited commercially (Ehlers and Goss, 2003).

The most important microbes for phosphorus uptake are mycorrhiza fungi that form associations with plant roots. The majority of plants establish such an association with certain types of soil fungi; this association is known as a *mycorrhiza*. Micorrhizas are generally mutualistic. Carbohydrate is passed from the plant to the fungus, and in return the fungus facilitates increased nutrient uptake, particularly of phosphorus, from the soil to the plant. Mycorrhizal fungi can also increase the availability of zinc to the root.

Interest in these symbioses has increased dramatically in recent years because of their potential benefit in agriculture, forestry and re-vegetation of damaged ecosystems. Some plants cannot become established or grow normally without an appropriate fungal partner. Even when plants can survive without mycorrhizas, those with 'fungus-roots' grow better on infertile soils and areas needing re-vegetation. The mechanism by which this occurs is a combination of increased surface area for adsorption from the soil solution and inward translocation of phosphorus from beyond zones of depletion around the roots. Without micorrhizas, the depletion zone of phosphorus around a root is ca. 1 mm, but micorrhizas can extend for 100 mm, thereby greatly increasing the zone from which phosphorus is adsorbed. In addition to the larger volume of soil explored, the kinetics of nutrient uptake may also be enhanced (Ehlers and Goss, 2003).

Micorrhizal fungi have also been shown to transport nitrogenous compounds between plants, but the most effective microbes for providing higher plants with nitrogen are rhizobia. These form symbioses with leguminous plants. The rhizobia enter the host plant through the root hair and travel into the cortex via an infection thread. The bacterium undergoes a transformation to produce many bacteriodes, and the growth and division of infected root cells leads to the formation of root nodules. For example, under ideal conditions, the symbiosis between soybean and rhizobia can fix about 500 kg N ha^{-1} y^{-1}.

6.2.2 Structure of the Root Tip

Root tips, the growing ends of individual root branches, have a striking appearance. They look fresh and white, rather like the colour of blanched asparagus. At some distance from the tip, the colour of the branch changes to light brown. This is the

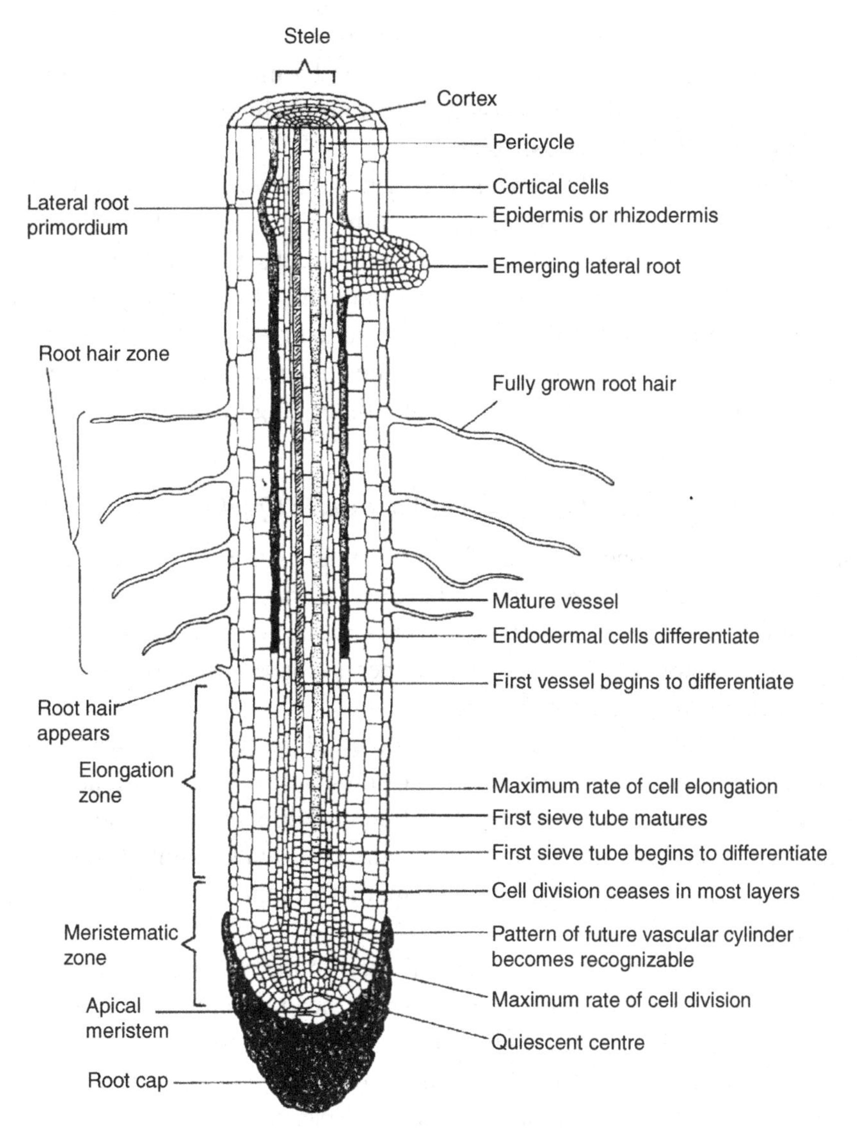

Figure 6.1 Longitudinal section of a root tip (Ehlers and Goss, 2003). Reproduced with permission of CAB International, Wallingford, UK.

older part of the branched root system, where the root epidermis has been formed. When the root tissue starts to decay, the colour turns to dark brown.

Depending on the species, the root tip may extend to several centimetres in length. The tip can be subdivided into four sections: the root cap, the meristematic zone, the zone of elongation and the root hair zone (Figure 6.1). The starting point of growth is the *meristematic zone*, which produces new cells by division. The cells at the bottom form the *root cap*. The cap is the protection shield of the growth tissue. It excretes a mucilage, lowering considerably the friction between the root cap and the surrounding soil particles. The slimy material protects the tip from desiccation. It supports the colonization of the immediate surroundings of the tip by microorganisms, forming the rhizosphere. The cells of the cap sense gravity, thus enabling the root to grow vertically downward (Ehlers and Goss, 2003).

The meristematic cells at the upper side generate embryonic cells backwards. Thus the root cap is being moved forwards through the soil as new cells are formed and then expand. It is in the zone of elongation that these cells expand, particularly growing in length, allowing the root system grow to depth. Thereafter the enlarged cells turn into the maturation zone, also named the root hair zone. Here the cells become specialized into tissues of different function. The epidermis and beneath it the cortex are formed, and inside is the stele, the vascular tissue (Figure 6.3).

6.2.3 Physiology of Root Water Uptake

Within root cells, the concentration of many solutes is greater than in the solution outside the root. Hence these nutrients have to be transported against an existing concentration gradient. Also in some cases ions are selectively excluded from cells. Therefore nutrient uptake is an active and energy-consuming process. The energy required is generated by cell metabolism.

On the other hand, the uptake of water by roots and the conduction within the plant does not rely on energy consumption. Water flows from sites with higher hydraulic head to sites with lower hydraulic head. Differences in hydraulic head cause water flow within the plant, from the root surface to the xylem in the central stele, from the stele to the various organs of the plant and finally from the leaves to the atmosphere (Ehlers and Goss, 2003).

The hydraulic head declines from the bulk soil to the root surface and drives the water towards the root. The higher the hydraulic head difference, the higher is the water uptake rate. Hainsworth and Aylmore (1986) measured for the first time the water content distribution around a single root by computer-aided tomography (Figure 6.2). The lower water contents measured near the root surface correspond to a lower hydraulic head.

Without hairs, roots may lose hydraulic contact with the soil. Either the root shrinks when the plant experiences water shortage or the soil contracts and separates from the root due to drying, a feature particularly associated with clayey soils. But also without any shrinkage of root or soil, the hydraulic contact will decrease, as the soil dries owing to root uptake and the wetted contact area between soil and root diminishes. Therefore root hairs are important for bridging the gap. In the language of hydraulics: root hairs lower the hydraulic resistance.

Water can enter the root interior through two pathways: via the root hairs of the rhizodermis cells (cellular pathway) or by entering into passage cells of the exodermis (apoplast pathway). Water arriving at the innermost concentric cellular layer of the

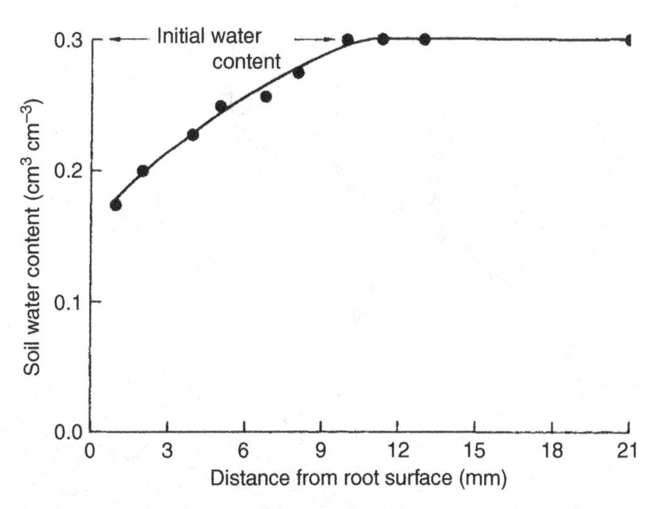

Figure 6.2 Water content distribution around a single root of radish after 10 hours of water uptake (Hainsworth and Aylmore, 1986).

cortex always has to enter the cell protoplasm. Here at the endodermis flow between neighbouring cell wands or intercellular air spaces is obstructed by suberization of the lateral cell walls, the barrier named *Casparian strip* (Figure 6.3). At the endodermis layer the plant has its last chance to modify the composition of solutes coming in with water before they are transported from the root. Nutrients in abundance and nonessential or even toxic ions can be excluded from passing onwards to the vascular system (Ehlers and Goss, 2003; Kirkham, 2005).

The hydraulic resistance of the cortex depends largely on the permeability of the cell membranes. And the permeability depends to a large extent on cell respiration. Respiration again relies upon temperature and oxygen supply. These interrelations indicate a phenomenon that plants may wilt in poorly drained or waterlogged soils. The plants show signs of water deficiency, while large amounts of soil moisture occur in the root zone!

For maintenance of root tissue permeability, most of crop plants need an oxygen supply to the roots via the pore system of the soil and an adequate soil temperature. In the soil, oxygen is transported from the soil surface to the oxygen respiring roots by diffusion (Cook, 1995; Bartholomeus et al., 2008). As long as some soil pores are drained, oxygen diffusion through the aerated pore system may satisfy oxygen requirements. But in soil with excessive water, oxygen diffusion is greatly impaired, as the diffusion coefficient of oxygen in water is 10^4 times smaller than in air. As a result, the oxygen supply rate to the roots is very much reduced. Table 6.1 lists the minimum volumetric air contents required in the top soil for potential root water uptake of different crops in a humid climate. It shows that bulb and beet crops require higher air contents than grain crops and grasses. In clay soil, a lower air content is

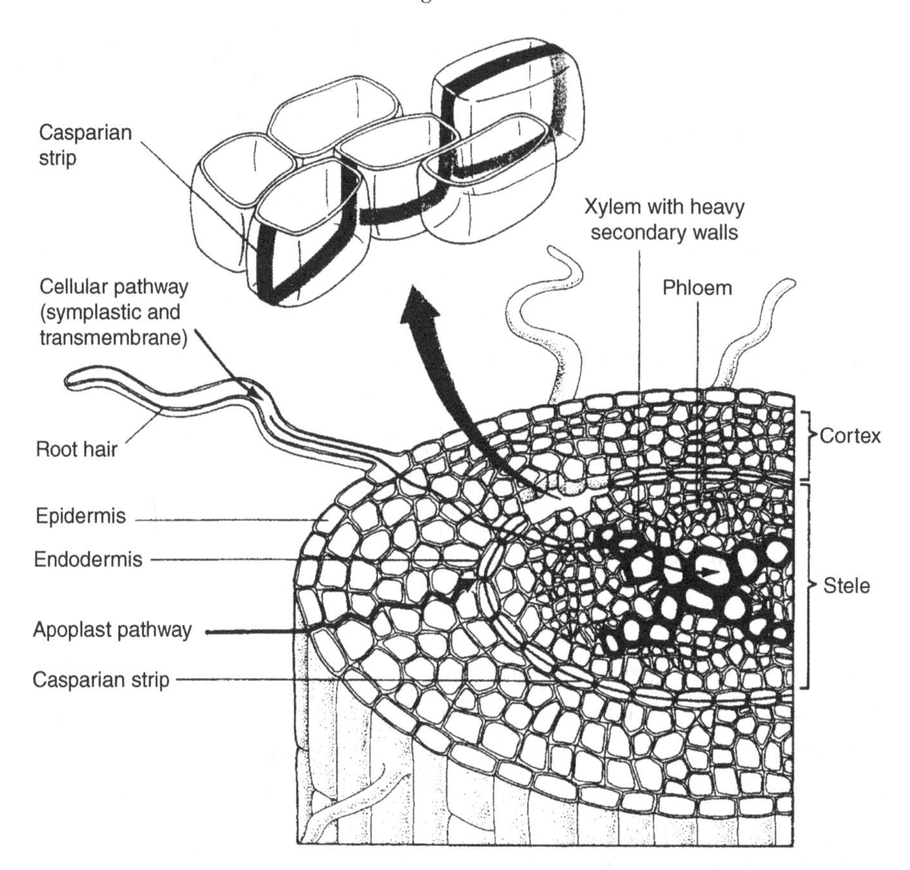

Figure 6.3 Possible pathways of water conduction in the root tip (Ehlers and Goss, 2003). Reproduced with permission of CAB International, Wallingford, UK.

allowed compared to loam and sand, as the macro pores in unsaturated clay are able to transport large amounts of air.

Plants living under submerged conditions have adapted to the limited oxygen supply from the soil system. In these plants, parenchymatic cells in the shoot and root tissue are only sparsely packed, leaving large air-filled spaces in between. This special tissue is called aerenchyma and serves the internal oxygen transport to the roots by diffusion. For instance, rice and reed are plants with effective aerating tissues. Wheat is better adapted to conditions of waterlogging than barley, as wheat can develop more porous root tissue in the event of flooding (Ehlers and Goss, 2003).

6.2.4 Modelling of Root Water Uptake

Root water uptake will be affected by atmospheric (potential transpiration rate), plant (root density, wilting point, radial and axial root resistances) and soil (retention function, hydraulic conductivity function) properties. Microscopic models intend to

Table 6.1 Minimum air content in the top 25 cm of different soil textures needed to have sufficient aeration for potential root water uptake in a humid climate

	Minimum volumetric air content ($m^3\ m^{-3}$) in top 25 cm		
Crop	Sand	Loam	Clay
Bulb crops	0.20	0.16	0.10
Beet crops	0.16	0.12	0.08
Grain crops	0.12	0.08	0.06
Grass	0.08	0.06	0.04

account for all these physical factors simultaneously. In macroscopic models fewer factors are considered, which makes them easier to apply but requires calibration of semi-empirical input parameters.

Microscopic Models

Microscopic models describe the convergent radial flow of soil water toward and into a representative individual root. The roots are considered as a line or narrow-tube sink uniform along its length. The root system as a whole can then be described as a set of such individual roots, assumed to be regularly spaced in the soil at definable distances that may vary within the soil profile (De Willigen et al., 2000). In such a geometry, water extraction at the roots generates a radial flow pattern. The water balance of a radial flow pattern results in the following continuity equation (Appendix D):

$$\frac{\partial \theta}{\partial t} = -\frac{q}{r} - \frac{\partial q}{\partial r} \tag{6.1}$$

where θ is soil water content ($m^3\ m^{-3}$), t is time (d), q is soil water flux density ($m\ d^{-1}$) and r is radial distance from the centre of the root (m). The soil water flux itself can be described by the Darcy equation in which the gravity component can be omitted (Chapter 4):

$$q = -k\frac{\partial h}{\partial r} \tag{6.2}$$

where k is hydraulic conductivity and h is soil water pressure head (cm).

When we solve Eq. (6.2) using realistic figures for extraction rates and soil hydraulic properties, we get a very strong gradient $\partial h/\partial r$ near the root surface (Figure 6.4). This is caused by the rapid decline of the hydraulic conductivity at lower h values and by the increasing flux density due to converging flow lines. To solve Eq. (6.1) numerically, researchers had to use gross simplifications (Gardner, 1960; Herkelrath et al., 1977; Feddes and Raats, 2004). However, Eq. (6.1) can be solved more

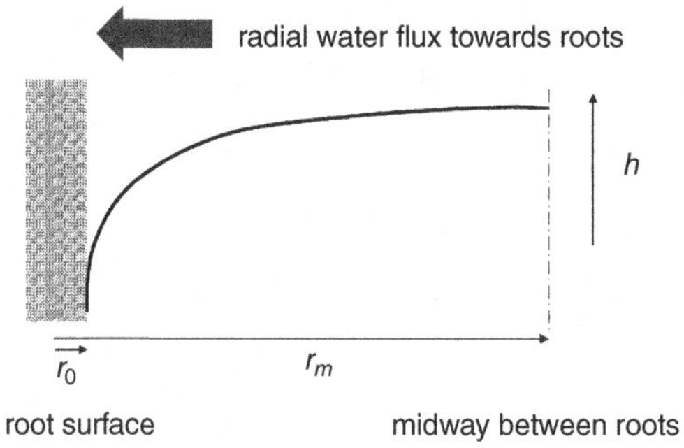

Figure 6.4 Strong gradients $\partial h/\partial r$ near root surface due to radial flow and low hydraulic conductivity.

accurately when we use the matric flux potential, instead of the soil water pressure head, as driving variable. The matric flux potential M ($m^2\ d^{-1}$) is defined as:

$$M = \int_{h_w}^{h} k(h)\mathrm{d}h \tag{6.3}$$

where h_w is the pressure head corresponding to plant wilting point. Inserting the matrix flux potential in the Darcy equation gives:

$$q = -k\frac{\partial h}{\partial r} = -\frac{\partial M}{\partial r} \tag{6.4}$$

When we numerically solve Eq. (6.1), resulting $M(r)$ profiles are more linear than $h(r)$ profiles, which is obvious from the linear character of Eq. (6.4). Use of the matric flux potential allows us to derive an analytical solution for microscopic root water extraction, and we might even upscale this approach to entire root zones (De Jong van Lier et al., 2008). The general theoretical background is explained below.

When the soil is relatively wet, root water extraction is not limited by soil hydraulic resistances and is equal to the potential transpiration rate T_p (Section 8.1). In this so-called *constant rate phase*, mass conservation yields:

$$\frac{\partial \theta_a}{\partial t} = -\frac{T_p}{D_r} \tag{6.5}$$

where θ_a is the average soil water content in the rhizosphere ($m^3\ m^{-3}$) and D_r is the thickness of the root zone (m). Numerical solution of radial soil water flow to roots showed that the change of water content with time $\mathrm{d}\theta/\mathrm{d}t$ is more or less independent of r. Therefore Eq. (6.5) can be generalized for any r:

$$\frac{\partial \theta}{\partial t} = -\frac{T_p}{D_r} \tag{6.6}$$

Equation (6.6) can be combined with Eqs. (6.1) to (6.3) to yield the following second-order differential equation:

$$-\frac{T_p}{D_r} = -\frac{q}{r} - \frac{\partial q}{\partial r} = \frac{\partial M}{r\partial r} + \frac{\partial^2 M}{\partial r^2} \tag{6.7}$$

Equation (6.7) can be solved for the governing boundary conditions (Appendix D):

$$M - M_0 = \frac{T_p}{2D_r}\left[\frac{r_0^2 - r^2}{2} + \left(r_m^2 + r_0^2\right)\ln\frac{r}{r_0}\right] \tag{6.8}$$

where r_0 is the root radius, and r_m is equal to the half mean distance between roots (Figure 6.4). The latter is related to the root length density R (m m^{-3}) by:

$$R = \frac{1}{\pi r_m^2} \quad \text{or} \quad r_m = \sqrt{\frac{1}{\pi R}} \tag{6.9}$$

Metselaar and de Jong van Lier (2007) showed by numerical analysis that $M(r)$ under limiting soil hydraulic conditions (or falling rate phase) has the same shape as under nonlimiting conditions and may be described with an expression equivalent to Eq. (6.8), with T_p replaced by the actual transpiration rate T_a and M_0 equal to the matric flux potential at permanent wilting point, which by definition is equal to zero (Eq. (6.3)). Therefore, in the falling rate phase,

$$M = \frac{T_a}{2D_r}\left[\frac{r_0^2 - r^2}{2} + \left(r_m^2 + r_0^2\right)\ln\frac{r}{r_0}\right] \tag{6.10}$$

To account for water uptake per soil layer, we apply an equation similar to Eq. (6.8), substituting the T_p/D_r term by the root water uptake per unit of soil volume at depth z, S_z (m^3 m^{-3} d^{-1}):

$$M_z - M_{0,z} = \frac{S_z}{2}\left[\frac{r_{0,z}^2 - r^2}{2} + \left(r_{m,z}^2 + r_{0,z}^2\right)\ln\frac{r}{r_{0,z}}\right] \tag{6.11}$$

in which the index z refers to layer-dependent parameters.

At a radial distance from the root surface r_a (m), the water content will be equal to the mean (bulk) soil water content in the rhizosphere, and M_a is the corresponding matric flux potential. Therefore a coefficient a_z can be defined as:

$$a_z = \frac{r_{a,z}}{r_{m,z}} \tag{6.12}$$

Numerical analysis showed that a_z has a relatively constant value of 0.53. Substituting M_a and r_a into Eq. (6.11), and incorporating Eq. (6.12) yields:

$$M_{a,z} - M_{0,z} = \frac{S_z}{2}\left[\frac{r_{0,z}^2 - a_z^2\, r_{m,z}^2}{2} + \left(r_{m,z}^2 + r_{0,z}^2\right)\ln\frac{a_z\, r_{m,z}}{r_{0,z}}\right] \tag{6.13}$$

Rewriting Eq. (6.13) results in a general root water extraction formulation that is valid for both the constant and falling rate phase and that can be applied at any depth in the root zone:

$$S_z = \frac{4\left(M_{a,z} - M_{0,z}\right)}{r_{0,z}^2 - a_z^2\, r_{m,z}^2 + 2\left(r_{0,z}^2 + r_{m,z}^2\right)\ln\frac{a_z\, r_{m,z}}{r_{0,z}}} \tag{6.14}$$

Integration of Eq. (6.14) over the root zone yields the total actual transpiration. The input data for this methodology consists of potential transpiration rate, plant wilting point, root length density profile and the soil hydraulic functions (retention function and conductivity function). The approach may include layers with different soil hydraulic properties and root densities. Most soil water will be extracted at locations with high root density and soil water pressure head. When at certain locations in the root zone soil water extraction is limited, other locations will automatically extract more soil water.

Question 6.1: The discussed microscopic approach can be used to quantify the effect of atmosphere, plant and soil on root water uptake. Which input parameters in Eq. (6.14) relate to atmosphere, which to plant and which to soil?

De Jong van Lier et al. (2006, 2008) applied the methodology to various soil types and atmospheric conditions. Figure 6.5 depicts results for a clay soil during a drying period with three root densities and two transpiration rates. In all cases both the constant and falling rate phases are clearly visible. In the case of higher transpiration rates and lower root densities, the falling rate phase starts earlier. The pressure head at the root surface, h_{root}, shows a diurnal fluctuation (lower values during day time), especially in case of low root densities.

In the above approach no hydraulic resistances inside the roots are considered. Noordwijk et al. (2000) and Heinen (2001) followed a similar approach in the soil, but included the radial hydraulic resistance within the roots. Such an approach no longer can be solved directly, but requires numerical iteration. Even more detailed, Javaux et al. (2008) and Schröder et al. (2009) use a three-dimensional numerical model in which all the hydraulic resistances in the root and soil system are made explicit. Their research model can be used to explore the complex feedback mechanisms between

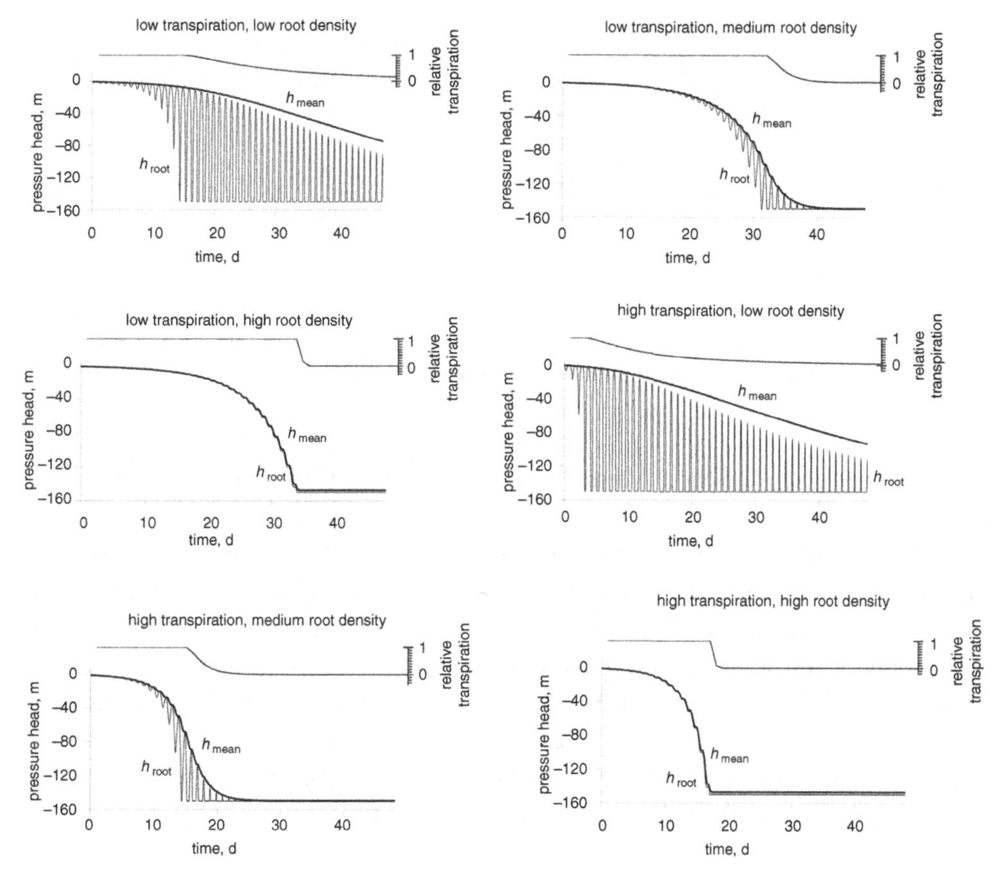

Figure 6.5 Simulated pressure head at root surface (h_{root}), mean pressure head (h_{mean}) and relative transpiration as a function of time for low and high potential transpiration rates and low, medium and high root density in a clay soil (De Jong van Lier et al., 2006).

plant and soil that occur during heterogeneous dry conditions in the root zone, and to verify simplifying assumptions in operational ecohydrological models.

Macroscopic Models

In the macroscopic approach, the entire root system is viewed as a diffuse sink that penetrates each soil layer uniformly, though not necessarily with a constant strength throughout the root zone. This approach disregards the flow patterns toward individual roots and thus avoids the geometric complications involved in analyzing the distribution of fluxes and potential gradients on a microscale. The major shortcoming of the macroscopic approach is that it is based on gross spatial averaging of the pressure and osmotic heads. The approach does not consider the decrease in pressure head and increase in concentration of salts at the immediate periphery of absorbing roots. However, the macroscopic approach has been very useful to model root water

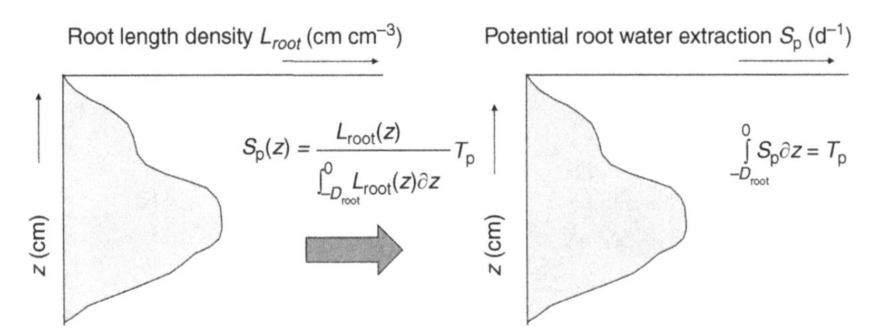

Figure 6.6 The distribution of the potential root water extraction rate, according to Eq. (6.15), is similar as the distribution of the root length density.

uptake in the past 30 years; we therefore elaborate on this approach (Feddes and Raats, 2004).

In the case of optimal soil water conditions, the plant transpiration T_p entirely depends on weather and plant characteristics. At daily intervals we may neglect water storage differences in the plant; thus the amount of transpiration should be equal to the amount of soil water extracted in the root zone. How is the extraction rate distributed over the rooting depth? The most common approach is to make the extraction rate proportional to the root length density, $L_{root}(z)$ (cm cm^{-3}). Keeping in mind that the integrated amount of extraction should be equal to T_p, we may derive the potential root water extraction rate at a certain depth, $S_p(z)$ (d^{-1}):

$$S_p(z) = \frac{L_{root}(z)}{\int_{-D_r}^{0} L_{root}(z)\partial z} T_p \tag{6.15}$$

With Eq. (6.15) the distribution of S_p with soil depth is the same as the distribution of L_{root}, which is illustrated in Figure 6.6. For practical reasons and because of feedback mechanisms in the root zone, many studies assume a homogeneous distribution of root length density with depth. In that case S_p simply becomes:

$$S_p(z) = \frac{T_p}{D_r} \tag{6.16}$$

Question 6.2: Consider a crop with a rooting depth of 80 cm and a potential transpiration rate T_p = 8 mm d^{-1}. The root density declines linearly with depth. Which potential root water extraction rate S_p do you expect at soil surface (z = 0)? And which S_p at z = –30 cm?

So far we considered root water uptake under optimal soil water conditions. Under nonoptimal conditions, that is, either too dry or too wet, S_p is reduced. As discussed

Table 6.2 Critical pressure head values $h_1 - h_4$ of the reduction factor for root water uptake α_{rw} (Figure 6.6) for some main crops (Wesseling, 1991)

Crop	h_1	h_2	$h_{3,\,high}$	$h_{3,\,low}$	h_4
Potatoes	–10	–25	–320	–600	–16 000
Sugar beet	–10	–25	–320	–600	–16 000
Wheat	0	–1	–500	–900	–16 000
Pasture	–10	–25	–200	–800	–8000
Corn	–15	–30	–325	–600	–8000

The value of $h_{3.high}$ applies to a high transpiration rate; the value of $h_{3.low}$ to a low transpiration rate.

at the microscopic approach, the root water uptake rate is limited by the product of hydraulic conductivity and gradient of soil water pressure head (Darcy's law). Most macroscopic root water uptake models simplify the reduction of root water uptake, by defining a dimensionless reduction factor as function of either soil water content, soil water pressure head or soil hydraulic conductivity (Feddes and Raats, 2004; Hupet et al., 2004). A much used reduction function is the one formulated by Feddes et al. (1978), which is shown in Figure 6.7. At soil water pressure heads above h_2 the root water uptake is reduced because of oxygen deficiency. Below h_3 water uptake is reduced because of too dry conditions. When the atmospheric demand for transpiration is higher, the reduction of transpiration rate will start at higher water contents. Therefore the parameter h_3 depends on T_p. For an indication of the parameter values $h_1 - h_4$, see Table 6.2. Although the parameters $h_1 - h_4$ are commonly defined with respect to crop type, they are to a certain extent also affected by soil texture.

In addition to water stress, plants may experience stress due to high salt concentrations, which cause a certain osmotic head. Thinking in head differences, it might seem logical to add the osmotic head to the pressure head in the soil, and derive the reduction factor as function of the total head. However, pressure head and osmotic head have entirely different impacts on the soil hydraulic conductivity, which plays a key role in root water uptake. In the case of a lower pressure head, the soil hydraulic conductivity will decrease. In the case of a higher osmotic head, the soil hydraulic conductivity is hardly affected. Therefore the concept based on differences in the sum of hydraulic and osmotic head can be applied only with additional parameters that account for the different impact on the hydraulic conductivity (Skaggs et al., 2006).

A pragmatic solution to derive root water uptake in the case of both water and salt stress, is by considering the effects of both stresses separately, and multiply dimensionless reduction factors (Cardon and Letey, 1992). We previously discussed the reduction factor for water stress (Figure 6.7). A simple and much used reduction factor for salt stress is depicted in Figure 6.8. Below a critical value for the soil water electrical conductivity no reduction occurs; above this threshold the reduction factor

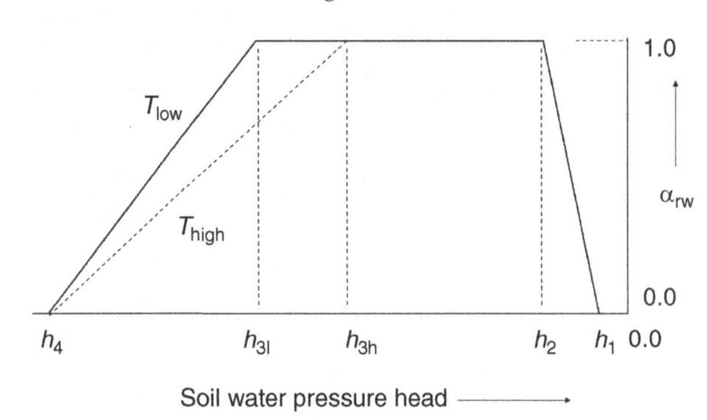

Figure 6.7 Dimensionless reduction coefficient for root water uptake, α_{rw}, as a function of soil water pressure head h and potential transpiration rate T_p (Feddes et al., 1978).

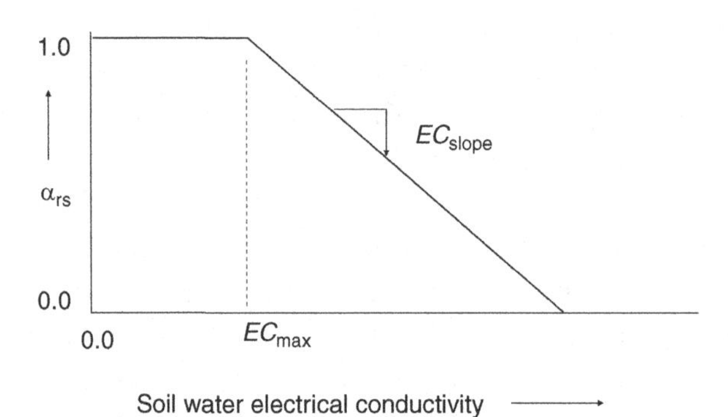

Figure 6.8 Dimensionless reduction coefficient for root water uptake, α_{rs}, as a function of soil water electrical conductivity EC (Maas and Hoffman, 1977).

declines linearly with the electrical conductivity of the soil water (Maas and Hoffman, 1977). The actual root water extraction rate $S(z)$ (d^{-1}) is then derived by multiplying the reduction factors for water (α_{rw}) and salt stress (α_{rs}) with the potential root water extraction rate:

$$S(z) = \alpha_{rw}(z)\,\alpha_{rs}(z)\,S_p(z) \tag{6.17}$$

Finally, the actual plant transpiration rate T_a (mm d^{-1}) is derived by integrating $S(z)$ over the rooting depth:

$$T_a = \int_{-D_r}^{0} S(z)\,\partial z \tag{6.18}$$

Let us take an example of an irrigated crop with combined water and salt stress. Consider a corn crop with a linear decline of the root density over a rooting depth of 80 cm. On a particular summer day potential transpiration $T_p = 8.0$ mm d^{-1}. Table 6.3

Table 6.3 Calculation of potential root water extraction (S_p), reduction factors for water (α_{rw}) and salt (α_{rs}) stress and actual transpiration ($\Sigma\, S\, \Delta z$) according to the macroscopic approach

Depth (cm)	h (cm)	EC_{sw} (dS m^{-1})	S_p (d^{-1})	α_{rw} (-)	α_{rs} (-)	S (d^{-1})	$S\, \Delta z$ (cm d^{-1})	$\Sigma\, S\, \Delta z$ (cm d^{-1})
0–20	−1200	2.0	0.0175	0.895	1.000	0.0157	0.314	0.314
20–40	−800	2.5	0.0125	0.947	1.000	0.0118	0.236	0.550
40–60	−300	5.5	0.0075	1.000	0.827	0.0062	0.124	0.674
60–80	−250	8.0	0.0025	1.000	0.655	0.0016	0.032	0.706

Details are in the text.

lists the h and EC_{sw} values, which are measured during this day at different soil depths. The question is which actual transpiration we may expect according to the described macroscopic approach. The sensitivity of corn to water and salt stress is expressed by $h_1 = -15$ cm, $h_2 = -30$ cm, $h_3 = -400$ cm, $h_4 = -8000$ cm, $EC_{max} = 3.0$ dS m^{-1}, and $EC_{slope} = 6.9\%$ per dS m^{-1}.

First calculate the potential root water extraction rate S_p, taking into account the triangular distribution of the root density with depth (Table 6.3). Next determine by linear interpolation the reduction factors α_{rw} and α_{rs} according to Figures 6.7 and 6.8. This gives the actual root water extraction rate S according to Eq. (6.17). Finally we can calculate the actual transpiration rate by integrating S over the rooting depth (Table 6.3). Note that, like in most irrigated soils, the water stress occurs in the top part of the root zone, while the salt stress occurs in the bottom part of the root zone. In this example the potential transpiration rate of 8.0 mm d^{-1} reduces to 7.1 mm d^{-1} because of water and salt stress.

Question 6.3: In the case of irrigated soils, why does the water stress often occur in the top part of the root zone and the salinity stress in the bottom part of the root zone?

6.3 Water Flow within the Plant

After passing through the endodermis of the root, the water enters the stele (Figure 6.1), where it is conducted through the cell tissue by osmosis, finally arriving at the xylem strands. The xylem is composed of a few living parenchyma cells and cells of large diameter that have lost their protoplasts. One of these types of cells is called the trachea or the vessel member. Strung together these member form a vessel (Figure 6.9). The vessels represent the 'hydraulic pipelines' for long-distance transport of water within the plant. They permit rapid conduction from the root through the stem axis to the leaf. The vessel members are joined together at their ends by open perforations or perforation plates (Figure 6.9), restricting to some degree the vertical

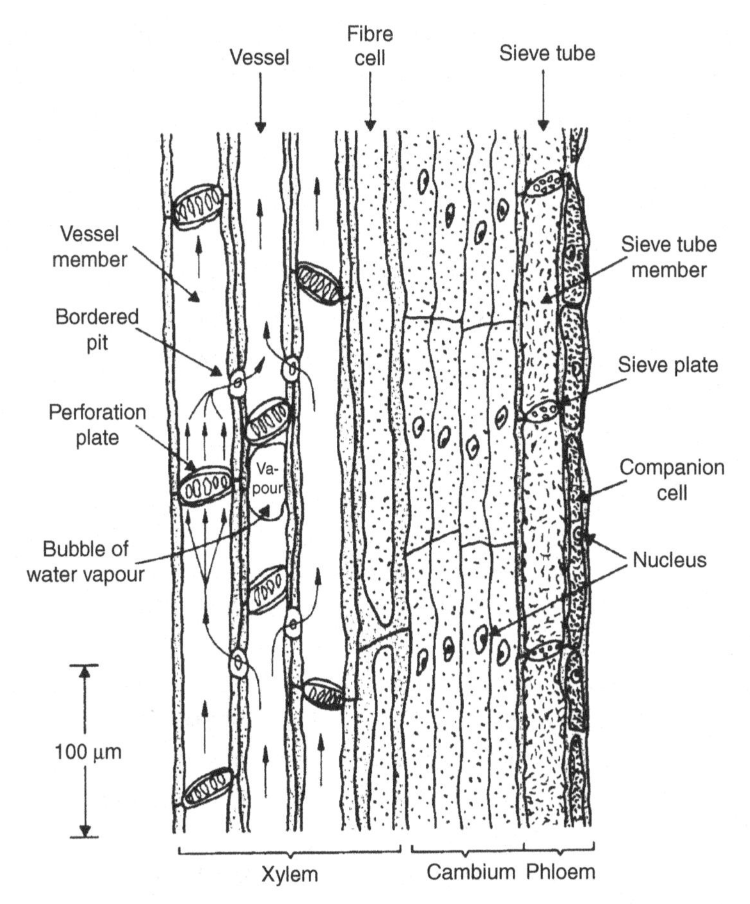

Figure 6.9 The vascular tissue of plants is composed of vessel tubes and sieve tubes. The sieve tubes serve for transport of organic compounds such as assimilates, whereas the vessel tubes convey water and minerals, as well as organic compounds, metabolized in the roots. Arrows within the vessels indicate the direction of sap flow. The pit density is in reality many times the number shown in the diagram(Ehlers and Goss, 2003). Reproduced with permission of CAB International, Wallingford, UK.

flow path by transverse constrictions. Bordered pits are inserted into the strengthened, lignified longitudinal walls of the vessels, which are passage openings in a horizontal direction (Figure 6.9). They allow the transverse transport of water into neighbouring cells, but may seal off the vessel by closing membranes, when by accident the vessel has dried out (Ehlers and Goss, 2003).

In the vessels, water is conducted by mass flow, like water through pipes, driven by a gradient in hydraulic head. The long-distance transport is carried out at high speed. The speed can be measured by heating the stem locally and measuring the temperature change at a point higher in the stem. In deciduous trees the velocity ranges from 1 to 40 m h^{-1}, and in herbaceous plants from 10 to 60 m h^{-1}. In the following paragraph we calculate the hydraulic head gradient necessary to attain these flow velocities.

The flow velocity v (m s^{-1}) in capillary tubes such as xylem vessels can be calculated with Poiseuille's law (Koorevaar et al., 1983):

$$v = -\frac{r^2}{8\eta}\frac{\partial H_p}{\partial x} \tag{6.19}$$

where r (m) is the radius of the tube, η is the dynamic viscosity (≈ 0.001 Pa s), H_p is the pressure equivalent of the hydraulic potential (Pa) and x is the flow direction. For example, take a sap flow velocity v of 10.8 m h^{-1} = 0.003 m s^{-1}. When r = 50 μm as in many trees, Eq. (6.19) yields 10 kPa m^{-1} = 0.1 bar m^{-1} for the equivalent pressure gradient. This pressure gradient is valid for an ideal tube with smooth walls. With the plant vessels one has to expect a higher flow resistance, caused by the roughness of vessel walls and presence of perforation plates. Therefore the pressure gradient has to be greater to attain v = 10.8 m h^{-1}, approximately 1.1 × 0.1 = 0.11 bar m^{-1}. Note that this gradient would allow water to flow through a horizontal pipe (Ehlers and Goss, 2003).

With plants growing upright, the gravitational head also has to be considered when calculating the gradient necessary for the suggested water flux in the vertical direction. The gradient for the gravitational head is 10 kPa m^{-1} = 0.1 bar m^{-1} of plant height. Therefore the total hydraulic head gradient in vertical trees is ca. 0.11 + 0.10 = 0.21 bar m^{-1}. For a 30 m high tree we calculate a hydraulic head difference of 6.3 bar between the soil surface and the canopy, and for a coastal sequoia from California of 100 m height or a karri tree of the same size from the south coast of Western Australia a difference of at least 21.0 bar. How can plants generate these giant hydraulic head differences?

Question 6.4: Calculate the hydraulic head loss (m) due to flow friction and due to gravity in case of a 20 m high tree. The xylem vessels have a radius r = 50 μm. The sap flow velocity v = 0.002 m s^{-1}. Increase the friction head loss according to Eq. (6.19) with a factor of 1.1 due to the roughness of vessel walls and the perforation plates.

For quite a long time it was thought that the hydraulic head difference in the vessels was caused by excess pressure in the roots. Today we know that the hydraulic head differences of up to 210 m, as calculated for the sequoia, are not caused by pressure but by pull. That is pull by a negative hydraulic head in the canopy. Meteorological factors control the relative air humidity in the intercellular spaces near the stomata. For instance at a relative air humidity of 98%, the hydraulic head will be as low as –135.7 m. The decline in hydraulic head within the soil–plant–atmosphere continuum is drawn in Figure 6.10 for four hypothetical situations. Case 1 depicts the hydraulic head decline when the soil is moist. Within the mesophyll cells of the leaf (DE), the hydraulic head stays much above the critical limit of about –200 m, below which the plant will start wilting. In case 2 the transpiration rate is greater at

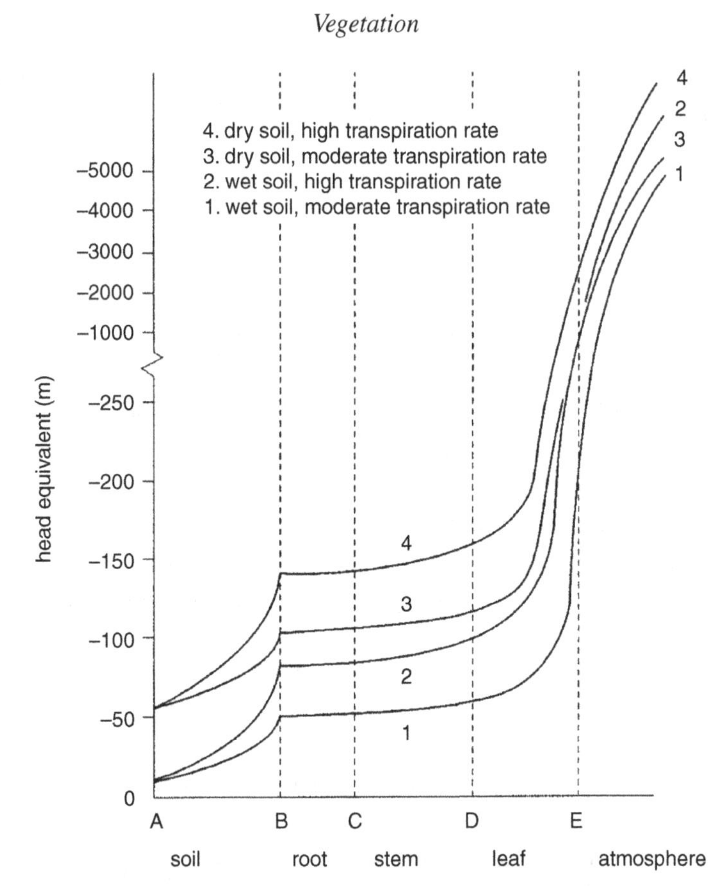

Figure 6.10 The potential decline in the soil–plant–atmosphere continuum at two soil moisture levels and slower and faster transpiration rates. (After Hillel, 1980)

an identical soil water hydraulic head, so that the critical hydraulic head within the leaf is almost attained. There is a similar situation just before wilting in case 3, when the available soil moisture content has been greatly depleted, the leaf hydraulic head is close to –200 m in some of the mesophyll cells, and the transpiration rate is also small. But finally, if the transpiration rate increases when soil moisture is in short supply, which is the situation in case 4, the leaf hydraulic head will fall below the critical level and the plant wilts.

Within the xylem sap a large negative pressure head of $-150 < h < -50$ m may exist. Because of wall thickening (Figure 6.9) the vessels will not collapse, but can withstand the pressure difference between inside and outside. Normal tissue cells with more elastic and unreinforced walls would break down more easily with the application of such pressures.

There is another problem with water transport through the plant. Air, dissolved in water, is released when the pressure head of the water becomes more negative. The gas forms bubbles, a process called *cavitation*. The bubbles will interrupt the water transport through the vessel (Figure 6.9). Water molecules in the liquid state are linked together by high intermolecular binding forces. A pulling force of more than 300 bar is necessary to separate water molecules from each other. Even in the highest trees the cohesion between water molecules should be sufficiently high to ensure intactness of the water threads within the vessels. But in reality the linkage can be broken by cavitation. The formation of 'gas seeds' breaks the water threads immediately.

Plants will protect themselves from disastrous consequences of cavitation by partial breaks of the conducting vessel tubes. These take the form of open perforations or *perforation plates* (Figure 6.9). Plants cannot avert the onset of cavitation. As soon as the water thread in a vessel is broken, the water flow will be diverted into neighbouring vessels by the bordered pits of a vessel below and above the obstructed vessel member (Figure 6.9). At the same time the expansion of the bubble from one vessel to the next is limited by the specific vessel structure. The holes at the perforations and the pits at the walls are good for liquid but not for gas transport. Because of the surface tension of water at the liquid–vapour interface, the holes will stop the air bubbles from escaping or steadily enlarging. During the night, when the transpiration rate falls to zero and the tension eases, the air bubbles can be dissolved again. Thus the damage is repaired and the diversion will be closed (Ehlers and Goss, 2003).

6.4 Transpiration, Photosynthesis and Stomatal Control

6.4.1 Transpiration

Plants take up liquid water, with nutrients dissolved in it, from the soil through their roots. The water is transported upward and most of it leaves the plant, as water vapour, whereas only a small fraction (about 1%) is used in the photosynthesis process. Losing this water is the unavoidable by-product of carbon exchange.

The water vapour leaves the plants through the stomata. These are small openings that occur mainly on the plant leaves, but to a lesser extent on other plant organs as well. Stomata are the main path way for the exchange of both CO_2 and water vapour between the plant and the atmosphere because the cuticle is rather impermeable for gases (Figure 6.11). In herbaceous plants stomata occur at both the upper and the lower side of leaves, whereas trees have stomata only at the lower side (Willmer and Fricker, 1996). The density and size of stomata varies considerably between plant species, but there is a roughly inverse relationship between stomatal density and stomate size that leads to a rather constant total area of stomata (Hetherington and Woodward, 2003). Typical stomatal densities are 200 per mm^2 (but ranging from 50 to 1000 per mm^2). The length of the guard cells, which regulate the stomatal opening, ranges from 20 to 80 μm. With a typical stomatal aperture of 6 μm, a fully open stomate has a pore area of about $2 \cdot 10^{-4}$ mm^2 (Franks and Beerling, 2009). The total area of the pores, when open, amounts to about 2–5% of the total leaf area (Willmer and Fricker, 1996).

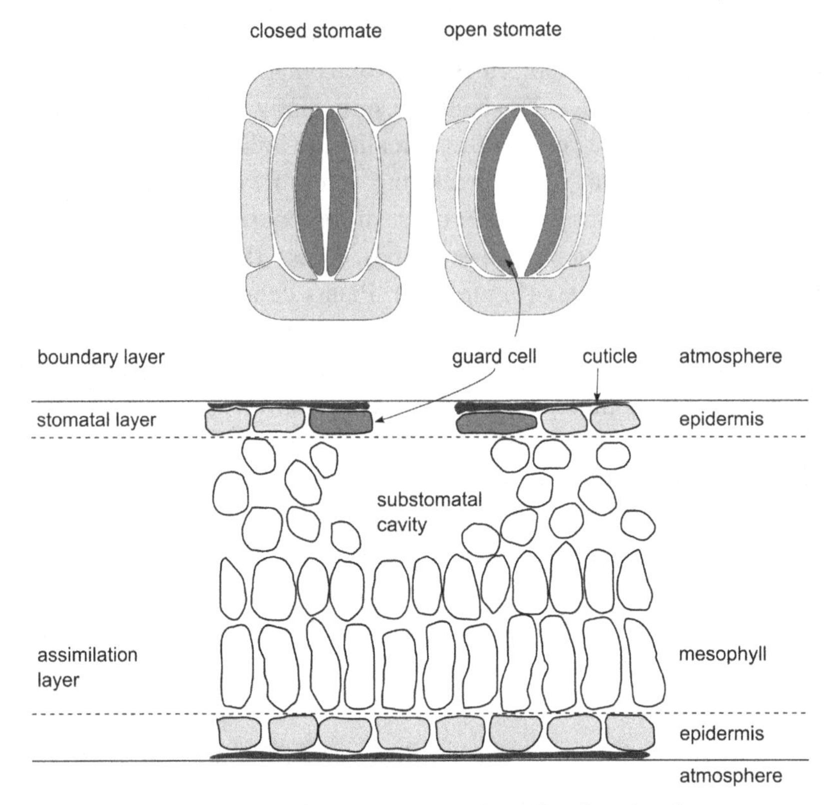

Figure 6.11 Schematic view of a stomate: top view of a closed and an open stomate (top) and side view of a stomate in a leaf (bottom). (After Konrad et al., 2008)

Water vaporization occurs within the leaf, namely in the intercellular spaces. From there the water vapour has to move within the leaf, leave the leaf through the stomate and finally has to escape from the air layer directly adjacent to the leaf, viz. the leaf boundary layer. If the vapour flux is formulated in terms of a potential difference and a resistance (compare Chapter 3), two resistances can be identified on this route: the variable stomatal resistance r_s and the boundary-layer resistance r_b. The latter is not only important as an obstacle to transport, but it also provides the link between the conditions in the air within the canopy and the conditions at the leaf surface (e.g., temperature, CO_2 concentration) as they are experienced by, and relevant for, the leaf (Goudriaan and Van Laar, 1978; Collatz et al., 1991). The boundary-layer resistance is dealt with in Section 6.6.4.

If we focus on the transport through the stomate (see Figure 6.12), the water vapour flux (transpiration) can be expressed as:

$$T = -\rho \frac{q_e - q_i}{r_s} \tag{6.20}$$

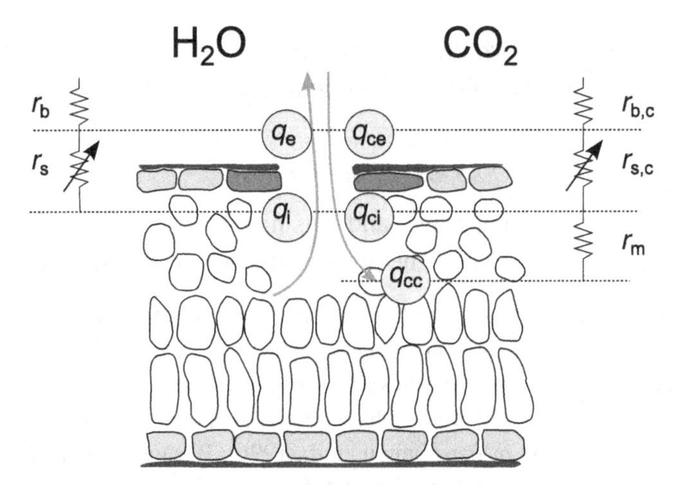

Figure 6.12 Pathways of water vapour and CO_2 out and into a leaf. Both encounter the boundary-layer resistance r_b and the variable stomatal resistance r_s. In addition, CO_2 has to pass a number of cell interfaces, reflected by the mesophyll resistance r_m. (After Willmer and Fricker, 1996)

where q_i and q_e are the specific humidity inside the substomatal cavity and just above the stomate, respectively, and r_s is the stomatal resistance (compare the aerodynamic resistances discussed in Chapter 3). Another parameter often used to express the effect of the stomata is the stomatal conductance g_s, which is simply the reciprocal of the stomatal resistance ($g_s = 1/r_s$) which gives g_s the units of a velocity. In plant physiology literature fluxes are often given as molar fluxes, rather than mass fluxes. Then g_s is used with units of mmol m^{-2} s^{-1} where at the same time the concentration is given as mole fractions.

The air inside the substomatal cavity is considered to be saturated with water vapour, and hence q_i is equal to the saturated specific humidity at the temperature of the leaf:

$$T = -\rho \frac{q_e - q_{sat}(T_s)}{r_s} \tag{6.21}$$

This implies that there is a clear link between transpiration and the temperature of the leaves. This notion is relevant in the context of the microclimate within the canopy, as the temperature may vary vertically (see Section 6.6). Furthermore, as the vegetation temperature is the outcome of the energy balance of the surface, the transpiration rate is related to the balance between radiative forcing and convective and evaporative cooling.

If the stomata are fully closed (and $r_s \rightarrow \infty$) there may still be some vapour transport through the cuticle. In that case the relevant resistance in Eq. (6.20) would be the replacement resistance for two parallel resistances: $(1/r_s + 1/r_{cut})^{-1}$. A typical value for the cuticular resistance is 4000 s m^{-1} (Leuning, 1995; Ronda et al., 2001).

6.4.2 Photosynthesis

Plants ingest CO_2 through their leaves and – using the energy from sunlight – fix this CO_2 in the form of carbohydrates. The process consists of two steps:

- The light-dependent process: formation of energy-storage molecules (ATP and NAPDH) using the energy of absorbed photosynthetically active radiation (PAR). PAR is part of the shortwave radiation, covering the wave length range of 0.4–0.7 μm. In terms of energy fluxes, PAR represents 40–50% of global radiation, depending on season, sky conditions and location (see review in Papaionnou et al., 1996). Because individual photons interact with the photosynthesis mechanism, PAR can also be expressed in terms of number of photons per unit of time and area. Typically, the ratio of PAR to global radiation is 2 μmol J^{-1} (Jacovides et al., 2007).
- The light-independent (or dark) process: the fixation of CO_2 into carbohydrates, using the energy supplied by ATP and NADPH. This process is called the Calvin cycle and is catalysed by the enzyme RuBisCO.

Photosynthesis mainly takes place in the mesophyll cells (see Figure 6.12). Three different mechanisms of photosynthesis can be distinguished:

- **C_3 carbon fixation**: Carbon fixation takes place in the mesophyll cells and the first carbohydrate produced in the fixation process is a three-carbon organic acid. This in turn is used to produce glucose. RuBisCO, which catalyzes the carbon-fixation is also able to catalyze oxygen fixation. The relative importance of carbon fixation and oxygen fixation depends on temperature: at higher temperatures the balance shifts to oxygen fixation, even to the point that more CO_2 is produced than taken up: net photorespiration. This makes C_3 plants unsuited to grow under hot conditions.
- **C_4 carbon fixation**: The CO_2 is fixated in two steps and two locations. The first carbohydrate produced – in the mesophyll cells – is a four-carbon organic acid. For this production the enzyme PEP carboxylase is used, which has a strong preference for CO_2 and also works at relatively low concentrations of CO_2. Subsequently, this product is transported to the bundle sheath cells (photosynthetic cells arranged around the veins of a leaf). In the latter cells, the reaction is reversed to produce CO_2 from the organic acid. Subsequently, this CO_2 is used in the regular Calvin cycle to produce glucose or other carbohydrates. Because in the bundle sheath cells the oxygen concentration is low, the fixation of oxygen by RuBisCO is nearly absent.
- **CAM carbon fixation.** The CO_2 is fixated in two steps at two different times. In CAM plants (crassulacean acid metabolism) the uptake of CO_2 and photosynthesis are asynchronous. During the night these plants open their stomata to ingest CO_2 that is stored in organic acids. During the day, CO_2 is extracted from the organic acids and used in a regular photosynthesis process. This arrangement, where stomata are closed when

the largest evaporative loss would occur, is beneficial for plants that are growing in arid conditions, such as succulents (Lambers et al., 2008).

The net CO_2 assimilation is the net result of the fixation of CO_2, the photorespiration (occurring when RuBisCO fixes oxygen rather than CO_2) and other respiratory processes that are required to provide energy for growth, maintenance, and transport, so-called dark-respiration (Lambers et al., 2008). As photorespiration is directly linked to the photosynthetic activity of RuBisCO, it is included in the gross assimilation rate A_g. Dark-respiration (R_d) is not completely independent of light conditions and shows a decay after the initiation of darkness (Byrd et al., 1992; Lambers et al., 2008). The net assimilation rate is then decomposed as:

$$A_n = A_g - R_d \tag{6.22}$$

The rate of CO_2 assimilation is determined by the most limiting part of the chain. For both the light-dependent and light-independent processes this may be the temperature as both processes depend on enzymes. These operate optimally within a certain temperature range only: a minimum temperature is needed to activate the enzymes whereas temperatures beyond a certain maximum will cause denaturation (changes in the three dimensional structure) (Gates, 1980). Furthermore, for the light-dependent process the amount of PAR supplied to the leaf may be limiting, whereas for the light-independent process the supply of CO_2 may hamper photosynthesis.

Figure 6.13 sketches each of these responses. At constant temperature and radiation input, the net assimilation rate initially increases with internal CO_2 concentration (within the leaf) until a plateau is reached at which the supply of CO_2 is no longer the limiting factor (Figure 6.13a). For low internal CO_2 concentrations photorespiration plus dark respiration dominate over gross photosynthesis, leading to a negative net assimilation rate. The CO_2 concentration at which the net assimilation is zero is called the CO_2 compensation point, denoted by Γ (Lambers et al., 2008). At constant CO_2 concentration and temperature an increase of absorbed PAR initially leads to an increase of the net photosynthesis rate (Figure 6.13b). But at high amounts of absorbed radiation the photosynthesis system becomes light saturated. The small negative net photosynthesis rate at zero absorbed PAR is due to respiration that is not related to the light-dependent process: dark respiration R_d. The point where $A_n = 0$ is called the light compensation point (LCP; Lambers et al., 2008) and the slope of the curve at the origin is called initial light use efficiency. In this slope the photorespiration is included: it shows the net effect of adding extra light where most of it is used to fix CO_2 and a small part is used to fix O_2, under the release of CO_2. The light response is different for leaves that have developed in full sunlight and leaves that are acclimated to shade (lower in the canopy): although the initial light use efficiency for both types of leaves is similar, sun-exposed leaves have

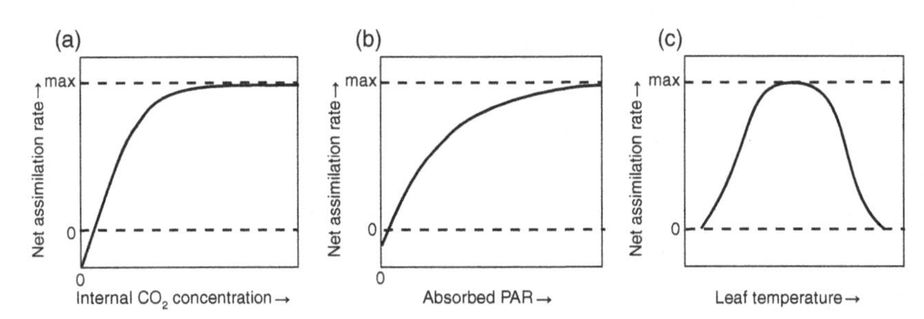

Figure 6.13 Dependence of photosynthesis rate on various environmental conditions (all other factors kept constant). (**a**) CO_2 concentration in the air space of the leaf. (**b**) The supply of photosynthetically active radiation. (**c**) Temperature. (After Gates, 1980)

higher maximum assimilation rates than shaded leaves (the plateau in Figure 6.13b is at a higher level). At the same time, sun-exposed leaves have a higher respiration rate, leading to a higher light compensation point: a downward shift of the curve in Figure 6.13b (Lambers et al., 2008). Finally, the temperature of the leaf is a limiting factor if it is either too low or too high for optimal functioning of the enzymes (Figure 6.13c). However, species may acclimate to the temperature regime of their habitat, leading to optimum temperatures that can range from below 10 °C to above 30 °C (Lambers et al., 2008).

The pathway for CO_2 is similar to that of water vapour, except for the fact that – provided that assimilation dominates over respiration – it is directed in the opposite direction. Besides, as the CO_2 is used for photosynthesis inside the cells (in contrast to water vapour that originates mainly from the intercellular space) it has to pass a number of extra interfaces, reflected by the mesophyll resistance r_m (see Figure 6.12) The r_m is sufficiently high to have a strong impact on the CO_2 uptake. Furthermore, r_m is highly dynamic, reacting to environmental factors (e.g., temperature and radiation) on timescales that are shorter than the reaction time of the stomatal resistance (seconds to minutes rather than tens of minutes; Flexas et al., 2008). The internal CO_2 concentration in the substomatal cavity is determined by the combination of the net CO_2 demand from the photosynthesizing cells (A_n) and the various resistances. It turns out that this interplay results in a q_{ci} that, at high light levels, has a rather constant relationship to the external CO_2 concentration (Zhang and Nobel, 1996).

If again we focus on the path through the stomata only, the net assimilation A_n can be written as:

$$A_n = -F_c = \rho \frac{q_{ce} - q_{ci}}{r_{s,c}} \quad (6.23)$$

where q_{ci} and q_{ce} are the specific CO_2 concentrations[1] inside the substomatal cavity and just above the stomate, respectively, and $r_{s,c}$ is the stomatal resistance for CO_2 transport. Note that we take A_n positive when the assimilation is positive. In that case the CO_2 flux F_c is negative as it is directed towards the leaf.[2]

Despite the similarity in pathways, the resistances for CO_2 transport differ from those for water vapour transport. The transport at this scale is due to molecular diffusion and the respective diffusion coefficients D_v and D_c are contained in the resistances: $r_s = l/D_v$ (where l is the diffusion path length). According to Graham's law, the molecular diffusion coefficient is inversely proportional to the molecular mass of the gas under consideration (Willmer and Fricker, 1996). Thus $D_c = \sqrt{M_v / M_c} D_v \approx \frac{1}{1.6} D_v$ and the relationship between the stomatal resistances for CO_2 and water vapour becomes $r_{s,c} = 1.6 r_s$. Consequently, Eq. (6.23) can be written as[3]:

$$
\begin{aligned}
A_n &= \rho \frac{q_{ce} - q_{ci}}{r_{s,c}} \\
&= \rho \frac{q_{ce} - q_{ci}}{1.6 r_s} = \frac{g_s}{1.6} \rho \left(q_{ce} - q_{ci} \right)
\end{aligned}
\tag{6.24}
$$

For a steady-state situation, the net CO_2 flux entering the leaf is equal to that entering the mesophyll cells. This means that, referring to Figure 6.12, the flux can also be written as:

$$
A_n = \rho \frac{q_{ci} - q_{cc}}{r_m} \tag{6.25}
$$

With the use of this expression we can look at the difference between C_3 and C_4 plants. Because C_4 plants use a more efficient enzyme in the first fixation step they can have a lower intracell CO_2 concentration q_{cc} than C_3 plants: $4 \cdot 10^{-6}$ kg kg^{-1} rather than $70 \cdot 10^{-6}$ kg kg^{-1}. At the same time the mesophyll resistance r_m is also lower in C_4 plants than in C_3 plants: 60 s m^{-1} rather than 140 s m^{-1} (Ronda et al., 2001). Thus for a given net uptake of CO_2, C_4 plants can maintain a lower internal CO_2 concentration because $q_{ci} = \frac{A_n}{\rho} r_m + q_{cc}$. A lower q_{ci} in turn means that for a given net CO_2 uptake

[1] Note that in literature CO_2 concentrations are often given as densities ($\rho_c = \rho q_c$) or as volume fractions ($f_c = \frac{M}{M_{CO_2}} q_c$, where M_{CO2} is the molar mass of CO_2). Volume fractions are usually given in parts per million by volume (ppmv). See also Appendix B.

[2] This sign convention is consistent with the notion that transport occurs generally down the gradient, in this case of CO_2.

[3] Here we ignore the fact that the diffusion of water vapour and CO_2 will interact, leading to a modification of the relationship between net assimilation and CO_2 concentration difference (see Von Caemmerer and Farquhar, 1981).

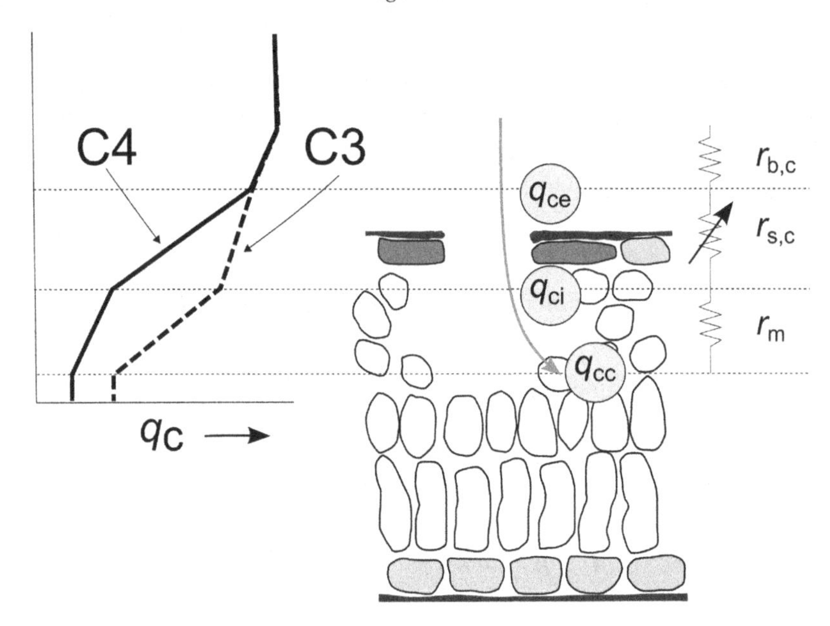

Figure 6.14 CO_2 concentration along the pathway from atmosphere to cell for C_3 and C_4 plants. Due to a lower intracell concentration (q_{cc}) and a smaller mesophyll resistance, C_4 plants maintain a lower CO_2 concentration in the stomata (q_{ci}) than C_3 plants. As a result, the stomatal resistance for a C_4 plant can be higher than that for a C_3 plant for a given assimilation rate (thus limiting water loss). Note that the slope of the q_c-line is proportional to the resistance at that location.

C_4 plants can have a higher stomatal resistance (see Figure 6.14) and hence a lower transpiration. This implies that the water use efficiency (mass of fixed CO_2 per mass of water lost) is significantly higher for most C_4 plants.

Question 6.5: Given the following data: $q_e = 9$ g kg^{-1}, $q_i = 12$ g kg^{-1} $q_{ce} = 385$ mg kg^{-1} and $q_{ci} = 300$ mg kg^{-1}, $\rho = 1.1$ kg m^{-3} and $r_s = 70$ s m^{-1}.

a) Determine the transpiration and assimilation fluxes E and A_n.
b) Determine the water use efficiency (using the definition of a plant physiologist: kg CO_2 fixed per kg water lost).
c) C_4 plants are able to maintain a lower internal CO_2 concentration (q_{ci}) than C_3 plants. Explain why this leads to higher water use efficiency (as defined earlier).
d) Explain what happens to the transpiration and the water use efficiency if the leaf temperature increases.

6.4.3 Stomatal Behaviour

In Section 6.4.1 it was shown that the exchange of both water vapour and CO_2 between plants and the atmosphere is – to an important extent – related to the stomatal resistance, and hence the stomatal aperture: the larger the aperture, the smaller the resistance. Inversely, the variation of stomatal aperture can be related to the assimilation rate and the transpiration rate.

To understand the photosynthesis-related part of stomatal response we rewrite Eq. (6.24) as an expression for the stomatal conductance:

$$g_{s,c} = g_{0,c} + \frac{1}{\rho}\frac{A_n}{q_{ce} - q_{ci}} = g_0 + \frac{1}{\rho}\frac{A_n}{q_{ce}\left(1 - \frac{q_{ci}}{q_{ce}}\right)} \tag{6.26}$$

where we used $g_s = 1/r_s$ and we added a residual conductance $g_{0,c}$, which becomes relevant at very low assimilation rates and is of the order of 0.24 mm s^{-1} (Leuning, 1995). For a given value of q_{ci}/q_{ce}, (which is rather constant, see above Eq. (6.23)), Eq. (6.26) predicts a linear variation of the conductance with net assimilation: if a larger CO_2 supply is needed to keep up with the demand of the photosynthesis system (resulting from a higher amount of radiation) the stomata are opened accordingly. Conversely, if the external CO_2 concentration increases the stomatal conductance will decrease.

The internal CO_2 concentration appears to be hardly affected by the assimilation rate (e.g., Wong et al., 1985; Leuning, 1995). Goudriaan et al. (1985) argue that the real conservative ratio is:

$$f = \frac{q_{ci} - \Gamma}{q_{ce} - \Gamma} \tag{6.27}$$

where Γ is the CO_2 compensation point (the internal CO_2 concentration at which net assimilation stops, which is of the order of q_{cc} as used in Eq. (6.25)) Γ is included to take into account situations where respiration is significant relative to net assimilation (low q_{ce} or low A_n).

The transpiration-related part of the stomatal response is such that the stomata are closed when the atmospheric demand for water vapour (expressed here as vapour pressure deficit $D = e_{sat}(T) - e$) increases. If the assimilation rate is not changed, the internal CO_2 concentration will decrease in order to maintain the necessary influx of CO_2 while stomata close. It turns out that the resulting response of the internal CO_2 concentration to the vapour pressure deficit at leaf level (denoted as $D_e = e_{sat}(T_e) - e_e$) can be expressed as (Zhang and Nobel, 1996; Jacobs et al., 1996):

$$\frac{q_{ci} - \Gamma}{q_{ce} - \Gamma} = f_{max}\left[1 - \frac{D_e}{D_0}\right] + f_{min}\frac{D_e}{D_0} = f_{max} - a_d D_e \tag{6.28}$$

where f_{max} is the maximum value of the concentration ratio (order of 0.9) at zero vapour pressure deficit and f_{min} is the minimum value that occurs when the stomata are fully closed (Ronda et al., 2001). D_0 is the vapour pressure deficit at which the stomata close. The right-hand side of (6.28) is an alternative way to denote this linear relationship by defining the slope a_d, which yields $D_0 = (f_{max} - f_{min})/a_d$. Γ (equivalent to q_{cc}) and a_d are plant-specific parameters. Usually, a distinction is made between C_3 and C_4 plants. The slope a_d is steeper for C_4 plants than for C_3 plants: C_4 plants react more strongly to dry air.

Combination of Eqs. (6.26) and (6.28) yields a model for the stomatal conductance that incorporates both the correlation to net assimilation and the response to transpiration:

$$g_{s,c} = g_{0,c} + \frac{1}{\rho} \frac{a_1 A_n}{(q_{ce} - \Gamma)\left(1 + a_2 \frac{D_e}{D_0}\right)} \tag{6.29}$$

where $a_1 = (1 - f_{max})^{-1}$ and $a_2 = (1 - f_{min})/(1 - f_{max}) - 1$ (this expression for a_2 is slightly different from the one given by Ronda et al., 2001). This shows that indeed the stomatal conductance decreases as the vapour pressure deficit increases.

Equation (6.29) can be simplified further by noting that the transpiration rate $T = 1.6 g_{s,c} \rho \left(q_{sat}(T_e) - q_e\right)$ and neglecting g_0, resulting in (Leuning, 1995):

$$g_{s,c} = \frac{a_1}{\rho\left(q_{ce} - \Gamma\right)} A_n - \frac{a_2}{\rho a_3 D_0} T \tag{6.30}$$

where we used $\left(q_{sat}(T_e) - q_e\right) = \frac{R}{R_v p} D_e = a_3 D_e$. Thus, stomatal conductance increases with an increasing assimilation rate (note that $A_n > 0$ for net uptake) and decreases with increasing transpiration. This expression can be rewritten to obtain the water use efficiency (WUE) at the leaf level:

$$WUE = \frac{A_n}{T} = \frac{1}{1.6 a_1 a_3} \left(q_{ce} - \Gamma\right) \left[\frac{1}{D_e} + \frac{a_2}{D_0}\right] \tag{6.31}$$

Clearly, a high vapour pressure deficit is detrimental for the WUE. Furthermore, an increase in the external CO_2 concentration will increase the WUE. Finally, WUE will differ between C_3 and C_4 plants through the CO_2 compensation point and the difference in sensitivity to vapour pressure deficit. The concept of WUE is sometimes employed to explain optimal stomatal responses: how do plants obtain the maximum assimilation while losing as little water as possible (e.g., Farquhar and Sharkey, 1982; Zhang and Nobel, 1996)?

The preceding analysis (in particular Eq. (6.29)) shows how the stomatal conductance is related to a number of variables related to the photosynthesis process

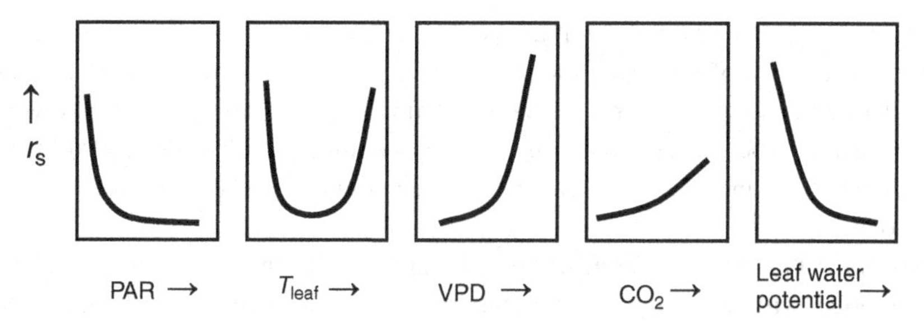

Figure 6.15 Schematic representation of the response of stomatal resistance to external and internal factors (after various graphs in Willmer and Ficker, (1996, and Oke, 1987). Note that the leaf water potential is negative, that is, larger potential means a less negative value (less water stress).

and the environment. The responses of stomata to internal and external factors[4] are summarized as follows (the responses of the stomatal resistance are given schematically in Figure 6.15):

- **Radiation**. Stomata need to be open only if photosynthesis can occur. Hence, stomata react on PAR. The stomatal reaction to light appears to be direct: the guard cells are sensitive to (even low levels of) light. CAM plants form an exception with respect to this response: they have their stomata closed during daylight.
- **Temperature.** There is a direct effect of temperature on the metabolism of the guard cells: with increasing temperature the metabolic activity increases until an optimum temperature is reached above which the activity decreases to prevent damage to the cells. The increased metabolism causes the stomata to open. An indirect effect of temperature is that temperature will increase respiration, which in turn will increase the internal CO_2 concentration in the leaf. This will cause the stomata to close. The optimum temperature (where the minimum resistance occurs) differs between plants but is of the order of 20 °C to 40 °C. At the same time, the photosynthesis process (that affects A_n in Eq. (6.29)) is also temperature dependent.
- **Atmospheric vapour deficit**. The vapour pressure deficit is an expression of the dryness of the air. If the air outside the leaves is very dry, the stomata are closed. Because a high vapour pressure deficit causes a high transpiration rate, the response to vapour pressure deficit could be interpreted as a response to high transpiration rates (Monteith, 1995). One proposed mechanism is that water evaporates from the cells in the leaves. The resulting decrease in turgor then sets up a feedback loop that reduces stomatal aperture to the extent that the turgor is restored. However, there are also indications that the guard cells or cells in their vicinity react directly to the vapour pressure deficit (Willmer and Fricker, 1996).
- **Internal CO_2 concentration.** The internal CO_2 concentration is a compromise between the need for a high concentration to supply the photosynthesis process with sufficient building material and the need for a low concentration (relative to the external

[4] Here, these factors are presented independently, but in practice it is difficult to separate them, as some factors will change simultaneously (Willmer and Fricker, 1996).

concentration) to maintain a high influx. At a given demand for CO_2 (determined by the available amount of radiation) the plant can regulate the internal CO_2 concentration by regulating the stomatal aperture. The regulation happens through the sensitivity of the guard cells to the internal CO_2 concentration (Willmer and Fricker, 1996). It turns out that the ratio of internal to external CO_2 concentration is rather conserved variable (see Eq. (6.26)).

- **External CO_2 concentration.** If the external CO_2 concentration is increased, the first-order effect will be that the CO_2 flux into the stomata is increased. If the influx becomes too large, the plant will no longer be able to use all the CO_2 for photosynthesis and the internal CO_2 concentration will increase. This in turn causes an increase of the stomatal resistance (Jarvis and Davies, 1998).
- **Leaf water potential.** When the leaf water potential decreases (becomes more negative) the stomata close to prevent further water loss. Because the leaf water potential is the net result of water uptake and transport towards the leaves and loss of water through transpiration, any process that influences one of these fluxes will have an effect on stomatal aperture (e.g., excessive evaporative demand, lack of root water uptake). Though maintaining turgor should be an important strategy for the plant, stomatal closure also seems to be modulated directly by signals from the roots (Davies and Zhang, 1991).

The responses sketched in the preceding text generally lead to a clear diurnal cycle in the stomatal resistance with low values during daylight and infinite values during the night. However, under dry and hot conditions the vapour pressure deficit may increase beyond its critical value around midday, leading to midday closure of the stomata. This prevents excessive water loss during the time of maximum temperatures and vapour pressure deficit.

When comparing the responses of the stomatal resistance to environmental factors in Figure 6.15 to the responses of the photosynthesis rates in Figure 6.13 it appears that they are related: the responses to CO_2 concentration, radiation and temperature are such that low resistances are linked to high net photosynthesis rates. This relationship – expressed also in Eq. (6.29) – is exploited in one of the models for the canopy resistance that is discussed in Section 9.2.4.

6.4.4 CO_2 Exchange at the Ecosystem Level

In Section 6.4.2 photosynthesis and respiration have been discussed on the level of individual leaves. However, the exchange of CO_2 between the atmosphere and the land surface is determined not only by the net photosynthetic activity of plants. Respiration by organisms in and on the soil is an important source of CO_2. This includes respiration by plant roots, as well as microbes and animals feeding on organic matter within and on top of the soil. The respiration by plants is called autotrophic respiration because the plants use their own carbohydrates, whereas the respiration by microbes and animals is called heterotrophic respiration.

The additional source of CO_2 implies that the total net exchange of CO_2 between an ecosystem and the atmosphere consists of the following terms:

$$\mathrm{NEE} = A_n - R_s = A_g - R_d - R_s \tag{6.32}$$

where NEE is the net ecosystem exchange, A_n is the net assimilation rate (often denoted as net primary production NPP), R_s is the soil respiration, R_d is the dark respiration and A_g is the gross assimilation (often denoted as gross primary production [GPP]). All fluxes have units of mass CO_2 per unit time per unit surface area. The NPP is comparable to the net photosynthesis discussed in Section 6.4.2, whereas NEE is the net total exchange, including soil respiration, as it would for instance be measured by an eddy-covariance system installed over that ecosystem (see Chapter 3). The NPP leads to the buildup of biomass: both above ground and below ground.

Because GPP is close to zero when there is no radiation input, and the dark respiration also decays to zero after sunset. Hence the soil respiration can be estimated from the night time NEE observations. For a given ecosystem, under given conditions of soil moisture and nutrition, the soil respiration is mainly a function of temperature. Hence, the diurnal cycle of temperature needs to be taken into account to translate night time NEE observations to day time estimates of R_s. A commonly used method is to take the respiration rate at 10 °C ($R_{s,10}$) as a reference and use an Arrhenius type of temperature dependence. Lloyd and Jackson (1994) suggest the following formulation, based on a large number of data sets:

$$R_s = R_{s,ref} e^{E_0\left(\frac{1}{T_{ref}-T_0} - \frac{1}{T-T_0}\right)} \tag{6.33}$$

where E_0 is a temperature sensitivity factor (in K), T_0 equals 227.13 K and T_{ref} is the reference temperature at which $R_{s,ref}$ is determined (e.g., 283.15 K or 10°C). Reichstein et al. (2005) show that E_0 is not a universal constant, but may be site specific. More importantly, they show that it strongly depends on the time period over which the temperature dependence is determined (if the period is too long, seasonal trends in soil conditions may erroneously be taken into account). They suggest a time window of the order of 15 days yielding values for E_0 that vary roughly between 100 and 250 K.

Because during nighttime no photosynthesis takes place, the surface is a source of CO_2. When conditions are very stable, vertical turbulent transport of CO_2 away from the surface may be absent. This will lead to a buildup of high CO_2 concentrations close to the surface. This effect can be enhanced when small-scale topography gives rise to density currents that horizontally transport CO_2 to the lowest points in the landscape (de Araújo, 2008).

Question 6.6: Given are soil respiration data from two nights: $4.0 \cdot 10^{-7}$ kg m^{-2} s^{-1} and $4.7 \cdot 10^{-7}$ kg m^{-2} s^{-1}. The soil temperatures were 15 and 20 °C during those nights.

a) Determine the temperature sensitivity E_0 for this data set (taking the data from the first night as the reference).
b) Determine the soil respiration when the soil temperature is 17 °C.

6.5 Dry Matter Production

There are two fundamental reasons for a close relation between dry matter production and water use by crop stands. First, both the processes of CO_2 assimilation and of H_2O transpiration strongly depend on radiant energy. Second, during water shortage both processes are in the same way reduced by stomatal control.

At the beginning of the 20th century agricultural scientists started to search for the relationship between water use and dry matter production. It turned out that experiments with plants growing in deep pots or containers were much easier to control than experiments in the field. In the field transpiration and evaporation could not be separated and the soil water balance was difficult to determine. In containers, however, evaporation could be prevented by covering the soil and an accurate soil water balance could be derived. As an example, Figure 6.16 shows the relationship between cumulative transpiration and accumulated dry matter for Kubanka wheat. In the experiments on which Figure 6.16 is based, different amounts of water were applied that produced the depicted different levels of biomass (Ehlers and Goss, 2003). Other examples with a proportional relation between transpiration and dry matter production are given by Kirkham (2005).

In field experiments soil evaporation is inevitably included in measured water use. Figure 6.17 shows data on water use and biomass from northeast Germany. Forage maize was grown in lysimeters with varying supplies of water, allowing determination of the evapotranspiration (Mundel, 1992). The intercept with the *x*-axis is 194 mm, and might be viewed as an estimate of the cumulative soil evaporation in the presence of plants.

We may define the *transpiration efficiency* or *water productivity*, WP_T (kg m^{-3}) as:

$$WP_T = \frac{DM_\mathrm{a}}{T_\mathrm{a}} \tag{6.34}$$

with DM_a the actual cumulative dry matter (kg m^{-2}) and T_a the cumulative actual transpiration (m). In the case of Kubanka wheat in the Great Plains (Figure 6.16) WP_T = 2.07 kg m^{-3}, while in case of forage maize in Germany (Figure 6.17) WP_T = 8.93 kg m^{-3}. There are two reasons for this striking difference of more than a factor 4: (1) the atmospheric evaporative demand, which was much higher in the experiments with Kubanka wheat; and (2) wheat is a C_3 crop and maize is a C_4 crop. At the leaf surface, stomates control the gas exchange of CO_2 and H_2O. The exchange of the two gases through the stomatal openings is a diffusion process. In the case of CO_2, the concentration in the atmosphere is fairly constant. However, in the case of H_2O, the air

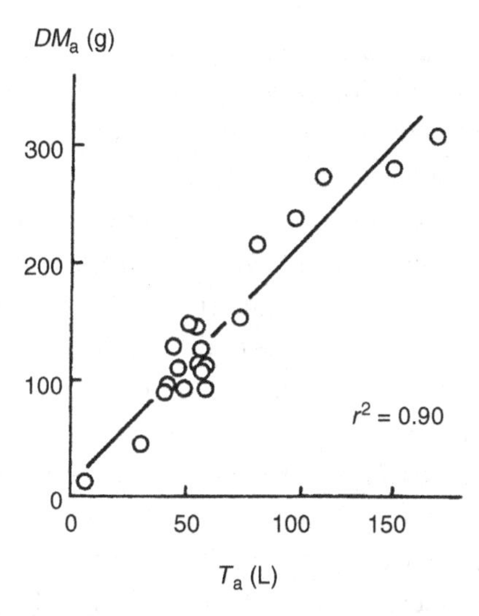

Figure 6.16 Cumulative dry matter DM_a of above-ground parts in Kubanka wheat (*Triticum durum*) grown in a single pot, as function of cumulative transpiration T_a. (Data from Briggs and Shantz, cited by de Wit, 1958.)

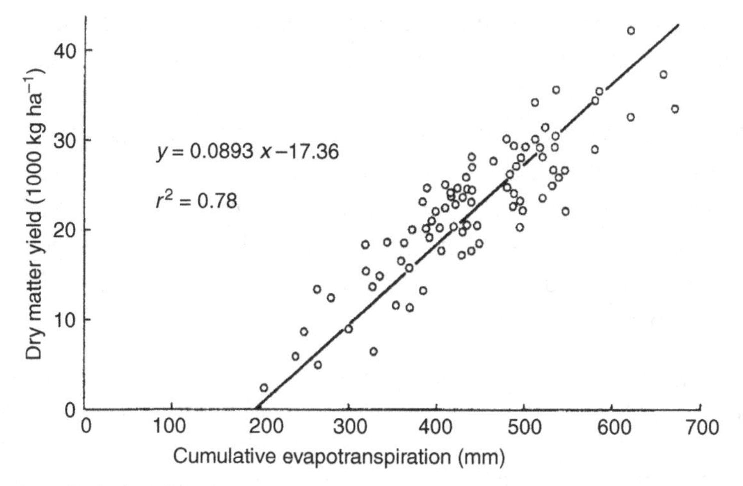

Figure 6.17 Relationship between cumulative evapotranspiration and dry matter yield of forage maize. The crop was grown in Paulinenaue, northeast Germany, on a lysimeter with a groundwater table 125 cm below the soil surface. (Based on Mundel, 1992.)

humidity varies during the course of a day and from day to day. This means that the vapour pressure deficit (VPD) of the air ($D_a = e_{sat}(T_a) - e_a$, here T_a is air temperature) is highly variable and therefore also the amount of water lost by diffusion. Taking this into account, Bierhuizen and Slayter (1965) incorporated a mean D_a (Pa) in the relationship between cumulative transpiration T_a (m) and dry matter DM_a (kg m^{-2}):

$$\mathrm{WP}_T = \frac{DM_a}{T_a} = 1000\frac{\mu}{D_a} \tag{6.35}$$

where the factor μ (Pa) is a measure for transpiration efficiency, which is more or less independent of the climatic conditions during crop growth. The factor 1000 depends on the units of T_a (m) and DM_a (kg m^{-2}). Note that the dependence of Eq. (6.35) on VPD is roughly consistent with the dependence of the leaf scale water use efficiency on VPD (Eq. (6.31)). An extensive literature review of Tanner and Sinclair (1981) confirms that Eq. (6.35) is very useful to relate water use and plant yield. Important for application of Eq. (6.35) is that the saturation deficit is calculated for the daytime, when the stomata are open, and not for the entire day. Ehlers and Goss (2003) provide μ values for a number of crops, which are listed in Table 6.4. As discussed earlier, C_4 crops are more effective in fixation of CO_2 within the leaf than C_3 crops. Therefore, at a certain light intensity, the CO_2 uptake rate and the photosynthesis of C_4 plants are much higher (Figure 6.18; compare with the sketch in Figure 6.13b). This causes also the higher water use efficiency of C_4 crops (Table 6.4).

Using the crop-specific factor μ and the average daylight saturation deficit during the growing period, we may calculate with Eq. (6.35) the amount of transpiration for an expected yield (provided no other stresses occur) or predict the yield for a given amount of water extracted by the roots. This is illustrated in Table 6.5 for locations in Germany and Colorado. When the climate becomes more arid, the amount of water for crop transpiration increases. In the case of Göttingen and Akron the transpiration amount more than doubles. That is only the water used for transpiration. Additional water will be required for evaporation and possibly for drainage. Conversely, from a fixed quantity of water, stored in the soil profile and replenished by precipitation or irrigation, only a comparatively small amount of dry matter can be attained in the dry climate with high saturation deficit of the air. Whereas in Göttingen 15 t ha^{-1} of wheat biomass can be produced from 300 mm of water, only 7 t ha^{-1} will be obtained in Akron (Table 6.5).

Question 6.7: Which water productivity (kg m^{-3}) and amount of transpiration (mm) do you expect for a wheat crop grown near Wageningen? The average saturation deficit amounts 1200 Pa, and a yield (dry matter above ground) of 16 t ha^{-1} is expected. Which yield do you predict if 300 mm of soil water is extracted by the roots?

At a specific location, we may apply Eq. (6.35) to potential and actual conditions, yielding:

$$\frac{DM_a}{DM_p} = \frac{T_a}{T_p} \tag{6.36}$$

Therefore a first approximation of the relative yield of dry matter, grain or other marketable products is the relative transpiration. Equation (6.36) is often used by hydrologists. However, at certain stages of crop development, such as pollination, marketable yield may be extraordinarily affected (Kirkham, 2005). Figure 6.19 shows a generalized relation between yield and adequacy of water at different stages

Table 6.4 Crop transpiration efficiency factor μ (Eq. 6.35) for various C_3 and C_4 crops

Crop	Type of CO_2 fixation	μ (Pa)
Sorghum	C_4	13.8
Maize	C_4	7.4–10.2
Wheat	C_3	3.1–6.2
Barley	C_3	4.0
Oat	C_3	2.9–4.2
Potato	C_3	5.9–6.5
Lucerne	C_3	4.3
Soybean	C_3	4.0
Pea	C_3	3.8
Fava bean	C_3	3.1

From Ehlers and Goss (2003).

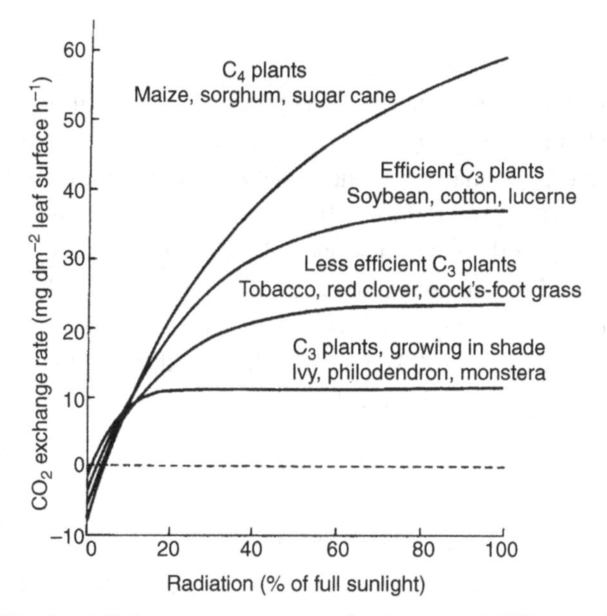

Figure 6.18 Idealized light response curves for leaves of different plant species. The dashed horizontal line marks the light compensation point LCP, where the CO_2 uptake rate due to photosynthesis is the same as the CO_2 release rate from respiration. (After Gardner et al., 1985).

of growth. The curve was developed for sugar cane in Hawaii, but shows a similar pattern at other crops. Kirkham (2005) summarizes moisture-sensitive periods for selected crops during which a water deficit depresses the economic yield much more than other periods.

Table 6.5 Application of Eq. (6.35) for three crops grown at two locations: Göttingen (Germany) and Akron (Colorado)

Crop	μ	Location	VPD	WP_T	Yield expectation	Transpiration	Predicted yield from 300 mm
	Pa		Pa	kg m^{-3}	t ha^{-1}	mm	t ha^{-1}
Maize	9.1	Göttingen	900	10.1	20	198	30.3
		Akron	1900	4.8		418	14.4
Wheat	4.5	Göttingen	900	5.0	15	300	15.0
		Akron	1900	2.4		633	7.1
Faba bean	3.1	Göttingen	900	3.4	10	290	10.3
		Akron	1900	1.6		613	4.9

For a given value of expected yield the transpirational water use is calculated. Also the yield is calculated at 300 mm water extracted by roots. The yield is total above-ground biomass. After Ehlers and Goss (2003).

The environmental conditions for plant production have an ironic aspect. In dry and arid areas with bright sunshine and high radiation for photosynthesis, the amount of soil water stored is normally low; nevertheless a comparatively large quantity of water is needed for the production of biomass. The scarceness of water limits the production level, while the dry atmosphere lowers the efficiency of transpirational water use. In wet and humid areas, however, frequently having less incoming radiation, the demand of water supply is less, as the efficiency of water use in production is greater (Ehlers and Goss, 2003). On the other hand, the actual supply may be so abundant that the surplus can have a harmful effect on plant growth and development.

6.6 Microclimate

The meteorological conditions within a vegetation may be very different from those above the vegetation. Because the processes in the plants react to the local conditions and because the conditions above the canopy are influenced by processes inside the canopy, it is worthwhile to discuss a number of aspects of the microclimate inside the canopy.

For any of those aspects the architecture of the vegetation is important. This not only entails the height of the vegetation, but also the size of the leaves, the orientation of the leaves and the vertical distribution of the leaves. The latter determines to a large extent at what height the active level of the vegetation is located (at what height most of the

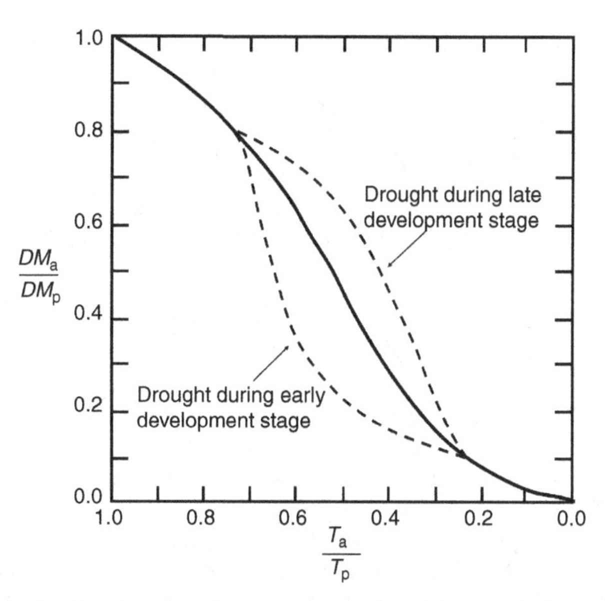

Figure 6.19 Generalized relation between relative yield and relative evapotranspiration, indicating the effect of stress timing (Kirkham, 2005).

radiation and momentum is absorbed; see Figure 6.20). For annual plants and deciduous trees and shrubs, the location of the active level, and the scalars for which it serves as the source or sink (heat, water vapour or CO_2), changes through the year owing to changes in the height of the vegetation and changes in the presence of leaves.

In contrast to the flow over a flat surface where sources and sinks are located only at the surface, in a canopy sources and sinks of radiation, momentum, heat, water vapour and various scalars can be located throughout the canopy. This implies that an extra term needs to be added to the budget equations for momentum and scalars (here exemplified by temperature; compare to Eq. (3.16)):

$$\frac{\partial \bar{u}}{\partial t} = -\frac{\partial \overline{u'w'}}{\partial z} + S_u$$
$$\frac{\partial \bar{\theta}}{\partial t} = -\frac{\partial \overline{w'\theta'}}{\partial z} + S_\theta \qquad (6.37)$$

where S_u and S_θ are the source/sink of momentum and heat respectively (e.g., drag on leaves, heating of the air by contact with hot leaves; in the case of CO_2 photosynthesis in the leaves would serve as a sink). Note that in Eq. (6.37) for simplicity only vertical gradients occur whereas in reality the conditions within a canopy will be highly nonhomogeneous in the horizontal as well.

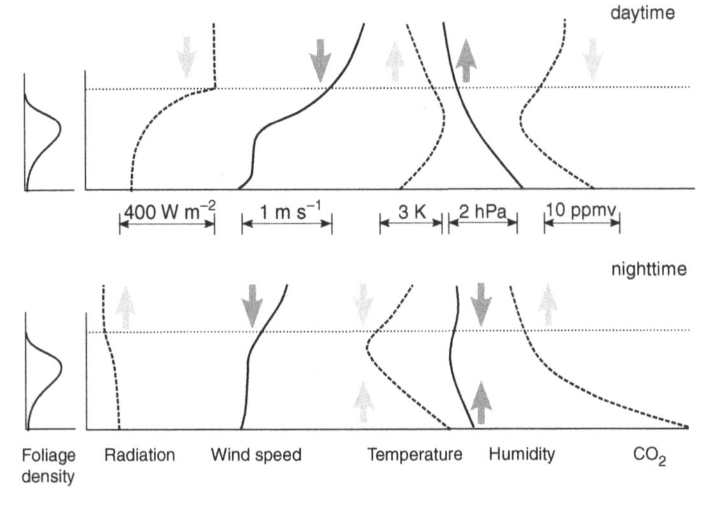

Figure 6.20 Canopy profiles of foliage density, net radiation, wind speed, temperature, humidity and CO_2 concentration during daytime (top) and night time (bottom). The arrows indicate the direction of transport. Note that in the daytime figure the in-canopy arrows are deliberately omitted owing to the possibility of counter-gradient transport (see later). (After Monteith and Unsworth, 2008.)

6.6.1 Radiation

Shortwave Radiation

Leaves absorb radiation, but they reflect and transmit radiation as well. The optical properties depend on the wavelength of the light, as can be seen in Figure 6.21. The strong absorption in the visible part is due to chlorophyll, with a small dip around 520 μm, which corresponds to green (thus leaves reflect green light effectively, hence their green colour). Absorption at wavelengths around 2 μm is due to absorption by water. The spectral properties of the leaves cause a change in the spectral composition of the light. Because, in the visible part of the spectrum, green is least absorbed, the light down in the canopy will be enriched with green, and will contain relatively little radiation that can be used for photosynthesis. The difference in reflectivity in the visible (r_{VIS}) and near-infrared (r_{NIR}) wavelength regions, which is specific for green vegetation, is exploited in the normalized difference vegetation index (NDVI):

$$\mathrm{NDVI} = \frac{r_{\mathrm{NIR}} - r_{\mathrm{VIS}}}{r_{\mathrm{NIR}} + r_{\mathrm{VIS}}} \tag{6.38}$$

The NDVI can be determined from remote sensing observations of reflectivity in the two wavelength bands and it can be linked to a number of vegetation properties, such as LAI and vegetation cover (see, e.g., Carlson and Ripley, 1997). In the case of sparse vegetation, the spectral properties of the underlying soil become important.

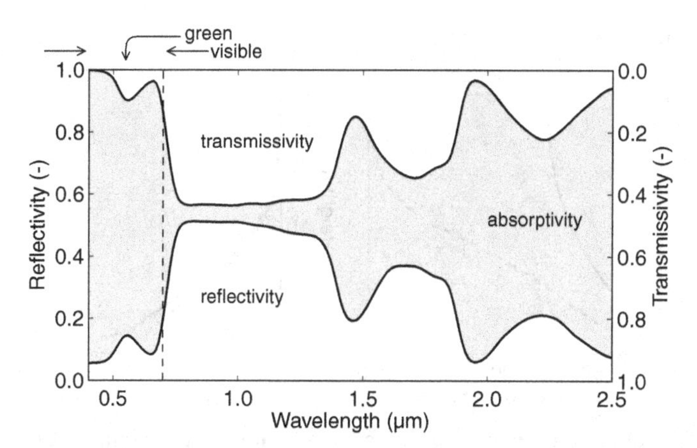

Figure 6.21 Spectral properties of a soybean leaf: reflectivity (bottom line), transmissivity (top line, with axis in opposite direction) and absorptivity (shaded area). (Data from Jacquemoud and Baret, 1990.)

This has given rise to a number of alternative vegetation indices (e.g., Enhanced Vegetation Index [EVI]; see Huete et al., 2002).

The attenuation of the total amount of short wave radiation can be approximated by Lambert–Beer's law:

$$K^{\downarrow}(z) = K^{\downarrow}e^{-aA(z)} \tag{6.39}$$

where the $K^{\downarrow}(z)$ is the shortwave radiation at a given height, $A(z)$ is the cumulative leaf area index (accumulated starting at the top, see, e.g., Figure 6.22) and a is an extinction coefficient. The latter depends not only on the canopy characteristics, but also on the characteristics of the radiation, in particular on the fraction of diffuse radiation and on the solar zenith angle.

Because leaves are not necessarily oriented horizontally, reflection of radiation will take place not only upward (out of the canopy) but also sideways and downward into the canopy. As a result, the proportion of diffuse radiation will be higher at the bottom of the canopy than at the top. Furthermore, if the incident radiation is mostly diffuse (on cloudy days) the radiation can penetrate deeper into the canopy because there will always be pathways that are not blocked by leaves. This can be seen in Figure 6.22, where on a clear day the radiation decreases sharply below 10 cm, whereas on an overcast day the decrease is more gradual. Of course on overcast days the total amount of radiation will be less than on clear days.

The vertical attenuation of PAR has a direct influence on the rate of photosynthesis in each layer. Here it should be noted that the light-use efficiency (amount of fixed carbon per amount of absorbed PAR) of leaves decreases above a certain radiation level (decreasing slopes in Figure 6.18). This implies that although on clear days the top of a vegetation receives a large amount of radiation, not all of this radiation

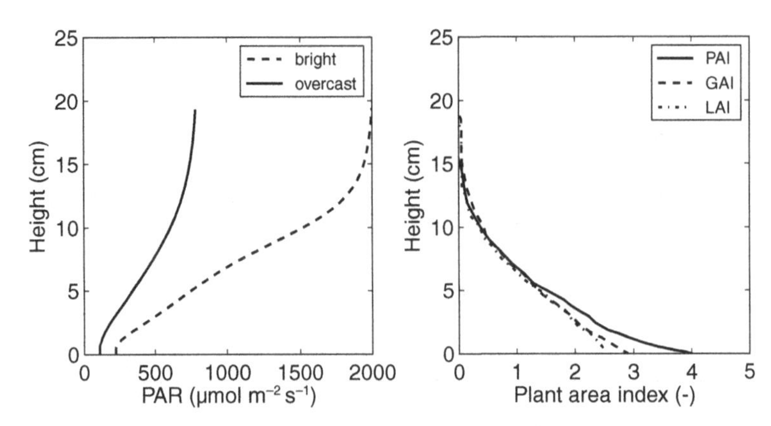

Figure 6.22 Radiation penetration in the vegetation of an Alpine pasture. (Right) Cumulative plant area index (PAI), green area index (GAI) and leaf area index (LAI). (Left) Profile of intensity of PAR for a clear and an overcast day. (Data from Tappeiner and Cernusca, 1989.)

is used for photosynthesis. Therefore, overcast conditions are relatively beneficial in terms of the total carbon fixation: the total amount of radiation is lower, but more evenly spread in the vertical. Related to the vertical variation of net assimilation, the stomatal resistance also varies with height (see Section 6.4.3). Furthermore, the variation in radiation affects the energy input to the leaf and hence it's temperature (see Section 6.6.4).

The attenuation described by Eq. (6.39) represents the (horizontal) mean variation of radiation with height. But owing to the porous nature of vegetation, the horizontal variation of downward radiation in the canopy can be large: certain locations may be exposed to direct sunlight, whereas others may be exposed only to multiply reflected light (see Figure 6.23). The penetration of radiation also changes continuously owing to the changes in azimuth and zenith angle of the sun, and to the motion of leaves and branches caused by wind. Furthermore, leaves are not necessarily oriented horizontally; hence the amount of radiation received by a leaf may be larger or smaller than the amount of radiation incident on a horizontal plane (this holds mainly in the upper part of the canopy where the fraction of direct radiation is still high).

Question 6.8: The attenuation as described in Eq. (6.39) is defined in terms of broadband shortwave radiation.

a) Suppose a global radiation of 800 W m^{-2} above a vegetation, an extinction coefficient (a in the equation) of 0.5 and values for A of 1.5 halfway up the canopy and A is 4 at the floor of the canopy. Compute the global radiation at the two levels (halfway and at the bottom).
b) From Figure 6.21 we can see that the transmission and reflection characteristics of leaves are different for different wavelengths. Consider radiation of 0.5 μm wave-

length and radiation in the near infrared of 1.0 μm. Assume the extinction coefficients for both wavelengths to be 0.7 and 0.3, respectively. Compute the change in spectral composition of the radiation at the same two levels as in question a) (where the spectral composition can be expressed as the ratio of the flux densities of radiation of 0.5 and 1.0 μm; above the canopies flux densities for both wavelengths are assumed identical).

Longwave Radiation

The effect of the leaves on longwave radiation inside the canopy is located mainly just below the top of the canopy. The top leaves receive incoming longwave radiation from the atmosphere (which has a relatively low temperature and emissivity). Most leaves below the top do not 'see' the sky and receive their incoming longwave radiation from the leaves above, which have a relatively high temperature and an emissivity close to one. Hence downward longwave radiation will on average be higher within the canopy than above it. Just as in the case of shortave radiation, the horizontal variation of downward longwave radiation can be large, depending on the fraction of sky the specific point in the canopy is exposed to (see Figure 6.23).

6.6.2 Air Temperature

The air temperature in a canopy will depend both on the temperature of the air above the canopy, the exchange of radiation (both the amount of radiation and the vertical distribution) and the exchange of heat with the soil. In Figure 6.20 a typical profile is shown. The highest temperature is located at the height of maximum foliage density, and hence maximum radiation absorption. Below that level, the temperature shows a stable stratification (temperature increases with height).

During nighttime, the reverse will happen. Maximum cooling will occur at the top of the vegetation, whereas some heat input from the soil will enter the air in the trunk space. As a result, the air below the foliage will be unstably stratified at night, giving rise to buoyancy-induced convection and strong mixing (Jacobs et al., 1995, 1996; Dupont and Patton, 2012).

6.6.3 Wind Speed

Within the canopy, the wind speed decreases strongly at the top of the canopy and stays rather constant with height below that. In vegetations that have a relatively open structure at the bottom (such as forests) the wind speed close to the ground may show a secondary maximum due to the fact that at that level the air encounters fewer obstacles.

Figure 6.23 Penetration of radiation from the sky through the crown of a pine forest to the forest floor. Photo courtesy of Bert Heusinkveld.

One important feature in the wind speed profile in Figure 6.20 should be noted: the inflection point just below the canopy top (point where the second derivative of $\overline{u}(z)$ changes sign). This inflection point is the cause of an instability that in turn causes large scale turbulent eddies that sweep into the canopy (carrying momentum and CO_2) and cause ejection of air (carrying heat and moisture). These large-scale structures are efficient in the transport of momentum and scalars, but are hardly related to the local gradients of those variables. Hence, transport counter to the gradient (rather than along the gradient, as assumed in standard K-theory; see Chapter 1) is an important feature of canopy turbulence (see Figure 6.24 and Finnigan, 2000).

6.6.4 *Leaf Temperature*

The temperature of the leaves is of vital importance for the biological processes taking place in the leaf. Furthermore, seen from above the canopy, the leaf temperature

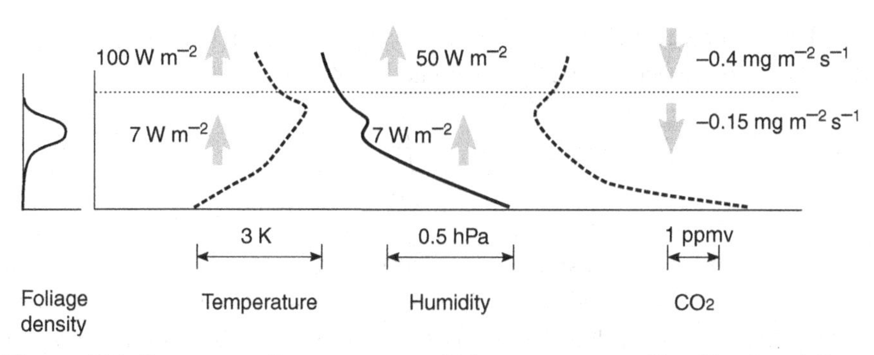

Figure 6.24 Counter-gradient transport within a canopy: profiles (lines) and fluxes (arrows) in a pine forest with a canopy height of 20 m (dashed line indicates canopy top): temperature and sensible heat flux (left), vapour pressure and latent heat flux (center) and CO_2 concentration and CO_2 flux (right). (Data from Denmead and Bradley, 1987.)

forms the surface temperature that determines the upwelling longwave radiation and sensible heat flux. To determine the temperature of a leaf, we need to consider its energy balance:

$$H_{\text{leaf}} = Q^*_{\text{leaf}} - L_{\text{v}} E_{\text{leaf}} \tag{6.40}$$

The net radiation is determined by the radiative fluxes leaving and entering the leaf (see Section 6.6.1) and the leaf evapotranspiration depends on the stomatal opening, leaf temperature and the ambient humidity. If we use a resistance law for the sensible heat flux ($H = -\rho c_{\text{p}} (T_{\text{a}} - T_{\text{leaf}}) / r_{\text{b,h}}$), the leaf temperature can be expressed as:

$$T_{\text{leaf}} = T_{\text{a}} + \frac{r_{\text{b,h}}}{\rho c_p} \left(Q^*_{\text{leaf}} - L_{\text{v}} E_{\text{leaf}} \right) \tag{6.41}$$

where $r_{\text{b,h}}$ is the laminar boundary-layer resistance for heat (similar, but not equal to the boundary-layer resistances for water vapour; see Figure 6.12). Though Eq. (6.41) clearly shows the factors that determined the leaf temperature, in principle it is an implicit expression for the leaf temperature: net radiation and transpiration themselves depend on the leaf temperature through the emitted longwave radiation and the saturated water vapour concentration in the stomata, respectively. Equation (6.41) shows that the degree to which the leaf temperature is coupled to the air temperature depends on the boundary-layer resistance and the energy input to the leaf. $r_{\text{b,h}}$ depends on the thickness of the laminar boundary layer: for thin boundary layers, the resistance is low. The boundary-layer thickness in turn depends on wind speed (the higher the wind speed, the thinner the boundary layer) and the distance over which the boundary layer can develop (larger leaves have thicker boundary layers). The boundary-layer resistance can be approximated as (Gates, 1980)[5]:

$$r_{b,h} = k_f \sqrt{\frac{\ell_f}{u_f}} \qquad (6.42)$$

where u_f is the wind speed outside the laminar boundary layer, ℓ_f is the length of the leaf in the direction of the air flow and k_f is of the order of 180 s$^{1/2}$ m^{-1}. Thus, a small value of $r_{b,h}$, and consequently strong coupling between leaf temperature and air temperature, occurs at high wind speeds and for small leaves.

Question 6.9: Consider an individual leaf in air with a temperature of 20 °C. The leaf is exposed to a net radiation of 400 W m^{-2}, transpires an amount of water equivalent 10^{-4} kg m^{-2} s^{-1} and has a boundary-layer resistance of 40 s m^{-1}.

a) Determine the leaf temperature (assume an air density of 1.20 kg m^{-3} and c_p = 1013 J kg^{-1} K^{-1}).
b) The plant is under water stress, partly closes its stomata, leading to a reduction of the transpiration to $0.3 \cdot 10^{-4}$ kg m^{-2} s^{-1}. Again determine the leaf temperature.
c) The same conditions apply as for question (a), but owing to a reduced wind speed the boundary-layer resistance has increased to 80 s m^{-1}. Determine the leaf temperature.

Question 6.10: In Figure 6.24 the fluxes of sensible heat, latent heat and CO_2 are shown at two levels, one within the canopy and one above. Explain for each of the quantities the difference in flux between the two levels.

6.6.5 Dew

Whereas during day time vegetation converts liquid water to water vapour, during night time the reverse may happen: water vapour condensates on the canopy surface. This happens if the temperature of the surface falls below the dew point of the air in and above the canopy. For dew to occur two things are needed: a sink of energy and a source of water vapour.

A sink of energy is needed because the latent energy released on condensation needs to be removed from the canopy. For condensation to occur, the other terms in the (simplified) surface energy balance need to act as a net sink ($Q^* - H - G < 0$). This

[5] The boundary-layer thickness δ can be approximated as $\delta = \sqrt{\nu \ell_f / u_f}$ where ν is the kinematic viscosity of air (order of $1.5 \cdot 10^{-5}$ m^2 s^{-1}). Assuming a linear temperature profile in the boundary layer, the heat flux is $H = -\rho c_p D_T (T_a - T_{leaf}) / \delta$, where D_T is the molecular diffusivity for heat. This defines $r_{b,h}$ as $\frac{\sqrt{\nu}}{D_T} \sqrt{\ell_f / u_f}$. It also shows that the boundary-layer resistance depends on the transported scalar because the molecular diffusivities for heat, water vapour and CO_2 are different ($D_T = 2.13 \cdot 10^{-5}$ m^2 s^{-1}, $D_v = 2.42 \cdot 10^{-5}$ m^2 s^{-1} and $D_c = 1.47 \cdot 10^{-5}$ m^2 s^{-1} at 20 °C; Gates, 1980).

will usually imply conditions where $Q^* < 0$ (cloud-free conditions) and H and G are not too strongly negative.

Two sources of water vapour are available.[6] First, water vapour can be extracted from the atmosphere (dewfall): a downward turbulent moisture flux. For this it is necessary that the stable stratification does not suppress turbulence completely and hence some wind is needed to maintain turbulence. On the other hand, if the wind is too strong, H will become strongly negative and of the order of Q^* so that the surface cannot cool below the air temperature. Thus dewfall can occur only for a limited range of wind speeds. The second source of water vapour can be the soil (dewrise). The soil can both be a direct source of water vapour (in the case of unsaturated soils water vapour diffusion will occur) or indirectly through evaporation from the soil surface. In the latter case the resulting dew is energetically neutral as far as the control volume that contains both the soil and the canopy is considered: first energy is consumed to evaporate water at the soil surface and subsequently latent energy is released on condensation. For a maize canopy, Jacobs et al. (1990) showed that the contribution of dewrise is at least an order of magnitude smaller than that of dewfall.[7]

Typical amounts of dewfall range from 0.05 to 0.5 mm per night (Xiao et al., 2009). The amount of dew need not be uniformly distributed over the depth of the canopy. The vertical distribution of dew found by Jacobs et al. (1990) in a maize crop can be seen in Figure 6.25. Despite the fact that the leaf area distribution changes significantly during the observation period, the peak of the dew formation always occurs at $z/h = 0.7$. This is probably due to the fact that under the conditions studied dewfall dominates and dew interception is concentrated in the top of the canopy. Furthermore, the location of maximum cooling (see Figure 6.20) is a tradeoff between the location of maximum foliage area (large cooling leaf area, deeper in the canopy) and maximum longwave cooling (per unit leaf area, top of the canopy).

Jacobs et al. (2006) estimate the contribution of dew to the water balance of a Dutch grassland to be 37 mm per year, or nearly 5% of the precipitation. The number of nights in which some dew occurs is 250 per year or nearly 70% of the nights. Whereas for temperate climates the contribution to the water balance is only limited, in arid conditions the annual dewfall can be an important source of water with a lower variability than rainfall (Zangvil, 1996).

Apart from the impact on the water balance, the wetness of leaves has other implications as well. First, fungal spores and other plant pathogens can develop in the layer of liquid water. The length of the period of leaf wetness is a critical parameter in this

[6] Another source of liquid water on leaves is guttation: water emerges from special pores due to the supply of water through the vascular system. If the air is close to saturation the guttation droplets will not evaporate but accumulate on the leaf. Because no phase change occurs, guttation cannot be accounted for by looking at latent heat fluxes (Hughes and Brimblecombe, 1994).

[7] For semi-arid conditions, with irrigated soils, the reverse may hold: there the soil will be the primary source of water vapour whereas the air is too dry to give a significant contribution (Weiss et al., 1989). Other atmospheric factors affecting the relative importance of soil and atmosphere factors related to atmospheric transport (stability, wind speed). Furthermore, the architecture of the canopy (height, bare soil fraction) will have an effect (Garrat and Segal, 1988).

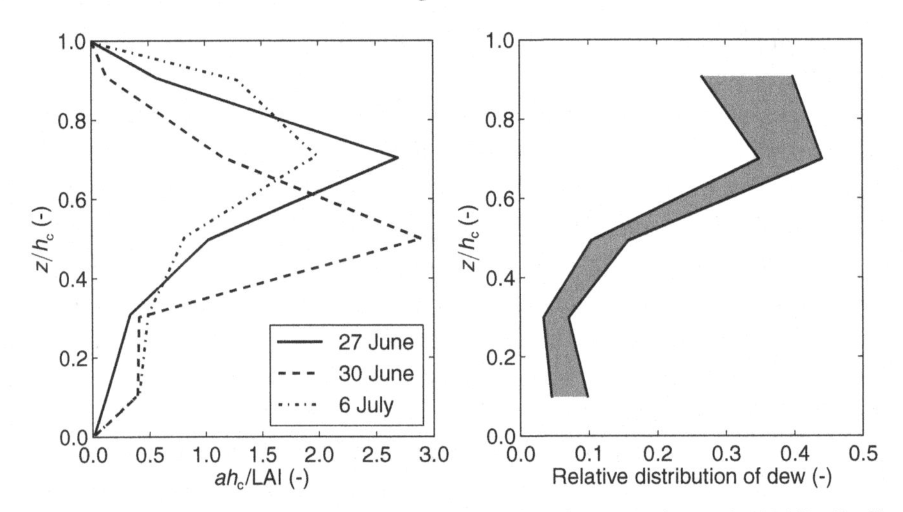

Figure 6.25 Dew formation in a maize canopy (data from Jacobs et al. (1990). (Left) Normalized leaf area distribution *ah/LAI* (*a* is the one-sided leaf area per unit volume; *h* is canopy height) for three dates. (Right) Normalized dew profiles (normalized with the total amount of dew) for the period June 22 to July 5; the area indicates the range from minimum to maximum value and total amounts of dewfall ranged from 0.05 to 0.25 mm per night.

development (e.g., Huber and Gillespie, 1992). Furthermore, surface wetness plays a role in the deposition of atmospheric trace gases that are well-soluble in water, such as NH_3 and SO_2 (Wichink Kruijt et al., 2008).

In Section 7.4 a method to quantify the amount of dewfall is discussed.

Question 6.11: Consider a maize canopy. The canopy has a height of 2 m and the air temperature within the canopy is 10 °C. A typical amount of dewfall for this night is 0.25 mm in one night.

a) Does all dewfall originate from the air inside the canopy? To answer this, assume that there is no exchange of water vapour with the soil or the overlying air (the canopy would be a closed box). Then how much dewfall would be possible if we assume that all water vapour in the air would condensate on the leaves? Assume the initial relative humidity of the air to be 100% (and assume p=101 300 Pa).
b) What net energy flux density ($Q^* - H - G$) would be needed to allow a dewfall of 0.25 mm in a night of 8 hours length?
c) If the aforementioned dewfall of 0.25 mm per night is extracted from a layer 100 m deep, which was initially isothermal and saturated at 10 °C, what will be the mean mixing ratio at the end of the night (ignore the change of temperature in the layer during the night)? Assume an air density of 1.24 kg m^{-3}.

6.7 Rainfall Interception

Interception is the process by which precipitation falls on vegetative surfaces, where it is subject to evaporation. Interception loss depends strongly on (1) vegetation type and stage of development, which should be well characterized by leaf area index; and (2) the intensity, duration, frequency and form of precipitation. The interception loss ranges from 10–40% of gross precipitation, depending on vegetation and climate (Dingman, 2002; Muzylo et al., 2009; Gerrits, 2010). Therefore, to simulate evapotranspiration and rainfall infiltration into the soil, we should quantify properly the amount of rainfall interception.

Rutter et al. (1975) presented a conceptual, physically based model for forests that proved to be very useful. Their model represents the interception process by a running water balance of rainfall input, storage and output in the form of drainage and evaporation (Figure 6.26). The canopy structure is described by the free throughfall coefficient r (-), the stemflow partitioning coefficient r_t (-), the canopy storage capacity S (mm) and the trunk storage capacity S_t (mm). The Rutter model estimates throughfall, stemflow and interception loss from input rainfall and evapotranspiration data. Essentially, it is based on the dynamic calculation of the water balance for the canopy and for the trunks through the equations:

$$
\begin{aligned}
(1-r-r_t)\int P\,\mathrm{d}t &= \int D\,\mathrm{d}t + \int E_{\mathrm{int,\,c}}\,\mathrm{d}t + \Delta C \\
r_t \int P\,\mathrm{d}t &= \int I_s\,\mathrm{d}t + \int E_{\mathrm{int,\,t}}\,\mathrm{d}t + \Delta C_t
\end{aligned}
\qquad (6.43)
$$

where P is the intensity of gross rainfall (mm d^{-1}), D is the drainage rate from the canopy (mm d^{-1}), $E_{\mathrm{int,\,c}}$ is the evaporation rate of water intercepted by the canopy (mm d^{-1}), ΔC is the change in canopy storage (mm), I_s is the stemflow (mm d^{-1}), $E_{\mathrm{int,\,t}}$ is the evaporation rate of water intercepted by the trunks (mm d^{-1}) and ΔC_t is the change of trunk storage (mm).

The evaporation rate from a saturated canopy E_{pot} is calculated using the Penman–Monteith equation with the canopy resistance set to zero (Chapter 7). When actual canopy storage C (mm) is less than canopy storage capacity S, evaporation rate is reduced in proportion to C/S. The rate of drainage from the canopy is usually calculated as:

$$
\begin{aligned}
D &= D_s \exp\left(b(C-S)\right) && \text{if} \quad C \geq S \\
D &= 0 && \text{if} \quad C < S
\end{aligned}
\qquad (6.44)
$$

where D_s is the drainage rate when $C = S$ and b is an empirical coefficient.

Modelling of stemflow and trunk evaporation follows closely the procedure previously used for the canopy. Evaporation from trunks is calculated as:

$$
\begin{aligned}
E_{\mathrm{int,\,t}} &= \varepsilon E_{\mathrm{pot}} && \text{if} \quad C_t \geq S_t \\
E_{\mathrm{int,\,t}} &= \varepsilon E_{\mathrm{pot}} \frac{C_t}{S_t} && \text{if} \quad C_t < S_t
\end{aligned}
\qquad (6.45)
$$

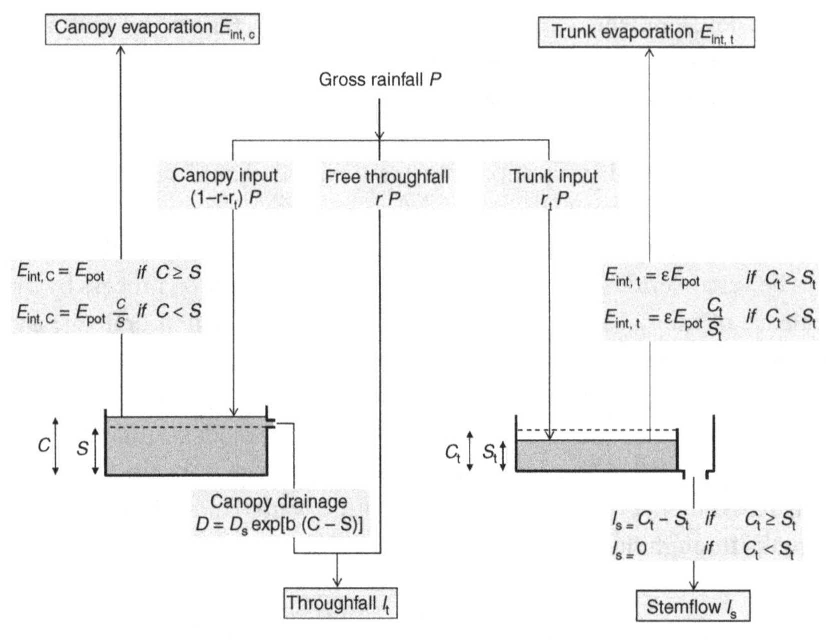

Figure 6.26 Scheme of the Rutter model (Rutter et al., 1975).

where ε is a constant describing the evaporation rate from saturated trunks as a proportion of that from the saturated canopy. Stemflow I_s for a given time step is calculated with the following equation:

$$\begin{aligned} I_s &= C_t - S_t \quad &\text{if} \quad C_t \geq S_t \\ I_s &= 0 \quad &\text{if} \quad C_t < S_t \end{aligned} \tag{6.46}$$

The model requires rainfall intensities at short time intervals, for example, every 10 minutes. Proper simulation of the canopy and trunk storage amounts requires a suitable numerical integration method or analytical integration of the model equations (Lloyd et al., 1988). The model has been developed for relatively closed canopies, particularly for the evaporative process, through the assumption that the canopy and trunk storages extend to the whole plot area. Valente et al. (1997) adapted the Rutter model for sparse forests.

Gash (1979, 1995) simplified the Rutter model and put forward a well-known analytical interception model. His model represents rainfall input as a series of discrete storms that are separated by intervals long enough for the canopy and stems to dry completely – this assumption is possible by the rapid drying of forest canopies. Each individual storm is then divided into three subsequent phases: canopy wetting-up, saturation and drying. For the first two of these phases, the actual rates of evaporation and rainfall are replaced by their mean rates for the entire period being modelled (Muzylo et al., 2009).

During wetting up, the increase of intercepted amount P_i is described by:

$$\frac{\partial P_i}{\partial t} = (1 - r - r_t) P_{mean} - \frac{P_i}{S} E_{mean} \tag{6.47}$$

where P_{mean} is the mean rainfall rate (mm d^{-1}), and E_{mean} is the mean evaporation rate of intercepted water when the canopy is saturated (mm d^{-1}). Integration of Eq. (6.47) yields the amount of rainfall that saturates the canopy, P_s (mm):

$$P_s = -\frac{P_{mean} S}{E_{mean}} \ln\left(1 - \frac{E_{mean}}{P_{mean}(1 - r - r_t)}\right) \quad \text{with } E_{mean} \le P_{mean}(1 - r - r_t) \tag{6.48}$$

Question 6.12: Derive Eq. (6.48).

For small storms ($P \le P_s$) the interception can be calculated from:

$$P_i = (1 - r - r_t) P \tag{6.49}$$

For large storms ($P > P_s$) the interception follows from:

$$P_i = (1 - r - r_t) P_s + \frac{E_{mean}}{P_{mean}} (P - P_s) \tag{6.50}$$

Figure 6.27 shows the Gash interception as function of rainfall amounts for typical values of a pine forest. The slope $\partial P_i / \partial P$ before saturation of the canopy equals $(1 - r - r_t)$; after saturation of the canopy this slope equals E_{mean}/P_{mean}.

In forests, the evaporation of intercepted water occurs at rates several times larger than for transpiration under identical meteorological conditions of a dry canopy (see section 7.2.4). Thus intercepted water disappears quickly and interception loss replaces tree transpiration only for short periods. Therefore accurate calculation of forest interception amounts is required to determine the total amount of evapotranspiration.

At short vegetation, atmospheric conductances are much lower than at forests, and interception loss occurs at rates comparable to transpiration. Thus for grasses and common agricultural crops, interception loss is to a large extent compensated by reduction in transpiration and makes little difference to cumulative evapotranspiration (Dingman, 2002). Therefore in the case of short vegetation a lower accuracy of interception amounts is required. Also, for short vegetation the distinction between canopy drainage and stem flow is less relevant. Von Hoyningen-Hüne (1983) and Braden (1985) performed a large number of lysimeter experiments to determine

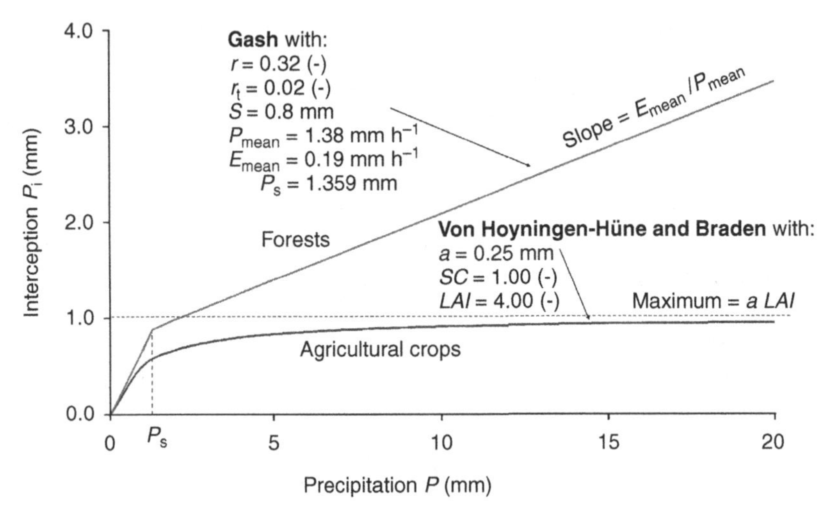

Figure 6.27 Interception as function of precipitation according to Gash (1979) and Hoyningen-Hüne (1983) and Braden (1985).

interception in agricultural crops. They proposed the following simple formula for canopy interception:

$$P_{\mathrm{i}} = a\,\mathrm{LAI}\left(1 - \frac{1}{1 + \frac{\mathrm{SC}\,P}{a\,\mathrm{LAI}}}\right) \tag{6.51}$$

where LAI is leaf area index (leaf m^2 soil m^{-2}), a is an empirical coefficient (m d^{-1}) and SC represents the soil cover fraction (-). For increasing amounts of precipitation, the amount of intercepted precipitation asymptotically reaches the saturation amount aLAI (Figure 6.27). In principle a must be determined experimentally, but a common value for ordinary agricultural crops is 0.25 mm d^{-1}. Equation (6.51) is based on daily precipitation values and yields daily interception amounts.

6.8 Summary

Water plays a key role in many plant physiological processes. As plants rely on the water available at their local spot, a proper analysis of root water uptake as affected by climate, soil texture, plant type and drainage condition forms the basis for many environmental studies. Also a close relation appears to exists among root water uptake, plant transpiration and vegetation growth. For analysis of root water uptake we described a physically based microscopic approach and a more empirical macroscopic approach. We discuss the hydraulic head decline in the soil–plant–atmosphere continuum and how plants cope with large negative pressure heads in their xylem vessels. Plants change their stomatal resistance to control the transpiration loss, in response to light intensity, leaf temperature, air humidity, internal CO_2 concentration and leaf water potential. The water use efficiency or water productivity appears to be affected mainly by daytime water vapour pressure deficit and plant type (C_3 vs. C_4 plants).

With respect to microclimate within the vegetation, important physical processes are the extinction of radiation, the leaf temperature as function of radiation, the heating and cooling of air temperature by leaves and soil, and the wind speed profile and its effect on turbulent eddies. During nighttime water vapour may condensate on the canopy surface as dew, which may affect the water balance, plant pathogen development and deposition of atmospheric trace gases. Rainfall interception may range from 10% to 40% of gross precipitation, depending on vegetation and climate. Rainfall interception modelling concepts for forests and agricultural crops are explained.

7

Combination Methods for Turbulent Fluxes

In Chapter 3 a number of methods were presented that can be used to determine the atmospheric fluxes of the surface energy and water balance: sensible and latent heat flux. Those methods were based solely on the use of data regarding wind, temperature and humidity: either the fluctuating parts of the signal (eddy-covariance method) or the mean values: vertical gradients or differences (similarity theory).

In this chapter we not only use our knowledge on the turbulent fluxes, as dealt with in Chapter 3, but also combine it with the energy balance equation (Chapters 1 and 2) and information on the vegetation (Chapter 6). First the Bowen ratio method is discussed in Section 7.1. Next the Penman–Monteith equation that describes the transpiration from vegetation is dealt with in Section 7.2. Finally, simplified estimates for evapo(-transpi)ration are given in Section 7.3 and dewfall (inverse evaporation) is discussed in Section 7.4.

The term "combination methods" in the title of this chapter has two different connotations:

- In general, the term "combination methods" refers to methods that combine the energy balance equation with information on turbulent transfer (all methods in this chapter).
- In a more restricted sense, the term "combination equation" refers to the Penman method (and derived methods) that combines the effects of two factors that determine evaporation (see Farahani et al., 2007). These are available energy (represented by the "radiation term") and atmospheric demand (ability of the atmosphere to remove water vapour, represented by the "aerodynamic term"). In this restricted sense, the term refers only to the methods discussed in Section 7.2.

7.1 Bowen Ratio Method

7.1.1 Sensible and Latent Heat Flux

The Bowen ratio is the – dimensionless – ratio of sensible heat flux and latent heat flux. If we apply this to the surface fluxes, the *surface* Bowen ratio is defined as:

$$\beta \equiv \frac{H}{L_v E}, \tag{7.1}$$

With this definition, in combination with the energy balance equation (Eq. (1.5)), the surface sensible heat flux and latent heat flux can be expressed as a function of available energy $Q^* - G$ and the Bowen ratio:

$$\begin{aligned} H &= \beta \frac{Q^* - G}{1+\beta} \\ L_v E &= \frac{Q^* - G}{1+\beta} \end{aligned} \tag{7.2}$$

But now the problem of determining the fluxes has been shifted to the determination of the Bowen ratio. For this we can make use of the expressions developed in Chapter 3 that link the turbulent fluxes to vertical differences in the concentration of the transported quantity:

$$H = -\rho c_p \frac{\overline{\theta}(z_{t2}) - \overline{\theta}(z_{t1})}{r_{ah}}, \quad L_v E = -\rho L_v \frac{\overline{q}(z_{v2}) - \overline{q}(z_{v1})}{r_{ae}}$$

where z_{t1}, z_{t2}, z_{v1}, and z_{v2} are the heights where temperature and specific humidity are measured, respectively, and r_{ah} and r_{ae} are the aerodynamic resistances for heat and water vapour transport, respectively. Now, provided that

- the location of the upper observations of temperature and humidity coincide ($z_{t2} = z_{v2}$);
- the locations of the lower observations of temperature and humidity coincide ($z_{t1} = z_{v1}$);
- the aerodynamic resistances for heat and water vapour are identical, which is a consequence of ($z_{t2} = z_{v2}$) and ($z_{t2} = z_{v2}$), in combination with the fact that the similarity relationships for heat and water vapour are supposed to be equal

the Bowen ratio can be determined as:

$$\beta = \frac{H}{L_v E} = \frac{c_p \left(\overline{\theta}(z_{t2}) - \overline{\theta}(z_{t1})\right) / r_{ah}}{L_v \left(\overline{q}(z_{v2}) - \overline{q}(z_{v1})\right) / r_{ae}} = \frac{c_p}{L_v} \frac{\Delta \overline{\theta}}{\Delta \overline{q}} \tag{7.3}$$

Hence, observations of temperature and humidity at two heights, in combination with soil heat flux and net radiation, suffice to determine the sensible and latent heat flux. No detailed measurements (or assumptions) regarding the nature of turbulence (e.g., with respect to r_a) are needed. Because the use of the surface energy balance is an essential part of the method, it is often referred to as Bowen ratio energy balance method (or BREB method).

There are a number of caveats related to the Bowen ratio method. First, the vertical differences in temperature and humidity can be quite small (and thus relative errors in β quite large):

- If measurements are not made sufficiently close to the surface (gradients are largest close to the surface)
- When the aerodynamic resistance is small (strong turbulence)
- When the fluxes are small (around sunrise and sunset, or under extremely wet or dry conditions)

Hence accurate instruments are needed with an accuracy of about 0.05 K for temperature, and about 0.02 g kg^{-1} for humidity (note that the error in the difference is already double the error in the observations at the individual levels). One way of eliminating systematic errors is to interchange the instruments periodically: for example, measure 5 minutes with sensor 1 on top and sensor 2 below, and the next 5 minutes with sensor 2 on top and sensor 1 below: combination of the two 5-minute averages will give a 10-minute average vertical difference without a systematic error (see, e.g., McCaughey et al., 1987).

Another problematic condition occurs when the available energy is close to zero (near sunrise and sunset or under cloudy conditions). Then the sensible and latent heat fluxes are of equal magnitude but opposite sign and hence the Bowen ratio will be close to –1. As a result the fluxes, as determined from Eq. (7.2), will become very sensitive to measurement errors resulting in unrealistic values.

Question 7.1:

a) What is the value of the Bowen ratio when the available energy equals zero ($Q^*-G=0$)?

b) What are the values of sensible and latent heat fluxes for this situation?

7.1.2 Trace Gases

The concept of a Bowen ratio can also be applied to the fluxes of trace gases, such as CO_2, N_2O, CH_4, etc. For some gases this is still the only way to infer vertical fluxes, as sensors that can be used for eddy-covariance measurements are not yet available. For other gases the fast and compact gas analysers have been developed only recently (e.g., Hendriks et al., 2008).

For a trace gas with a surface flux F_X one would define the 'mass Bowen ratio' as the ratio of the unknown flux to a known mass flux (e.g., the water vapour flux):

$$\beta_X \equiv \frac{F_X}{E} = \frac{\Delta \overline{q_X}}{\Delta \overline{q}}, \qquad (7.4)$$

where the last equality sign is based on the assumption that the aerodynamic resistances for water vapour and trace gas X are equal. Then, by measuring simultaneously at the same height the concentrations of water vapour and the gas under consideration, the mass Bowen ratio can be determined. In combination with a separate observation of E (either with the Bowen ratio method, or otherwise) this will give the trace gas flux F_X.

Note that although Eq. (7.4) is an elegant definition for a mass Bowen ratio that seems dimensionless, strictly speaking it is not. The dimensions of q_X are kilograms of gas X per kilogram air, and those of q are kilograms of water vapour per kilogram of air. Hence the dimensions of β_X are kilograms of gas X per kilogram of water vapour. This ratio is not dimensionless. Therefore, a Bowen ratio involving, for example, F_X and the sensible heat flux (and hence the vertical differences of q_X and temperature) would be an equally valid approach (and useful if an independent observation of H is available and an observation of L_vE is not).

Question 7.2: Given the following observations:

Quantity	Value at 5 m height	Value at 2 m height	Value at surface	Unit
Net radiation		500		W m^{-2}
Surface soil heat flux			50	W m^{-2}
Air temperature	28.5	28.8		°C
Density of dry air (ρ_d)	1.155	1.154		kg m^{-3}
Water vapour density (ρ_v)	$10.0 \cdot 10^{-3}$	$10.5 \cdot 10^{-3}$		kg m^{-3}
CO_2 density (ρ_c)	$600 \cdot 10^{-6}$	$597 \cdot 10^{-6}$		kg m^{-3}

a) Compute the sensible and latent heat fluxes using the Bowen ratio method. (Use c_p = 1013 J kg^{-1} K^{-1}).
b) Compute the CO_2 flux density using the mass Bowen ratio.

7.2 Penman–Monteith Equation

For the Bowen ratio method, observations at two levels are needed, which are often not routinely available. The Penman–Monteith equation is a way to use observations at only one level, and treat the lowest level (the surface) only implicitly (i.e., no surface temperature or humidity information is needed).

First the Penman equation (valid for wet surfaces) will be derived (Section 7.2.1). Then the concept of the 'big-leaf' approach is presented, and used to derive the Penman–Monteith equation, which describes the transpiration from dry vegetation (Section 7.2.2)

7.2.1 Penman Derivation

Penman (1948) derived a well-known expression for the evaporation from wet surfaces. Here we follow a formal derivation, using the knowledge gathered in Chapters 2 and 3. The starting point is the energy balance equation, and the resistance expressions for sensible and latent heat fluxes:

$$Q^* - G = H + L_v E \tag{7.5}$$

$$H = -\rho c_p \frac{\overline{\theta}(z) - \theta_s}{r_{ah}} \tag{7.6}$$

$$L_v E = -\rho L_v \frac{\overline{q}(z) - q_s}{r_{ae}} \tag{7.7}$$

where θ_s and q_s are the potential temperature and specific humidity at the surface, and r_{ah} and r_{ae} are the aerodynamic resistances for heat and moisture transport. Now, we make a number of small modifications to Eqs. (7.5) to (7.7):

- Because the upper observation level will be relatively close to the ground (order of metres), the difference between potential temperature and regular temperature are small, and θ is replaced by T.
- The specific humidity will be replaced by the vapour pressure,[1] which is more commonly used in this context: $q \approx \frac{R_d}{R_v}\frac{e}{p} = \frac{c_p}{L_v \gamma} e$ where γ is the psychrometric constant (Pa K^{-1}), R_d, and R_v are the gas constants for dry air and water vapour (J kg^{-1} K^{-1}), respectively (see Appendix B).
- The indication of the observation height is dropped, and the values related to air are indicated by $\overline{T}_a$ and $\overline{e}_a$, and the surface values by T_s and e_s.

This gives the following new versions for Eqs. (7.6) and (7.7):

$$H = -\rho c_p \frac{\overline{T}_a - T_s}{r_{ah}} \tag{7.8}$$

$$L_v E = -\rho \frac{c_p}{\gamma} \frac{\overline{e}_a - e_s}{r_{ae}} \tag{7.9}$$

Let us assume that we have observations for the available energy $Q^* - G$, the aerodynamic resistances r_{ah} and r_{ae} (where the dependence of the resistances on stability, hence

[1] Note, that the derivation of the Penman equation is equally well possible when using specific humidity as the humidity variable (in that case the slope of the saturated vapour pressure curve s, used later on, will indicate dq_{sat}/dT rather than de_{sat}/dT; see, e.g., van Heerwaarden et al. (2009))

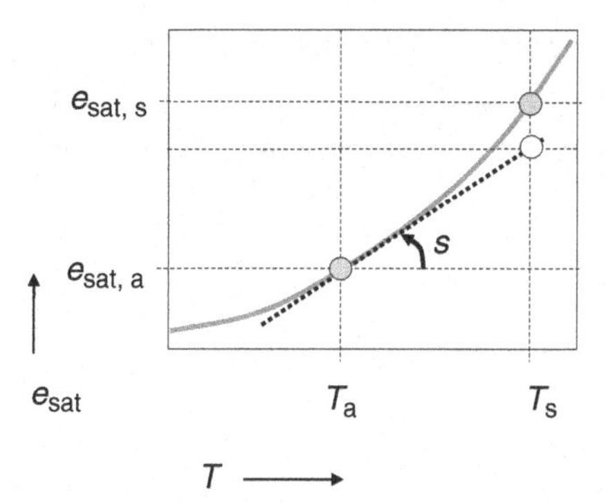

Figure 7.1 Linearzation of the saturated vapour pressure curve, as used in the Penman derivation. Grey points are actual combinations of e_{sat} and T, whereas the white point is the combination of e_{sat} and T at the surface, as implied by the linearization.

H, is temporarily neglected) and the temperature and humidity at a given height. Then the system of equations (Eqs. (7.5), (7.8) and (7.9)) cannot yet be solved because there are four unknowns (H, L_vE, T_s and e_s) and only three equations. However, given the fact that we consider a wet surface, the vapour pressure at the surface can be assumed to be equal to the saturated value at the surface temperature (see Appendix B). Thus:

$$e_s = e_{sat}(T_s) \tag{7.10}$$

Now there are as many equations as unknowns and the system can be solved. But, this system has a nonlinear component, since the saturated vapour pressure curve is a non-linear function of temperature, and thus the system cannot be solved explicitly, but only iteratively. The main trick of the Penman derivation is that we can eliminate the surface temperature altogether. To this end, the surface vapour pressure is estimated from the saturated vapour pressure at the observation level z:

$$e_{sat}(T_s) \approx e_{sat}(\overline{T}_a) + s(\overline{T}_a)\left[T_s - \overline{T}_a\right] \tag{7.11}$$

where $s(T)$ is the slope of the saturated vapour pressure curve, $\frac{\mathrm{d}e_{sat}}{\mathrm{d}T}$ (see Appendix B).

As can be seen in Figure 7.1, this linearization is only an approximation, but for situations in which the difference between air temperature and surface temperature is not too large (i.e., small sensible heat flux) it is sufficient. Combining Eqs. (7.11) and (7.9) results in the following expression for the latent heat flux:

$$\begin{aligned} L_vE &= -\frac{\rho c_p}{\gamma}\left[\frac{\bar{e} - e_{sat}(\overline{T}_a)}{r_{ae}} + s(\overline{T}_a)\frac{\overline{T}_a - T_s}{r_{ae}}\right] \\ &= -\frac{\rho c_p}{\gamma}\frac{\bar{e}_a - e_{sat}(\overline{T}_a)}{r_{ae}} + \frac{s(\overline{T}_a)}{\gamma}\frac{r_{ah}}{r_{ae}}H \end{aligned} \tag{7.12}$$

Finally, if we assume the aerodynamic resistances for water vapour and heat to be equal and combine Eq. (7.12) with the energy balance Eq. (7.5) (viz., $H = Q^* - G - L_v E$) we obtain (after some serious rearrangement) the Penman equation for evaporation from a wet surface:

$$L_v E = \frac{s(Q^* - G)}{s + \gamma} + \frac{\frac{\rho c_p}{r_a}\left(e_{sat}(\overline{T}_a) - \overline{e}_a\right)}{s + \gamma} \tag{7.13}$$

where we have omitted the explicit temperature dependence of s and e_{sat} for reasons of clarity. Furthermore, the aerodynamic resistances for heat and water vapour have been denoted simply by r_a (which is *not* equal to the aerodynamic resistance for momentum, due to differences in roughness length and stability dependence). Although the focus of the Penman equation is on evaporation, the sensible heat flux can simply be obtained as the residual from the energy balance: $H = Q^* - G - L_v E$.

The first term on the right-hand side of Eq. (7.13) is called the *radiation* term because it describes the evaporation due to energy input by radiation. The second term is called the *aerodynamic* term because it depends explicitly on the turbulent transport and the atmospheric moisture conditions through $e_{sat}(\overline{T}_a) - \overline{e}_a$ (vapour pressure deficit [VPD]). Referring to the derivation of the Penman equation, the two terms can also be interpreted as follows. The first term is proportional to the evaporation that is due to the fact that the temperature of the surface deviates from the air temperature as a result of net energy input: a temperature contrast between surface and air results in a contrast in $e_{sat}(T)$ between surface and air. The second term is proportional to the evaporation that would occur if the surface temperature would be equal to the air temperature.

The Penman equation shows that even if there is not net input of energy by $(Q^* - G)$ evaporation can proceed: in that case the energy required for evaporation will be extracted from the air through a negative sensible heat flux.

Question 7.3: To circumvent the need for an iterative solution of a system of three equations, the Penman equation uses the linearization given in Eq. (7.11). Determine the error in the estimation of $e_{sat}(T_s)$ due to the linearization for a situation with an air temperature of 20 °C, an aerodynamic resistance of 30 s m^{-1}, and a sensible heat flux of:

a) 0 W m^{-2}
b) 100 W m^{-2}
c) 300 W m^{-2}

Assume an air density of 1.2 kg m^{-3}. Hint: from Eq. (7.8) one can obtain the *real* surface temperature for a given sensible heat flux and hence the *real* saturated vapour pressure at the surface. Compare this with the linearization.

A number of conclusions regarding evaporation from a wet surface can be drawn, based on this equation:

a) Evaporation increases with increasing energy input (radiation term).
b) Evaporation increases with decreasing aerodynamic resistance, that is, under conditions of high wind speed, or strong convective turbulence (aerodynamic term).
c) Evaporation increases with increasing water vapour pressure deficit, thus with increasing dryness of the air (aerodynamic term).
d) The relative importance of the radiation term and the aerodynamic term depends (at a given values of $Q^* - G$, VPD and r_a) on temperature: at higher temperatures, the radiation term becomes increasingly important due to the increase of s with temperature. On the other hand, at a given temperature, sunny conditions will favour the radiation term, whereas windy conditions and/or dry air will favour the aerodynamic term.

The conclusions drawn under points (a) to (c) could be considered as common knowledge: if one wants to dry the laundry outside, this works best when the sun shines (a), when there is sufficient wind (b) and when the air is dry (c).

Some remarks need to be made regarding the *application* of the Penman equation:

- The Penman equation has a sound physical basis and as such it *describes* the evaporation well (except for the small effect of the linearization in Eq. (7.11)): if the input data have been measured over the wet surface for which one wants to determine the evaporation, the Penman equation should give a correct answer. But one should be cautious when using the Penman equation to *predict* evaporation: if one would use observations at a given location, not taken over the wet surface one is interested in (e.g., because the wet surface is not yet there), the calculated evaporation will be in error, because the observed $Q^* - G$, T, VPD and r_a will not be representative of the conditions over a wet surface (e.g., the observed temperature may be too high, the net radiation may be too low, due to a too high surface temperature or a different albedo).
- The aerodynamic resistance depends on wind speed *and* stability (see Chapter 3). Hence, to determine the evaporation from observed $Q^* - G$, VPD, temperature and wind speed (and assumed roughness lengths for momentum and heat) the stability needs to be determined as well. As this depends mainly on H (and to some extent on $L_v E$) an iterative procedure needs to be used to find the correct combination of sensible heat flux, evaporation and aerodynamic resistance.
- Although water bodies (e.g., ditches, lakes, etc.) are wet surfaces as well, the application of the Penman equation to those surfaces needs to be done with care. First, usually no observations above the water body are available, but only observations over land surfaces near the water body. Hence the first point above is applicable (difference in surface temperature, roughness lengths, albedo). Second, the storage of heat in the water takes over the role of the soil heat flux in $Q^* - G$. This storage may be due to exchange of energy at the water surface, but also due to the penetration of solar radiation down to some depth in the water.

Question 7.4: Consider the Penman equation.
a) What happens with $L_v E$ if air is saturated?
b) What happens with $L_v E$ if there is no net energy input ($Q^* - G = 0$)?
c) What happens with r_a if there is no net energy input?

d) What happens with $L_v E$ if the temperature increases (at given relative humidity)
e) Under which conditions is it possible that $L_v E$ is larger than the available energy input ($Q^* - G$)?

7.2.2 Penman–Monteith Derivation

Monteith (1965) and Rijtema (1965) independently proposed how to extend the Penman method to vegetated surfaces, resulting in what is nowadays called the Penman–Monteith method.

To make the link from a wet surface to a vegetated surface, the concept of the 'big leaf' is introduced. The vegetation is simplified to one single leaf, with one idealized stomatal cavity (see Figure 7.2, and compare to Section 6.4 in Chapter 6). All water vapour is assumed to originate from the stomata, and hence the Penman–Monteith methods aims to describe the process of transpiration, not evaporation. Furthermore, the method is – strictly speaking – limited to surfaces that are fully covered by vegetation: a mixture of vegetation and bare soil (sparse vegetation) cannot be dealt with, as this would involve two different pathways for water vapour (from the soil and from the plants) and two different surface temperatures. Shuttleworth and Wallace (1985) provide an example of how the Penman–Monteith method could be extended to deal with evapotranspiration from sparse canopies: both transpiration and soil evaporation.

Nevertheless, the water vapour flux described by the Penman–Monteith method is here referred to as evapotranspiration (hence including both transpiration and evaporation from soil and intercepted water). Where appropriate, it is indicated which component of evapotranspiration is described with the Penman–Monteith equation.

As compared to the wet surface discussed in Section 7.2.1, the transport path for heat has not changed: transport takes place from a surface with temperature T_s to the atmosphere with temperature $\overline{T}_a$ at a certain height. But water vapour does not originate at the surface of the leaf, but from within the stomatal cavity. This leads to two important assumptions:

- The air within the stomatal cavity is saturated with water vapour, at the surface temperature T_s, that is, $e_s = e_{sat}(T_s)$. This appears to be a sound assumption, even under conditions of considerable water stress (Ball, 1987).
- The transport from within the leaf to the surface of the leaf experiences a separate resistance, the canopy resistance[2] r_c. This resistance acts in series with the aerodynamic resistance r_a. Although the introduction of a canopy resistance is both mathematically simple and conceptually appealing (it looks like a stomatal resistance; see Figure 7.2), it shifts the complexity of the determination of evapotranspiration largely toward the specification of the canopy resistance (see Section 7.2.3).

[2] The canopy resistance is also frequently called 'surface resistance', and indicated as r_s. But to prevent confusion with the stomatal resistance, we use the term canopy resistance and symbol r_c here.

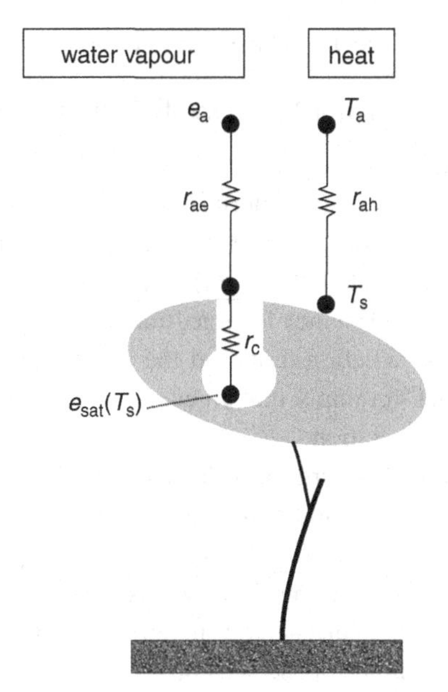

Figure 7.2 The 'big leaf' approach: vegetation replaced by one leaf, with one stomatal cavity. Water vapour transport experiences an extra resistance, r_c (canopy resistance). Note that the aerodynamic resistances are those for scalar transport, not for momentum.

Now the derivation of the Penman equation can be revisited to include the extra resistance. Equation (7.9) becomes:

$$L_v E = -\rho \frac{c_p}{\gamma} \frac{\overline{e}_a - e_{sat}(T_s)}{r_{ae} + r_c} \tag{7.14}$$

Subsequently, Eq. (7.12) becomes

$$\begin{aligned} L_v E &= -\frac{\rho c_p}{\gamma}\left[\frac{\overline{e}_a - e_{sat}(\overline{T}_a)}{r_{ae} + r_c} + s(\overline{T}_a)\frac{\overline{T}_a - T_s}{r_{ae} + r_c}\right] \\ &= -\frac{\rho c_p}{\gamma}\frac{\overline{e}_a - e_{sat}(\overline{T}_a)}{r_{ae} + r_c} + \frac{s(\overline{T}_a)}{\gamma}\frac{r_{ah}}{r_{ae} + r_c}H \end{aligned} \tag{7.15}$$

Finally, in combination with the energy balance equation, this gives the Penman–Monteith equation (along the same derivation as for Eq. (7.13)):

$$L_v E = \frac{s(Q^* - G)}{s + \gamma\left(1 + \frac{r_c}{r_a}\right)} + \frac{\frac{\rho c_p}{r_a}\left(e_{sat}(\overline{T}_a) - \overline{e}_a\right)}{s + \gamma\left(1 + \frac{r_c}{r_a}\right)} \tag{7.16}$$

The Penman–Monteith equation is very similar to the Penman equation, except for the factor $1+\frac{r_c}{r_a}$ in the denominator.[3] This modification has the following effects:

- The canopy resistance is the main determining factor for the partitioning of available energy between latent heat flux and sensible heat flux (Figure 7.3a). For small values of the canopy resistance most (but not all) available energy is used for transpiration, whereas when the canopy resistance is large, transpiration will be reduced (because the denominator of both the radiation term and the aerodynamic term will increase). But, whereas $r_c = 0$ s m^{-1} does not imply that all energy is used for transpiration, $r_c \rightarrow \infty$ *does* imply the absence of transpiration.
- The partitioning of available energy also depends on the aerodynamic resistance (Figure 7.3b). But this influence is modified by the canopy resistance in peculiar way. If the canopy resistance is zero or low, a reduced r_a (e.g., due to higher wind speed) always leads to a higher evapotranspiration. This seems trivial, as a more efficient exchange of water vapour between the surface and the air should favour evapotranspiration. But the same argument could hold for the sensible heat flux. However, for a fixed available energy, only one of the two fluxes can increase and the other has to decrease accordingly. So why does the evapotranspiration 'benefit' from a lower aerodynamic resistance if the canopy resistance is low? And why does the reverse happen when the canopy resistance is high: the sensible heat flux increases with decreasing r_a?

  The key to this problem is the surface temperature, which influences both H and L_vE (through $e_{sat}(T_s)$; see Eq. (7.14)). A lower r_a leads to a stronger coupling between surface and atmosphere. Hence, a smaller contrast in temperature and humidity between the surface and the atmosphere is sufficient to yield the same fluxes of sensible and latent heat: the surface temperature will be closer to the air temperature.

  - If r_c is small, the change in total resistances for heat (r_a) and moisture ($r_a + r_c$) will be nearly identical. Then the impact of a change in r_a on H and L_vE depends solely on the related changes in $T_a - T_s$ and $e_a - e_{sat}(T_s)$, respectively (see Eqs. (7.8) and (7.14)). Because the latter can approximated as VPD + $s(T_a - T_s)$, we can see that the relative change in the contrast in moisture between atmosphere and surface will always be less than the relative change in temperature contrast. Hence L_vE benefits most from the decreased aerodynamic resistance.
  - In the case of a large r_c, the changes in the total resistances for heat (r_a) and moisture ($r_a + r_c$) will no longer be identical. The relative decrease in $r_a + r_c$ will be much less than that in r_a. If we assume $r_a + r_c$ to be nearly constant, while the surface-to-atmosphere moisture contrast decreases due to the decrease in surface temperature, the evapotranspiration will *decrease* with decreasing aerodynamic resistance (in favour of an increase of the sensible heat flux).

[3] Often $\gamma\left(1+\frac{r_c}{r_a}\right)$ is denoted by γ^*, a psychrometric constant that takes into account that the resistances for moisture and heat transfer differ.

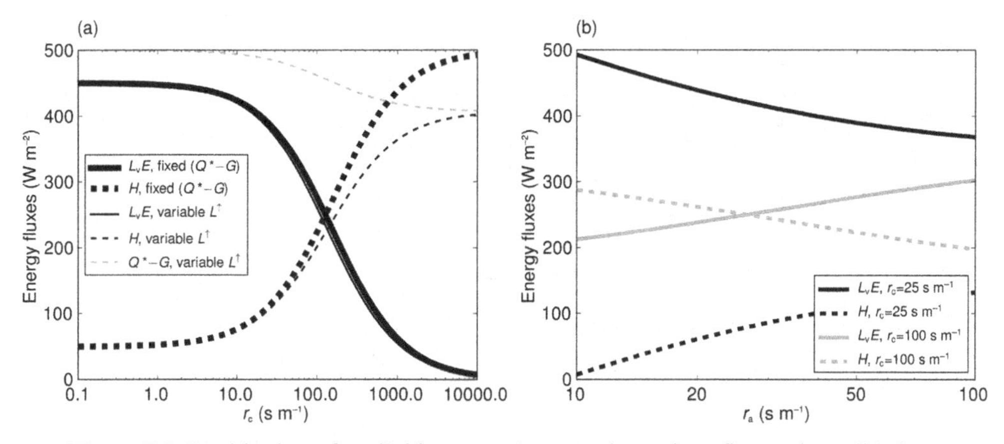

Figure 7.3 Partitioning of available energy between latent heat flux and sensible heat flux as a function of canopy resistance **(a)** and aerodynamic resistance **(b)**. Fluxes determined with the Penman–Monteith equation, with a fixed available energy of 500 W m^{-2}, T_a = 20 °C, RH = 60%. In **(a)** r_a is set to 50 s m^{-1}. Because the surface temperature increases when transpiration decreases, the longwave emission will increase with increasing r_c (the effect of this is shown in thin lines). The effects of changes in stability on r_a have not been taken into account.

> To summarize: in the case of a decrease in r_a, the type of energy that is most easily available is extracted extra from the surface. If the surface is wet (or at least has a low r_c), latent heat is most readily available, so evapotranspiration will benefit from a lower r_a, at the expense of the sensible heat flux. But for higher r_c the situation will be reversed, as sensible heat is more easily available than latent heat (release of water to the atmosphere is hampered by stomatal closure).

The partitioning between latent and sensible heat flux usually varies through the day because available energy and latent heat flux are out of phase (Figure 7.4a). The latent heat flux generally peaks a few hours later than the available energy and hence the sensible heat flux peaks *before* the available energy. This observation can be explained using the Penman–Monteith equation as follows. Whereas the radiation term in the Penman–Monteith equation is in phase with the available energy (provided r_a and r_c are constant), the aerodynamic term generally has its maximum later. This delay in the peak of the aerodynamic term is caused mainly by the fact that the vapour pressure deficit peaks late in the afternoon, which is closely linked to the fact that the maximum temperature usually occurs late in the afternoon (and hence the maximum in $e_{sat}(T_a)$; Figure 7.4c, d). In some cases VPD also increases by a decrease in the vapour pressure due to entrainment of dry air (see Section 7.3.1). The phase shift between radiation term and aerodynamic term also implies that the relative contribution of both terms to the total evapotranspiration varies through the day, where the contribution of the aerodynamic term gradually increases (Figure 7.4b).

Regarding the predictive powers of the Penman–Monteith equation the same disclaimer holds as for the Penman equation: the Penman–Monteith equations is only a

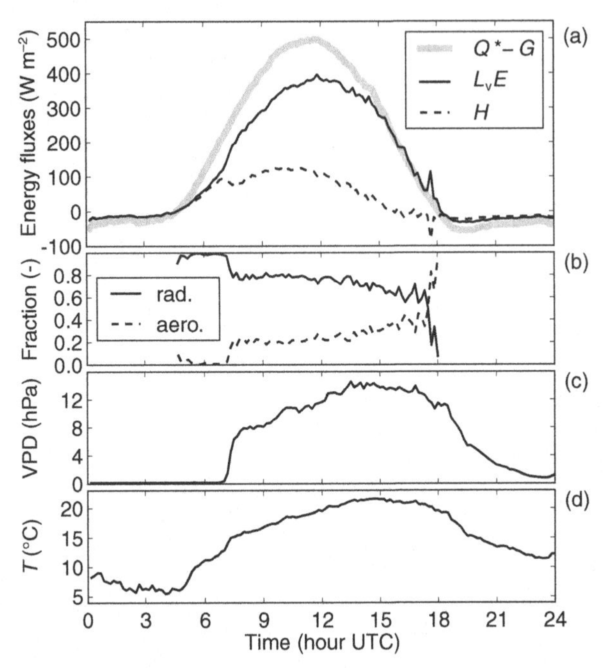

Figure 7.4 Diurnal cycle of latent and sensible heat flux (**a**) as determined with the Penman–Monteith method with z_0 = 0.03 m and r_c = 30 s m$^{-1.}$ (**b**) Relative contribution of radiation and aerodynamic term to total evapotranspiration, for conditions when $(Q^* - G) > 0$ only Diurnal cycle of driving variables vapour pressure deficit (**c**), air temperature (**d**) and wind speed (**e**)**.** (Data from Haarweg Meteorological station, May 23, 2007)

descriptive equation. Only if it is used with observations that are related to the surface and situation for which one would like to compute the evapotranspiration will it give the correct answer.

Another note regarding the use of the Penman–Monteith equation for computations of evapotransporation is related to the surface temperature. Although it is no longer visible in the expression (through Penman's linearization) it still plays a role. The surface temperature is a complex resultant of the various heating processes (for daytime: incoming radiation) and cooling processes (for daytime: upwelling radiation, soil heat flux and turbulent fluxes of heat and water vapour). As the surface temperature affects many of these processes ($L^{\uparrow}$, G, H and L_vE), it will adjust itself until the terms of the energy balance actually balance (see also Figure 7.3). When one *prescribes* $Q^*- G$ in the computation of evapotranspiration, one implicitly assumes to know the resulting surface temperature beforehand. This is true when observed values for Q^* and G are used, or when Q^* and G have a relatively fixed relationship to incoming radiation (as is the case for well-watered vegetated surfaces).

Although the Penman–Monteith equation is designed to describe transpiration, it can also describe the evaporation of water from the surface of vegetation (e.g., after rain or dewfall). In that case there is no stomatal control and the canopy resistance is set to zero. The Penman–Monteith equation then reduces to the Penman equation. The latter can also be used to model dewfall, as discussed in Section 7.4.

7.2.3 Canopy Resistance

Despite the fact that the big-leaf model treats a vegetation as one single leaf, the canopy resistance is an integrated property of an entire vegetation. It is related to, but not identical to, the stomatal resistance discussed in Chapter 6 (see, e.g., Jarvis and McNaughton, 1986, and Baldocchi et al., 1991). It is related in the sense that the same responses to internal (water potential) and external factors (radiation, temperature and vapour pressure deficit) do appear. But the canopy resistance is different from the stomatal resistance in the sense that the first is integrated over many leaves, possibly covering more than 1 m^2 of leaf per m^2 of ground area. Furthermore, some of the environmental factors that affect the stomatal resistance vary with height within the canopy (e.g., radiation).

The canopy resistance can be inferred from observed latent heat fluxes, provided that all other variables occurring in the Penman–Monteith equation are known from observations as well (the separation between transpiration and evaporation, possibly from the soil, is a special point of attention in this respect; see Kelliher et al., 1995). An example is shown in Figure 7.5 for a forest in the Netherlands. It is clear that the canopy resistance shows a similar response to environmental conditions as the stomatal resistance discussed in Chapter 6:

- The resistance increases with vapour pressure deficit: the vegetation tends to close its stomata under dry conditions.
- The resistance decreases with increasing global radiation: the plants only open their stomata when light is available that is needed to perform photosynthesis.

In modelling, the minimum canopy resistance ($r_{c,min}$) is often taken as a starting point. This is the resistance under optimal conditions: sufficient light and soil water, moderate temperatures and high relative humidity. Kelliher et al. (1995) report values of 30 to 50 s m^{-1} (these values may include soil evaporation). They suggest that the difference in minimum resistance is not so much dependent on the height (or leaf area index) of the vegetation, but more on whether the vegetation is a natural vegetation (higher minimum resistance) or agricultural crop (lower minimum resistance). They find a more or less fixed relationship between the stomatal resistance and canopy resistance: $r_c \approx 1/3\, r_s$.

Typical midday values for the canopy resistance reported for low vegetation are 30–50 s m^{-1} and for high vegetation 60–100 s m^{-1} (Kelliher et al., 1995; Monteith and Unsworth, 2007). The higher values for high vegetation are due to the fact that trees generally show a stronger response (increase in resistance) to the vapour pressure

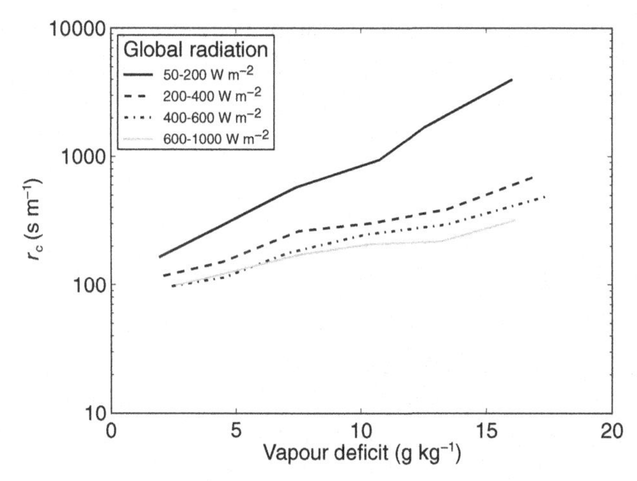

Figure 7.5 Observed canopy resistance of a Douglas fir stand in the Netherlands, as a function of vapour deficit ($q_{sat}(T) - q$, comparable to the vapour pressure deficit $e_{sat}(T) - e$) and global radiation. (Data from Bosveld and Bouten, 2001)

deficit of the air than low vegetation (see Chapters 6 and 9). This difference is particularly visible at midday when VPD is highest. Furthermore, broadleaf trees tend to show resistances that are somewhat lower than those of needleleaf tree (see, e.g., Dolman, 1986). Models that describe the variation of the canopy resistance due to variations in external (meteorological) and internal (plant) factors are dealt with in Chapter 9.

Question 7.5: In Chapter 6 the *stomatal* resistance was discussed, whereas in this chapter the related *canopy* resistance is used.

a) If a vegetation has a leaf area index (LAI) larger than 1, will the canopy resistance be higher or lower than the stomatal resistance?
b) The stomatal resistance depends on the amount of photosynthetically active radiation (PAR; see Figure 6.15). If two types of vegetation have the same LAI, but the first has one layer of leaves, and the second has a number of layers, shading each other, which of the vegetation types will have the highest canopy resistance?

Question 7.6: Consider the transpiration from a short grass vegetation. Use the Penman–Monteith equation in your answers. Given are the following observations:

Quantity	*Value*	*Unit*
Q^*	250	W m^{-2}
G	22	W m^{-2}
T_a (2 m)	15	°C
e_a (2 m)	14	hPa
r_a	50	s m^{-1}
r_c	60	s m^{-1}
P	1013	hPa

a) Compute the latent heat flux (use $c_p = 1013$ J kg^{-1} K^{-1} and $\rho = 1.22$ kg m^{-3}).
b) Compute the sensible heat flux.
c) Compute the surface temperature.

7.2.4 Analysis of Evapotranspiration from Different Surface Types

Using the Penman–Monteith equation, some hydrologically very relevant differences in evapotranspiration from different surfaces can be explained. We consider grass and a forest. The important differences are the differences in roughness (and hence aerodynamic resistance) and the difference in canopy resistance. We do not consider the difference in albedo (affecting net radiation) and differences in soil heat flux. In Table 7.1 the relevant resistances for both surfaces are given, for the situation that the canopy is dry as well as when it is wet (as a result of, e.g., rain or dew fall). Also given is the evapotranspiration for all cases (computed using the Penman–Monteith equation and decomposed in the radiation term and the aerodynamic term).

First we focus on the situation that both grass and forest are dry. In that case grass produces significantly more transpiration than forest. This is due mainly to the radiation term: the difference between surface temperature and air temperature is much larger for grass than for forest because forest is more closely coupled to the air temperature. This is a result of the low aerodynamic resistance for forest. The result of the higher surface temperature of grass is that the saturated vapour pressure in the stomata will be higher, leading to a larger humidity contrast between surface and atmosphere. The difference in the aerodynamic term between the two surfaces is only marginal because for forest the lower in r_a in the numerator is nearly compensated by the lower of r_a in the denominator. The r_a has a dominant effect on the value of the denominator because ($1/r_a$) it is multiplied with a large r_c. One could wonder how forest is able to get rid of the energy input, if both the latent heat flux and the surface temperature are low. But the smaller vertical temperature contrast (approximately half of that for grass) is more than compensated by the smaller aerodynamic resistance (one quarter of that for grass).

For a wet canopy the situation is reversed. Water is readily available at the surface of the canopy and hence there is no stomatal control for the evaporation ($r_c = 0$). In this case the radiation term is identical for both surfaces and hence the only difference in evaporation can come from the aerodynamic term. The fact that the aerodynamic resistance of forest is only 25% of that of grass directly translates in an aerodynamic term for forest that is four times the value for grass. The total evaporation for grass is slightly larger than the available energy, but for forest the evaporation amounts to 2.24 times the available energy. Thus a significant amount of energy needs to be extracted from the air (negative sensible heat flux).

To summarize: owing to the low aerodynamic resistance of forests, the variations in evapotranspiration due to variations in the canopy resistance are magnified as compared to surfaces with higher aerodynamic resistances. The water loss depends not

Table 7.1 Typical values for aerodynamic resistance and canopy resistance for low and high vegetation (grass and forest) and the resulting contributions of the radiation and aerdynamic term for dry and wet vegetation

	Grass		Forest	
	Dry	Wet	Dry	Wet
r_c (s m^{-1})	50	0	100	0
r_{ah} and r_{av} (s m^{-1})	40	40	10	10
r_c/r_a (-)	1.25	0	10	0
L_vE (radiation term, W m^{-2})	247	343	83	343
L_vE (aerodynamic term, W m^{-2})	124	195	140	780
$L_vE/(Q^* - G)$	0.74	1.08	0.45	2.24
H (W m^{-2})	129	–38	277	–623

Available energy (Q^*– G) has been set to 500 W m$^{-2,}$ ambient temperature to 20 °C and relative humidity to 60%.

only on the vegetation type, but also on the duration of periods that the vegetation is wet. If a forest is located in a climate with frequent, light, rains the canopy will be wet for most of the time, and water loss will be large as compared to grass. If, on the other hand, the same amount of rain falls in infrequent but intense events, the canopy will be dry most of the time, and a forest would lose less water than grassland. The magnitude of the interception reservoir is also relevant in this respect, as it determines how long evaporation from a wet surface can continue before the liquid water on the canopy is depleted (see Chapter 6). These differences in evapotranspiration between grass and forests have important consequences for water management (e.g., which part of the precipitation is available for ground water recharge or runoff).

Question 7.7: Evaporation is different for different surface types and surface conditions. The table below gives surface properties for various surface types and conditions.

Give typical values or qualitative indications (e.g., high, low, not relevant) for the surface properties mentioned in the first column. Do this for each of the surface types listed in the header, that is, wet and dry forest, wet and dry grassland and lakes.

Also explain in one sentence for each surface property the origin of the differences between the different surface types.

	Forest (dry)	Forest (wet)	Grassland (dry)	Grassland (wet)	Lake
Canopy resistance, r_c					
Aerodynamic resistance, r_a					
Roughness length z_0					
Albedo r					
Evaporation of intercepted water					

7.3 Derived Evapotranspiration Models

7.3.1 Equilibrium Evaporation

If we consider an extensive surface well supplied with water, we would expect that at a certain stage the air above the surface would become saturated with water vapour. As a result the aerodynamic term of the Penman equation would become zero and the evaporation from the surface would be equal to the radiation term only. If energy would continue to be supplied, the evaporation would continue and one would intuitively expect the air to become oversaturated (i.e., $e_a > e_{sat}(T_a)$) due to the continuous supply of water vapour. One could imagine this scenario to occur, for instance, when air is advected over the ocean over long distances.

However, the surface supplies not only water vapour to the air, but heat as well: $H = (Q^* - G)\left[1 - \frac{s}{s+\gamma}\right]$. Hence the air temperature increases so that $e_{sat}(T_a)$ increases as well. It can be shown that the increase of e_a and $e_{sat}(T_a)$ occurs at the same pace so that $e_a - e_{sat}(T_a)$ remains zero once saturation is reached.

This remaining evaporation is called the equilibrium evaporation:

$$L_v E_{eq} \equiv \frac{s}{s+\gamma}(Q^* - G) \tag{7.17}$$

In practice, however, it appears that the air above wet surfaces hardly ever becomes completely saturated.[4] This lack of saturation of the air is due to the fact that the atmospheric boundary layer (ABL), is not a closed box. The air above the ABL is warm and dry. Hence, at the top of the ABL warm and dry air is entrained from the free atmosphere into the boundary layer leading to a warming and drying of the boundary layer (see Figures 3.1 and 3.2). This entrainment of dry air is equivalent to a loss of humidity *from* the boundary layer (see Figure 7.6). This drives the water vapour content away from saturation. As a result, the aerodynamic term of the Penman (or likewise Penman–Monteith) equation always plays a role.[5]

Question 7.8: Consider the situation in the right part of Figure 7.6.

a) Explain the difference in the development of the water vapour pressure with the case *with* a lid (no loss of water vapour at the top, left part of Figure 7.6).
b) In Figure 7.6 the flux of water vapour at the top of the box is sketched to be equal at all x-locations. Explain that the flux in reality will change with location (and how does it change?).

[4] *If* saturation occurs (e.g., fog formation), this will not be due to evaporation from the underlying surface only, but some form of cooling or advection of moisture must play an additional role.

[5] De Bruin (1983) was one of the first to describe quantitatively the importance of the ABL on surface evapotranspiration.

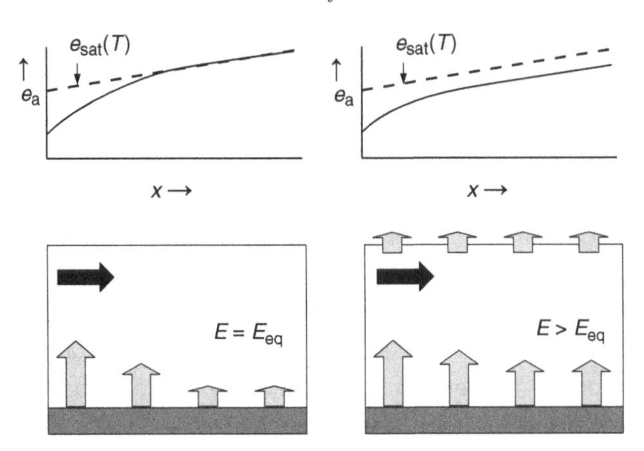

Figure 7.6 The concept of equilibrium evaporation for a closed box (left) and an atmospheric boundary layer that exchanges water vapour with the free troposphere through entrainment of dry air (right). The air flows in the direction of the black arrow. The grey arrows signify evaporation from the surface or loss of water vapour through entrainment (only at the right). The saturated vapour pressure (top) increases as air flows over the surface because heat is added to the air as well.

Question 7.9: Compute the Bowen ratio for a situation where equilibrium evaporation takes place for two conditions assume $\gamma = 0.66$ hPa K^{-1}:

a) Air temperatures is 5 °C.
b) Air temperature is 25 °C.
c) Which of the two cases has the highest evaporation?

Question 7.10: Show that for the closed box in Figure 7.6 (left) indeed the VPD ($e_a - e_{sat}(T_a)$) does not change in time if the evaporation is equal to the equilibrium evaporation.

An indirect effect of the entrainment of dry air is that a negative feedback will occur with the surface evapotranspiration: more dry air entrainment leads to a drier boundary layer, which leads to an increase of surface evapotranspiration (through the aerodynamic term). This increase in evapotranspiration in turn counteracts – partly – the effect of the dry air entrainment by moistening the boundary layer (see feedback loop 3 in Figure 7.7 and van Heerwaarden et al., 2009). Other negative feedbacks that limit the variation in the aerodynamic term are the following. If evapotranspiration becomes limited, the sensible heat flux increases, which leads to direct heating of the ABL, a subsequent increase of the VPD and hence an increase of evapotranspiration (loop 1.1). The ABL is also heated by the entrainment of warm air from the free troposphere, which leads to an increase of evapotranspiration through increased VPD as well (loop 1.2). Finally, direct moistening of the

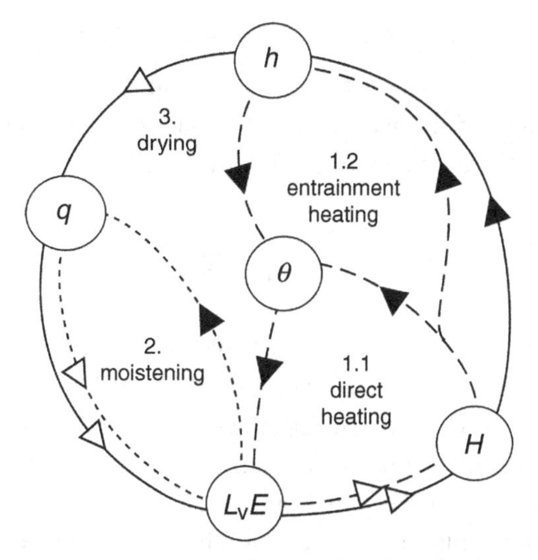

Figure 7.7 Feedbacks between surface fluxes *LE* and *H* and the temperature (θ) and humidity (q) in the boundary layer and boundary-layer height (h). Solid arrows are positive feedbacks, open arrows negative feedbacks. Line styles indicate different feedbacks: heating from surface and through entrainment (long dashes), moistening from surface (short dashes) and drying through entrainment (solid line). (From van Heerwaarden et al., 2009; used with permission from John Wiley & Sons)

ABL by evapotranspiration limits VPD, and thus the magnitude of the aerodynamic term (loop 2).

Even though the air does not become saturated, due to dry air entrainment, the concept of an equilibrium evaporation can still be defined: it is the evaporation that leads to a stationary VPD (rather than VPD = 0). Its value solely depends on the speed of entrainment in combination with the contrast in temperature and humidity between the boundary layer and the free atmosphere above. Together these determine the rate of entrainment heating and entrainment drying, respectively, both of which affect VPD (see Figure 7.7). In equilibrium the surface evapotranspiration would exactly cancel the effect of entrainment on VPD. The equilibrium evaporation does not depend on the surface conditions. But whether or not the surface evapotranspiration actually reaches this equilibrium value (in the course of a diurnal cycle) depends mainly on the magnitude of the canopy resistance and aerodynamic resistance: higher resistances delay the approach to equilibrium, in many cases beyond the duration of daytime conditions (see van Heerwaarden et al., 2009).

Despite the utility of the concept of equilibrium evapotranspiration in an entraining boundary layer, we refer – in the remainder of the book – to the classical expression in Eq. (7.17) as the equilibrium evapotranspiration.

7.3.2 Priestley–Taylor Equation

Priestley and Taylor (1972) were the first to recognize from *observations* that the actual evapotranspiration from well-watered surfaces (water or low vegetation) was generally higher than the equilibrium evaporation:

$$L_v E = \alpha_{PT} \frac{s}{s+\gamma}(Q^* - G) \tag{7.18}$$

where α_{PT} is the Priestley–Taylor coefficient which is of the order of 1.26 for well-watered surfaces. For a wet surface, where the Penman Eq. (7.13) would be a good estimate of the actual evaporation, the Priestley–Taylor equation implies that the aerodynamic term of the Penman equation is proportional to the radiation term: the aerodynamic term equals 26% of the radiation term. Fixing the proportionality between the two terms of course omits some variability that occurs in reality. But, as shown in the previous section, feedbacks in the ABL make that variations in the aerodynamic term will be damped. Furthermore, because the aerodynamic term is only of the order of one quarter of the radiation term, deviations from this strict proportionality (due to variations in aerodynamic resistance or VPD) do not influence the total evaporation greatly.

For vegetated surfaces, the situation is more complex, because then we need to compare the Penman–Monteith estimate for evapotranspiration (Eq. (7.16)) with the Priestley–Taylor estimate (Eq. (7.18)). In that case not only r_a and VPD play a role, but more importantly the canopy resistance.

Figure 7.8 shows the observed dependence of α_{PT} on the canopy resistance for typical mid-latitude summer conditions over short vegetation. Indeed for low values of the canopy resistance (wet or well-watered surfaces) the value of α_{PT} is of the order of 1.1 to 1.2, whereas it decreases with increasing r_c.

Despite the fact that the Penman–Monteith equation is not a predictive equation it can still be useful for sensitivity analyses, provided that the results are treated with caution. In the present context one interesting feature that can be investigated with the Penman–Monteith equation is the dependence of the Priestley–Taylor coefficient on wind speed (or aerodynamic resistance). Therefore Figure 7.8 also shows the modelled dependence of α_{PT} on canopy resistance, for a number of values for the aerodynamic resistance. The tendency of the modelled α_{PT} vs. r_c is the same as for the observations. Furthermore, for low values of the canopy resistance the evaporation (and hence α_{PT}) *increases* with decreasing aerodynamic resistance (i.e., with higher wind speed), as one would expect. But, for high values of r_c the evaporation *decreases* with decreasing r_a. This behaviour is in accordance with the analysis related to Figure 7.3b. The value of the canopy resistance at which α_{PT} is independent of the aerodynamic resistance $\left(\text{i.e., } \frac{\partial L_v E}{\partial r_a} = 0\right)$ happens to be the point where $\alpha_{PT} = 1$

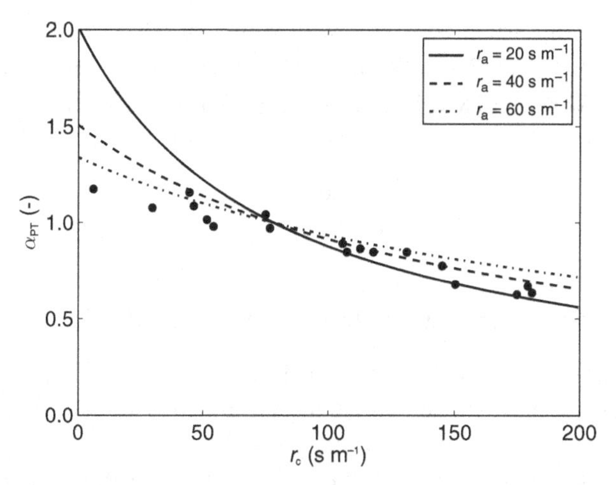

Figure 7.8 Priestley–Taylor α_{PT} as a function of canopy resistance (**a**). Symbols show observations obtained in 1977 at Cabauw (The Netherlands) at local noon for sunny conditions. (Data from DeBruin, 1983). Lines show α_{PT} for a range of aerodynamic resistances, as calculated from the Penman–Monteith equation (with $Q^* - G = 500$ W m^{-2}, RH = 50% and $T_a = 25\,°C$).

and hence L_vE equals L_vE_{eq}. The Bowen ratio at this point is $\beta = \frac{\gamma}{s} = \beta_{eq}$ and the corresponding canopy resistance can be derived to be $r_{c,eq} = \left(1+\beta_{eq}\right)\frac{\rho c_p}{\gamma}\frac{e_{sat}(\overline{T}_a)-\overline{e}_a}{Q^*-G}$ (see Jacobs and de Bruin, 1992). This $r_{c,eq}$ is typically of the order of 50 s m^{-1} which is representative of well-watered low vegetation. Hence, for well-watered surfaces, the sensitivity of evapotranspiration to aerodynamic resistance (and thus to wind speed) is rather limited.

Question 7.11: Why does the Priestley–Taylor coefficient α_{PT} decrease with increasing canopy resistance (see Figure 7.8). Use the Penman–Monteith equation in your answer.

Question 7.12: Explain why at high values of the canopy resistance the evaporation decreases with decreasing aerodynamic resistance (see Figures 7.3b and 7.8).

7.3.3 Makkink Equation

Already in 1957 (15 years before the paper of Priestley and Taylor) Makkink (1957) found that the evapotranspiration of grass for Dutch summer conditions could be estimated as:

$$L_vE = c_1\frac{s}{s+\gamma}K\downarrow + c_2 \tag{7.19}$$

where c_1 and c_2 are empirical constants with values of 0.9 and 30 W m^{-2} respectively (for daily mean values). DeBruin (1987) proposed a slightly simplified version of Eq. (7.19) which is now generally referred to as the Makkink equation:

$$L_v E = 0.65 \frac{s}{s+\gamma} K \downarrow \tag{7.20}$$

Although the original Makkink equation predates the Priestley–Taylor equation, formally it could be derived from it (DeBruin and Lablans, 1998):

- For daily mean values the soil heat flux can be neglected (in first approximation the same amount of energy enters the soil during the day as leaves the soil at night). Hence in the Priestley–Taylor equation $Q^* - G$ can be replaced by Q^*.
- For well-watered surfaces the net radiation is roughly 50% of the global radiation. This fixed ratio is due to the well-defined albedo of grass and the moderate difference between air temperature (important in determining $L^{\downarrow}$) and surface temperature (determining $L^{\uparrow}$).

These two steps yield Eq. (7.20) from Eq. (7.18) with a proportionality constant of 0.63 rather than 0.65 (which is close enough given the empirical nature of the Makkink equation).

Question 7.13: Make a table listing all methods to estimate evapotranspiration presented in this chapter. For each method, list the assumptions/restrictions, the required input data and advantage/disadvantage of the method.

7.4 Dewfall

Because the Penman–Monteith equation is based on a combination of the surface energy balance and a description of atmospheric heat and moisture transport, it should in principle be able to reproduce dewfall as well. As dewfall generally occurs under conditions when the amount of incoming solar radiation is small or zero, the stomata will be closed and water vapour is transported to the surface of leaves only. If we assume that there is some initial thin layer of liquid water on the vegetation, and if the dynamics of transport of radiation, heat and water vapour *within* the canopy is ignored, the Penman equation can be used to determine the amount of dewfall D (dewfall occurs if $D < 0$):

$$L_v D = \frac{s(Q^* - G)}{s+\gamma} + \frac{\frac{\rho c_p}{r_a}\left(e_{sat}(\overline{T}_a) - \overline{e}_a\right)}{s+\gamma} \tag{7.21}$$

Whereas both terms are positive in the case of daytime evaporation, in the case of dewfall the first term will be negative (taking care of the removal of the energy released upon condensation), whereas the second term will be positive (drier air will suppress the formation of dew through evaporation of part of the formed dew). Figure 7.9 shows an example of the skill of Eq. (7.21) in predicting the dewfall on grassland.

The upper limit for dew formation will be reached when the aerodynamic term in Eq. (7.21) (which counteracts dewfall) is zero. The so-called potential dewfall then is:

$$L_v D_{pot} = \frac{s(Q^* - G)}{s + \gamma} \tag{7.22}$$

For example, for a night with a mean air temperature of 10 °C, mean $Q^* - G$ of –40 W m^{-2}, the potential dewfall rate will be 0.032 mm h^{-1}, or 0.26 mm during a night of 8 hours.

There are two ways for the aerodynamic term in Eq. (7.21) to become zero:

- The air is saturated, resulting in $e_{sat}(\overline{T}_a) - \overline{e}_a = 0$. But if the air is so saturated that fog occurs, the radiative cooling will be suppressed, leading to a low value for the potential dewfall.
- The aerodynamic resistance tends to infinity. This occurs when the wind speed is zero or so low that stability effectively suppresses turbulence completely. In that case all turbulent transport is absent: the supply of water vapour to the surface will vanish and the production of dew will cease quickly once all water vapour close to the surface has condensed.

Thus, favourable conditions for dewfall occur when the air is close to saturation and the wind speed is low, but not so low that turbulence is suppressed.

It should be noted that for situations other than low canopies or wet soils, the modelling of dewfall (and dewrise) will be more complicated. For higher canopies (higher than about 35 cm) the energy balance needs to be treated in a number of separate layers (e.g., Jacobs et al., 1990, 2005). Consideration of the energy balance of a control volume rather than a surface is needed (see Chapter 1). For dry soils the assumption of a wet surface will no longer be valid and the model needs to be adjusted accordingly. Furthermore, in the case of sloping terrain the effect of differences in insolation (and subsequent longwave cooling at night) between slopes needs to be taken into account, as well as the possibility of katabatic drainage flows leading to local cooling in depressions (see Jacobs, 2000).

Question 7.14: Estimation of dewfall using the Penman equation.

a) Which conditions are favourable for dewfall to occur (consider time of day, cloudiness, water vapour content in the air and wind speed; explain for each variable why this is a favourable condition).
b) For dewfall to occur, the air does not need to be saturated. Use the Penman–Monteith equation to derive an expression for the maximum vapour pressure deficit (VPD) at which dewfall can still occur.

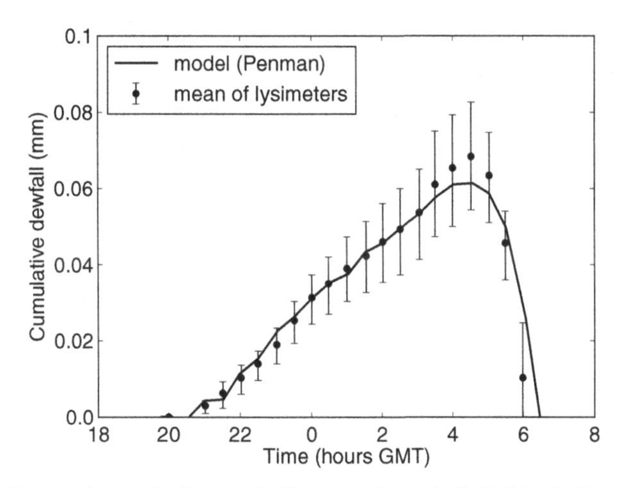

Figure 7.9 Comparison of observed (dots) and modelled (black line) cumulative dewfall for a single night over grass. (Data from Jacobs et al., 2006)

c) Given the following observations: $Q^* = -60$ W m^{-2}, $G = -10$ W m^{-2}, $T_a = 15$ °C, $e_a =$ 15 hPa, $p = 1013$ hPa and $r_a = 100$ s m^{-1}. Assume $L_v = 2.47 \cdot 10^6$ J kg^{-1}, $\rho = 1.22$ kg m^{-3}, $c_p = 1013$ J kg^{-1} K^{-1}. What is the dewfall rate in mm h^{-1}?

7.5 Summary

With the surface-energy balance and the formulations for the turbulent fluxes a number of combination methods can be developed. The Bowen ratio method uses observations of temperature and humidity at two heights, in combination with observations of the available energy to derive the sensible and latent heat flux. The method can be extended to other scalars, if their vertical concentration difference is measured in conjunction with either temperature or humidity differences.

The Penman method combines the available energy with expressions for the turbulent fluxes based on observations at two levels: one atmospheric level and one at the surface. The surface is supposed to be wet. The flux expression contains an aerodynamic resistance that expresses the ease with which heat and water vapour are transported between the surface and the atmosphere. Because the air at the surface is saturated, the vertical humidity difference can be determined from the humidity at the atmospheric level and the temperature at the surface. To arrive at a closed-form equation, the surface temperature is eliminated through a linearization. The Penman equation consists of two terms: the radiation term and the aerodynamic term.

The Penman–Monteith method extends the Penman method to vegetated surfaces by including an extra resistance, the canopy resistance, in the pathway of humidity transport. For practical applications one needs to model the canopy resistance (discussed in Chapter 9), or assume a value for it. The relative magnitudes of canopy resistance and aerodynamic resistance are an important factor in determining evapotranspiration from different surfaces (rough vs. smooth and dry vs. wet).

Whereas the radiation term of the Penman equation is determined mainly by the energy input at the surface, the aerodynamic term is strongly coupled to the atmospheric conditions (in particular vapour pressure deficit). Through various feedback mechanisms, the magnitude of the aerodynamic term relative to the radiation term has a limited range. This fact is employed in Priestley–Taylor method, which uses the radiation term of the Penman equation (also called equilibrium evaporation) as the starting point and adds the aerodynamic term as a fixed fraction. The Makkink method is a further simplification that replaces the available energy by global radiation.

As dewfall is the inverse process of evaporation from a wet surface, it can be well described with the Penman method.

8
Integrated Applications

This chapter shows how the knowledge from the previous chapters can be combined to understand and manage processes at the land–atmosphere interface. First, attention is paid to the estimation of crop water requirements using the crop factor method and to the direct measurement of evapotranspiration using lysimeters. Then it is shown how in a semiarid region the water productivity of irrigated crops can be studied and improved. Finally, the response of different vegetation types (grass and forest) to heat wave conditions is studied.

8.1 Crop Water Requirements

Evapotranspiration determines to a large extent the hydrological cycle and the environmental conditions near the soil surface. There is a direct relation between the ratio of actual to optimal transpiration and the ratio of actual to optimal crop yield. Irrigation water requirements are determined by the amount of evapotranspiration relative to the amount of natural rainfall and readily available soil moisture. Groundwater recharge and soil salinization also depend largely on the amount of evapotranspiration. In the context of agricultural practice the water required to grow a crop does not only include the water loss due to evapotranspiration, but also the water needed to leach salts and to compensate for nonuniform application of the water (Allen, 1998).

This section focuses on the application of the methods developed in the previous chapter in the determination of crop water requirements.

8.1.1 Definitions of Terms and Units

A confusing variety of terms regarding evaporation, transpiration and evapotranspiration exists. To reduce the confusion we will list here the definitions as they are used here (where we mainly follow Moors (2002)).

- The total evapotranspiration E[1] consists of:
  - Transpiration (T): the part of the total water vapour flux that enters the atmosphere from the soil through the vegetation (stomata and cuticula).
  - Evaporation of intercepted water (E_{int}): evaporation of water that has been intercepted by plants.
  - Soil evaporation (E_{soil}): evaporation of water from the soil (the soil may either be saturated or partly dry).
- Potential evapotranspiration E_{pot} is the *theoretical* evapotranspiration that would occur if a given vegetation, completely covering the soil, is exposed to prevailing meteorological conditions (without itself affecting the meteorological conditions). The term 'potential evaporation' sometimes leads to confusion when no reference is made to a specific type of vegetation. In that case referring to '*the* potential evapo(transpi-)ration' becomes useless.
- Reference evapotranspiration E_{ref}: is the *theoretical* evapotranspiration that would occur if a well-defined, theoretical vegetation, completely covering the soil is exposed to prevailing meteorological conditions (without affecting the meteorological conditions).

Furthermore, in this context a short discussion of units is needed. Whereas meteorologists consider evapotranspiration as an energy term, for practical applications the depth of the water layer that evaporates is important (usually in mm d^{-1}). To arrive from energy flux densities (in W m^{-2}) to fluxes in mm d^{-1}, the following steps are needed:

- A flux density of 1.0 W m^{-2} that continues for 1.0 day (86 400 s) amounts to a total energy flux of 86 400 J m^{-2} d^{-1}.
- With this amount of energy one can evaporate 86 $400/L_v$ kilograms of water per square metre in one day. Taking $L_v = 2.45 \cdot 10^6$ J kg^{-1} (strictly speaking, L_v is temperature dependent) the mass of water evaporated by this amount of energy (1.0 W m^{-2} during one day) is approximately $3.53 \cdot 10^{-2}$ kg m^{-2} d^{-1}.
- This mass of water corresponds to $3.53 \cdot 10^{-2}$ kg/ρ_w cubic meters of water (ρ_w is the density of water: 1000 kg m^{-3}). This is $3.53 \cdot 10^{-5}$ m^3 m^{-2} d^{-1}.
- This volume per square meter of surface per day is equivalent to a layer of water of $3.53 \cdot 10^{-5}$ m d^{-1}, or $3.53 \cdot 10^{-2}$ mm d^{-1}.

We could also start at the other end and pose the question what daily mean flux density is needed to evaporate 1 mm of water in one day. The answer is (1 mm d^{-1})/ ($3.53 \cdot 10^{-2}$ mm d^{-1}/W m^{-2}), which equals 28.4 W m^{-2}.

Question 8.1: The latent heat of vaporization is temperature dependent (see Appendix B). Which temperature (air temperature, surface temperature, or another temperature) should be used to compute the value of L_v that is needed to convert the evaporation in terms of a mass-flux into evaporation in terms of an energy flux. Explain your answer.

[1] Note that the literature on crop water use evapotranspiration is often denoted by *ET*, rather than *E*.

8.1.2 Factors Affecting Evapotranspiration

Factors related to weather, crop characteristics, management and environmental aspects affect evaporation and transpiration. The weather factors have been incorporated in the evapotranspiration models developed in Chapter 7. The other factors are summarized here.

Roughly three types of crop parameters that affect evapotranspiration can be identified:

- Parameters that affect the meteorological conditions, in particular the albedo (affecting the available energy through Q^*) and the roughness (affecting the aerodynamic resistance).
- Parameters that affect the ease with which water vapour is released by the plants, that is, the canopy resistance. The canopy resistance is in part determined by the type of plant, but also by the leaf area index (LAI), which changes in time, and the condition of the plants (water stress, illness).
- Parameters that affect the ease with which the plants can extract water from the soil (in particular rooting depth).

The management factors that affect evapotranspiration are multifaceted. Important characteristics of the root zone that influence the evapotranspiration rate are the water content, the salinity levels and the nutrient concentrations. In Section 6.2.3 the effect of water excess, water shortage and salinity on plant transpiration were discussed. In Section 9.1.3 the effect of water shortage on soil evaporation is considered. The nutrient concentrations *indirectly* affect the evapotranspiration rate by the fact that they affect crop growth and hence the LAI, crop height and rooting depth. Management aspects as irrigation, nutrient application, mulching, water harvesting, intercropping, drainage, leaching and conservation tillage directly affect the soil status with respect to water, salts and nutrients.

8.1.3 Crop Factor Method: General Structure

The identification of different factors that affect the evapotranspiration of an actual crop has led to the development of the so-called crop factor method, which dates back to Van Wijk and De Vries (1954) and Jensen (1968). The general idea of the crop factor method is the assumption that the weather factors and the other factors (related to crop and management) can be strictly separated. This leads to

$$E = K_c E_{ref} \tag{8.1}$$

where K_c is a dimensionless crop factor (or crop coefficient) and E_{ref} is the reference evapotranspiration.

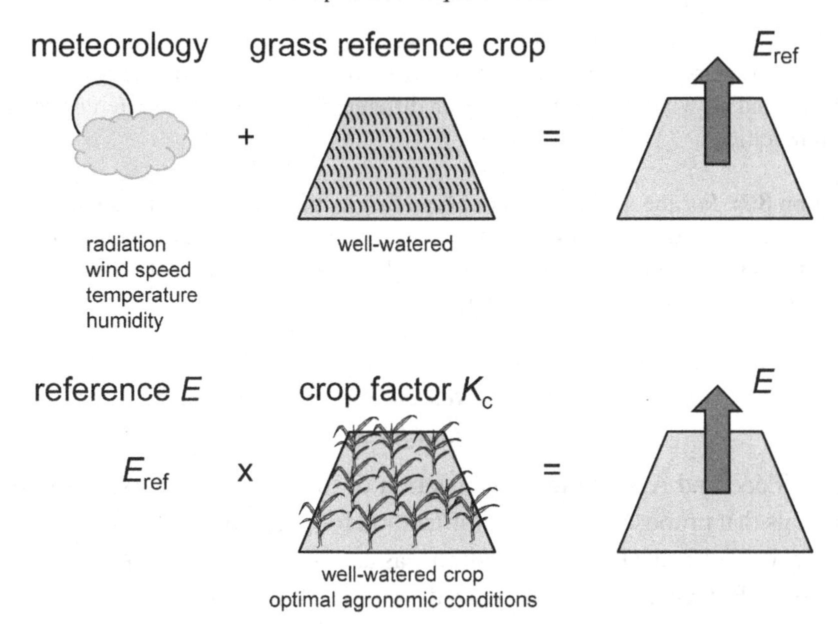

Figure 8.1 Structure of the crop factor method: meteorology and properties of the reference crop determine reference evapotranspiration. Crop type and stage determine optimal evapotranspiration for that crop. (Adapted from Allen et al., 1998.)

In fact, Eq. (8.1) serves as a *definition* of the crop factor, but no claim is made yet about its value: $K_c \equiv \frac{E}{E_{ref}}$, which implies that the crop factor is specific for a specific choice of the reference evapotranspiration. The idea is that all variation of E with crop type or management practice can be contained in the crop factor K_c (see Figure 8.1).

One should make a clear distinction between the *development* of the crop factor method and the *use* of it:

- The *development* of the crop factor method entails the performance of a large number of field experiments, for different crops (and possibly different management practices). In those experiments the actual evapotranspiration E needs to be measured, alongside the input variables that are needed to compute the reference evapotranspiration E_{ref}. From E and E_{ref} the crop factor can be computed (which will vary through the growing season). This finally yields a tabulated collection of crop factors for various crops, for various crop growth stages (initial, flowering, maturity, full senescence).
- The *use* of the crop factor method (e.g., to determine water requirements for a crop) entails the calculation of the reference evapotranspiration from observed meteorological data and the selection of the appropriate crop factor from the tabulated collection.

Question 8.2: Explain why a crop factor (or set of crop factors) is linked to a specific definition of the reference evaporation (in other words: why do crop factors for a specific crop and growing stage differ between different definitions of the reference evapotranspiration)?

Question 8.3: For the determination of crop factors the actual evapotranspiration of a crop should be measured, alongside the input variables needed for the calculation of the reference evapotranspiration. Under what conditions (i.e., above what type of surface) should those input variables for E_{ref} be measured?

8.1.4 Crop Factor Method: Penman–Monteith Equation for E_{ref}

The FAO (Food and Agriculture Organization of the United Nations) has published two manuals that propose the crop factor method for practical application. In Doorenbos and Pruitt (1977) the Penman equation was used to determine the reference evapotranspiration. But later it was recognized that the Penman–Monteith equation was better suited to determine E_{ref} (Allen et al., 1998). Here we discuss the latter method. Note that the method has been developed to be used mainly for estimations of evapotranspiration on daily to decadal (10-daily) basis.

Reference Evapotranspiration: A Hypothetical Crop

First the evapotranspiration of a reference crop E_{ref} (mm d^{-1}) is calculated according to the Penman–Monteith equation. This hypothetical reference crop is defined as a full cover crop with height $h_c = 0.12$ m, a fixed canopy resistance $r_c = 70$ s m^{-1} and an albedo $r = 0.23$. If the diurnal cycle of E_{ref} needs to be resolved canopy resistances of 50 s m^{-1} and 200 s m^{-1} are to be used for daytime and nighttime, respectively. The roughness lengths for momentum and heat are 0.012 m and 0.0012 m respectively, and the displacement height is 0.08 m. This hypothetical crop closely resembles an extensive surface of green grass of uniform height, actively growing, completely shading the ground and with adequate water. Because generally not all data required to compute E_{ref} with the original Penman–Monteith equation is available, Allen et al. (1998) list a range of empirical methods to deal with lack of data:

- The aerodynamic resistance is calculated from observed wind speed and the roughness length of the reference crop, but without stability corrections. This could be warranted by the fact that under well-watered conditions the stability correction is relatively small.[2]
- For daily and 10-daily calculations, the soil heat flux is generally neglected. For calculations with time steps of less than one day empirical relationships with net radiation are

[2] But the daily mean aerodynamic resistance is not necessarily identical to the aerodynamic resistance calculated with the mean wind speed and mean (near neutral) stability (see Chapter 3, on the Schmidt paradox).

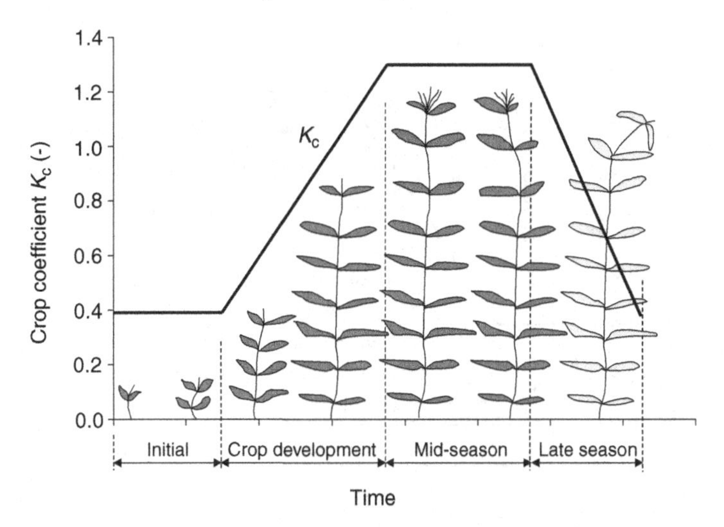

Figure 8.2 Generalized crop coefficient curve for the single-crop coefficient.

used (see Chapter 2). For periods longer than 10 days, an empirical relationship with air temperature is used (based on the assumption that the soil temperature follows the air temperature on that time scale).

- The net radiation is not based on observed values, but rather on empirical estimates. Extraterrestrial radiation is reduced to global radiation using empirical relationships with sunshine duration (similar to those given in Appendix A). For the net longwave radiation an empirical approximation is used that incorporates air temperature, humidity and cloudiness.

Furthermore, Allen et al. (1998) discuss the problem when the weather data needed to compute E_{ref} have *not* been measured in environmental conditions that correspond to the definition of reference evapotranspiration. This may happen for instance, if the crop factor method is to be used to *plan* an irrigation system in an arid region. The weather data used will refer to conditions where the irrigation scheme is not yet there, and therefore will reflect hotter and drier conditions than will actually occur once the irrigation system is in place.

Single-Crop Coefficient

The optimal evapotranspiration of the actual crop E (mm d^{-1}) is simply calculated by Eq. (8.1).

The crop growing season is divided in four stages: initial, development, mid season and late season (Figure 8.2). Allen et al. (1998) provide extensive data on the K_c values as a function of growing stage and for a large number of crops. Because of the simplifying assumptions this single-crop coefficient approach can be used only for long-term water balances and basic irrigation scheduling.

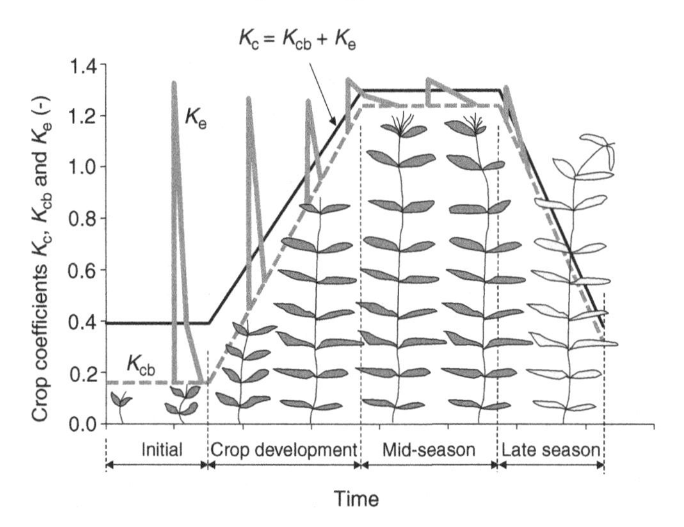

Figure 8.3 Crop coefficient curves showing the basal K_{cb}, soil evaporation K_e and the corresponding single $K_c = K_{cb} + K_e$ curves.

Dual Crop Coefficient

One of the main disadvantages of the single-crop coefficient is that two processes (transpiration and soil evaporation) have to be covered by one crop coefficient. After all, the relationship between actual soil evaporation and E_{ref} has nothing to do with plant characteristics. Therefore Allen et al. (1998) proposed to use for more accurate and detailed studies (e.g., studies on daily basis) dual-crop coefficients:

$$E = \left(K_{cb} + K_e\right) E_{ref} \tag{8.2}$$

where K_{cb} is the basal crop factor and K_e the soil evaporation coefficient. K_{cb} is defined as the ratio E/E_{ref} when the soil surface is dry. K_e describes the evaporation component of E. Figure 8.3 illustrates the methodology. The value of K_{cb} is smaller than the value of K_c in Figure 8.2 because the latter includes the average soil evaporation. If the soil is wet following rain or irrigation K_e may be large. However, the sum of K_{cb} and K_e can never exceed a maximum value as determined by the total energy amount available for evapotranspiration. K_e decreases sharply when the top soil dries out. The corresponding smoothed K_c (i.e., $K_{cb} + K_e$) curve is also shown in Figure 8.3 and illustrates the effect of averaging K_{cb} and K_e over time. The estimation of K_e requires a daily water balance computation for the water content in the top soil. Allen et al. (1998) describe in detail the procedure to determine both K_{cb} and K_e. Compared to the single-crop coefficient approach, the dual-crop coefficient approach is more suitable to analyse daily irrigation scheduling or other research studies in which daily variations in soil surface wetness affect evapotranspiration and soil water fluxes.

Extra Crop Coefficients for Nonstandard Conditions

The method of Allen et al. (1998) also provides a way to deal with nonstandard or suboptimal conditions. For example, in the case of insufficient soil moisture, the crop factor (K_c or K_{cb}) is multiplied with a water stress coefficient (equal to 1 for sufficient soil moisture, and less than 1, depending on the degree of water shortage). Nonstandard conditions that are dealt with are, for example, situations in which the vegetation cover is less than would occur under optimal conditions, or situations in which the management is different than standard.

Question 8.4: Figure 8.3 refers to a crop that is irrigated in regular intervals.

a) How can you identify from Figure 8.3 the moments in which irrigation is applied?
b) Explain the rapid increase and decrease in time of K_e.
c) Why does the relative importance of K_e in the total crop factor decrease during the 'crop development' stage?
d) Why are there no peaks of K_e in the 'late season' growing stage?

8.1.5 Crop Factor Method: Makkink Equation for E_{ref}

Instead of the Penman–Monteith equation, other methods can be used to compute the reference evapotranspiration as well. In the Netherlands the Makkink equation is used to determine E_{ref}. The use of a less sophisticated method to determine E_{ref} for temperate conditions is warranted given the fact that – especially for the growing season – evapotranspiration is largely driven by radiation. DeBruin and Stricker (2000) showed that during the growing season the Makkink method is equivalent to the Penman–Monteith equation (where the skill depends on which variables have been measured and which have been approximated empirically) and compares well with observed evapotranspiration from well-watered grass (the reference crop).

Furthermore, the advantage of the Makkink method is that only observations of global radiation and temperature are needed. Observations of wind speed and humidity, which may be more easily disturbed by the exact location of the weather station, are not needed. The limited amount of input data also makes the Makkink method suitable to derive reference evapotranspiration from satellite data (Schüttemeyer et al., 2007; DeBruin et al., 2010).

In Appendix E the crop factors for use with the Makkink equation for conditions in the Netherlands are given for a variety of vegetations.

Question 8.5: In the Netherlands the Royal Netherlands Meteorological Institute (KNMI) reports the Makkink reference evapotranspiration. Below are given the climatological monthly averages (period 1971–2000) for station De Bilt (station 280) (centre of the Netherlands) in mm/month. In addition, the precipitation (P) is also given.

	J	F	M	A	M	J	J	A	S	O	N	D
E_{ref}	7.9	15.1	31.4	54.5	82.9	86.7	91.5	80.2	48.2	27.1	11.0	6.2
P	67.0	47.5	65.4	44.5	61.5	71.7	70.0	58.2	72.0	77.1	81.2	76.8

Crop factors for use with the Makkink reference evapotranspiration can be found in Appendix E.

a) Compute the (climatological mean) optimal evapotranspiration for grass for the period April–September (determine the monthly mean crop factor for each month from the three decadal values).
b) Compute the (climatological mean) optimal evapotranspiration for potatoes for the period April–September.
c) Can each of the crops (grass and potatoes) grow on the precipitation that falls in each of the months during the growing season in an average year (i.e., is the precipitation sufficient to sustain optimal evapotranspiration)?
d) Suppose that 120 mm of water is stored in the root zone of each of crops on March 31. Is the precipitation in each of the months sufficient to let the crop grow unstressed?

8.2 Evapotranspiration Measurement: Lysimeters

In previous chapters a number of methods to *measure* evapotranspiration have been dealt with. The eddy-covariance method, as well as the profile method (using similarity relationships) have been dealt with in Chapter 3. In Chapter 7 the Bowen ratio method has been described. Here we discuss a measurement technique that does not consider the turbulent flux in the atmosphere, but rather determines the evapotranspiration as a residual of the water balance of a well-defined soil column.

The water balance of a soil column accounts for all incoming and outgoing fluxes of a soil profile and has been discussed in Section 4.2. The actual evapotranspiration E can be calculated when all the other fluxes and the change of soil water storage are known. This means that all errors in the other fluxes and soil water change will be reflected in the estimate of E.

In the soil water balance method it can be difficult to quantify the drainage or deep percolation flux D. In the case of deep groundwater levels this term is equal to the percolation flux. In the case of shallow groundwater levels, D includes both percolation and capillary rise. Despite the effort of many researchers, until today no practical device could be developed to measure soil water fluxes. For a proper evaluation of D therefore lysimeters should be used. A lysimeter is an isolated undisturbed soil column, typically 0.5–2.0 m in diameter, with or without a crop, in which apart from the evapotranspiration all terms of the water balance can be assessed. The lysimeter permits the measurement of drainage or makes it zero.

Figure 8.4 shows a nonweighable lysimeter, in which a bottom porous plate is used to apply a soil water pressure head corresponding with that in the field at the

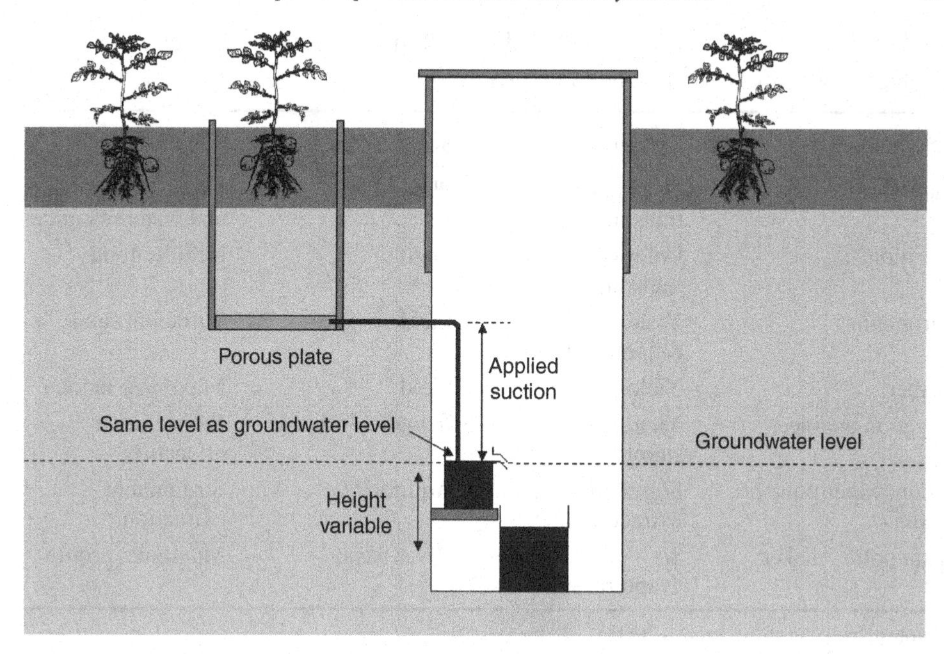

Figure 8.4 Possible setup of a nonweighable lysimeter.

same depth. The difference in soil water storage might be measured with TDR sensors (Section 4.11.3) or a neutron probe in preinstalled access tubes. In weighable lysimeters the change of soil water storage can be assessed directly from the difference in weight. Though difficult and expensive to install, lysimeters have been used widely to test different evapotranspiration formula (Aboukhaled et al., 1982).

Microlysimeters of 5–10 cm diameter and 5–10 cm high can be used to measure evaporation from bare soil (Boast and Robertson, 1982). The thin-walled cylinders are pushed into the field soil and carefully lifted from their place. Next the cylinder bottom is closed water-tight and the mass of the cylinder including the moist soil is determined. The cylinder is replaced in the field with its top even with the surrounding field, leaving it exposed to environmental conditions representative for the bare soil. After a period of time (typically 1–2 days) the cylinder is weighed again. The weight loss equals the amount of evaporation. Microlysimeters of 5 cm high cannot be used for periods longer than 2 days, because after this time the closed bottom of the cylinder starts to restrict the soil evaporation in comparison with the surrounding field. Recently, automatic microlysimeters (where the weighing is done in the lysimeter itself) have been developed (Heusinkveld et al., 2006).

Question 8.6: Give the water balance of a nonweighable lysimeter and explain how the evapotranspiration is determined from that water balance.

Table 8.1 Some examples of stakeholders and their targets in the water productivity framework as related to agriculture

Stakeholder	Definition	Scale	Target
Plant physiologist	Dry matter/ transpiration	Plant	Utilization of light and water resources
Nutritionist	Calorie/ transpiration	Field	Healthy food
Agronomist	Yield/ evapotranspiration	Field	Sufficient food
Farmer	Yield/supply	Field	Maximize income
Irrigation engineer	Yield/irrigation supply	Irrigation scheme	Proper water allocation
Groundwater policy maker	$/groundwater extraction	Aquifer	Sustainable extraction
Basin policy maker	$/ evapotranspiration	River basin	Maximize profits

Adapted from Molden et al. (2003).

Question 8.7: Consider a microlysimeter with a diameter of 10 cm, a depth of 5 cm, filled with sandy soil with a porosity of 40% and a water content of 30% (i.e., ¾ of the pores is filled with water).

a) What is the weight of the contents of this lysimeter (use Table 2.2)?
b) The microlysimeter is used during a night in which the average dewfall is 0.03 mm h^{-1}. Suppose that we want to keep track of the dewfall every five minutes. What is the resolution for the weight measurement needed to attain this resolution?
c) What is the needed resolution (answer of question b) relative to the total weight?

8.3 Water Productivity at Field and Regional Scale

8.3.1 Introduction

In an increasing number of regions the claims for fresh water by agriculture, industries, households and nature reserves exceed the amounts of fresh water available, thus demanding a better management of fresh water. Because irrigated agriculture is by far the biggest consumer of fresh water, increasing water productivity in irrigated agriculture is a logical way to save water (IWMI, 2007). Water productivity (WP) relates to the value or benefit derived from the use of water. Definitions of WP are not uniform and change with the background of the researcher or stakeholder involved (Table 8.1). For example, obtaining more kilograms of dry matter per unit of water transpired is a key issue for plant breeders. However, at basin level, policymakers may wish to maximize the economic value of the irrigation water used (Molden et al., 2003).

This case study presents nominal WP values based on the yield/evapotranspiration ratio of an irrigated basin in a semiarid region of India and combines field measurements, remote sensing and simulation models. An important reason to include the field scale is that many choices with regard to crop and water management, which directly affect WP, are made by the farmer. Also much of our scientific knowledge on crop-soil-water interactions applies to field scale processes. The regional scale is important as many decisions on water management and agricultural policies are made at this level. Another reason to consider regional scale is that water management in one region may affect other regions in the catchment. For instance, reduced groundwater recharge upstream, will result in reduced groundwater availability downstream. To evaluate options for improvement of WP both at field and regional scale, we applied a physical field scale crop and soil model. The paragraphs that follow provide a summary; details can be found in Van Dam and Malik (2003).

8.3.2 Sirsa District

The study area, Sirsa District, is located in the western part of Haryana State, India, and covers ca. 4270 km^2 (Figure 8.5). The soil texture in Sirsa District varies from sand to sandy loam, with a belt of silty loam to silty clay loam along the Ghagger River, which flows from east to west through the central part of the district. The climate of the region can be defined as subtropical, semiarid and continental with monsoon (July–September). Average annual rainfall in Sirsa District varies from 100 to 400 mm, which represents only 10–25% of the reference evapotranspiration (Jhorar et al., 2003).

The temperature conditions in Sirsa District allow growing of crops throughout the year. However, farmers generally grow two crops per year: a *rabi* crop (winter, from October to April) and a *kharif* crop (summer, from April to October). Crop production is very limited without irrigation, even in the summer. Since the mid-1950s, the Bhakra irrigation system has been distributing the surface irrigation water among the farmers in Sirsa District. The canal water distribution among farmers follows the Warabandi system, which means that they receive canal water amounts in proportion to their land holdings. The limited canal water supply in Sirsa District forces farmers to extract groundwater for supplementary irrigation. Groundwater quality determines the amounts of groundwater used for irrigation. Groundwater quality in the northern and the southern parts is generally poor compared to the central and southwestern parts of the district. Therefore, in the period 1990–2000, the northern and southern parts of Sirsa District experienced a rise in groundwater levels (in some parts +10 m), whereas groundwater levels declined in the central parts (in some parts –7 m).

Water management in Sirsa District is thus complex owing to low and erratic rainfall, canal water scarcity, high evaporative demands, sandy soils with low water holding capacity, marginal to poor groundwater quality and rising and declining groundwater levels. Marketable yield in the farmer fields is considerably less than at the

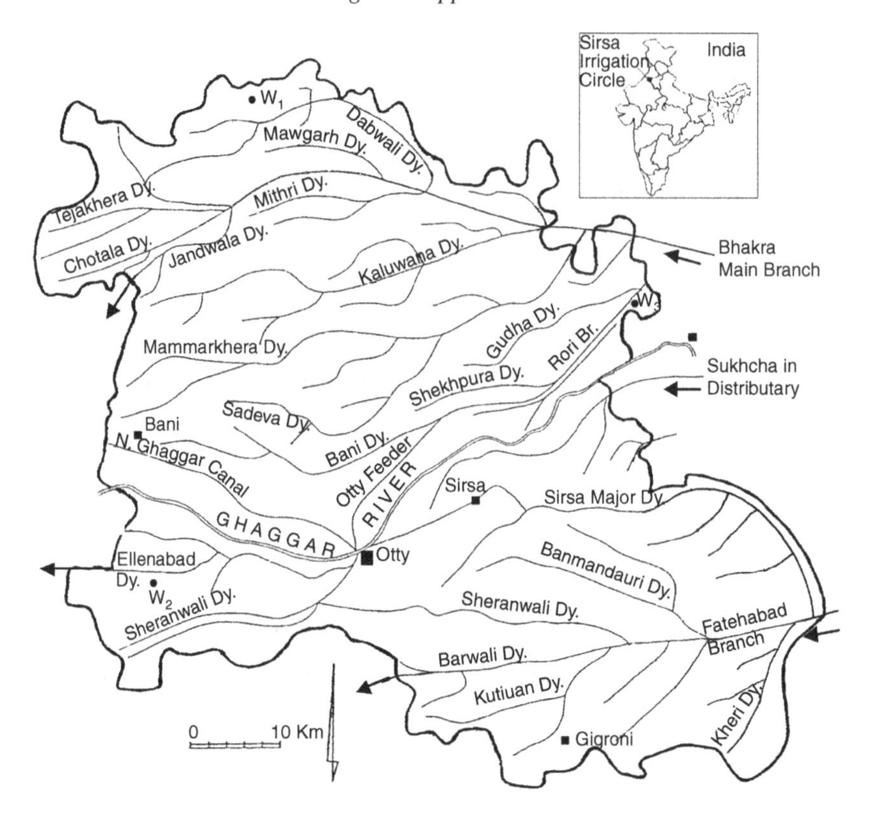

Figure 8.5 Location and canal network of Sirsa District.

experimental stations. These water management problems made Sirsa District a suitable pilot area for a water productivity analysis.

8.3.3 *Modelling Tools*

The ecohydrological model SWAP (Soil, Water, Atmosphere, Plant; Chapter 9) has been used to simulate water and salt transport in the soil and crop growth in relation to weather and irrigation data. In addition, the satellite image processing model SEBAL (Surface Energy Balance Algorithm for Land) has been used. SEBAL calculates actual and potential evapotranspiration rates from cropped and bare land (Bastiaanssen et al., 2005). The key input data for SEBAL consist of satellite images with spectral radiance in the visible, near-infrared and thermal infrared part of the spectrum. SEBAL computes a complete radiation and energy balance along with the resistances for momentum, heat and water vapour transport for every individual pixel. The resistances are a function of physical conditions near the soil surface, such as soil hydraulic head (and thus soil moisture and soil salinity), wind speed and air temperature. Satellite radiances are converted first into land surface characteristics, such as surface albedo, leaf area index, vegetation index and surface temperature. These land

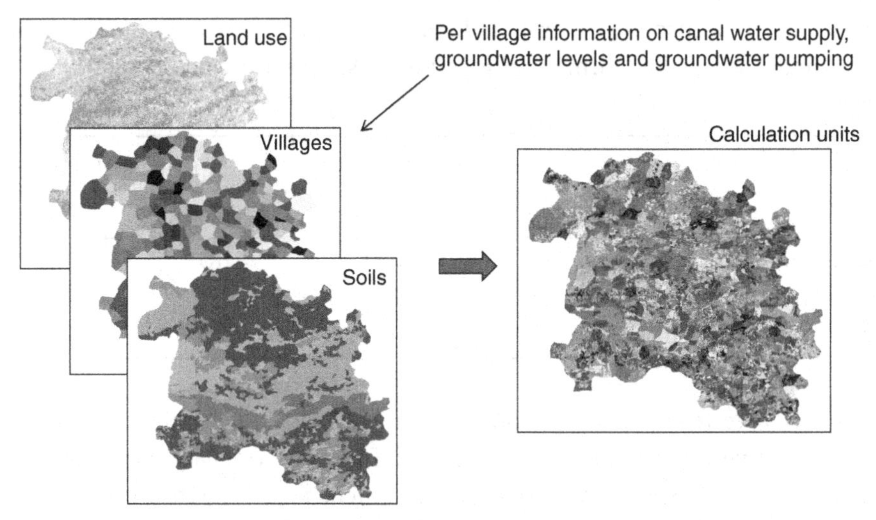

Figure 8.6 Overlays with ArcView to derive unique combinations of land use, irrigation management and soils for the regional analysis in Sirsa District.

surface characteristics can be derived from different types of satellites. First, instantaneous evapotranspiration is computed, which is subsequently scaled up to 24 hours and longer periods. In addition to satellite images, the SEBAL model requires daily average data on wind speed, humidity, solar radiation and air temperature.

8.3.4 Measurements

To run SWAP, we collected data at experimental stations and in farmer fields. Trials at experimental stations in the area were used to calibrate input parameters of the main crops (Bessembinder et al., 2003). A total of 24 farmer fields with different crops, soils, groundwater levels and canal water allocation were monitored to identify the yield gap between experimental stations and farmer fields. Regional geographical data were collected and digitized to perform a regional analysis with distributed modelling. Overlays of maps on land use, topography (villages) and soils, each with a grid size of 30 m, yielded 2404 unique combinations for entire Sirsa District (Figure 8.6).

8.3.5 Yield Gap

Crop yield is much higher at the experimental stations, in comparison to the farmer fields (Table 8.2). To arrive at 7.4 tons of wheat ha^{-1}, farmer yields should increase with 61%. In case of cotton, the farmer yields should increase with 38%. Also water productivity is much higher at the experimental stations: in case of wheat 67%, in case of cotton 62%. This means that with the same amount of water, the yield of wheat may increase with 67% and of cotton with 62%. The main differences between the experimental station and

Table 8.2 Evapotranspiration, crop yield and water productivity as measured at experimental stations and farmer fields in Sirsa District from October 2000 to October 2001

Item	Farmer fields		Experimental stations	
	Wheat	Cotton	Wheat	Cotton
Evapotranspiration (mm)	299	609	287	525
Crop yield (tons ha^{-1})	4.6	2.1	7.4	2.9
Water productivity (kg m^{-3})	1.54	0.34	2.58	0.55

farmer fields were related to crop management: proper land levelling, optimal sowing time, and strict pest, weed and disease control. Thus with ordinary crop management measures the water productivity in Sirsa District can be raised significantly.

8.3.6 Crop Yields at Field Scale

Table 8.3 contains the mean crop yields as derived by distributed SWAP modelling, by remote sensing (SEBAL) and as measured at the farmer fields. Despite their different approach, both SWAP and SEBAL result in similar average yields for wheat, rice and cotton. Therefore the calibrated SWAP model could be used to explore various management options at farmer fields. For instance, the effect of deficit irrigation, higher salinity levels and various sowing dates on the final crop yield and water use could be investigated (Bessembinder et al., 2003).

The standard deviation of the yield is underestimated by remote sensing, in comparison to distributed modelling and measurements in farmer fields. This can be explained by the difference in resolution between remote sensing (30–1100 m) and distributed modelling and measurements (1–30 m).

8.3.7 Water Productivity at a Regional Scale

SWAP and SEBAL determine *ET* and crop yield in entirely different ways: agrohydrological modelling versus analysis of remote sensing images. Therefore, it is interesting to compare the water productivity values as derived by distributed modelling and remote sensing (Table 8.4). The mean WP values for the main crops are very close. The WP standard deviation is larger for the results derived from distributed modelling than for the results derived from remote sensing due to the higher resolution of distributed modelling (Singh et al., 2006a).

8.3.8 Scenario Analysis

To realize the dual objectives of meeting the growing food demands and restricting water use, water management in Sirsa District should aim at higher crop yields per

Table 8.3 Mean crop yields and standard deviations (fresh matter, tons ha^{-1}) of wheat, rice and cotton as obtained by distributed SWAP-WOFOST modelling, remote sensing (SEBAL) and field measurements at farmer fields in Sirsa District during the agricultural year 2001–2002

Method	Wheat		Rice		Cotton	
	Mean	SD	Mean	SD	Mean	SD
SWAP-WOFOST	4.8	1.0	3.5	2.5	2.0	0.5
SEBAL	4.4	0.3	3.7	1.1	2.2	0.3
Field measurements	4.5	1.5	—	—	2.1	1.1

After Singh et al. (2005a).

Table 8.4 Water productivity WP_{ET} (kg m^{-3}) as estimated with SWAP-WOFOST (distributed modelling) and with SEBAL (remote sensing) in Sirsa District during the agricultural year 2001–2002

	Wheat		Rice		Cotton	
	Mean	SD	Mean	SD	Mean	SD
SWAP-WOFOST	1.37	0.20	0.47	0.30	0.36	0.05
SEBAL	1.22	0.07	0.43	0.19	0.31	0.04

The evapotranspiration is calculated during the entire growing season.

unit water consumed. At the same time, the irrigated agriculture should be sustainable. This means a higher water productivity, less groundwater rise and lower salinity levels in the northern commands and less groundwater level decline in the central commands. Therefore we evaluated three scenarios (Table 8.5) and compared them to 'business as usual' for a 10-year period.

Scenario 1 mimics the reference situation with the crop and water management as measured during the agricultural year 2001–2002. Scenario 2 quantifies the impact of improved crop cultivars, cultivation and nutrient, weed, pest and disease management. These developments are expected to increase crop yields by about 15%. Scenario 3 targets the rising groundwater levels in the northern parts of Sirsa District. The seepage losses from the conveyance system amounted about 40% of the net canal inflow. By canal lining and proper canal maintenance it should be able to reduce the seepage amount by 25–30%. Scenario 4 was formulated to divert canal water from the northern regions with rising groundwater levels to the central and southern commands with declining groundwater levels.

The simulation results showed that improved crop management (Scenario 2) increases water productivity with about 12%. Lower seepage losses in the irrigation canals (Scenario 3) reduced the current salinization in the area with 35%. Canal water

Table 8.5 Alternative water management scenarios that were analysed for Sirsa District

Scenario	Description	Required action
1	Reference situation	Business as usual
2	Increased crop yields (15%)	Improved crop varieties, better nutrient supply, effective pest and disease control
3	Reduced seepage losses (25–30%)	Lining and improved maintenance of irrigation canals
4	Canal water reallocation (15%)	Divert canal water from northern parts to central parts

reallocation (Scenario 4) increased the uniformity of the groundwater recharge substantially. Ideally, these scenarios should be combined to increase water productivity and improve sustainability in Sirsa District (Singh et al., 2006b).

8.3.9 Satellite Data Assimilation

In this study, we used remote sensing and simulation modelling separately to derive WP values. Remote sensing of *ET*, yield and WP may also be used to calibrate plant and soil parameters of the crop and soil models. For instance, Jhorar et al. (2004) used remotely sensed evapotranspiration to calibrate soil hydraulic parameters. Another way to benefit from both the information produced by generic simulation models and remote sensing is by so-called data assimilation (Walker and Houser, 2001; Schuurmans et al., 2003; Pauwels et al., 2007; Vazifedoust et al., 2009). In this method simulation models are updated with remote sensing information whenever an observation is available. While adjusting the model state variables, both model errors and measurement errors by remote sensing are taken into account. In this integration of simulation models and remote sensing data, all information sources are optimally used to improve regional water productivity analysis and crop yield prediction.

8.4 Response to Heat Wave Conditions of the Energy and Water Balance of Grassland and Forests

Observations suggest that in western Europe the length and intensity of prolonged periods of above-average warm conditions in summer (heat waves) have increased in the past century (Della-Marta et al., 2007). Land–atmosphere interactions may play an important role (Seneviratne et al., 2006). Those interactions may be different for different land use types. To understand those differences, the response of the energy partitioning in the energy balance of grass and forest will be compared for two different conditions: normal summer days are compared to exceptionally hot days as occurring during heat waves. Heat waves are defined here as a period of at least 5 days in which the maximum temperature exceeds the climatological value for that date by at least 5 K (Frich et al., 2002).

The rationale for this analysis is twofold. First, it is intended to show the mechanism behind the difference in response to changing atmospheric conditions between grassland and forest. Second, the differences in partitioning may influence the intensity of the heat wave itself.

The data set as well as part of the analysis in this section is based on Teuling et al. (2010).

8.4.1 Data

To study the actual response of different land use types, direct flux observations are needed. Teuling et al. (2010) provide a synthesis of flux observations from 30 stations (grass or forest) in western Europe, covering the summer months (June–August) in the period 1997–2006 (data sets per station varying from 2 to 10 years). The data have been obtained from the FLUXNET data set (Baldocchi et al., 2001; Baldocchi, 2008). The data encompass not only the turbulent surface fluxes of heat and water vapour, but net radiation and soil heat flux as well.

To study the response of the surface energy balance, the climatological energy balance is constructed for both land use types separately: the median fluxes for all grass/ forest stations, for the time period 9–13 UTC from days without heat wave conditions. Subsequently, the median anomaly in the energy balance terms is determined for the 2003 and 2006 heat waves in western Europe. Here the median observations will be used as if they were representative observations of two single composite stations: one grass station and one forest station.

To analyse further the normal and heat wave behaviour of the energy balance, we a need to derive a number of variables that have not been directly observed: air temperature and canopy resistance. We start from the resistance expressions for the sensible as used in the derivation of the Penman–Monteith equation in Chapter 7:

$$H = -\rho c_p \frac{T_a - T_s}{r_a} \tag{8.3}$$

Assuming typical values for the aerodynamic resistance for grass and forest (40 and 10 s m^{-1}, respectively), and deriving the surface temperature from the upwelling longwave radiation, the air temperature can be determined using Eq. (8.3).

Next it is assumed that the relative humidity above both surfaces is 70%. With the derived air temperature this gives the vapour pressure and with that the total resistance (aerodynamic + canopy resistance) can be derived from

$$L_v E = -\rho \frac{c_p}{\gamma} \frac{e_a - e_{sat}(T_s)}{r_a + r_c} \tag{8.4}$$

(which was used before in Chapter 7 in the derivation of the Penman–Monteith equation).

8.4.2 Energy Balance during Normal Summers

Figure 8.7a shows the components of net radiation and the energy balance for the composite grass and forest station. The downwelling radiation fluxes are nearly equal for both sites, which simplifies the analysis as both sites are exposed to the same radiative input. However, the grass and forest *do* differ in albedo (0.10 vs. 0.18) and surface temperature (21.6 °C vs. 18.8 °C, assuming a surface emissivity of 0.96), leading to a difference in net radiation of more than 10%. The partitioning of the available energy over sensible and latent heat flux is consistent with the findings in Chapter 7: grass allocates a larger proportion of the energy to evapotranspiration than the forest (evaporative fraction: 53% vs. 38%; see Table 8.6). The relatively low evapotranspiration for the forest is due to the strong coupling of the surface temperature to the air temperature (resulting from the low aerodynamic resistance). As a result, the gradient of water vapour between the stomata (saturated air at leaf temperature) and the air is lower than for grass. This lower gradient is only partially compensated by the lower aerodynamic resistance.

The air temperatures above grass and forest (derived using Eq. (8.3)) appear to be close to each other and the surface-to-air temperature difference is smaller for the forest due to the small aerodynamic resistance (strong coupling to the air). Furthermore, the computed air temperature is roughly consistent with the climatological median maximum temperature of the stations, 20.6 °C (note that the maximum temperature occurs later than the time period used here: 9–13 UTC, which could explain the discrepancy of about 2.5 °C).

Using Eq. (8.4) the total resistance ($r_a + r_c$) can be determined, which leads – with the assumed aerodynamic resistance values – to canopy resistances of 76 and 83 s m^{-1} for grass and forest, respectively. Given the crudeness of the analysis and the assumptions made, these values are close to the expected values (see Chapter 7).

The analysis shows that under normal summer conditions grass and forest supply approximately the same amounts of water vapour to the atmosphere, but forests supply more sensible heat than grass.

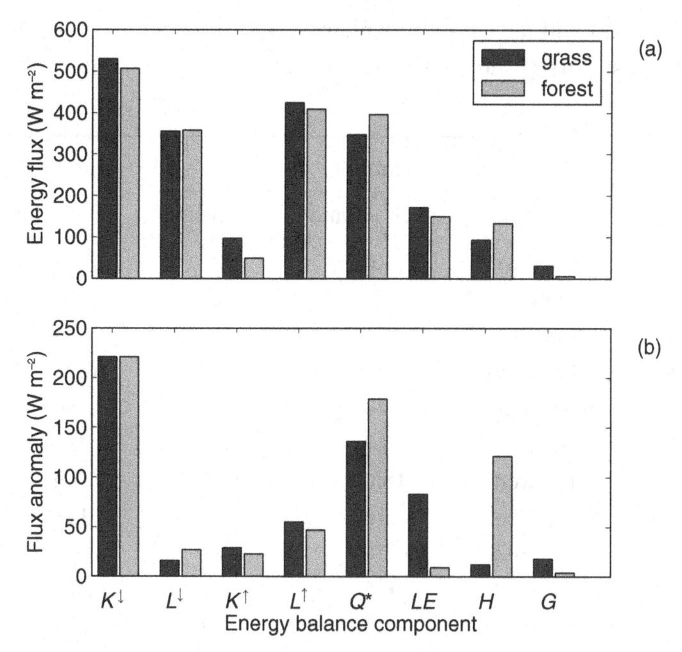

Figure 8.7 Components of net radiation and the surface energy balance for grass and forest sites during summer months (June–August) in western Europe. (**a**) Climatological values of the fluxes during non-heat wave conditions. (**b**) Anomalies of the components during heat wave conditions. (After Teuling et al., 2010)

8.4.3 Energy Balance during Heat Wave Conditions

In Figure 8.6b the anomalies of the energy balance terms are shown for heat wave conditions. The first obvious anomaly is in the incoming shortwave radiation, as heat wave conditions are usually related to clear sky conditions. Apart from the extra supply of short wave radiation, there is also a slight increase in the incoming longwave radiation (somewhat larger for forest than for grass). This is probably the net effect of a decrease in $L^{\downarrow}$ due to a decrease in cloud cover and an increase in $L^{\downarrow}$ due to a higher temperature and humidity content of the atmosphere (see Chapter 2). The upwelling short wave radiation increases more for the grass than for the forest, consistent with the higher albedo for grass. Finally, the upwelling longwave radiation increases stronger for the grass than for the forest, indicating that the difference in surface temperature between grass and forest increases as compared to normal conditions (the grass being warmer). The net effect of all four terms is that the forest has an extra 179 W m^{-2} available, whereas the net radiation of the grass increases by 136 W m^{-2}, thus further increasing the disparity between both surfaces in terms of available energy.

The response of both surfaces to this extra supply of energy is completely opposite. Whereas the grass allocates nearly all extra energy input to the evapotranspiration,

Table 8.6 Observed and derived atmospheric conditions and surface properties for grass and forest during 'normal' conditions (climatology) and during heat wave days

Quantity	Source	Normal		Heat wave	
		Grassland	Forest	Grassland	Forest
T_s (°C)	Observed	21.6	18.8	30.9	26.9
H (W m^{-2})	Observed	93	133	105	254
L_vE (W m^{-2})	Observed	171	149	254	158
$L_vE/(H + L_vE)$ (-)	Observed	0.65	0.53	0.71	0.38
r_a (s m^{-1})	Assumed	40	10	40	10
RH (-)	Assumed	0.7	0.7	0.7	0.7
T_a (°C)	Derived	18.6	17.7	27.4	24.8
$r_a + r_c$ (s m^{-1})	Derived	116	93	138	158
r_c (s m^{-1})	Derived	76	83	98	148
VPD (hPa)	Derived	6.3	6.0	10.8	9.3
ΔT_a (°C in 4 h) in BL of 750 m	Derived	1.5	2.1	1.7	4.0

The values for the aerodynamic resistance r_a and relative humidity were assumed; the other variables have been derived (see the text for details).

the forest only increases the sensible heat flux. What could be the explanation for these opposite responses? Table 8.6 (two rightmost columns) provides the data for an analysis of this behaviour, where the same assumptions were used as for the normal conditions. The first difference between normal data and the heat wave conditions is the increase in air temperature of about 8 °C, indicating that indeed heat wave conditions occurred (maximum temperatures more than 5 K above climatology). As a result of this increase in air temperature, and assuming an unchanged relative humidity, the vapour pressure deficit increases significantly, and hence the evaporative demand of the atmosphere does as well. Whereas the grass completely yields to this demand (with only a small increase in canopy resistance), the forest responds strongly to the increased demand by nearly doubling its canopy resistance. This response is found not only in the data analysed here, but is – at least qualitatively – consistent with results of, for example, Kelliher et al. (1993). The data clearly show that due to the strong aerodynamic coupling of forests to the atmosphere (and subsequent cooling) they do not need much evaporative cooling to keep the surface temperature in bounds.

The analysis shows that under heat wave conditions forests remain more conservative than grass as it comes to the use of water. But the reverse side of that medal is that forests supply more heat to the atmosphere, thus increasing the intensity of a heat wave. If the sensible heat fluxes given Table 8.6 are fed into a boundary layer with a mean depth of 750 m, during 4 hours, and assuming no entrainment of heat[3], the boundary layer would heat up by 1–2 K under normal conditions, with stronger heating for the forest. For heat wave conditions the difference in heating over grass and forest is much more pronounced: for grass the heating rate increases only slightly, whereas over the forest it nearly doubles. One could argue that the higher sensible heat flux over the forest will give rise to a deeper boundary layer owing to stronger convection. The deeper boundary layer would lead to a smaller temperature increase, but this growth would also entail a larger entrainment of warm air.

8.4.4 Temporal Development of the Energy and Water Balance

The previous analysis considered only the partitioning of available energy under conditions that were assumed to be not limited by soil moisture: the differences in the energy balance were solely due to the different atmospheric forcing. However, as time progresses and soil moisture is not replenished by rainfall, the soil will dry out and the transpiration may be reduced below its unstressed values.

To study the development of surface fluxes during the dry down process a simple soil water balance model is used, based on the Warrilow model introduced in Chapter 4. This model consists of the following components and parameter choices (which are realistic but arbitrary):

- Unstressed daily evapotranspiration is based on the observations given in Table 8.6. To convert the average fluxes during the 9–13 UTC period to a daily evapotranspiration the fraction of the total flux that occurs between 9 and 13 UTC is assumed to be 0.3 (the data in Teuling et al. (2010) show a range for this fraction of 0.3–0.4). Unstressed surface fluxes are supposed to be constant from day to day.
- Available energy is assumed to be constant from day to day: a reduction in evapotranspiration will translate in an equal increase in sensible heat flux.
- The rooting depth of grass and forest is set to 0.4 m and 0.6 m, respectively (the depth of the layer in which 80% of the roots are located; Zeng, 2001).
- The volumetric soil moisture content at which reduction of evapotranspiration starts to occur is set to 0.2 and evapotranspiration stops at a water content of 0.1.
- The initial soil moisture content is set at 0.3.

With these ingredients, the development of the surface fluxes (9–13 UTC averages) as shown in Figure 8.8 can be simulated. As long as the soil moisture content is above the critical value, evapotranspiration is controlled by the atmospheric forcing, and hence constant in the present simple model. For the climatological conditions evapotranspiration of grass and forest are nearly identical, and hence the difference in timing of

[3] If the entrainment of heat from the atmosphere above the boundary layer would be taken into account, the heating would be approximately 20% higher.

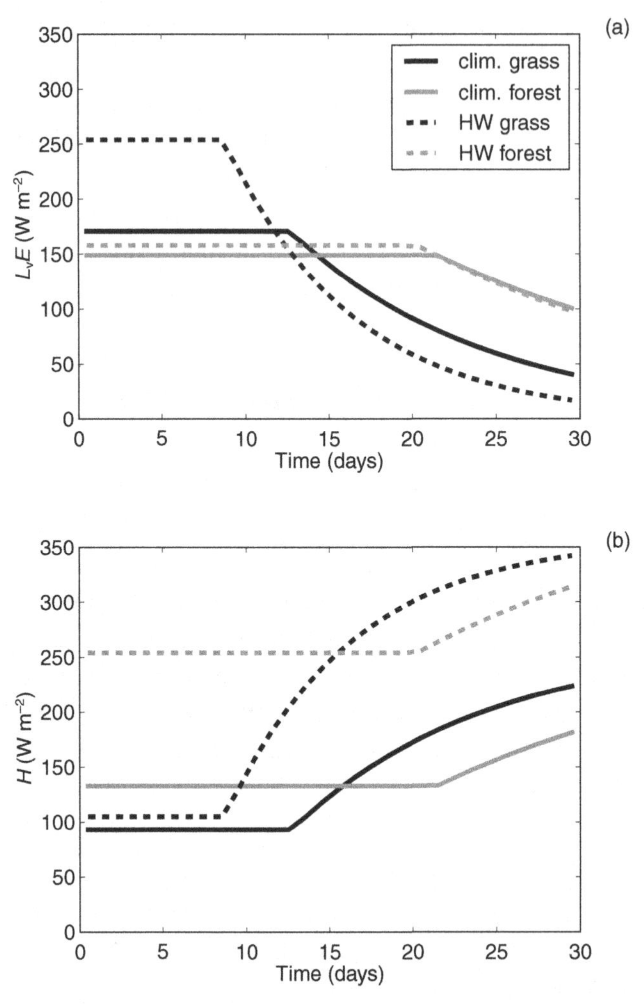

Figure 8.8 Temporal development of latent heat flux **(a)** and sensible heat flux **(b)** during a 30-day dry-down. Comparison of climatological conditions (clim.) and heat wave conditions (HW) for grass and forest. Unstressed fluxes are identical to those given in Table 8.6. Grass and forest differ not only in unstressed flux, but also in rooting depth.

the first signs of reduction of evapotranspiration is due only to the shallower rooting depth of the grass. From day 12 onward the evapotranspiration starts to fall and stays below the value of the forest. The evapotranspiration decreases approximately logarithmically in time (see Teuling et al., 2006). Through the constraint of the energy balance, a decrease in the evapotranspiration leads to an increase in sensible heat flux (Figure 8.8b). Because the initial sensible heat flux of the grass was below that of the forest, it takes some additional days before the sensible heat flux of the grass exceeds that of the forest (approximately 2 weeks after the start of the dry down).

The heat wave conditions start out with a large disparity in evapotranspiration between the grass and the forest. The forest loses water at nearly the same rate as under climatological conditions, showing a slight reduction after only about 20 days. On the other hand, the grass depletes soil moisture much quicker, leading to reduced evapotranspiration after 9 days. Although the sensible heat flux of grass under unstressed conditions is only slightly higher for the heat wave situation than for normal conditions, the quicker soil moisture depletion leads to an earlier and stronger rise in the sensible heat flux. Within 16 days the grass shows a larger sensible heat flux than forest. Thus, although forests tend the aggravate the heat wave initially, grass dries out so strongly that after approximately 2 weeks grass starts to be the major supplier of heat to the atmosphere.

9

Integrated Models in Hydrology and Meteorology

This chapter shows how the methods discussed in the previous chapters are applied in hydrological and meteorological models. The SWAP (Soil, Water, Atmosphere, Plant) model is an example of a field-scale ecohydrological model (Section 9.1). In Section 9.2 various aspects of land–surface models as used in weather and climate models are discussed.

9.1 SWAP

9.1.1 Introduction

SWAP simulates transport of water, solutes and heat in the vadose zone (Kroes et al., 2008; Van Dam et al., 2008). The model includes vegetation growth, as affected by meteorological and hydrological conditions. The upper boundary of the model domain is a plane just above the canopy. The lower boundary corresponds to a plane in the shallow groundwater (Figure 9.1). In this model domain the transport processes are predominantly vertical; therefore SWAP is a one-dimensional, vertical directed model. The flow below the groundwater level may include lateral drainage fluxes, provided that these fluxes can be prescribed with analytical drainage formulas. The model is very flexible with regard to input data at the upper and lower boundaries. At the top general data on rainfall, irrigation and evapotranspiration are used. For frost conditions a simple snow storage module has been implemented and soil water flow will be impeded when soil temperature descends below zero. To facilitate temporal detailed studies on surface runoff and diurnal transpiration fluxes, evapotranspiration and rainfall data can be specified at daily and shorter time intervals. At the model lower boundary, various forms of head and flux based conditions are used.

In the horizontal plane, SWAP's main focus is the field scale. At this scale most transport processes can be described in a deterministic way, as a field generally can be represented by one microclimate, one vegetation type, one soil type and one drainage condition. Also cultivation sequences and farmer management decisions apply to the field scale. Both the physical characterization and the cultivation practices make

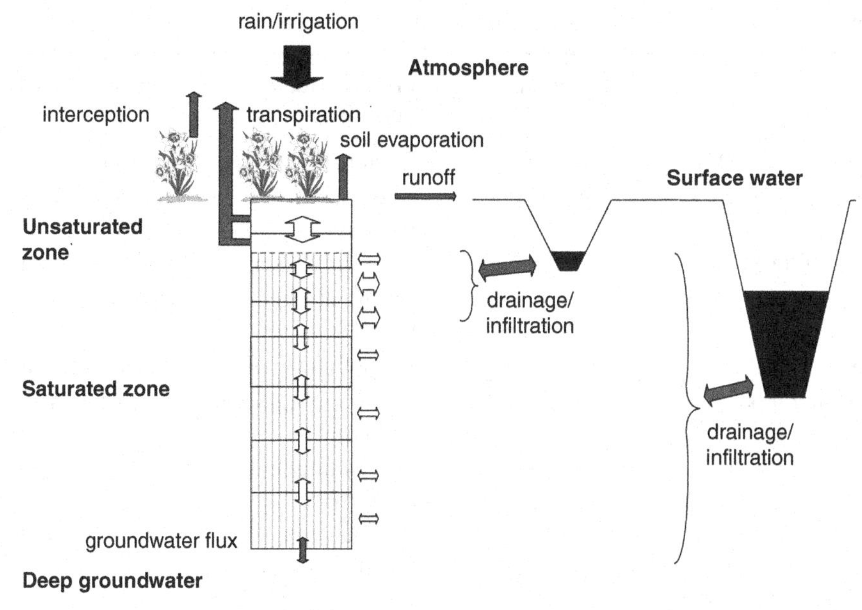

Figure 9.1 Scheme of water flow processes in SWAP.

the field scale very relevant. For broader management or policy studies, the catchment or regional scale might be important. Up-scaling from field to regional scale can be accomplished by simulating the enclosed fields parallel, such as illustrated in Chapter 8.

Table 9.1 lists a number of typical studies with SWAP that appeared in scientific literature. Current developments with multidimensional physically based models and integrated hydrological frameworks will further improve our analysis of water and solute movement in soils. However, because of their flexibility, accessibility and speed, the coming decade one-dimensional models as SWAP will be functional to explore new flow and transport concepts, to analyse laboratory and field experiments, to select viable hydrological management options, to perform regional studies within geographical information systems and to illustrate atmosphere-vegetation-soil interactions for education and extension.

In the next paragraphs we discuss specific modelling features of SWAP. Subsequently we address soil water flow, solute transport, heat flow and crop growth.

9.1.2 Soil Water Flow

Combination of Darcy's law and the principle of mass conservation results in the versatile Richards' equation for soil water flow, as discussed in Chapter 4. To solve Richards' equation numerically for arbitrary field conditions, we need to know the soil hydraulic functions $\theta(h)$ and $k(h)$, the actual root water extraction rate and the

Table 9.1 Typical hydrological studies with SWAP in recent scientific literature

Citation	Location	Primary study objective	Unique features used
Droogers et al. (2000)	Turkey	Regional irrigation	Water productivity
Sarwar et al. (2000)	Pakistan	Sustainable irrigation	Dynamic drainage design
Wolf et al. (2003)	Netherlands	Nutrient transport	Regionalization
Bethune and Wang (2004)	Australia	Water balance predictions	Macropore flow
Utset et al. (2004)	Spain	Evapotranspiration	Irrigation scheduling
Hupet et al. (2004)	Belgium	Maize transpiration and growth	Parameter calibration
Jhorar et al. (2004)	India	Soil hydraulic parameters	Remote sensing data
Droogers et al. (2004)	Worldwide	Adaptation to climate change	Crop yield prediction
Ritsema et al. (2005)	Australia	Preferential flow and transport	Dual modelling approach
Crescimanno and Garofalo (2005, 2006)	Italy	Irrigation with saline water	Macropore flow
De Jong van Lier et al. (2006, 2008)	Canada	Root water uptake during drought	Microscopic root concept
Van Walsum and Groenendijk (2007)	Netherlands	Regional three-dimensional modelling	Generation meta-functions
Bartholomeus et al. (2008)	Netherlands	Natural vegetation type	Oxygen shortage stress
Vazifedoust et al. (2009)	Iran	Regional crop yield	Assimilation of satellite data
Schaik et al. (2010)	Spain	Runoff	Macropore flow

boundary conditions (initial, top and bottom). Although the basic assumptions of Richards' equations are very straightforward, it is less easy to derive a reliable solution of Richards' equation that can be used for general field conditions. Numerical problems may arise due to the high nonlinearity of the $\theta(h)$ and $k(h)$ relations, distinct soil layering and rapid changes from wet to dry conditions and vice versa in the top soil. We discuss how SWAP addresses these numerical problems.

Discretization of the soil water flow equation should occur both in space and time, as depicted in Figure 9.2. In fact we did perform such a discretization in Chapter 4, when calculating the root water extraction using tensiometer measurements. The numerical scheme employed in SWAP is based on finite difference, which means

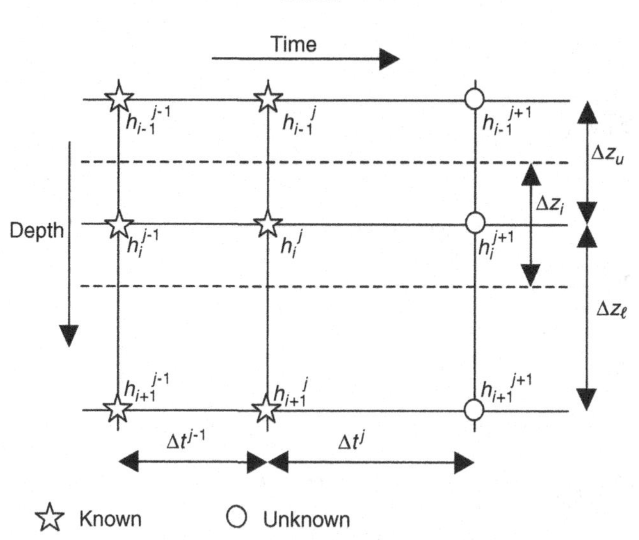

Figure 9.2 Spatial and temporal discretization used to solve Richards' equation.

that the vertical column is divided in compartments (don't confuse with natural soil layers!) with calculation nodes in the centre. Both the time and space steps may vary in length. The subscript i is used for the node number (increasing with depth) and superscript j for the time level (increasing with time). At a certain time level j all the state variables (h, k and θ) are known in the nodes depicted as a star in Figure 9.2. The task of the numerical scheme is to calculate the new state variables at time level $j + 1$ (depicted as open circles).

We may calculate the new state variables by solving for each compartment the water balance. A first approximation is given in Figure 9.3. All the values of the state variables at time level j are known. We want to calculate h_i^{j+1}. We may use the Darcy fluxes at the top and bottom of the compartment at time level j. The new pressure head h_i^{j+1} is provided by the mass balance, as depicted in Figure 9.3. This scheme is a so-called explicit finite difference scheme, which will work if the time steps are small. At larger time steps this explicit scheme becomes unstable. The reason is that the water fluxes at time level j are used, although in fact the average water fluxes during time step Δt^j should be used. At larger time steps and longer simulation periods this may result in substantial errors.

The finite difference scheme can be made stable if we use the fluxes at the new time level. Our starting point is again Richards' equation (Chapter 4):

$$C(h)\frac{\partial h}{\partial t} = \frac{\partial\left[k(h)\left(\frac{\partial h}{\partial z}+1\right)\right]}{\partial z} - S(z) \tag{9.1}$$

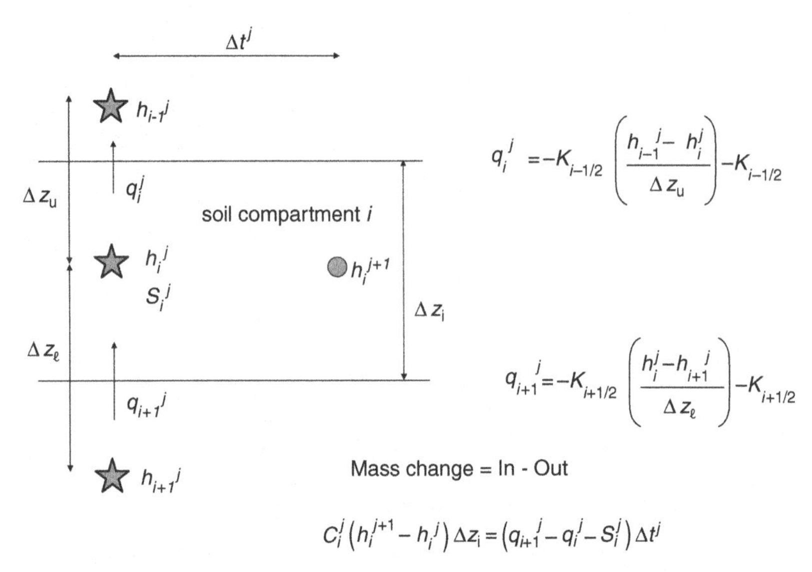

Figure 9.3 Straightforward numerical discretization of Richards' equation.

where C is the differential water capacity (m^{-1}), h the soil water pressure head (m), k the hydraulic conductivity ($m\ d^{-1}$) and S the root water extraction rate (d^{-1}). A straightforward, finite difference scheme with fluxes at the new time level $j+1$ and the differential water capacity halfway the new and old time level ($j+1/2$), is:

$$C_i^{j+1/2}\left(h_i^{j+1} - h_i^j\right) = \frac{\Delta t^j}{\Delta z_i}\left[k_{i-1/2}^j\left(\frac{h_{i-1}^{j+1,p} - h_i^{j+1,p}}{\Delta z_u}\right) + k_{i-1/2}^j - k_{i+1/2}^j\left(\frac{h_i^{j+1,p} - h_{i+1}^{j+1,p}}{\Delta z_\ell}\right) - k_{i+1/2}^j\right] - \Delta t^j S_i^j \tag{9.2}$$

We call this scheme implicit, as the new pressure head h_i^{j+1} is a function of itself and can be solved only by iteration (the subscript p indicates the iteration step). Although this scheme may work well for ordinary field conditions, it is not accurate to simulate rapid hydrological events such as intensive rain showers on dry soils or fast fluctuations of the groundwater table near the soil surface. In such cases, numerical errors originate from two main sources: the averaging of the hydraulic conductivity k between the nodes, and the averaging of the water capacity C during the time step. Let's see how we can address these error sources.

To determine the average k between the nodes, different methods can be used, for example, arithmetic, geometric and harmonic. If we view the system as a series of layers with different k, the harmonic average would seem the most appropriate (Chapter 4). Especially at sharp wetting fronts, the averaging method may have a large impact on the calculated soil water flux, as illustrated in Question 9.1.

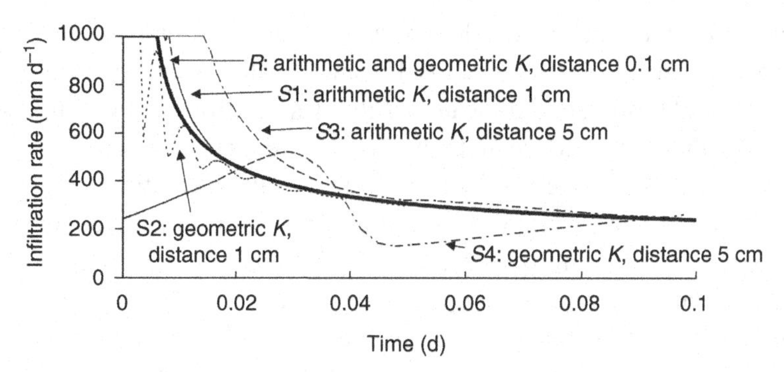

Figure 9.4 Infiltration rate of sand in case of intensive rain at a dry soil as simulated with geometric and arithmetic averages of hydraulic conductivity k at nodal distances of 1 and 5 cm.

Question 9.1: Consider two adjacent nodes near the infiltration front in a sandy soil. The upper node is in the wetted part and its state variables have the values $h_{i-1} = 0$ cm, $\theta_{i-1} = 0.431$ cm^3 cm^{-3} and $k_{i-1} = 9.65$ cm d^{-1}. The lower node is still ahead of the wetting front with state variables $h_i = -100$ cm, $\theta_i = 0.260$ cm^3 cm^{-3} and $k_i = 0.12$ cm d^{-1}. The vertical distance between the nodes is 10 cm. Which soil water flux would you calculate between both nodes if the hydraulic conductivity is arithmetically averaged? Which soil water flux in the case of a geometric average of the hydraulic conductivity? And in the case of a harmonic average?

Above exercise shows that the arithmetic average inclines to the largest k, resulting in high fluxes, while the geometric and harmonic average tend to the lowest k, resulting in low fluxes. If we would use different methods of averaging for runoff calculations, the results may deviate by a factor of 20 or more! The most suitable method for averaging has been evaluated for extreme hydrological events with SWAP (Van Dam and Feddes, 2000). One of the extreme events was an intensive rain shower of 100 mm in 0.1 d on a dry sand soil with $\theta = 0.1$.

Figure 9.4 shows the calculated infiltration rates at the soil surface during the 0.1-day period for various nodal distances and methods of averaging. The bold continuous line is the theoretical infiltration curve. Initially the rain may infiltrate at a rate of 1000 mm d^{-1} into the dry sand soil. At $t = 0.008$ d, h at the soil surface becomes zero. Gradually the infiltration rate declines, ultimately reaching the value of the saturated hydraulic conductivity of the top soil. The total amount of infiltration is 39 mm out of 100 mm of rainfall, the remaining amount is runoff. As Figure 9.4 shows, use of arithmetic averages results in larger hydraulic conductivities and thus larger infiltration fluxes than use of geometric averages.

In case of $\Delta z_i = 5$ cm, arithmetic averages of k seriously overestimate the infiltration rate (total = 47 mm) whereas geometric averages seriously underestimate the infiltration rate (total = 27 mm). The very steep wetting front due to low geometric k-averages causes infiltration rate oscillations when the geometric average is used. When smaller nodal distances are used (1 cm instead of 5 cm) the infiltration fluxes calculated by arithmetic and geometric averaging approach the theoretical curve. At $\Delta z_i = 0.1$ cm, they converge towards the same solution, as the spatial differences of k become so small that every averaging method yields the same result. Harmonic means (not shown in Figure 9.4) underestimate the mean k at the wetting front and the infiltration rate even more than the geometric mean. In the case of arithmetic averages with $\Delta z_i = 1$ cm, the calculated infiltration rate is very close to that of the theoretical infiltration curve (40 compared to 39 mm), while the infiltration rate with geometric averages deviates more (37 compared to 39 mm). Based on this and other cases, SWAP applies arithmetic averaging of k and maximum nodal distances at the soil surface of 1 cm.

The second main error source concerned the temporal averaging of water capacity C. In Question 9.2 the problem is illustrated.

Question 9.2: Suppose at $t = j$ a node in a sandy subsoil has the following state variables: $h_i^j = -50$ cm, $\theta_i^j = 0.26210$ cm^3 cm^{-3} and $C_i^j = \partial\theta / \partial h = 0.00278$ cm^{-1}. The subsoil becomes more dry, and at $t = j + 1$ the node shows the following state variables: $h_i^{j+1} = -52$ cm, $\theta_i^{j+1} = 0.25660$ cm^3 cm^{-3} and $C_i^{j+1} = \partial\theta / \partial h = 0.00272$ cm^{-1}. The compartment of this node is 5 cm thick. How large is the real difference in water storage of this compartment? Which water storage difference would you calculate with the expression $C_i^j\,(h_i^{j+1} - h_i^j)\,\Delta z_i$? And which with the expression $C_i^{j+1}\,(h_i^{j+1} - h_i^j)\,\Delta z_i$?

Question 9.2 shows the mass balance problem due to temporal averaging of water capacity $C = \partial\theta/\partial h$. A very elegant solution to this problem was published by Celia et al. (1990). Instead of applying during a time step:

$$\theta_i^{j+1} - \theta_i^j = C_i^{j+½}\left(h_i^{j+1} - h_i^j\right) \tag{9.3}$$

where $C_i^{j+½}$ denotes some kind of average water capacity during the time step, we may use the water content estimate at the new time level, $\theta_i^{j+1,p-1}$, in the iterative solution:

$$\theta_i^{j+1} - \theta_i^j = C_i^{j+1,p-1}\left(h_i^{j+1,p} - h_i^{j+1,p-1}\right) + \theta_i^{j+1,p-1} - \theta_i^j \tag{9.4}$$

where the superscript p is the iteration level and $C_i^{j+1,p-1}$ is the water capacity evaluated at the pressure head value of the last iteration, $h_i^{j+1,p-1}$. At convergence, the term $(h_i^{j+1,p} - h_i^{j+1,p-1})$ will be small, which eliminates effectively remaining inaccuracies in the evaluation of C. Implementation of Eq. (9.4) results in a perfect water balance, also at larger time steps!

The implicit finite difference solution of Richards' equation that is currently applied in SWAP therefore reads:

$$C_i^{j+1,p-1}\left(h_i^{j+1,p} - h_i^{j+1,p-1}\right) + \theta_i^{j+1,p-1} - \theta_i^j =$$

$$\frac{\Delta t^j}{\Delta z_i}\left[K_{i-½}^j\left(\frac{h_{i-1}^{j+1,p} - h_i^{j+1,p}}{\Delta z_u}\right) + K_{i-½}^j - K_{i+½}^j\left(\frac{h_i^{j+1,p} - h_{i+1}^{j+1,p}}{\Delta z_\ell}\right) - K_{i+½}^j\right] - \Delta t^j S_i^j \tag{9.5}$$

This numerical scheme is mass conservative and stable. To solve accurately evaporation and infiltration fluxes at the soil surface, the compartment thickness should be maximum 1 cm. Application of Eq. (9.5) to each compartment results in a set of n equations with n unknown pressure heads, which can be solved efficiently. The time step is based on the number of iterations required to solve the set of equations. At a large number of iterations, the time step is decreased; at a small number of iterations, the time step is increased.

9.1.3 Top Boundary Condition Hydrology

Measurement of reliable evapotranspiration fluxes is far from trivial and strongly varies with the local hydrological conditions. Therefore SWAP simulates evapotranspiration fluxes from basic weather data with the Penman–Monteith equation (Chapter 7) or from reference evapotranspiration data. Application of the Penman–Monteith equation requires daily values of air temperature, net radiation, wind speed and air humidity. In case these data are not available, popular alternative evapotranspiration formulas can be used, such as Priestly–Taylor (1972), Makkink (Makkink, 1957; Feddes, 1987) and Hargreaves et al. (1985). The Priestly–Taylor and Makkink methods require only air temperature and solar radiation data. The method of Hargreaves requires solely air temperature data. These alternative methods calculate a reference evapotranspiration flux that is generally defined for a hypothetical grass cover of 12 cm high, with an albedo of 0.23 and a canopy resistance of 70 s m^{-1}. To derive the fluxes for the actual crop, so-called crop and soil factors are used (Chapter 8).

In general the daily water fluxes passing through a canopy are large compared to the amounts of water stored in the canopy itself (Chapter 6). On a daily basis we may assume soil water extraction by roots to be equal to plant transpiration. Whereas root water extraction occurs throughout the root zone, soil evaporation occurs at the soil surface. Owing to the steep gradient of water contents and pressure heads near the soil surface, during drying conditions evaporation fluxes decline more rapidly than transpiration fluxes. Once water has infiltrated into the soil and the soil surface has become dry, soil evaporation fluxes become small. Water harvesting, in which fields are left fallow during one or several seasons, is based on this phenomenon. Because of the different physical behaviour of the transpiration and evaporation process, SWAP simulates these processes separately.

SWAP calculates three quantities with the Penman–Monteith equation:

- Evapotranspiration rate of a wet canopy, completely covering the soil
- Evapotranspiration rate of a dry canopy, completely covering the soil
- Evaporation rate of a wet, bare soil

These quantities are obtained by using the appropriate values for canopy resistance, crop height and reflection coefficient. In the case of a wet canopy or a bare wet soil,

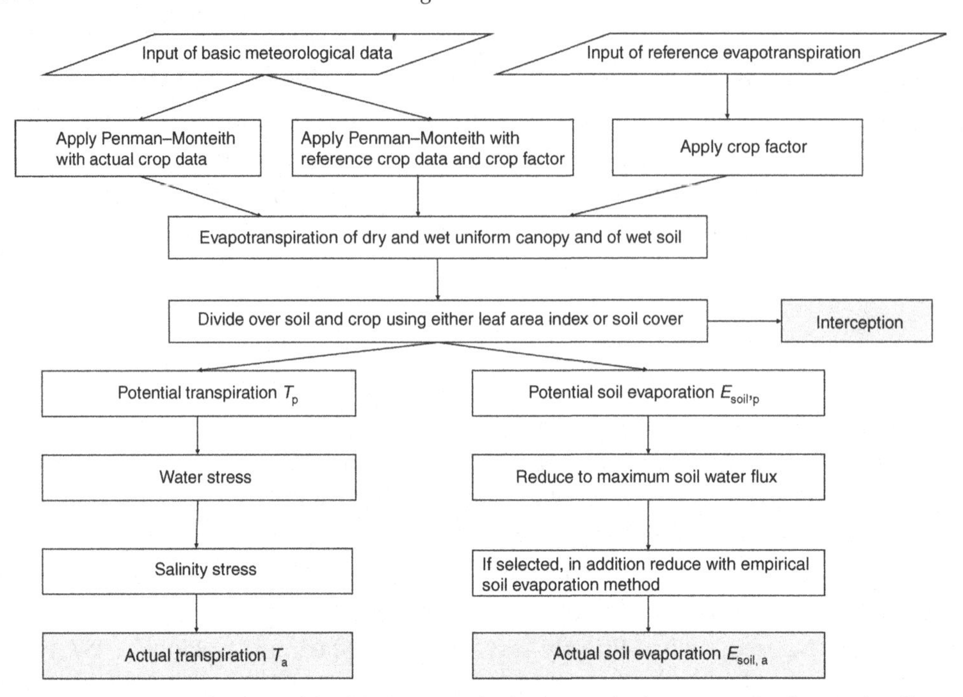

Figure 9.5 Method used in SWAP to calculate actual plant transpiration and soil evaporation of partly covered soils from basic meteorological input data.

the canopy resistance is set to zero and only the aerodynamic resistance applies. In the case of a dry crop with optimal water supply in the soil, the canopy resistance is equal to its minimum value and varies between 30 s m^{-1} for arable crops to 150 s m^{-1} for trees. For a dry and wet crop, the actual crop heights are used, while for bare soil 'crop height' is zero. The reflection coefficient in case of a wet or dry crop equals 0.23, while for a bare soil the value 0.15 (-) is assumed.

Figure 9.5 shows an overview of the top boundary procedure followed in SWAP. Both the use of the Penman–Monteith method and the use of a reference evapotranspiration rate with crop factors are allowed. After calculation of the evapotranspiration flux of the dry and wet canopy and the wet soil, the potential plant transpiration T_p and soil evaporation $E_{soil,\,p}$ fluxes are derived by taking into account the amounts of rainfall interception and the leaf area index (LAI) or soil cover. In the case of agricultural crops interception is calculated with the methods of Von Hoyningen-Hüne (1983) and Braden (1985) and in the case of forests with the method of Gash (1979, 1995) (Chapter 6).

Subsequently, T_p in combination with the root length distribution over the root zone, is used to derive the maximum root water extraction rates at various depths. The actual root water extraction rates are calculated taking into account reductions due to oxygen deficiency, water deficiency or salinity excess. For oxygen and water deficiency, the reduction function of Feddes et al. (1978) is used. For saline conditions, the reduction function of Maas and Hoffman (1977) is employed. In the case of simultaneous water and salt stress, SWAP multiplies both reduction factors. Integration over the depth of actual root water extraction rates yields the actual transpiration rate (Chapter 6).

Reduction of $E_{soil,p}$ for dry soil conditions occurs in two ways. SWAP calculates the maximum upward soil water flux near the soil surface, using Darcy's equation, the prevailing soil hydraulic functions and the actual soil water status. In addition, we employ empirical reduction functions based on Black (1969) or Boesten and Stroosnijder (1986). Although from a physical point of view the maximum soil water flux based on Darcy should suffice, the resulting flux generally overestimates the evaporation rates of dry soils. Probably the soil hydraulic functions change close to the soil surface because of splashing rain, crust formation and plant residues. Therefore SWAP determines the actual evaporation rate as the minimum of $E_{soil,p}$, the maximum Darcy flux and a selected empirical reduction function.

Question 9.3: The top boundary conditions for rainfall and evapotranspiration described in the preceding text can be applied at time intervals of days or shorter. For which purpose would you prefer shorter time intervals?

9.1.4 Bottom Boundary Condition Hydrology

The following options are offered to prescribe the bottom boundary condition:

1. Specify the groundwater level or soil water pressure head as function of time.
2. Specify the bottom flux as function of time.
3. Specify the bottom flux as function of groundwater level.

Measurements of groundwater levels are relatively easy and are often used during model calibration with experimental data. However, when alternative scenarios have to be simulated, groundwater levels may change, and therefore cannot be prescribed anymore. Prescribed bottom fluxes to simulate experiments are an attractive option, as fixed bottom fluxes may increase the accuracy of simulated soil moisture profiles and solute leaching. Unfortunately, despite considerable efforts no reliable and practical soil water flux measurement devices have been developed until now. Nevertheless, situations in which the bottom flux can be prescribed occur when a soil layer with a low permeability is present in the subsoil, or when the seepage flux is more or less constant and known.

For scenario analysis a more general boundary condition should be used, such as soil water flux as function of groundwater level or bottom pressure head. Such relations in general require calibration with field data.

When the groundwater level is relatively deep, we may assume a zero gradient of the soil water pressure head at the bottom of the soil profile, so-called *free drainage*. Application of Darcy's law gives for such a case:

$$q = -k(h)\left(\frac{\partial h}{\partial z} + 1\right) = -k(h)(0+1) = -k(h) \tag{9.6}$$

In the case of lysimeter experiments where *free outflow* occurs at the lysimeter bottom, SWAP will assume zero flow as long as $h \leq 0$ at the lysimeter bottom. As soon as h tends to become larger than zero, SWAP will fix h at zero and calculate the bottom flux.

Question 9.4: Imagine two columns of identical soils. Column 1 is taken to the laboratory for leaching experiments, and free outflow applies at the bottom. Column 2 is still undisturbed in the field, where the groundwater level is so deep that free drainage conditions apply at the bottom of the column. We irrigate both soil columns with the same amount of water. At column 1, drainage starts 5 hours after water application. Six hours after water application, which column contains more water, column 1 or column 2?

9.1.5 Lateral Drainage

In the saturated part of the soil column, SWAP simulates lateral drainage and bottom fluxes separately. Drainage fluxes refer to groundwater flow to or from the local drainage system. Bottom fluxes refer to water fluxes at the soil profile bottom, which in general are governed by regional groundwater flow and less by local water management (Figure 9.6). In many soil water flow models, bottom fluxes include the drainage fluxes. SWAP can be used in the same way, by omitting the drainage component. The feature of defining the lateral drainage flux separately allows the evaluation of surface water management and drainage design alternatives.

SWAP offers three methods to calculate the drainage flux density q_{drain} (m d^{-1}):

1. A linear relation between groundwater level ϕ_{gwl} (m) and q_{drain}:

$$q_{\text{drain}} = \frac{\phi_{\text{gwl}} - \phi_{\text{drain}}}{\gamma_{\text{drain}}} \tag{9.7}$$

where ϕ_{drain} is the drain level (m) and γ_{drain} is the drainage resistance (d). Simultaneous drainage fluxes to various drainage levels can be calculated, which are superimposed to derive the total drainage flux. This is depicted in Figure 9.7. Note that in the figure the assumption is made that water level $\phi_{\text{drain},1}$ is maintained in dry periods, resulting in infiltration. At the higher drainage levels, no water infiltration is assumed.

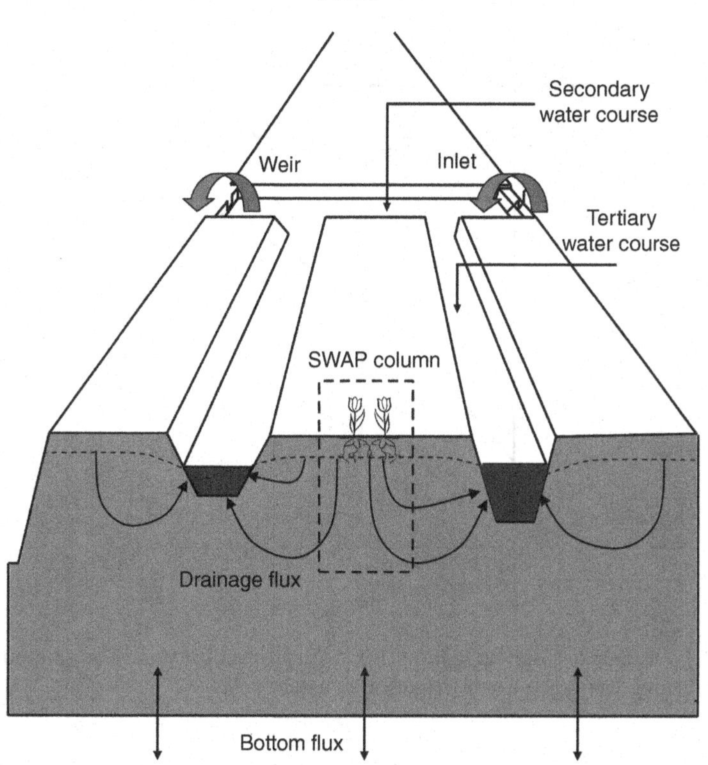

Figure 9.6 Superposition of lateral drainage and regional groundwater flow with respect to SWAP column.

2. A tabular relation between groundwater level and drainage flux. This option is useful in the case of drainage media at various levels, which cause a decreasing drainage resistance when the groundwater level increases. This situation gives a similar shape of the relation between drainage flux and groundwater level as depicted in Figure 9.7, but here no separate drainage levels and resistances need to be defined.
3. Various analytical drainage equations, which have been extensively described by Ritzema (1994). An example is the Hooghoudt equation. Consider groundwater flow towards a ditch or subsurface drain, as depicted in Figure 9.8. For steady-state conditions we may derive:

$$q_{\text{drain}} = R = \frac{4k_{\text{t}}m^2 + 8k_{\text{b}}dm}{L^2} \quad \text{with } d = \frac{L}{8\left[\dfrac{\left(L - D\sqrt{2}\right)^2}{8DL} + \dfrac{1}{\pi}\ln\left(\dfrac{D}{r_{\text{d}}\sqrt{2}}\right)\right]} \tag{9.8}$$

where q_{drain} is the drainage flux (m d^{-1}), R is recharge (m d^{-1}), m is the maximum height of the saturated region above drain level (m), k_{t} is the saturated conductivity in this region (m d^{-1}), D is the thickness of the aquifer below drain level (m), k_{b} is the saturated

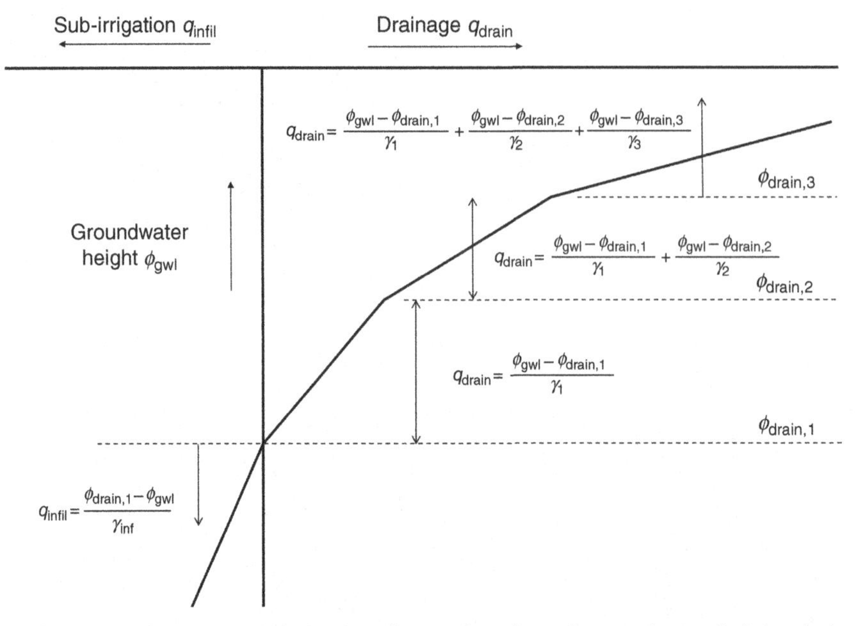

Figure 9.7 Drainage or sub-irrigation flux as function of groundwater height, drainage level φ and drainage or sub-irrigation resistance γ.

conductivity in this region (m d^{-1}), L is the drain spacing (m) and r_d is the hydraulic radius of the drainage canal (-). The variable d is an 'equivalent thickness', which is smaller than the thickness of the aquifer D and which expresses the hydraulic head loss due to convergent streamlines near the drainage canal. If the required drainage flux is known, Equation (9.8) can be solved iteratively to find the drain spacing L.

Question 9.5: Consider a field where the relation between drainage flux and groundwater level is similar as depicted in Figure 9.7. The three drainage resistances are: γ_1 = 1000 d, γ_2 = 500 d and γ_3 = 250 d. The three drainage levels are situated at: $\phi_{drain,1} = -3.0$ m, $\phi_{drain,2} = -2.0$ m and $\phi_{drain,3} = -1.0$ m with respect to soil surface. To solve this question, use method the first method from the above list:

a) How large is the drainage flux (mm d^{-1}) at a groundwater level 2.5 m below the soil surface?
b) How large is the drainage flux (mm d^{-1}) at a groundwater level 1.5 m below soil surface?
c) How large is the drainage flux (mm d^{-1}) at a groundwater level 0.5 m below soil surface?

Question 9.6: At an orchard high groundwater tables occur due to an aquitard at 4 m depth. We want to install subsurface drains such that the maximum groundwater level is 0.80 m below soil surface at a design discharge of 10 mm d^{-1}. The drains have a hydraulic radius of 5 cm and will be installed at a 2.0 m depth. The saturated

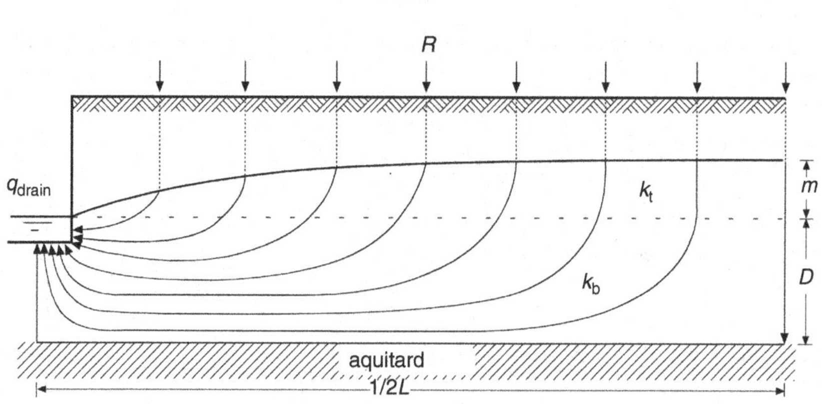

Figure 9.8 Hydrological scheme for the Hooghoudt drainage equation.

conductivity above the drain level is 0.8 m d^{-1}, below the drain level 1.2 m d^{-1}. What is the design drain spacing L?

9.1.6 Solute Transport

SWAP focuses on the transport of salts, pesticides and other solutes that can be described with relatively simple kinetics: convection, diffusion, dispersion, root uptake, Freundlich adsorption and first-order decomposition. The physical backgrounds of these processes are described in Chapter 5.

In the case of more advanced pesticide transport, such as volatilization and kinetic adsorption, SWAP can be used in combination with the model PEARL (Pesticide Emission Assessment at Regional and Local scales; Leistra et al., 2000; Tiktak et al., 2000). For detailed nutrient transport (nitrogen and phosphorus), SWAP can be used in combination with the model ANIMO (Agricultural Nutrient Model; Rijtema et al., 1997; Kroes and Roelsma, 1997).

As discussed in Chapter 5, we may derive a general transport equation for dynamic, one-dimensional, convective-dispersive mass transport, including nonlinear adsorption, linear decay and proportional root uptake in the vadose zone:

$$\frac{\partial(\theta c + \rho_b Q)}{\partial t} = -\frac{\partial(qc)}{\partial z} + \frac{\partial\left[\theta D \frac{\partial c}{\partial z}\right]}{\partial z} - \mu(\theta c + \rho_b Q) - K_r S c \tag{9.9}$$

where c is the solute concentration in the soil water (kg m^{-3}), ρ_b is the dry soil bulk density (kg m^{-3}), Q is the solute amount adsorbed (kg kg^{-1}), D is the effective diffusion coefficient (m^2 d^{-1}), μ is the decomposition parameter (d^{-1}), K_r is the preference factor for solute uptake by plant roots (-) and S is the water uptake by roots (d^{-1}).

SWAP solves this equation numerically, with an explicit, central finite difference scheme:

$$\frac{\theta_i^{j+1} c_i^{j+1} + \rho_b Q_i^{j+1} - \theta_i^j c_i^j - \rho_b Q_i^j}{\Delta t^j} =$$

$$\frac{q_{i-1/2}^j c_{i-1/2}^j - q_{i+1/2}^j c_{i+1/2}^j}{\Delta z_i} + \frac{1}{\Delta z_i}\left[\frac{\theta_{i-1/2}^j D_{i-1/2}^j \left(c_{i-1}^j - c_i^j\right)}{1/2\left(\Delta z_{i-1} + \Delta z_i\right)} - \frac{\theta_{i+1/2}^j D_{i+1/2}^j \left(c_i^j - c_{i+1}^j\right)}{1/2\left(\Delta z_i + \Delta z_{i+1}\right)}\right] - \tag{9.10}$$

$$\mu_i^j \left(\theta_i^j c_i^j + \rho_b Q_i^j\right) - K_r S_i^j c_i^j$$

where the superscript j denotes the time level, subscript i the node number and subscripts $i-1/2$ and $i+1/2$ refer to linearly interpolated values at the upper and lower compartment boundary, respectively. The scheme is explicit, as the unknown solute concentration c_i^{j+1} is a function of known variables at the former time level j. To ensure stability of the explicit scheme, the time step Δt^j should meet the criterion (Van Genuchten and Wierenga, 1974):

$$\Delta t^j \leq \frac{\Delta z_i^2 \, \theta_i^j}{2 D_i^j} \tag{9.11}$$

This stability criterion applies to nonsorbing substances and is therefore also safe for sorbing substances.

In general convective water fluxes dominate solute transport. To simulate solute transport, the input data in addition to water flow should specify the solute concentrations in rain-, irrigation-, canal-, and groundwater and the initial solute concentrations.

9.1.7 Heat Flow

Soil temperature affects many physical, chemical and biological processes in the top soil. Examples are the surface energy balance, soil hydraulic properties, decomposition rate of organic compounds and growth rate of roots. SWAP calculates the soil temperatures either analytically or numerically. In Chapter 2 a general analytical solution is discussed; here we explain a general numerical method based on De Vries (1963). The general flow equation for soil heat can be written as (Chapter 2):

$$C_s \frac{\partial T}{\partial t} = \frac{\partial \left(\lambda_s \frac{\partial T}{\partial z}\right)}{\partial z} \tag{9.12}$$

where C_s is the volumetric soil heat capacity (J m^{-3} K^{-1}), T is the soil temperature (K), and λ_s is the thermal conductivity (J m^{-1} K^{-1} s^{-1}). SWAP employs a fully implicit finite difference numerical scheme to solve Eq. (9.12):

$$C_{\mathrm{s}}^{j+1}\left(T_i^{j+1}-T_i^j\right)=\frac{\Delta t^j}{\Delta z_i}\left[\lambda_{i-1/2}^{j+1/2}\,\frac{T_{i-1}^{j+1}-T_i^{j+1}}{\Delta z_{\mathrm{u}}}-\lambda_{i+1/2}^{j+1/2}\,\frac{T_i^{j+1}-T_{i+1}^{j+1}}{\Delta z_{\ell}}\right] \tag{9.13}$$

where the superscript j denotes the time level, the subscript i is the node number, $\Delta z_{\mathrm{u}} = z_{\mathrm{i+1}} - z_{\mathrm{i}}$ and $\Delta z_{\mathrm{l}} = z_{\mathrm{i}} - z_{\mathrm{i+1}}$. As the coefficients C_{s} and λ_{s} are not affected by the soil temperature itself, Eq. (9.13) is a linear equation, which can be solved efficiently.

Both volumetric heat capacity and thermal conductivity depend on the soil components quartz, clay mineral, organic matter, water and air. The volumetric soil heat capacity C_{s} can be calculated as weighted mean of the heat capacities of each component:

$$C_{\mathrm{s}}=\sum_{i=1}^{n} f_{\mathrm{i}}\,C_{\mathrm{i}} \tag{9.14}$$

where f is the volume fraction ($\mathrm{m^3\ m^{-3}}$), C is the volumetric heat capacity ($\mathrm{J\ m^{-3}\ K^{-1}}$) and n *is* the number of soil components. Table 9.2 gives values of C for the different soil components.

Table 9.2 also lists thermal conductivity values, which are largest for sand and clay, an order smaller for organic matter and water, and again an order smaller for dry air. Because the thermal conductivity of air is much smaller than that of water or solid matter, a high air content (or low water content) corresponds to a low thermal conductivity.

The components that affect λ_{s} are the same as those affecting C_{s}. However, the variation in λ_{s} is much larger. In the range of soil wetness normally experienced in the field, C_{s} may undergo a threefold or fourfold change, whereas the corresponding change in λ_{s} may be hundredfold or more. As discussed in Chapter 2, thermal conductivity is sensitive to the sizes, shapes and spatial arrangements of the solid particles. In the case of dry soil, the addition of a small amount of water increases the contact area between soil particles considerably, and therefore the thermal conductivity increases rapidly (see Fig. 2.21). At larger water contents this increase becomes less pronounced (see Fig. 2.22).

Farouki (1986) gives an overview of various methods to calculate the thermal conductivity as function of soil moisture content. SWAP employs the method of De Vries (1963), which compares well to laboratory measurements (Ochsner et al., 2001). In this method the soil is considered a continuous liquid or gaseous phase in which soil and respectively gas or liquid 'particles' are dispersed. In the case of a 'wet' soil ($\theta > \theta_{\mathrm{wet}}$) liquid water is assumed to be the continuous phase and the thermal conductivity is given by:

$$\lambda_{\mathrm{s}}=\frac{k_{\mathrm{q}}f_{\mathrm{q}}\lambda_{\mathrm{q}}+k_{\mathrm{c}}f_{\mathrm{c}}\lambda_{\mathrm{c}}+k_{\mathrm{o}}f_{\mathrm{o}}\lambda_{\mathrm{o}}+\theta\lambda_{\mathrm{w}}+k_{\mathrm{a}}f_{\mathrm{a}}\lambda_{\mathrm{a}}}{k_{\mathrm{q}}f_{\mathrm{q}}+k_{\mathrm{c}}f_{\mathrm{c}}+k_{\mathrm{o}}f_{\mathrm{o}}+\theta+k_{\mathrm{a}}f_{\mathrm{a}}} \tag{9.15}$$

Table 9.2 Basic soil constituent data to calculate the composite soil heat capacity and thermal conductivity according to the method of De Vries (1963)

	Constituent				
	Sand	Clay	Organic	Water	Air
Volumetric heat capacity C_i (J cm^{-3} K^{-1})	2.128	2.385	2.496	4.180	1.212
Thermal conductivity λ_i (J cm^{-1} d^{-1} K^{-1})	7603	2523	216	492	Variable
Shape factor g_i (-)	0.144	0.00	0.50	0.144	Variable
Weight factor k_i (water continuous) (-)	0.2461	0.7317	1.2602	1.0000	Variable
Weight factor k_i (air continuous) (-)	0.0143	0.6695	0.4545	0.1812	1.0000

where k is a weighting factor and the subscripts denote the soil components: quartz (q), clay mineral (c), organic matter (o), water (w) and air (a). The weighting factors depend on the shape and orientation of the soil components and the ratio of the conductivities. De Vries assumed the particles to be spheroids whose axes are randomly oriented in the soil. In that case the weighting factor for soil component i can be calculated by:

$$k_i = \frac{1}{3}\sum_{j=1}^{3}\left[1+\left(\frac{\lambda_i}{\lambda_0}-1\right)g_j\right]^{-1} \tag{9.16}$$

where the subscript zero refers to the continuous fluid (air for dry soil, and water for moist soil), and g_j represents the shape factor for the ith component with $g_1 + g_2 + g_3 = 1$. We may assume g_1 and g_2 to be equal. Therefore only one shape factor must be estimated for each soil component and Eq. (9.16) can be written as:

$$k_i = \frac{0.66}{1+g_i\left(\frac{\lambda_i}{\lambda_0}-1\right)} + \frac{0.33}{1+(1-2g_i)\left(\frac{\lambda_i}{\lambda_0}-1\right)} \tag{9.17}$$

The shape factors should be calibrated and common values are given in Table 9.2. The weighting factors based on these shape factors and calculated with Eq. (9.17) are also listed in Table 9.2.

Question 9.7: Derive the weight factor k_i for sand with water as continuous phase, using Eq. (9.17) and the shape factors listed in Table 9.2.

For 'dry' soil ($\theta < \theta_{dry}$) air is considered as the continuous phase and the thermal conductivity is determined as:

$$\lambda_s = 1.25\frac{k_q f_q \lambda_q + k_c f_c \lambda_c + k_o f_o \lambda_o + k_w \theta \lambda_w + f_a \lambda_a}{k_q f_q + k_c f_c + k_o f_o + k_w \theta + f_a} \tag{9.18}$$

where all weighting factors k_i are defined with respect to air (Table 9.2).

The procedure is quite sensitive to the shape factor for air, which appears to depend on the air content itself. In wet soil ($\theta > \theta_{wet}$), the air shape factor is given by:

$$g_a = 0.333 - \frac{\phi - \theta}{\phi}(0.333 - 0.035) \tag{9.19}$$

where ϕ is soil porosity ($m^3\ m^{-3}$). In dry soil ($\theta < \theta_{dry}$), the air shape factor follows from:

$$g_a = 0.013 + \frac{\theta}{\theta_{dry}}\left(g_{a,dry} - 0.013\right) \tag{9.20}$$

where $g_{a,dry}$ is the value of Eq. (9.19) at $\theta = \theta_{dry}$. In this way g_a varies between 0.013 ($\theta = 0$) and 0.333 ($\theta = \phi$).

The thermal conductivity of air-filled pores is considered to be the sum of λ_{da} and λ_v, where λ_{da} is the thermal conductivity of dry air (22 J cm^{-1} d^{-1} K^{-1}) and λ_v accounts for heat transfer across the air-filled pores by water vapour. Above the critical water content θ_{wet} the air-filled pores are assumed to be saturated with water vapour, and λ_v is assumed to be 64 J cm^{-1} d^{-1} K^{-1}. Below θ_{wet} we assume that λ_v decreases linearly with water content to a value of zero for oven-dry soil:

$$\lambda_a = \lambda_{da} + \lambda_v = 22 + \frac{\theta}{\theta_{wet}} 64 \text{ J cm}^{-1}\text{ d}^{-1}\text{ K}^{-1} \tag{9.21}$$

In the case that neither water nor air can be considered as the continuous phase ($\theta_{dry} < \theta < \theta_{wet}$), λ_s is found by interpolation between values at the wet and dry limits:

$$\lambda_s(\theta) = \lambda_s\left(\theta_{dry}\right) + \frac{\lambda_s\left(\theta_{wet}\right) - \lambda_s\left(\theta_{dry}\right)}{\theta_{wet} - \theta_{dry}}\left(\theta - \theta_{dry}\right) \tag{9.22}$$

The values of θ_{dry} and θ_{wet} are commonly taken as 0.02 and 0.05 respectively.

Question 9.8: Consider a sandy soil with volume fractions $f_q = 0.55$, $f_c = 0.08$, and $f_o = 0.02$. Calculate the soil thermal conductivity λ_s for wet ($\theta = 0.25$) and dry ($\theta = 0.02$) conditions.

With respect to boundary conditions, in SWAP at the soil surface either the daily average air temperature T_{avg} or measured soil surface temperatures can be used. At the bottom of the soil profile either soil temperatures can be specified or $q_{heat} = 0.0$ can be selected. The latter option is valid for large soil columns, with negligible heat fluxes at the bottom.

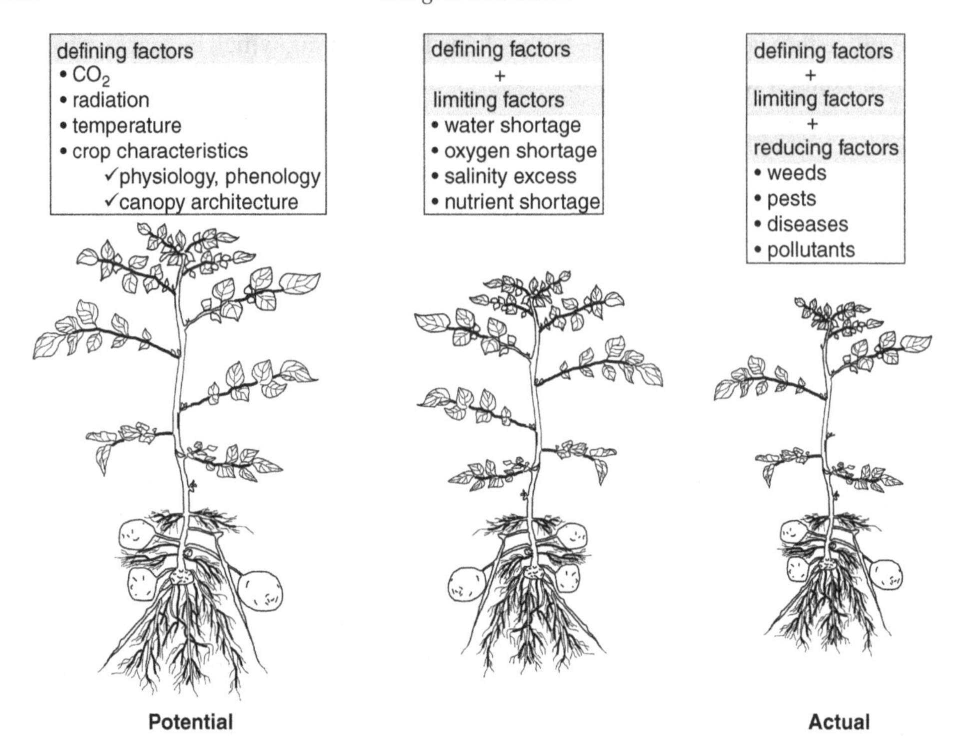

Figure 9.9 A hierarchy of growth factors, production situations and associated production levels (Van Ittersum et al., 2003).

9.1.8 Crop Growth

Three groups of growth factors (Figure 9.9) may be distinguished to obtain a hierarchy of production levels in crop production. Growth-defining factors determine the potential production that can be achieved in a given physical environment for a specific plant species. The main growth defining factors are radiation intensity, carbon dioxide concentration, temperature and crop characteristics. Their management, at least in open fields, is only possible through tactical decisions such as sowing date, sowing density and breeding. To achieve the potential production the crop must be optimally supplied with water and nutrients and fully protected against weeds, pests, diseases and other factors that may reduce growth.

Growth-limiting factors comprise shortage of water, oxygen and nutrients and excess of salts. Combined with the crop and climate characteristics, these factors determine a theoretical production level for a plant species in a given physical environment. Here, management can be used to control availability of water, oxygen, nutrients and salts, and may increase production towards potential levels.

Growth-reducing factors hamper growth further and comprise biotic factors such as weeds, pests and diseases, and abiotic factors such as pollutants and aluminium toxicity. Crop protection aims to limit the influence of these growth-reducing factors.

In the actual production situation, the productivity achieved will be the result of a combination of growth-defining, -limiting and -reducing factors (Van Ittersum et al., 2003).

In the 1980s a wide range of scientists in Wageningen became involved in the development and application of crop models. A comprehensive overview of the development and application of 'Wageningen' crop models is given in Van Ittersum et al. (2003). We will discuss the WOFOST (WOrld FOod STudies model; Supit et al., 1994), which has been incorporated in the SWAP model, and aims to calculate the crop production level as determined by climate and crop factors and limited by water and oxygen shortage or salt excess.

Figure 9.10 shows a flow diagram of the WOFOST model components. Plant dry matter production results from the photosynthesis process, in which CO_2 from the air is converted into *carbohydrates* $(CH_2O)_n$ according to the overall reaction:

$$CO_2 + H_2O + \text{solar energy} \rightarrow CH_2O + O_2 \qquad (9.23)$$

This process is known as CO_2 assimilation. For each kg of CO_2 absorbed, 30/44 kg of CH_2O is formed, where the numerical values represent the molecular weights of CH_2O and CO_2.

Canopy photosynthesis is calculated from the absorbed amount of photosynthetically active radiation (PAR; wavelength 400–700 nm) and the photosynthesis-light response of single leaves. Use of average illumination intensities of the leaves in the calculations would overestimate assimilation because of the convex assimilation-light response. Temporal and spatial variation in illumination intensity over the leaves therefore has to be taken into account. First, the instantaneous radiation flux at the top of the canopy at a certain time of day is derived from measured daily global radiation. A distinction is made between diffuse skylight and direct sunlight because of the large difference in illumination intensity between shaded leaves receiving only diffuse radiation, and sunlit leaves, receiving both direct and diffuse radiation. The assimilation rate in a leaf layer is calculated for sunlit and shaded leaves separately. Daily crop assimilation is obtained by integrating these assimilation rates over the leaf layers and over the day (Goudriaan, 1986).

The potential photosynthesis might be reduced by water or oxygen shortage or salt excess. The reduction is assumed to be proportional to the reduction in transpiration, as discussed in Chapter 6.

Part of the daily production of assimilates is used to provide energy for the maintenance of the existing biomass (*maintenance respiration*). Maintenance respiration is related to the standing biomass and to the metabolism intensity. Higher temperatures accelerate the turnover rates in plant tissues and hence the cost of maintenance. An increase in temperature of 10 °C increases maintenance respiration by a factor of about 2 (Penning de Vries and Van Laar, 1982).

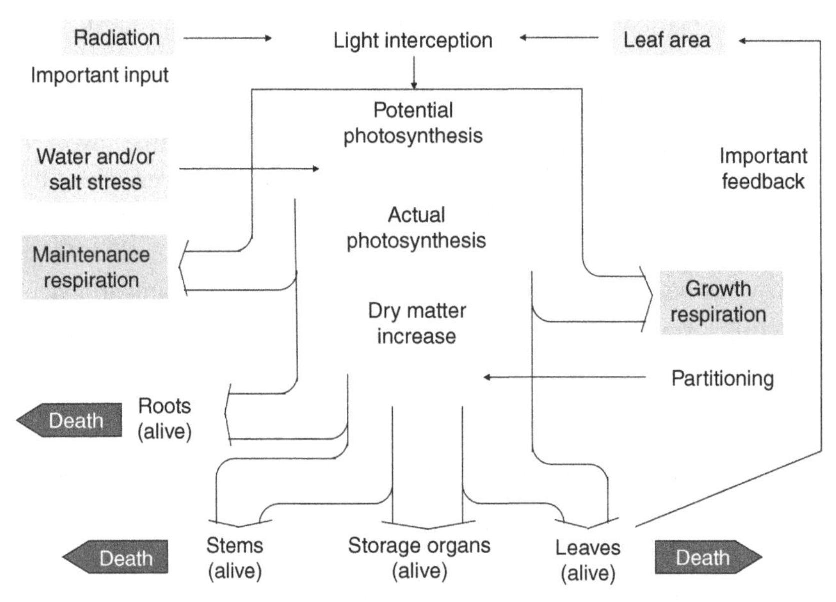

Figure 9.10 Schematization of crop growth processes incorporated in WOFOST.

The remaining carbohydrates are converted into structural plant material such as cellulose, proteins, lignin and lipids. In this conversion process some of the weight of carbohydrates is lost as *growth respiration*. The magnitude of growth respiration is determined by the composition of the end product formed. Thus the weight efficiency of conversion of primary carbohydrates into structural plant material varies with the composition of that material. In the model, crop specific conversion factors are used for leaf, storage organ, stem and root biomass.

As previously mentioned, a crop not only accumulates weight, but it also passes through successive crop development stages. The dry matter produced is therefore partitioned amongst the various plant organs in a distribution pattern that depends on the crop development stage. Crop development is characterized by the order and rate of appearance of vegetative and reproductive plant organs. The major environmental conditions influencing crop development are temperature and day length. Figure 9.11 provides an example of the partitioning of dry matter produced to the different plant organs. Initially the main part of the assimilation products will go to the roots and leaves. During flower initiation most of the dry matter produced goes to the stem. In the reproductive part almost all new dry matter will go to the storage organ.

The dry weights of the roots, leaves, stem and storage organs is obtained by integrating their growth rates over time. During the development of the crop, a part of the living biomass dies due to senescence. WOFOST takes daily time steps, which means that the calculation time is small. Obvious, solar radiation is the motor behind crop growth. The calculated LAI on a certain day is input for the photosynthesis process at the next day (Figure 9.10). Therefore a strong positive feedback exists among

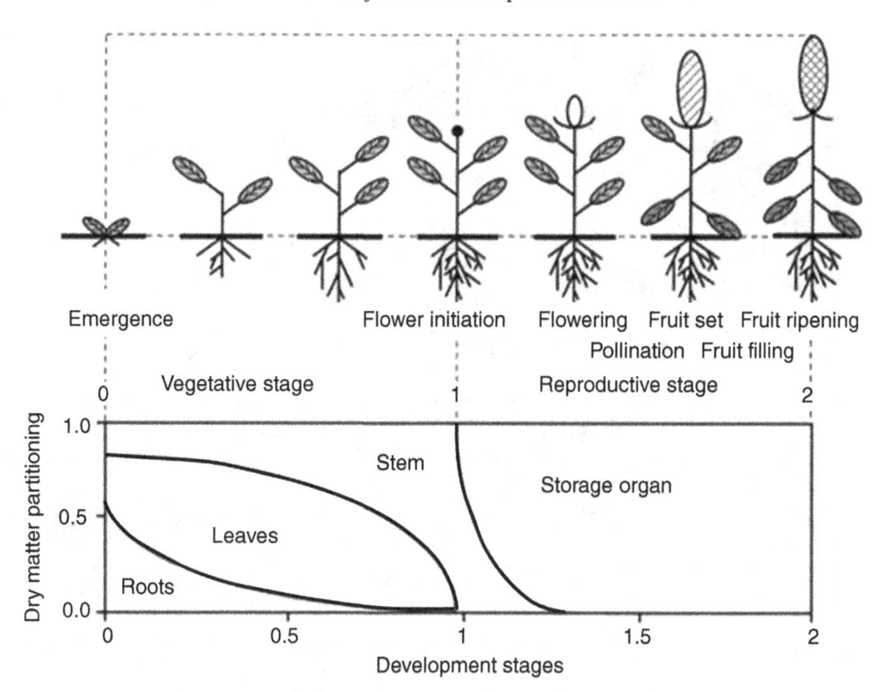

Figure 9.11 General overview of development stages in a crop's life cycle and the accompanying changes in dry matter partitioning (Lövestein et al., 1995).

radiation, leaf area and light interception. This means that proper input of daily solar radiation and CO_2 assimilation rate is very important for accurate simulation of crop growth! An overview of all input data required to simulate crop growth is listed in Table 9.3.

9.2. The Land-Surface in Atmospheric Models

Models of the atmosphere are always bounded by Earth's surface. At that boundary the models need to be provided with proper boundary conditions. To that end, each atmospheric model (which includes both numerical weather prediction models and climate models) are in need of a land-surface model (LSM).

9.2.1 The Role of LSMs in Atmospheric Models

In principle, atmospheric models simply solve the conservation equations for momentum (wind speed in two horizontal directions), heat and mass (mainly water vapour) (see Figure 9.12):

- The windspeed is affected by the pressure gradient, Coriolis force and the advection of momentum from neighbouring grid cells.

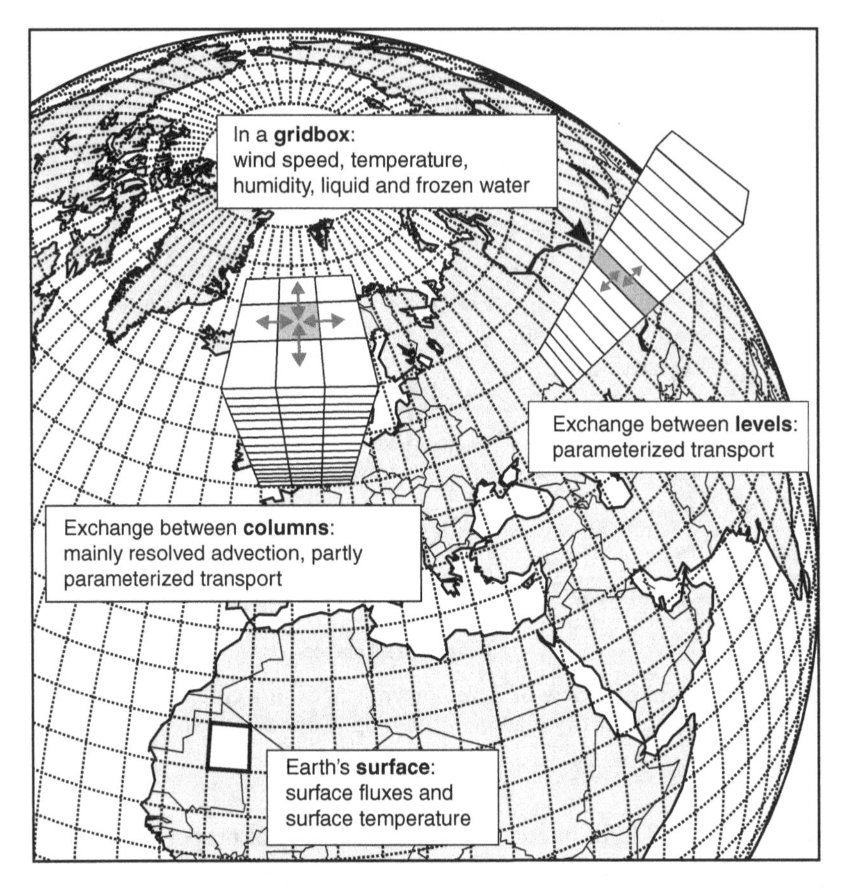

Figure 9.12 Atmospheric grid boxes in which one value for wind speed and direction, temperature, humidity and liquid and frozen water is defined are vertically embedded in atmospheric columns which in turn are horizontally connected. Earth's surface is the lower boundary for each column. Note that the size of the grid boxes in this figure (5 by 5 degrees) is much larger than used in present-day weather models (approx. 16 km in the ECMWF model in 2012).

Table 9.3 Input data required to simulate crop growth with WOFOST

Category	Input data
Weather	Temperature, solar radiation, wind speed, humidity, rainfall
Irrigation	Irrigation timing, gift amounts, irrigation water salinity
Crop	Emergence date, criteria crop development stage, light use efficiency, maximum assimilation rate, growth and maintenance respiration factors, sensitivity to water and salinity stress, dry matter partitioning as function of crop development, organ death rates, root density profile as function of crop development, rainfall interception
Soil moisture	Initial soil moisture, drainage, soil hydraulic functions

- Temperature changes because of horizontal advection of heat from neighbouring cells.
- Moisture and other scalars change because of horizontal advection from neighbouring cells as well.

But much more happens than horizontal exchange alone: many processes occur at scales that are much smaller than can be resolved on the grid of an atmospheric model (which has typical horizontal distances between cells of 10–100 km and vertical distances of 10–500 m). The processes that cannot be resolved encompass:

- Turbulent exchange of momentum, heat and moisture (in particular vertical exchange)
- Cloud formation and precipitation
- Radiation processes

Because these processes cannot be explicitly resolved, they have to be parameterized. This means that they have to be expressed in terms of variables that *are* known in the model. For example, radiation transfer through clouds needs to be calculated from information on the liquid water content within a grid box only, whereas in reality the radiation transfer would depend not only on moisture content but on the structure of the clouds as well. In most atmospheric models all of these processes are applied within each vertical column separately. So, as far as the parameterized processes are concerned, the atmosphere consists of a large collection of parallel, independent columns.

Whereas most grid boxes in an atmospheric model only have other grid boxes as their neighbour, the grid boxes at the lower end of the atmospheric column have Earth (either land surface, water or ice) as its lower boundary.

At this lower boundary momentum, heat and water vapour are exchanged between the land surface and the atmosphere (see Figure 9.13). This exchange will affect the values of horizontal wind speed, temperature and humidity in this grid box (e.g., a positive evapotranspiration will increase the specific humidity in the box). Thus, the role of an LSM in an atmospheric model is to provide the correct exchange of momentum, heat and water vapour between the surface and the atmosphere.

Two major complications arise in the specification of those fluxes, in particular over land:

- On the scale of a single vegetation unit (i.e., maize field or a forest) the fluxes originate from a number of different sources (e.g., soil and leaf). Those fluxes are regulated both by the external atmospheric conditions and by the state of the land surface (in particular vegetation cover and soil moisture content). The various levels of complexity in which this can be dealt with in LSMs is discussed in Section 9.2.3.
- On the scale of a single grid box the properties of the land-surface are not homogeneous: within a square of say 50×50 km one will find, cities, grassland, forests and open water). Methods to deal with this large-scale heterogeneity are discussed in Section 9.2.5.
- Apart from the fact that the *properties* of the land-surface vary within a grid box, the *elevation* may vary as well: at the grid-scale the model can only follow the mean elevation

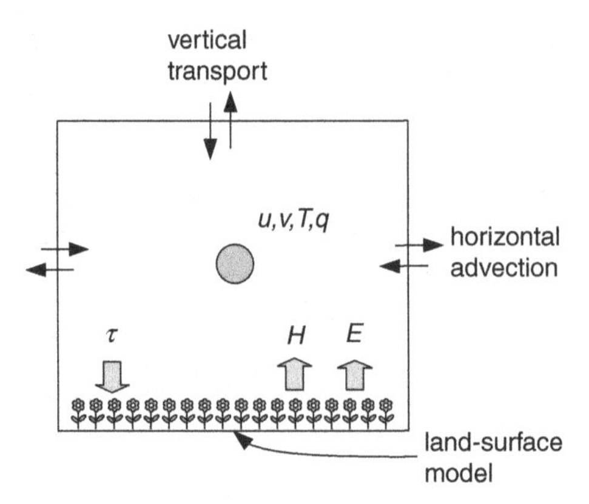

Figure 9.13 Role of land-surface model in lowest grid box: momentum is transported to the surface, heat and moisture is exchanged with the surface (here positive fluxes during daytime). Exchange with neighbouring cells through horizontal advection and vertical (turbulent) transport. The dot indicates the grid point within the box where velocity, temperature and humidity are stored.

of the terrain, whereas smaller mountains, hills and valleys will not be visible. This subgrid orography has two effects. First, it modifies the effective roughess length (for momentum): the atmosphere not only feels the roughness of the surface itself (grass, trees etc.) but also the roughness due to variations in elevation. In this way the momentum exchange between the surface and the atmosphere is affected (Wan and Porté-Agel, 2011). Second, orography may generate orographic drag due to gravity waves excited by the variations in terrain height. This orographic drag affects momentum transport throughout the atmosphere (Jiménez and Dudhia (2012) provides an example of how this drag is dealt with in NWP models).

Question 9.9: Consider Figure 9.13 and suppose that there is no horizontal advection from or into the grid box for each of the quantities.

a) What happens with the values of u, v, T and q in the grid box if the shear stress, heat flux and moisture flux at the surface have the sign as given by the arrows in the figure (assume the fluxes at the top of the grid box have the same direction, but are smaller in magnitude than the surface flux).
b) If the dimensions of the gridbox are Δx, Δy and Δz in the two horizontal and vertical directions, respectively and the heat flux at the top and bottom of the gridbox are indicated by $H(\Delta z)$ and $H(0)$, give an expression for the change in time of the temperature in the gridbox.

9.2.2 General Structure of a LSM

Any LSM needs to provide the turbulent fluxes of momentum, heat and water vapour (and possibly other trace gases) to the atmospheric part of the model. Furthermore, the upwelling longwave radiation (dependent on surface temperature) and reflected shortwave radiation are fed into the model.

Those fluxes need to be modelled (decribed) in terms of *variables* that are available in the atmospheric model in combination with *parameters* that characterize the surface. Such a description is called a parameterization.

For a LSM the relevant *forcing variables*, providing the boundary upper conditions for the land-surface model, are:

- The radiative forcing (incoming shortwave and longwave radiation)
- The liquid water forcing (precipitation)
- The atmospheric state in the lower gridbox (wind speed, temperature and humidity)

To determine the fluxes, the LSM also needs to keep track of some internal variables (among others):

- The soil conditions (temperature and water content)
- The state of the surface (e.g., snow cover, amount of intercepted water)

The list of *model parameters* very much depends on the level of complexity of the LSM, but will at least contain albedo, surface emissivity, roughness length, as well as information on the amount and type of vegetation (including properties related to stomatal behaviour). These parameters are usually based on land-use classifications (based on remote sensing data) where for each land-use class a set of parameters is given. In some cases these parameters are time-dependent, for example, seasonal variations in albedo or vegetation fraction. In more complex models (see Section 9.2.3) some of these parameters become variables within the LSM (and thus are modelled themselves).

Provided that the lowest grid point of the model is located at a height within the surface layer (or constant flux layer), surface layer similarity can be used to compute the fluxes of momentum, heat and water vapour using:

1. Wind speed, temperature and humidity known at that the lowest atmospheric grid point
2. The roughness lengths for momentum and heat
3. The surface values for wind speed (equal to zero), temperature and humidity. Usually the surface specific humidity is replaced by $q_{sat}(T_s)$ in combination with a canopy resistance (see Chapter 7 and Section 9.2.4).

The similarity relationships presented in Chapter 3 can be used to combine the information under items (1) to (3) to compute the fluxes. To arrive at the values needed under item (3), the energy balance needs to be computed. This generally involves the solution of a number of coupled problems:

- Net radiation is coupled to the surface by an albedo, surface emissivity (both relatively constant) and a highly variable surface temperature.
- Soil heat flux depends on the surface temperature as well (as well as on the soil temperatures).
- The partitioning of the available energy ($Q^* - G$) over sensible and latent heat flux determines the surface temperature (e.g., high evaporation gives a low surface temperature).

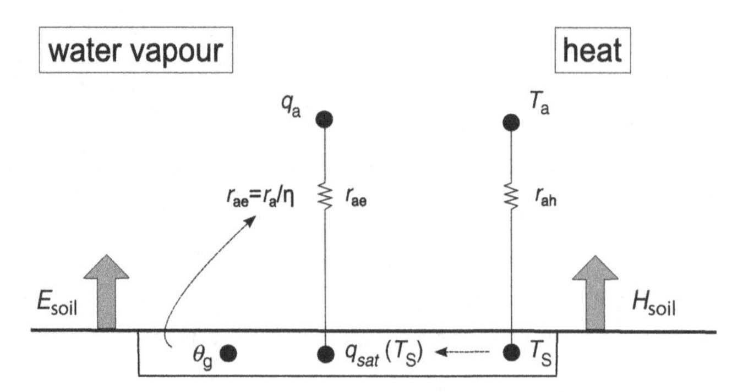

Figure 9.14 LSM without vegetation: the Manabe (1969) model. T_s is the temperature of the soil layer and θ_g is the volumetric moisture content of the soil (which has maximum moisture content $\theta_{max.}$). The aerodynamic resistance for water vapour is modified based on soil moisture availability, expressed as $\eta = \theta_g / \theta_{max}$. Note that $q_{sat}(T_s)$ is not an independent model variable as it directly depends on T_s.

> This partitioning is a strong function of the canopy resistance. Because water availability in the root zone is an important determining factor in the canopy resistance, not only the energy balance of the surface needs to be tracked, but also the water balance.

LSMs are both used in weather forecast models and in climate models. But given the longer integration times of a climate model, in those models more emphasis is placed on the correct long-term behaviour (e.g., the soil should not dry out too far; Hagemann et al., 2004).

9.2.3 Modelling of Vegetation

Roughly four levels of complexity can be distinguished in the treatment of vegetation in land-surface models (see also Sellers et al., 1997 and Pitman, 2003).

LSMs Without Vegetation (First Generation)

The model of Manabe (1969), the first LSM to be implemented in a climate model, is the archetype of this group (see Figure 9.14). In this model the fluxes of both heat and water vapour solely originate from the soil surface. The evaporation is regulated by modifying the aerodynamic resistance with a factor that depends on the relative saturation of the (only) soil layer. If the soil dries out, the resistance for water vapour transport increases, thus reducing evaporation. In the original Manabe (1969) model the soil heat flux is ignored, as it did not simulate the diurnal cycle. In later applications this simplification has been replaced by including a force-restore method for the soil temperature or with a multilayer soil model for temperature (as in, e.g., the MM5/WRF model; Chen and Dudhia, 2001). Although on time scales of years to months this simple scheme provides appropriate fluxes, on diurnal time scales larger discrepancies will occur (Desborough, 1999; Trier et al., 2006).

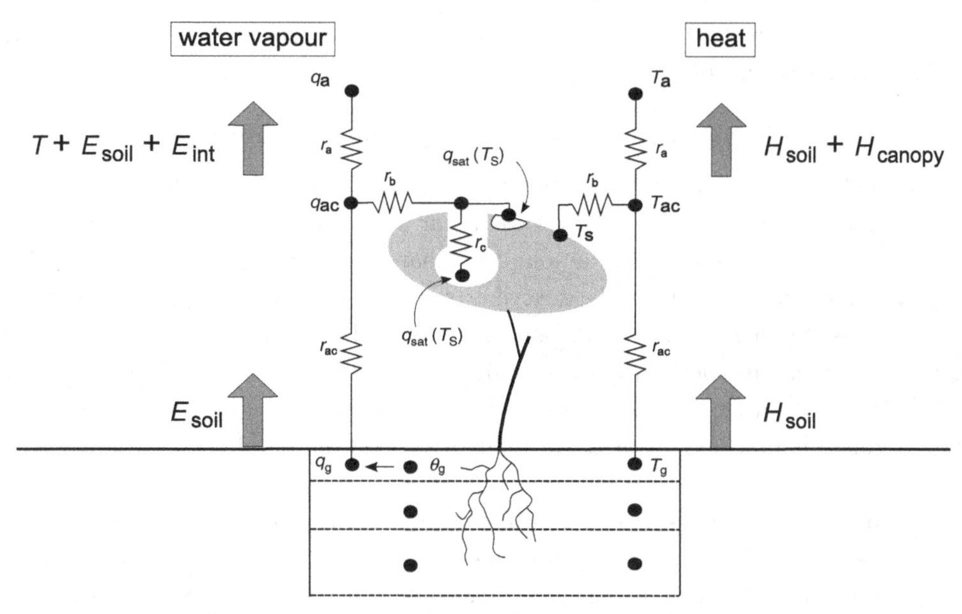

Figure 9.15 LSM with a single vegetation layer and multiple soil layers. The air temperature and humidity at the canopy level (T_{ac} and q_{ac}) are coupled to the air above through the aerodynamic resistance r_a, to the vegetation (through boundary-layer resistance r_b and canopy resistance r_c) and to the soil temperature and humidity (T_g and q_g) through the in-canopy aerodynamic resistance r_{ac}. The liquid water reservoir (rainfall interception or dew) is directly coupled (through r_b) to the canopy air, as there is no stomatal control.

LSMs with Empirical Stomatal Control (Second Generation)

Deardorff (1978) was the first to develop a LSM that incorporated vegetation, which has been further developed by many others into models of varying complexity (see, e.g., Pitman, 2003, for an overview).

In these LSMs the vegetation consists of one (or more) layer that covers a fraction σ_f of the ground, thus exposing a fraction $1-\sigma_f$ of bare soil. Figure 9.15 provides a sketch of the way evapotranspiration and sensible heat flux are parameterized for a single-layer model. To allow for separate fluxes from the soil and the vegetation layer, an extra atmospheric layer is introduced: the canopy air. Both the soil surface, the vegetation layer and the atmosphere above, exchange heat and water vapour with this layer.

An essential property of the vegetation layer in these models is that the transpiration from the plants is regulated by a canopy resistance, which depends on environmental conditions such as solar radiation and temperature, as well as on soil moisture content (see Section 9.2.4). The water needed for the transpiration flux is extracted from one or more soil layers depending on the vertical distribution of the roots. The latter in turn depends on the type of vegetation to be modelled. The surface of the vegetation may be wet due to rainfall interception or dew. This water can evaporate directly, without stomatal control.

LSMs Based on Plant Physiology (Third Generation)

In the 1990s the advance in plant-physiological knowledge made its way to land-surface models. Because CO_2 follows – as far as assimilation is concerned – the same pathway as water vapour a sound mechanistic description of the assimilation processes helps to properly describe transpiration. In particular, the factors that limit assimilation will limit transpiration as well. Hence, the plant-physiology approach affects the modelling of the canopy resistance in particular (see Section 9.2.4). The advantage of the plant-physiology as compared to the purely empirical approach is that less empiricism is involved and the number of parameters is smaller.

In some models, the simulated assimilation (as needed in the modelling of r_c) is used to assign the fixed carbon to particular parts of the plants. In this way some of the plant-parameters (such as leaf area index) used in LSMs are not prescribed but predicted by the model itself.

LSMs with Adaptive Vegetation (Fourth Generation)

On the time scale of numerical weather prediction the vegetation (cover and type) can be assumed to be nearly invariable. However, for long-term simulations with climate models (or Earth system models in general) the changes in the vegetation need to be taken into account. For simulations of the current climate, the known seasonal variation of the vegetation cover can be prescribed. But for simulations of future development the vegetation needs to be allowed to respond to a changing simulated climate (including changing CO_2 concentrations and temperature).

Two concepts are used to model the adaptation of vegetation to changing environmental conditions (Levis, 2010). The first group of models are so-called equilibrium vegetation models (EVMs). In these models the occurrence of a certain biome (combinations of plant types) at a certain location is determined by the local climate (among others precipitation, temperature, CO_2 concentration) and possibly the soil type. Some EVMs use plant-functional types rather than predefined combinations of plants. A plant-functional type represents a broad class of plants that has distinct characteristics (e.g., physiologically and morphologically) from the other classes: for example, grasses versus tropical rain forest trees. The use of plant-functional types allows the model to compose its own biomes, depending on the climate conditions.

The second group of models are dynamic (global) vegetation models (DGVMs). In these model not only the relationship between local conditions (climate and soil) are taken into account, but vegetation dynamics as well: for example, succession between different plant types, competition, disturbances in the form of wild fire.

Table 9.4 LSM model components responsible for variation of surface fluxes (in particular transpiration and CO_2 flux) on different time scales; also indicated which model variable causes this variation and which generation of LSM incorporated this variation for the first time

Timescale	Change in LSM parameter/variable	Relevant atmospheric model variable	First LSM generation
Minute	Surface temperature	Radiation	First
	H_2O concentration in stomata (due to change in surface temperature)	Radiation	Second, third
	CO_2 concentration in stomata (due to change in assimilation)	Radiation	None
Hour	Canopy resistance	Radiation	Second, third
	Surface temperature	Air temperature	
		Air humidity	
Day	Canopy resistance	Radiation	Second, third
	Surface temperature	Air temperature	
		Air humidity	
		Soil moisture	
Week	Canopy resistance	Radiation	Second, third
		Air temperature	
		Soil moisture	
Month	Canopy resistance	Radiation (cumulative)	Third, fourth
	Vegetation fraction	Air temperature	
		Soil moisture	
Year	Vegetation type	Air temperature Soil moisture	Fourth

Furthermore, DGVM's may include modelling of nutrient dynamics (e.g., Gerber et al., 2010).

The adaptive vegetation models produce their own amount and type of vegetation, but the physical description of transpiration and assimilation is similar to that in third-generation models. Through the inclusion of the process of assimilation, the models can determine the seasonal variation in biomass.

Table 9.4 summarizes the various time scales involved in land-surface processes and the parts of the LSM that is responsible to model this variation. It is clear that most second- and third-generation LSMs are capable of simulating variations in surface fluxes of water vapour (and for some third- generation models CO_2 fluxes) at time scales up to weeks. This is sufficient for weather forecast models, but for climate models one needs fourth-generation models in order to include the long-term dynamics of the vegetation.

9.2.4 Canopy Resistance

In the previous section, the canopy resistance was left unspecified. Some basic characteristics of the stomatal resistance (the resistance for individual stomata) have been dealt with in Chapter 6. Furthermore, the concept of a canopy resistance has been introduced in Chapter 7.

The main requirements for a model of the canopy resistance is that it should react to external factors in a way similar to the reactions of the stomatal resistance (see Section 6.4.3). Furthermore, one would expect that the modelled canopy resistance decreases when the amount of leaf area increases, as more leaf area provides more parallel pathways for water vapour transport. However, because the microclimate (temperature, radiation and humidity) varies vertically within a canopy and the variation of the orientation of leaves exposes them to different amounts of radiation, the dependence on leaf area may not be a simple one (Jarvis and McNaughton, 1986; Baldocchi et al., 1991; Ronda et al., 2001).

Whereas in the meteorological context it is logical to express the stomatal control on transpiration in terms of a resistance, the plant-physiology literature more often uses the reciprocal of the resistance: the conductance. Here we use both terminologies interchangeably. We discuss two approaches to model the canopy conductance: the empirical Jarvis–Stewart approach and the plant-physiology-based A–g_s (or Ball-Berry) approach.

Jarvis–Stewart Approach

The general structure of the so-called Jarvis–Stewart approach (Jarvis, 1976 and Stewart,1988) is based on a minimum stomatal resistance (i.e., the stomatal resistance under optimal conditions) that is modified by a number of empirical response functions plus a scaling from a single square meter of leaf to a canopy:

$$r_c = \frac{r_{s,min}}{\text{LAI}} f_1(K^{\downarrow}) f_2(D_{q,a}) f_3(T_a) f_4(\bar{\theta}) \qquad (9.24)$$

where $r_{s,min}$ is the minimum *stomatal* resistance; LAI is the leaf area index (surface area of leaves per surface of ground); and f_1, f_2, f_3 and f_4 are response functions for the influence of global radiation, vapour deficit $D_{q,a}$ (here in terms of specific humidity: $(q_{sat}(T_a) - q_a)$), air temperature T_a and soil moisture ($\bar{\theta}$ is soil moisture content averaged over the part of the soil column where roots are present, not to be confused with potential temperature). The proportionality of r_c to LAI^{-1} can be understood when considering r_c as the replacement resistance for a number of parellel resistances (where the number of resistances is LAI, and the magnitude of those resistances is $r_{s,min}$).

A wide range of formulations exists for the response functions, differing both in the exact shape and in the parameter values involved. As an example, we here discuss the functions as given by Chen and Dudhia (2001).[1] Note that the functions have similar shapes as the dependence of stomatal resistance on environmental factors as sketched in Figure 6.15).

The radiation response function f_1 is formulated as follows:

$$f_1(K^{\downarrow}) = \frac{1+f}{r_{\mathrm{s,min}} / r_{\mathrm{s,max}} + f}, \quad \text{where } f = 0.55 \frac{K^{\downarrow}}{\mathrm{PAR}_{\mathrm{limit}}} \frac{2}{\mathrm{LAI}} \tag{9.25}$$

where $r_{\mathrm{s,max}}$ is the cuticular resistance of the leaves (i.e., the resistance when the stomata are fully closed, of the order of 5000 s m^{-1}). The factor 0.55 in f is the fraction of global radiation that is photosynthetically active (somewhat higher than the values given in Chapter 6). $\mathrm{PAR}_{\mathrm{limit}}$ is the level of PAR at which the resistance is roughly doubled (equal to 30 W m^{-2} for trees and 100 W m^{-2} for low vegetation). Finally, the LAI is included in f because the positive effect of radiation on stomatal opening decreases below the top of the canopy due to the extinction of radiation.[2] At low values of global radiation f_1 tends to $r_{\mathrm{s,max}} / r_{\mathrm{s,min}}$ (so that the actual canopy resistance becomes $r_{\mathrm{s,max}} / \mathrm{LAI}$) whereas for high levels of radiation the function tends to one.

The response of the canopy resistance to specific humidity deficit (or in other models: VPD) is taken as a linear dependence:

$$f_2(D_{q,\mathrm{a}}) = 1 + h_{\mathrm{s}} D_{q,\mathrm{a}} = 1 + h_{\mathrm{s}} \left(q_{\mathrm{sat}}(T_{\mathrm{a}}) - q_{\mathrm{a}} \right) \tag{9.26}$$

where h_{s} depends on the vegetation type but has a typical value of 40 to 50 (kg kg^{-1})$^{-1}$. In some land-surface models h_{s} is nonzero only for high vegetation. The dependence of canopy resistance on vapour deficit is in fact a dependence on transpiration rate: plants limit their water loss if atmospheric demand for water vapour becomes too large (Leuning, 1995; Monteith, 1995; and Chapter 6).

The temperature dependence of the canopy resistance is parameterized as a parabolic function around the reference temperature T_{ref}:

$$f_3(T_{\mathrm{a}}) = \left[1 - a_T (T_{\mathrm{ref}} - T_{\mathrm{a}})^2 \right]^{-1} \quad \text{for } T_{\mathrm{ref}} - a_T^{-1/2} < T_{\mathrm{a}} < T_{\mathrm{ref}} + a_T^{-1/2} \tag{9.27}$$

The value for the reference temperature commonly applied is 25 °C, but in reality this should depend on plant species. The usual value for the parameter a_T is 0.0016 K^{-2} which corresponds with a temperature range where Eq. (9.27) can be used from T_{ref} – 25 K to T_{ref} + 25 K. Beyond this range f_3 needs to be set to an arbitrary large value. Eq. (9.27) with the given value for a_T implies that the canopy resistance doubles when the temperature differs from the reference temperature by around 18 K.

[1] Note that here we present the reciprocal of the functions of Chen and Dudhia (2001), due to the fact that in Eq. (9.24) we multiply with the response functions rather than divide by them.

[2] The factor 2 in f seems to be inconsistent with the original two-layer formulation of Dickinson et al. (1986) which would give a factor close to 1.

The last response function describes the dependence of the canopy resistance on soil moisture availability. This is used as a proxy for the leaf water potential (see Chapter 6). The empirical function f_4 is given by:

$$\left[f_4(\bar{\theta})\right]^{-1} = \begin{cases} 0 & \text{if } \bar{\theta} < \theta_{\text{pwp}} \\ \dfrac{\bar{\theta} - \theta_{\text{pwp}}}{\theta_{\text{fc}} - \theta_{\text{pwp}}} & \text{if } \theta_{\text{pwp}} \leq \bar{\theta} < \theta_{\text{fc}} \\ 1 & \text{if } \bar{\theta} \geq \theta_{\text{fc}} \end{cases} \tag{9.28}$$

where $\bar{\theta}$ is the root zone mean water content, θ_{pwp} is the volumetric soil moisture content at permanent wilting point and θ_{fc} is the water content at field capacity (note: the reciprocal of f_4 is given to show the similarity to the Warrilow model used to model the reduction of evapotranspiration; see Section 4.2). The method of determining the root zone mean water content $\bar{\theta}$ differs between models. In some models the volumetric water content is weighted with the soil layer thickness, whereas in others the fraction of the total root length in each layer is taken into account as an extra weighting. The distribution of roots over the different soil layers may differ between vegetation types (e.g., trees have deeper rooting systems than grass). It appears that the results in terms of fluxes can be sensitive to the choice of root density distribution (e.g., Desborough, 1997 and Ek and Holtslag, 2005).

The time scales at which each of the response functions discussed in the preceding text is active differs (see, e.g., Schüttemeyer et al., 2006). The radiation response function mainly follows the diurnal cycle as do the temperature and VPD responses. But temperature and VPD exhibit variations with longer time scales as well, including the annual cycle. Generally, the soil moisture response shows a variation on the time scale of weeks and more because soil moisture does not vary rapidly (except when a dry period is followed by significant rain).

Question 9.10: Consider the expression for the canopy resistance in Eq. (9.24) (and the expressions for the response functions).

a) Explain why LAI is the nominator of the expression in Eq. (9.24).
b) Verify that the expressions in Eqs. (9.25)–(9.28) are similar to the responses of the *stomatal* resistance as given in Figure 6.15.

A–g_s Approach

The A–g_s approach makes use of the fact that the stomata are the gateway not only for water vapour, but for CO_2 as well. The stomata have to strike a balance between optimal CO_2 uptake and minimal water loss. Thus some of the stomatal behaviour is related to photosynthesis, and other parts to water loss. Many variants exist for models of the stomatal behaviour based on plant physiology. These concepts have been transferred to the meteorological applications by, for example, Collatz et al. (1991),

Jacobs (1994, 1996) and Calvet et al. (1998). Here we discuss mainly the A–g_s model as presented by Ronda et al. (2001), which is largely based on Jacobs (1994).

The starting point is the expression for the stomatal conductance developed in Chapter 6, Eq. (6.29). However, this expression is valid only for high light intensities, as at low light intensities the ratio of internal to external CO_2 concentration is no longer constant (Jacobs, 1994). If Eq. (6.29) would be applied in low light conditions, A_g would be less than R_d, then A_n would be negative, and the computed conductance would be negative as well. Therefore, Ronda et al. (2001) pragmatically replaced A_n by A_g to obtain the correct behaviour[3]:

$$g_{s,c} = g_{0,c} + \frac{1}{\rho} \frac{a_1 A_g}{(q_{ce} - \Gamma)\left(1 + a_2 \frac{D_e}{D_0}\right)} \tag{9.29}$$

where a_1 and a_2 are given in Section 6.4.3. Recall that the variables with subscript 'e' (external) are defined just outside the stomata, not at some reference level above the vegetation.

Apart from the plant-specific parameters, this model for stomatal conductance model needs to be complemented with a model for A_g. For this, use is made of a more or less mechanistic model for the gross assimilation rate A_g (i.e., net assimilation plus dark respiration: $A_n + R_d$). This model is based on a quantification of the two limiting factors for assimilation: supply of CO_2 and supply of PAR (Jacobs et al., 1996; see also Chapter 6). If neither of the factors is limiting, the maximum net assimilation (or maximum primary production) $A_{n,max}$ is attained (the plateau in Figure 6.13a and b).

For the CO_2-limited case at low q_{ci} the net assimilation is given by $A_n = \rho g_m (q_{ci} - \Gamma)$ where g_m is the mesophyll conductance. g_m determines the initial slope in Figure 6.13a. For higher internal CO_2 concentrations the actual net assimilation rate A_n is related to q_{ci} through the following interpolation between CO_2-limited and CO_2-unlimited conditions (note that it is the net assimilation, not the gross assimilation that is limited by CO_2 supply):

$$A_{n,c} = A_{n,max}\left(1 - \exp(-\frac{\rho g_m (q_{ci} - \Gamma)}{A_{n,max}}\right) \tag{9.30}$$

where the extra subscript 'c' indicates that $A_{n,c}$ is the CO_2-limited net assimilation (in the literature often denoted as A_m)

For the light-limited case at low light intensities, the net assimilation is linearly related to the absorbed PAR: $A_n = \epsilon I_{PAR} - R_d$, where ϵ is the initial light use

[3] Jacobs (1994) uses an interpolation between the situation at high-light conditions where A_n can be used, and low-light conditions where A_g should be used to prevent a negative g_s. Furthermore, he subtracts the CO_2 transport through the cuticula (with conductance $g_{0,c}$) from the assimilation rate, thus eliminating $g_{0,c}$ from Eq. (9.29).

Table 9.5 Parameter values for the A–g_s model of Ronda et al. (2001)

Parameter	Function of	Plant type	
		C_3	C_4
ϵ_0(mg J^{-1})		0.017	0.014
Γ (mg kg^{-1})	T	68.5	4.3
g_m (mm s^{-1})	T	7.0	17.5
$A_{n,max}$ (mg m^{-2} s^{-1})	T	2.2	1.7
a_d (k Pa^{-1})		0.07	0.15
f_{max} (-)		0.89	0.85
f_{min} (-)	g_0, g_m	0.23	0.18
D_0 (kPa)	f_{max}, f_{min}, a_d	9.4	4.5

The values for Γ, g_m and $A_{m,max}$ are in principle temperature-dependent. Here the values at 298 K are given. Note that the value for a_d for C_4 plants differs from the one in Ronda et al. (2001), which was erroneous (Ronda, pers. comm., 2012).

efficiency, that is, the amount of CO_2 fixed at a given input of PAR, at low levels of PAR. The initial light use efficiency determines the initial slope in Figure 6.13b. The initial light use efficiency in turn depends on the maximum light use efficiency (ϵ_0) as $\epsilon = \epsilon_0 (q_{ce} - \Gamma)/(q_{ce} + 2\Gamma)$. Taking Eq. (9.30) to represent the maximum attainable assimilation rate under the given CO_2 conditions, the actual gross assimilation can be determined from the following interpolation:

$$A_{g,cl} = (A_{n,c} + R_d)\left[1 - \exp\left(-\frac{\epsilon I_{PAR}}{A_{n,c} + R_d}\right)\right] \tag{9.31}$$

where the subscript 'cl' denotes that this is the CO_2 and light-limited gross assimilation. R_d is parameterized as a fraction of the net assimilation at maximum light availability: $R_d = 0.11 A_{n,c}$.

The aforementioned model for A_g contains a number of parameters that are dependent on temperature owing to the fact that the associated chemical processes are temperature-dependent. In that way, the temperature response of the model is naturally incorporated (for the values at reference temperature, see Table 9.5; temperature dependences are given in Ronda et al., 2001). The parameters also differ between plant species (in particular C_3 vs. C_4 plants).

The soil moisture response of the A–g_s model is incorporated empirically, based on a reduction of gross assimilation rate. Ronda et al. (2001) suggest:

$$A_{g,clw} = A_{g,cl}\left[2 f_\theta(\bar{\theta}) - f_\theta^{\,2}(\bar{\theta})\right] \tag{9.32}$$

where the subscript 'w' denotes water limitation. f_θ is identical to $[f_4(\bar{\theta})]^{-1}$ used in the Jarvis–Stewart method (see Eq. (9.28)).

In the end, the A–g_s approach contains the same stomatal responses to environmental factors as the Jarvis–Stewart approach, but in the A–g_s method they are parameterized in a way that is more closely related to the physiology of plants. The radiation, temperature and the external CO_2 concentration enter the model through the model for the net assimilation. The vapour pressure deficit determines the ratio of internal to external CO_2 concentration. Soil moisture – empirically – affects the reduction of the gross assimilation. Owing to the close relation between the A–g_s method and the physiology of plants, synergistic interactions between different responses are implicitly incorporated in A–g_s models, rather than that all responses act independently, as in the Jarvis-Stewart approach. Furthermore, some of the parameters in A–g_s models are related to widespread characteristics of the photosynthesis process and thus vary little between plant species (Jacobs, 1994).

Up to this point only the stomatal conductance was dealt with. To scale this to a canopy conductance one would need to integrate over all layers of a canopy where all environmental factors close to the leaf may vary (radiation, temperature, humidity and external CO_2 concentration). Ronda et al. (2001) take into account the vertical exponential decay of PAR through the canopy, but use single values for leaf temperature and vapour pressure deficit. More elaborate methods can be found in, for example, Goudriaan (1977) and Sellers (1984).

Question 9.11: Assimilation increases with increasing internal CO_2 concentration and light input.
a) Determine the initial slope of the CO_2 response curve (Eq. (9.30)).
b) Determine the initial slope of the light response curve (Eq. (9.31)).

Question 9.12: Given the following observations at leaf level: I_{PAR} = 300 W m^{-2}, D = 1.5 kPa, q_{ce} = 5.77 · 10^{-4} kg kg^{-1} (corresponding to 380 ppmv). Leaf temperature is 298 K. Do the following calculations for a C_3 plant (see Table 9.5 for plant parameters).
a) Calculate the internal CO_2 concentration.
b) Calculate the CO_2-limited net assimilation $A_{n,c}$.
c) Calculate the radiation-limited gross assimilation $A_{g,cl}$.
d) Calculate the stomatal conductances for CO_2 and water vapour.
e) Calculate the stomatal resistance for CO_2 and water vapour.

Question 9.13: Given a vegetation with LAI equal to 2. Assume that half of the leaf area is exposed to I_{PAR} =300 W m^{-2} and the other half to I_{PAR} =100 W m^{-2}. The other environmental conditions are identical to those given in Question 9.12.
a) Calculate the stomatal conductance for water vapour for the leaf area exposed to I_{PAR} =100 W m^{-2}.
b) Calculate the canopy conductance for water vapour for this canopy.

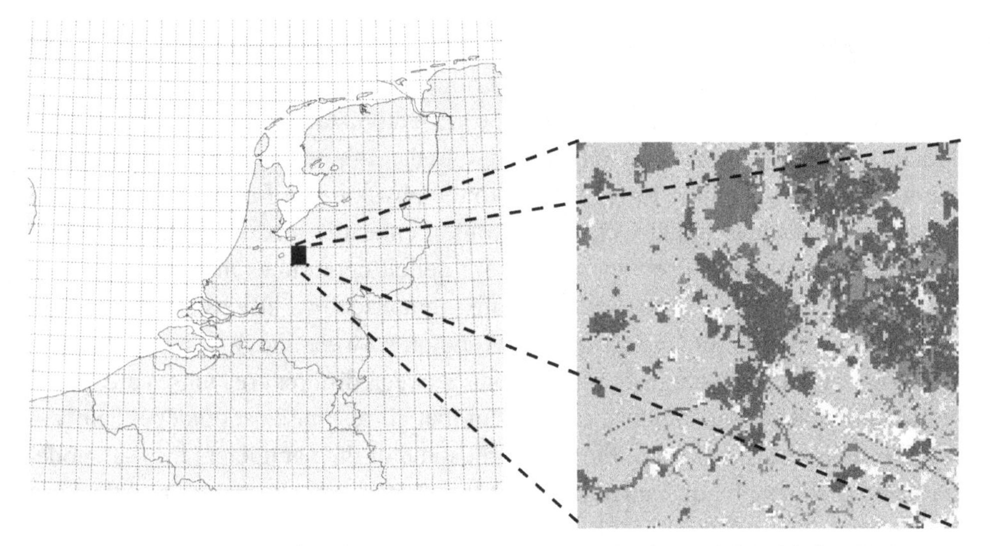

Figure 9.16 Typical grid of a high resolution atmospheric model (grid size 22 km) overlaid on the Netherlands. Enlargement: land use within a single grid box.

9.2.5 Surface Heterogeneity

The land surface is strongly heterogeneous, both in its properties (i.e., albedo, soil type etc.) and in its state (snow cover, soil moisture content, soil temperature). Within seemingly homogeneous landscapes with a single land use type, heterogeneity already may play a role (e.g., variations in LAI). But in the current context variations of land use type (e.g., forest vs. urban environment) and vegetation type within a single model grid cell is our major concern (see Figure 9.16). Each of these land-use types will have a different energy and water balance and hence will deliver different fluxes to the atmosphere.

There are two main categories of methods to deal with this heterogeneity:

- The surface within a grid cell is treated as one single land-use type. The parameters for that surface are constructed from the parameters of the different surface types within the grid. Various methods for the construction of effective parameters exist (e.g., Chehbouni et al., 1995; Arain et al.,1997; Intsiful and Kunstmann, 2008). The simplest averaging scheme is to use the parameters of the dominant land-use type.
- The surface within a grid cell is subdivided into a number subdivisions, each with its own set of parameters.
  - The least spatially explicit form of this method is the statistical approach in which only the statistical distribution of the parameters is given. With these distributions of parameters a distribution of fluxes can be constructed from which the grid-scale mean fluxes can be determined (e.g., Avissar, 1992; Bonan et al., 1993).
  - The next step is to group similar parts of the landscape into a single land-use type (e.g., low vegetation, open water): although some land surface parameters may vary

within such a land-use type (e.g., LAI, albedo) a single value is assigned that is supposed to be representative for that land use type. Then for each of these land use types ('tiles' or 'clumps') the LSM produces fluxes which are averaged to a grid-averaged flux based on the relative abundance of the given land-use type (e.g., Bonan, 1995; Viterbo and Beljaars, 1995).

- In the so-called mosaic approach the grid cell is subdivided into regular smaller units. Each of these units can have its own unique combination of parameters, as far as spatial information with sufficient detail is available, for example, from remote sensing or soil maps. Each of the mosaic elements will produce its own flux, thus contributing to the grid-averaged flux (e.g., Avissar and Pielke, 1989; Seth et al., 1994; Ament and Simmer, 2008).

Here the tile method is discussed in some more detail. The range of tiles distinguished varies between models. As an example of a tiled LSM, the Tiled ECMWF Surface Scheme for Exchange over Land (TESSEL), is discussed (Viterbo and Beljaars, 1995; van den Hurk et al.,2000). The following are the eight tiles that are used, including the characteristics that distinguish them:

1. Low vegetation (without snow): large aerodynamic resistance, small canopy resistance, shallow rooting zone
2. High vegetation (without snow): small aerodynamic resistance, canopy resistance sensitive to VPD, deep rooting zone
3. Liquid water in the interception reservoir: no stomatal control
4. Bare soil (without snow): extraction solely from upper soil layer, resistance depends on soil moisture content
5. Snow on bare soil or low vegetation: change in albedo, phase change before evaporation
6. High vegetation with snow beneath: higher roughness than snow on low vegetation, albedo mostly determined by vegetation that protrudes through snow; phase change before evaporation
7. Sea-ice: difference in albedo and roughness compared to open water;
8. Oceans and lakes: low albedo, conservative surface temperature.

The relative proportions of low vegetation, high vegetation, bare soil and open water do not change in time. Those fractions are taken from a global database. But the interception reservoir and snow cover *are* dynamic. If rain fills the interception reservoir, the proportions of low and high vegetation are reduced in favour of the proportion taken up by the interception reservoir. Furthermore, whether or not each of the vegetated or bare soil tiles is covered with snow *does* change in time. This implies that the model needs to keep track of the snow cover accurately.

The tiling affects only the surface. *Above* the surface, all tiles are coupled to the same atmospheric values for wind, temperature and humidity, valid for that particular grid box. And *below* the surface all tiles share the same soil parameters and state variables (soil moisture and temperature).

The parameters that make each of the tiles different are[4]:

- A number of vegetation-related parameters: LAI, vegetation coverage, parameters related to the canopy resistance (see Section 9.2.4), roughness length and the vertical root distribution (needed to determine where the vegetation extracts the soil moisture)
- The surface albedo which depends on snow cover
- The skin layer conductivity (see Section 2.3.6 on the soil heat flux for vegetated surfaces)

Those differences in parameters results in a situation in which each tile has its own surface temperature. The vegetation-related parameters are selected for the dominant high vegetation type and the dominant low vegetation type within the grid box. For example: within a grid box both 'tall grass' and 'irrigated crops' may be present as low vegetation types. But if the fraction of 'tall grass' is higher than that of 'irrigated crops' the vegetation-related parameters for the low-vegetation tile will be assigned as if all low vegetation in the grid box is 'tall grass'.

The water vapour transport, with the resistances involved, is depicted in Figure 9.17. The surface value of the specific humidity equals the saturated value at the surface temperature of the specific tile ($q_{s,i} = q_{sat}(T_{s,i})$, where i signifies the index of the tile). For *all* tiles the moisture transport encounters the aerodynamic resistance, whereas *some* tiles have an additional resistance: canopy resistance, snow resistance and soil resistance. The latter is modelled as:

$$r_{soil} = r_{soil,min} f_4(\theta_1) \tag{9.33}$$

where $r_{soil,min}$ is the resistance when sufficient water is available (taken equal to 50 s m^{-1}), θ_1 is the water content of the upper soil layer and f_4 is identical to the formulation in the canopy resistance (i.e., Eq. (9.28)).

For heat transport the picture will be similar, except for the fact that in that case only the aerodynamic resistance needs to be used, in combination with the surface temperature of the tile under consideration.

Apart from the tiles discussed in the preceding text, some models also include an urban tile, as urban surfaces have very distinct properties: pavement that prevents infiltration, large roughness, little or no vegetation, heat capacity of buildings, anthropogenic heat production (Arnfield, 2003). Because the spatial extent of urban regions is generally limited, the inclusion of an urban tile makes sense only for simulations at such resolution that the urban areas make a significant contribution to the total surface. Models that make use of only one dominant land-use type per grid cell need an urban surface parameterization only once the resolution becomes finer than the size of urban areas. See Grimmond et al. (2011) for an overview of current urban land surface models.

[4] For clarity we leave out one extra parameter, which is the fraction of shortwave radiation that directly reaches the ground surface (between/under the vegetation).

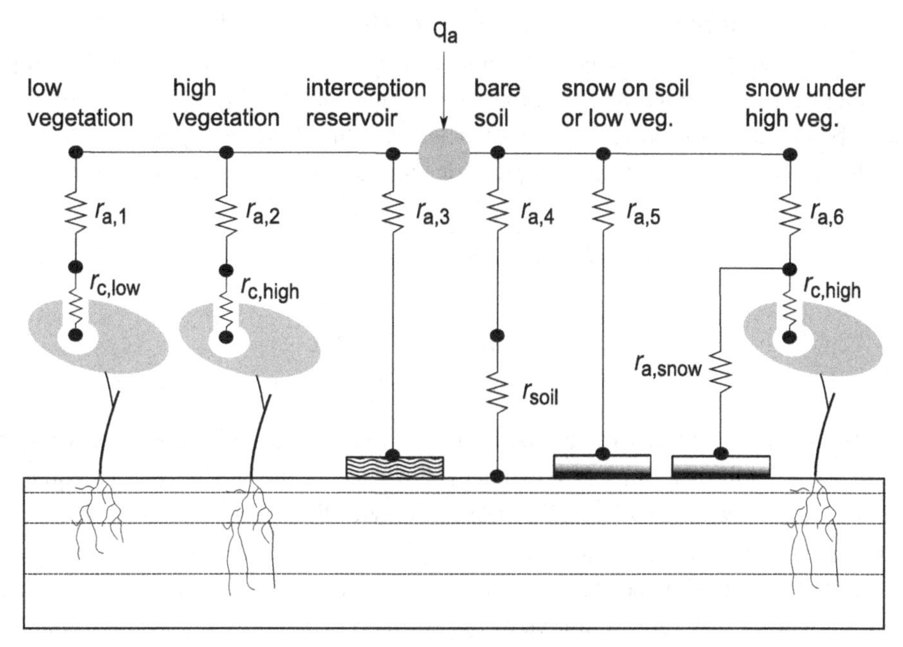

Figure 9.17 Schematic representation of water vapour transport in TESSEL for the six land tiles. Some tiles evaporate without an addition resistance, some have a canopy resistance, and snow below high vegetation has a parallel resistance for evaporation from snow and a canopy resistance. (After ECMWF, 2009)

Question 9.14: Make a sketch similar to Figure 9.17, but now for heat transport (i.e., the upper node is not for q_a, but for T_a). Pay special attention to the resistances.

Question 9.15: Explain for each of the land tiles depicted in Figure 9.17 why they are used in the model (i.e., in what respect do the tiles differ that makes it important to distinguish them).

9.2.6 Coupling to the Atmosphere and the Soil

In the previous section, the coupling of the land surface to the atmosphere has been dealt with schematically, but how does it work in practice? Again, TESSEL is used as an example. To simplify the discussion, in the rest of this section, the processes related to snow, snow melt and soil freezing will not be dealt with.

Coupling to the Atmosphere

For each of the tiles (with index i), the surface energy balance is:

$$\underbrace{(1-r_i)K^{\downarrow}+\varepsilon\left(L^{\downarrow}-\sigma T_{s,i}^{4}\right)}_{Q^*}-\underbrace{\Lambda_{veg,i}\left(T_{s,i}-T_{soil,1}\right)}_{G}=H_i+L_vE_i \tag{9.34}$$

Note that all tiles share the same soil temperature (the subscript '1' refers to the first soil layer). The surface temperature is not the temperature of the first soil layer but the temperature of a skin layer that has no heat capacity. Because this skin layer cannot store heat, it needs to be in immediate equilibrium with the energy flux. This energy balance Eq. (9.34) is supplemented with the following expressions for the sensible and latent heat flux:

$$
\begin{aligned}
H_i &= -\rho c_\mathrm{p} \frac{T_\mathrm{a} - T_{\mathrm{s},i}}{r_{\mathrm{a},i}} \\
L_\mathrm{v} E_i &= -\rho L_\mathrm{v} \frac{q_\mathrm{a} - q_\mathrm{sat}\left(T_{\mathrm{s},i}\right)}{r_{\mathrm{a},i} + r_{\mathrm{extra},i}}
\end{aligned}
\tag{9.35}
$$

where the nature of the extra resistance $r_{\mathrm{extra},i}$ depends on the tile (either a canopy resistance, or a soil evaporation resistance). For each of the tiles the set of equations given by Eqs. (9.34) and (9.35), in combination with the similarity relationship from Chapter 3, is solved, giving values for the sensible and latent heat flux for each tile. The weighted sum of those fluxes from separate tiles is passed to the atmospheric model as the total flux representative of the entire grid box. Details on the implementation of the coupling between a tiled land surface model and the atmospheric column can be found in Best et al. (2004).

Coupling to the Soil

In principle each tile in a tiled LSM could have its own soil properties. However, there is no reason why soil properties should vary with land use. Therefore, TESSEL uses a single soil type within a grid box, for all tiles. Between grid cells, the soil hydraulic properties vary: the properties related to the dominant soil type within the grid box are assigned to all tiles (as of the introduction of HTESSEL; Balsamo et al., 2009). However, all grid boxes have identical soil thermal properties (heat capacity and thermal conductivity).

The evolution of the soil temperature is determined from the solution of the heat diffusion equation (Eq. (2.31)). The soil in TESSEL is divided into four layers with a thickness that increases with depth (7 cm, 21 cm 72 cm and 189, i.e., a total soil column of 289 cm). The thicknesses have been chosen such that, for forcings with a frequency between 1 day and 1 year (see Chapter 2), the phase and amplitude of the soil temperature at each depth is close to those that would be obtained if a large number of layers would be used. The upper soil layer represents the diurnal cycle, the second layer represents variations at the timescale of 1 day to 1 week, the third layer variations between 1 week and 1 month and the deepest layer represents variation with a period longer than 1 month (Viterbo and Beljaars (1995). The upper boundary condition of the soil column is the surface soil heat flux (see Eq. (9.34)), whereas at the lower boundary the flux is taken zero.

Vertical soil moisture transport in the soil is governed by Richards' equation (Chapter 4), which is solved for the same layers as used for the soil temperature. The sink term in Richards' equation is the uptake of soil moisture by the roots. The total transpiration flux is extracted from those soil layers where roots are present, in proportion to the root density (which depends on the type of vegetation in the tile). If in a given soil layer the soil moisture content is below permanent wilting point, no water is extracted from that particular layer. The upper boundary condition of the soil column is precipitation diminished by interception and surface runoff (occurring if rainfall exceeds the infiltration capacity). At the lower boundary free drainage is allowed.

The energy balance is heavily modified if snow is present on the surface. First, the albedo of the surface increases. Second, the coupling between the surface temperature and soil temperature is decreased (see Section 2.3.8). The exact conductivity depends on the snow density, which changes in time (ECMWF, 2009). Finally, direct sublimation of snow may occur, but at temperatures above the freezing point of water there will also be snow melt. The timing of this melt has a large impact on the simulated surface energy balance (see Balsamo et al., 2011).

Question 9.16: The thicknesses of the soil layers in TESSEL are based on the period of the temporal variations they should be able to represent. TESSEL uses a loamy soil. Assume that the soil properties of a loamy soil are halfway in between those of a sandy soil and a clay soil (see Table 2.2).

a) Determine the damping depth of a dry loamy soil for forcings with a period of 1 day, 1 week, 1 month and 1 year.
b) Determine the damping depth of a saturated loamy soil for forcings with a period of 1 day, 1 week, 1 month and 1 year.
c) Compare those damping depths with the thicknesses of the soil layers as used in TESSEL.

9.2.7 The Role of Observations

Observations play a crucial role in the development, testing and use of LSMs. In the development phase observations are used to determine the underlying physical relationships needed to describe transport processes. Examples are the flux-gradient relationships (Chapter 3), parameterizations for the unsaturated soil-hydraulic conductivity (Chapter 4) and the empirical relationships that describe the regulation of the canopy resistance (Chapters 6 and 9).

Then, LSMs need a large number of parameters that describe properties of the surface: for example, albedo, soil properties, vegetation fraction, minimal stomatal resistance, and so forth. Some models include a model or parameterization for some of these parameters (e.g., dynamic vegetation, albedo that depends on vegetation fraction). Otherwise, these parameters are assumed to be immutable (or at most slowly varying on a seasonal time scale). In that case they are usually derived from remote-sensing observations as those are the only means to obtain this information on a global scale at sufficient spatial resolution (e.g., Hall et al., 1995; Masson et al., 2003).

Subsequently, LSMs are tested in two ways: off-line and on-line. In off-line testing the model is decoupled from the full atmospheric model: the variables that are normally provided by the atmospheric model are taken from observations. These *forcings* include precipitation, incoming longwave and shortwave radiation, wind speed, temperature and humidity. Furthermore, certain *parameters* are based on observations, such as albedo and roughness length. After the LSM has been run in this way, observed fluxes are used as a *validation* of the fluxes produced by the model (e.g., sensible and latent heat flux, net radiation). Whereas developers of models usually use their own data sets for a first validation, coordinated validation exercises are also used to intercompare the skills of different LSMs for a common set of observations. Examples of these activities are PILPS (Project for Intercomparison of Land-surface Parametrization Schemes: Henderson-Sellers et al., 1995), GSWP (Global Soil Wetness Project: Dirmeyer et al., 2006) and initiatives linked to a certain region such as West Africa (Boone et al. 2009) or surface types like cities (Grimmond et al., 2011).

Off-line testing does not provide insight in the effect that a certain LSM has on the overall behaviour of the model. Therefor on-line tests are also needed, which might reveal effects on, for example, precipitation, cloud formation, soil moisture depletion and runoff (e.g., Balsamo et al. 2009). For on-line tests not only surface fluxes can be used for validation, but, for example, screen-level temperature, humidity, wind speed and precipitation as well. For operational weather models these variables happen to be important variables to assess the skill of a model.

In the operational use of LSMs observations play a role as well. In the context of weather forecasting, each forecast needs to start with a correct initial state of the model. For LSMs this initial state may comprise soil temperatures and moisture content at various depths as well as the amount of snow cover and intercepted water. Without additional information, this initial state could be carried over from a previous forecast, but any error in that forecast will remain, or even amplify, in the new forecast. Therefore, similarly to what is done for the atmospheric part of weather models, data assimilation can be used to correct the initial state of the LSM and bring it as close to reality as possible. One of the methods that is used is based on the fact that screen level air temperature and humidity reflect in part the energy partitioning in the surface energy balance: a low Bowen ratio will lead to low temperatures and high humidity contents, whereas a high Bowen ratio will give warm, dry air. In turn, the Bowen ratio is strongly influenced by the amount of available soil moisture. Thus, an error in the soil moisture content will lead to an error in the screen level temperature and humidity and can hence be detected by comparing the forecast of T and RH with the observed values for the same moment. Based on the discrepancy between the two, the soil moisture can be adjusted (see, e.g., Giard and Bazile, 2000; Drush and Viterbo, 2007).

Furthermore, directly observed variables such as snow cover or top-soil soil moisture content can be used to correct the initial model state (see, e.g., Mahfouf, 2010).

Appendix A

Radiation

This appendix summarizes (without much commentary) basic equations related to radiation in the atmosphere.

A.1 Radiation Laws

Planck's law: monochromatic hemispherical emissive power of a black body (in W m^{-2} µm^{-1})

$$M_{b\lambda} = \frac{c_1 \lambda^{-5}}{e^{\frac{c_2}{\lambda T}} - 1} \tag{A.1}$$

in which $c_1 = 3.74 \cdot 10^8$ W µm^{-4} m^{-2} and $c_2 = 1.439 \cdot 10^4$ µm K; T is the absolute temperature of the black body and λ is the wavelength in µm.

Direct consequences of Eq. (A.1) are Stefan–Boltzmann's law and Wien's displacement law:

$$M_b = \sigma T^4 \tag{A.2}$$

$$\lambda_m T = 2898 \ \mu\text{m K} \tag{A.3}$$

where σ is the constant of Stefan–Boltzmann (5.67 · 10^{-8} W m^{-2} K^{-4}) and λ_m is the wavelength at which maximum emission takes place.

Hemispherical monochromatic emissive power of a general (non-black) body is given by:

$$M_\lambda = \varepsilon_\lambda M_{b\lambda} \tag{A.4}$$

where ε_λ is the monochromatic emissivity. The total hemispherical emissive power of a non-black body is given by:

$$M = \int_0^\infty \varepsilon_\lambda M_{b\lambda} \mathrm{d}\lambda \tag{A.5}$$

For a so-called grey body, the emissivity is equal for all wavelengths ($\varepsilon_\lambda = \varepsilon$), and hence Eq. (A.5) becomes:

$$M = \varepsilon \int_0^\infty M_{b\lambda} \mathrm{d}\lambda = \varepsilon \sigma T^4 \tag{A.6}$$

Kirchhoff's law states that, under equilibrium conditions, the monochromatic emissivity and monochromatic absorbtivity (α_λ) are equal:

$$\alpha_\lambda = \varepsilon_\lambda \tag{A.7}$$

A.2 Solar Radiation: Instantaneous

The amount of solar radiation at Earth's surface is determined by the amount of radiation at the top of the atmosphere and by the composition of the interfering atmosphere. The radiation at the top of the atmosphere is given by:

$$K_0^\downarrow = \overline{I_0} \left(\frac{d_{\mathrm{Sun},0}}{d_{\mathrm{Sun}}} \right)^2 \cos(\theta_z) \tag{A.8}$$

where $\overline{I_0}$ is the solar constant (flux density of solar radiation at the mean distance from Sun to Earth). This value varies with period of 11 years and an amplitude of about 1 W m^{-2}. Here we use a value of 1365 W m^{-2} for the solar constant, though recent research suggests a value of 1361 W m^{-2} (Kopp and Lean, 2011). Furthermore $d_{\mathrm{Sun},0}$ is the *mean* (over a year) distance between the Sun and Earth, d_{Sun} is the *actual* distance between the Sun and Earth (depending on the date) and θ_z is the solar zenith angle (angle between solar beam and the normal to Earth's surface) which depends on the location, date and time.

The eccentricity correction factor can be approximated as:

$$\left(\frac{d_{\mathrm{Sun},0}}{d_{\mathrm{Sun}}} \right)^2 = 1.000110 + 0.034221 \cos\Gamma + 0.001280 \sin\Gamma + 0.000719 \cos 2\Gamma + 0.000077 \sin 2\Gamma \tag{A.9}$$

with $\Gamma = 2\pi(d_\mathrm{n} - 1)/365$, where d_n is the day number of the year (January 1 equals 1). Γ is sometimes called the day angle.

The expression for the solar zenith angle is:

$$\cos\theta_z = \sin\delta \sin\phi + \cos\delta \cos\phi \cos\omega \tag{A.10}$$

where δ is the declination of the sun, φ the latitude (defined positive at the Northern Hemisphere) and ω the hour angle.

An approximate equation for the declination (in radians) is:

$$\begin{aligned}\delta = (&0.006918 - 0.399912\cos\Gamma + 0.070257\sin\Gamma - 0.006758\cos 2\Gamma \\ &+ 0.000907\sin 2\Gamma - 0.002697\cos 3\Gamma + 0.00148\cdot\sin 3\Gamma)\end{aligned} \tag{A.11}$$

The hour angle ω (in radians) is the angle between the observer's meridian and the solar meridian (zero at noon, positive in the morning):

$$\omega = \frac{2\pi}{24}(-t_{\text{solar}} - E_t) + \pi = \frac{2\pi}{24}\left(-\left(t_{\text{UTC}} + \eta\frac{24}{2\pi}\right) - E_t\right) + \pi \tag{A.12}$$

where t_{solar} is the local solar time (in hours, 12 at local solar noon), t_{UTC} is the time (UTC) in hours, η is the longitude in radians (east positive) and E_t is the equation of time (which corrects for the shift through the year of the exact moment of solar noon). E_t (in hours) is given by:

$$\begin{aligned}E_t = 3.8197(&0.000075 + 0.001868\cos\Gamma - 0.032077\sin\Gamma \\ &- 0.014615\cos 2\Gamma - 0.04089\sin 2\Gamma)\end{aligned} \tag{A.13}$$

A.3 Solar Radiation: Daily Values

The hour angle at sunrise is:

$$\omega_s = \cos^{-1}(-\tan\phi\tan\delta) \tag{A.14}$$

The day length is $2\omega_s$ (radians), or expressed in hours, $N_d = 2\frac{24}{2\pi}\omega_s$.

The daily mean shortwave radiation at the top of the atmosphere, on a plane parallel to Earth's surface, is:

$$\overline{K_0^{\downarrow}}^{24} = \frac{\overline{I_0}}{\pi}\left(\frac{d_{\text{sun},0}}{d_{\text{sun}}}\right)^2 [\omega_s \sin\phi\sin\delta + \cos\phi\cos\delta\sin\omega_s] \tag{A.15}$$

with ω_s in radians!

Appendix B

Thermodynamics and Water Vapour

This appendix summarizes some basic atmospheric thermodynamics and moisture variables. An overview of some physical constants is given in Table B.3.

B.1 Some Basic Thermodynamics

Air is a mixture of gases. The fractions of most constituents are rather constant, except for water vapour, which is highly variable (see Table B.1).

The equation of state for a perfect gas reads:

$$p = \rho \frac{R^*}{M} T = \rho R T \tag{B.1}$$

where p is pressure, ρ is the density, T is the temperature, R^* is the universal gas constant 8314 J kmol^{-1} K^{-1}, M is the molar mass of the gas and R is the specific gas constant. Because air is a mixture of gases, first the molar mass of this mixture needs to be determined. If the composition is given in terms of volume fractions ($f_{v,i}$) and given the fact that in a gas a mole of any gas occupies the same volume, the molar mass of the mixture can be determined simply as:

$$M = \sum_i f_{v,i} M_i \tag{B.2}$$

With the data from Table B.1, this yields for dry air (air without water vapour): $M_d = 28.976 \text{ kg mol}^{-1}$.

Based on Dalton's law, which states that the total pressure is the sum of the partial pressures of the constituents, it can be derived that the molar mass of a mixture with constituents with molar masses M_i and mass fractions $f_{m,i}$ is:

$$M = \left(\sum_i \frac{f_{m,i}}{M_i} \right)^{-1} \tag{B.3}$$

Table B.1 Composition of atmospheric air up to 105 km altitude

Gas	Molecular mass (kg kmol^{-1})	Volume fraction (relative to dry air)
Nitrogen (N_2)	28.013	78.08 %
Oxygen (O_2)	32.000	20.95 %
Argon (Ar)	39.95	0.93 %
Water vapour (H_2O)	18.02	0–5 %
Carbon dioxide (CO_2)	44.01	389 ppmv
Neon (Ne)	20.18	18 ppmv
Helium (He)	4.00	5 ppmv
Methane (CH_4)	16.04	1.87 ppmv
Krypton (Kr)	83.80	1 ppmv
Hydrogen (H_2)	2.02	0.5 ppmv
Nitrous oxide (N_2O)	56.03	0.32 ppmv
Ozone (O_3)	48.00	0–0.1 ppmv

From Wallace and Hobbs (2006). CO_2 data: global mean value for 2010. Source: Dr. Pieter Tans, NOAA/ESRL: www.esrl.noaa.gov/gmd/ccgg/trends. CH_4 and N_2O data: Mace Head (Ireland) October 2009–September 2010. Source: AGAGE network, Prinn *et al.* (2000).

With $M_{\mathrm{d}} = 28.976$ kg mol^{-1}, the specific gas constant for dry air becomes $R_{\mathrm{d}} = 287$ J kg^{-1} K^{-1}. Hence, using (B.1), the density of dry air can be defined as:

$$\rho_{\mathrm{d}} = \frac{M_{\mathrm{d}}}{R^*}\frac{p}{T} = \frac{p}{R_{\mathrm{d}}T} \tag{B.4}$$

The specific gas constant for water vapour is $R_{\mathrm{v}} = 462$ J kg^{-1} K^{-1}. Based on Eq. (B.3) the dependence of the molar mass of air that includes water vapour can be derived to be:

$$M = \left(\frac{q}{18} + \frac{1-q}{28.976}\right)^{-1} \tag{B.5}$$

where q is the specific humidity (defined in Section B.4). Hence the specific gas constant for moist air is:

$$R = \frac{R^*}{M} = R^*\left(\frac{q}{18} + \frac{1-q}{28.976}\right) = \mathrm{R}_{\mathrm{d}}\left(1 + 0.61q\right) \tag{B.6}$$

The specific heats of air at constant volume (denoted by a subscript v, not to be confused with the subscript v used for water vapour) and constant pressure are related as:

$$c_p - c_v = R \tag{B.7}$$

For dry air the values for the specific heats are: c_{vd} = 717 J kg^{-1} K^{-1} and c_{pd} = 1004 J kg^{-1} K^{-1} at a temperature of 285 K (Forsythe, 1954). The temperature dependence is small (1–2 % over the range 250–310 K). For moist air the specific heat at constant pressure is a combination of c_{pd} and the specific heat at constant pressure for water vapour c_{pv} (1849 J kg^{-1} K^{-1}):

$$c_p = c_{pd}\left(1-q\right) + c_{pv}q = c_{pd}\left(1+0.84q\right) \tag{B.8}$$

Then the value for c_v follows from (B.7) and (B.8):

$$c_v = c_{pd}\left(1+0.84q\right) - R_d\left(1+0.61q\right) = c_{vd} + 0.61q\left(1.4c_{pd} - R_d\right) \tag{B.9}$$

Deviations of R, c_p and c_v from their dry air values is of the order of 1–3% depending on the actual specific humidity. In many applications these deviation are small enough to be neglected and hence the dry air values of the thermodynamic properties of air can be used. However, the effect of water vapour on density, and more specifically density fluctuations, is in many cases not negligible as it affects buoyancy (see Chapter 3).

Finally, the first law of thermodynamics reads:

$$\begin{aligned} dq &= c_v dT + p d\alpha \\ &= c_p dT - \frac{dp}{\rho} \end{aligned} \tag{B.10}$$

where dq is a differential amount of heat added to the system, which is used both to increase the temperature by a differential increment dT and to change the specific volume by an amount $d\alpha$ (the specific volume is the inverse of the density: $\alpha = \rho^{-1}$). To arrive at the second equality, use has been made of Eq. (B.7).

B.2 Hydrostatic Equilibrium

The pressure at a certain height is determined by the weight of the air above that level. Hence the pressure decreases with height. In the case of hydrostatic equilibrium (that is, no vertical acceleration) the vertical pressure gradient exactly balances the weight of the air. For an infinitesimal height increment dz this is:

$$dp = -g\rho(z)dz \tag{B.11}$$

where g is the acceleration due to gravity. For an isothermal atmosphere (temperature constant with height) this yields (combination of Eqs. (B.1) and (B.11)):

$$p(z) = p(0)e^{-\frac{gz}{RT}} \tag{B.12}$$

B.3 Potential Temperature

For an adiabatic process ($dq = 0$ in Eq. (B.10)) the combination of Eq. (B.10) with the equation of state (B.1) yields a relationship between an infinitesimal temperature change and an infinitesimal pressure change:

$$\frac{dT}{T} = \frac{R}{c_p}\frac{dp}{p} \tag{B.13}$$

Integration of (B.13) between a reference pressure p_0 and pressure p yields the definition of the potential temperature:

$$\theta \equiv T\left(\frac{p_0}{p}\right)^{\frac{R}{c_p}} \tag{B.14}$$

Although the temperature changes during an adiabatic process, the potential temperature does not change (i.e., is a conserved variable). Combination of Eqs. (B.13) and (B.11) yields an expression for the temperature change with height for an adiabatic processes in a hydrostatic equilibrium:

$$\frac{dT}{dz} = -\frac{g}{c_p} \tag{B.15}$$

which is called the dry adiabatic lapse rate.

B.4 Measures of Water Vapour Content

In the Table B.2 various measures of water vapour content are summarized, indicating their symbol, name, unit and an indication of their use.

As all variables given in the table relate to the same amount of water vapour in air, they should all be related. Those relationships are explored in the text that follows.

Because water vapour pressure is the partial pressure of water vapour, it is directly related to the absolute humidity:

$$e = \rho_v R_v T \tag{B.16}$$

Table B.2 Overview of different variables used to indicate the amount of water vapour in air

Symbol	Name	Unit	Description/remark
ρ_v	Absolute humidity	kg m^{-3}	Water vapour density; often used in sensors based on absorption of radiation
e	Water vapour pressure	Pa	Partial pressure of water vapour (often used in models for evaporation)
r	Mixing ratio	kg kg^{-1}	Mass of water vapour as a fraction of mass of dry air: $r = \frac{\rho_v}{\rho_d}$ (conserved for adiabatic processes)
q	Specific humidity	kg kg^{-1}	Mass of water vapour as a fraction of mass of moist air $q = \frac{\rho_v}{\rho} = \frac{\rho_v}{\rho_d + \rho_v}$ (conserved for adiabatic processes)
T_v	Virtual temperature	K	Temperature that *dry* air should have to have the same density as air with a given moisture content: $T_v \approx T(1+0.61q)$. Used when the buoyancy effect of water vapour is relevant.
RH	Relative humidity	—	Vapour pressure as a fraction of the vapour pressure at saturation for a given temperature: $RH = \frac{e}{e_{sat}(T)}$ (see Eq. (B.20) for a definition of e_{sat}). Easily measured with hair hygrometers. Relevant since many natural materials are sensitive to relative humidity.
T_d	Dew point temperature	K	Temperature to which air needs to be cooled (at constant pressure) to reach saturation, i.e. where $e = e_{sat}(T_d)$.Relevant to predict effect of cooling at constant pressure. Can be measured with a dew point mirror.
T_w	Wet bulb temperature	K	The temperature to which the wet bulb of a psychrometer will cool when exposed to air with a given moisture content (heat will be extracted to provide energy for evaporation). Can be measured with a psychrometer.

Through the definitions, mixing ratio and specific humidity are directly related:

$$q = \frac{r}{r+1} \tag{B.17}$$

Because both r and q are much less than 1, for many practical applications q and r can be assumed to be identical. The fact that both q and r are conserved variables for adiabatic processes is due to the fact that both the total density, and the density of water vapour, change at the same rate when the pressure or temperature changes. As a result, their ratio is constant under those changes.

The relationship between the mixing ratio and the vapour pressure is derived from Dalton's law. Using the equation of state for dry air and water vapour yields:

$$r = \frac{R_d}{R_v} \frac{e}{p-e} \tag{B.18}$$

The ratio of the specific gas constant for dry air and water vapour is an important number in meteorology. It equals 0.621 and is close to 5/8. The relationship between q and e is:

$$q = \frac{R}{R_v} \frac{e}{p} \approx \frac{R_d}{R_v} \frac{e}{p} \approx \frac{5}{8} \frac{e}{p} \tag{B.19}$$

For the definition of relative humidity the saturated vapour pressure e_{sat} is needed. The saturated vapour pressure is the water vapour pressure in a gas that is in equilibrium (at a given temperature) with the liquid phase: the number of molecules that leave the liquid phase equals the number of molecules that leave the gas phase and thus rejoin the liquid phase. The empirical approximations for $e_{sat}(T)$ proposed by WMO (2008) is (see Figure B.1):

$$e_{sat}(T) = e_{sat,0} \exp\left[\frac{a(T-273.15)}{b+T}\right] \tag{B.20}$$

where $e_{sat,0}$ is the e_{sat} at 0 °C ($e_{sat,0}$ = 611.2 Pa.). The value of the constants a and b depends on the surface over which the saturated vapour pressure needs to be determined. For water surfaces a = 17.62 K^{-1} and b = –30.03 K, whereas for ice surfaces the values are a = 22.46 K^{-1} and b = –0.53 K Generally, a subscript 'w' or 'i' is used to indicate whether saturated values over water or ice are referred to. Here, we always refer to the saturated vapour pressure over water and omit the subscript 'w' (i.e., where e_{sat} is written, $e_{sat,w}$ is intended). Note that Eq. (B.20) was originally stated with the temperature in °C.

In evaporation theory the derivative of $e_{sat}(T)$ to temperature is used. From Eq. (B.20) this can be determined to be:

$$s(T) \equiv \frac{de_{sat}}{dT} = e_{sat}(T) \frac{a(b+273.15)}{(b+T)^2} \tag{B.21}$$

In Figure B.1 both $e_{sat}(T)$ and $s(T)$ are depicted.

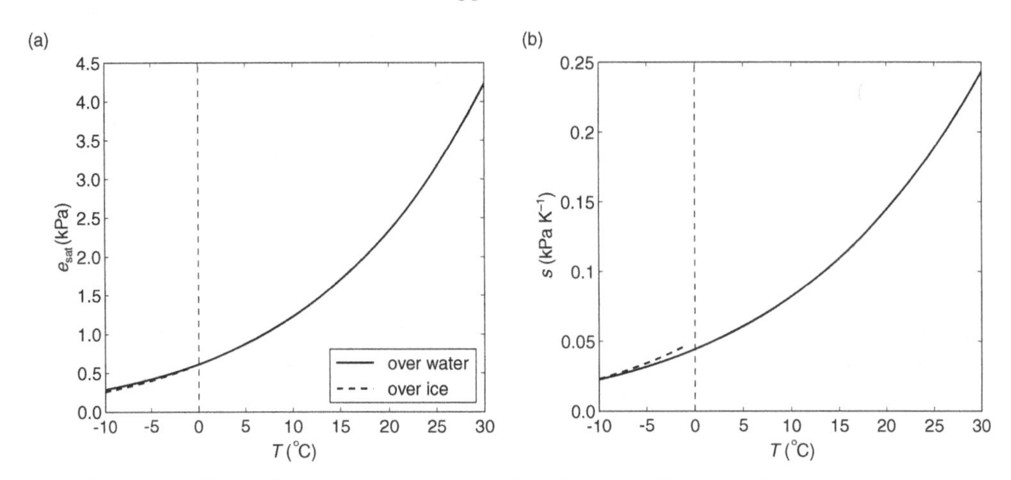

Figure B.1 Saturation vapour pressure **(a)** and slope of saturated vapour pressure curve **(b)**.

The wet bulb temperature as indicated by a psychrometer is related to the vapour pressure as:

$$e = e_{\mathrm{sat}}(T_{\mathrm{w}}) - \gamma(T - T_{\mathrm{w}}) \tag{B.22}$$

where γ is the psychrometer constant:

$$\gamma = \frac{R_{\mathrm{v}}}{R_{\mathrm{d}}} \frac{c_{\mathrm{p}}}{L_{\mathrm{v}}} p \tag{B.23}$$

This definition of the psychrometer constant assumes that the transfer of moisture from the wet bulb is exactly as efficient as the transfer of heat to the wet bulb; in practice this is not the case and hence for real calculations regarding psychrometers, the real psychrometer constant needs to be used, which may depend on ventilation rate (see Monteith and Unsworth, 2008). Also note that the psychrometer constant is not constant since it depends on pressure (and to a lesser extent on temperature, through the temperature dependence of L_{v}; see below and Table B.3).

B.5. Latent Heat of Vaporization

The temperature dependence of the latent heat of vaporization L_{v} of water vapour is approximately given by $2501000(1-0.00095(T-273.15))$ J kg^{-1} which is a good approximation over the temperature range 273–313 K (i.e., 0–40 °C) (relative to the data in Haynes, 2011, their table 6–5).

Table B.3 Overview of thermodynamic properties of dry and moist air, including their dependence on temperature, water vapour content and pressure

Constant	Dependent on			Value	Unit
	T	q	p		
R^*				8314	J kmol^{-1} K^{-1}
R_v				462	J kg^{-1} K^{-1}
R_d				287	J kg^{-1} K^{-1}
R		•		$287(1+0.61q)$	J kg^{-1} K^{-1}
c_{pd}				1004	J kg^{-1} K^{-1}
c_{vd}				717	J kg^{-1} K^{-1}
c_p		•		$1004(1+0.84q)$	J kg^{-1} K^{-1}
c_v		•		$717(1+0.37q)$	J kg^{-1} K^{-1}
L_v	•			$2501000(1-0.00095(T-273.15))$	J kg^{-1}
γ	•	•	•	$65.5\dfrac{1+0.84q}{1-0.00095(T-273.15)}\dfrac{p}{101300}$	Pa K^{-1}
e_{sat} (over water)	•			$611.2\ \exp\left[\dfrac{17.62(T-273.15)}{-30.03+T}\right]$	Pa
s (over water)	•			$e_{sat}(T)\dfrac{4284}{(-30.03+T)^2}$ (with e_{sat} over water)	Pa K^{-1}
e_{sat} (over ice)	•			$611.2\ \exp\left[\dfrac{22.46\ (T-273.15)}{-0.53+T}\right]$	Pa
s (over ice)	•			$e_{sat}(T)\dfrac{6123}{(-0.53+T)^2}$ (with e_{sat} over ice)	Pa K^{-1}

Appendix C

Dimensional Analysis

Some problems in natural sciences are too complex to describe with fundamental laws. In those cases, dimensional analysis is an important tool to find the dependence of a certain variable in a flow (e.g., a concentration gradient) on other quantities. Dimensional analysis consists of four steps:

1. Find the relevant physical quantities that (may) determine the quantity of interest.
2. Make dimensionless groups out of the quantities selected in step 1.
3. Do an experiment (or analyse existing data) in which all quantities selected in step 1 are measured.
4. Find a relationship between the dimensionless groups made in step 2, and calculated with the data of step 3. If all goes well, the dimensionless groups show a universal relationship that can also be used for other, *similar* situations.

Below, we briefly focus on the four steps, and we take as an example the vertical gradient of the mean horizontal wind speed $\left(\frac{\partial \bar{u}}{\partial z}\right)$, under conditions where buoyancy plays a role.

C.1 Choose Relevant Physical Quantities

The selection of relevant quantities requires insight into the problem, and some expert judgement. But if one selects too few quantities, the relationships found in step 4 will not be universal: they will differ from one experiment to another. On the other hand, if too many quantities are selected, it will turn out that the relationships found in step 4 will not depend on the irrelevant quantities.

For the example at hand, the relevant quantities are:

- The wind speed gradient $\frac{\partial \bar{u}}{\partial z}$ itself
- Height above the ground, z (this determines the size of turbulent eddies)

- The shear stress, $\overline{u'w'}$ (for two reasons: a) the shear stress vertically exchanges momentum, and hence influences the gradient of $\overline{u}$; b) the shear stress is part of the shear production term in the TKE equation (3.10)
- The buoyancy term in the TKE equation: $\frac{g}{\overline{\theta}_V}\overline{w'\theta_V'}$

C.2 Make Dimensionless Groups

The construction of dimensionless groups follows a straightforward recipe. This is called Buckingham's pi-theorem.

1. Determine the number of dimensionless groups n. This depends on the number of selected quantities m (in our case four) and the number of fundamental dimensions r in the following way: $n = m - r$. The number of fundamental dimensions requires some explanation. For the present subject, the units of all variables can be expressed as a combination of the fundamental dimensions of mass (M), length (L), time (T) and temperature (Θ). For example, the SI units of force is a Newton, which is equal to kg m s^{-2}, or in general terms $[M\,L\,T^{-2}]$. If there is only one dimensionless group, we know beforehand that it should be constant. The value of the constant still needs to be determined experimentally.
2. For each dimensionless group one so-called key quantity needs to be selected, such that all key quantities together contain all fundamental dimensions that are present in the physical quantities selected in step 1.
3. Each dimensionless group is the product of a key quantity and the remaining quantities, each raised to some power. This power should be chosen such that the entire dimensionless group is indeed dimensionless.

Returning to our example, the number of selected quantities is four. The number of fundamental dimensions of the four quantities is only two (L and T; the Θ contained in the virtual heat flux is cancelled by the division by $\overline{\theta}_V$). Hence the number of groups is $4 - 2 = 2$.

In the selection of the key quantities there is some arbitrariness, but we will choose $\frac{\partial \overline{u}}{\partial z}$ and $\frac{g}{\overline{\theta}_V}\overline{w'\theta_V'}$. This yields the following expressions for the dimensionless groups (which are identified by a capital pi, Π):

$$\Pi_1 = \left[\frac{\partial \overline{u}}{\partial z}\right][z]^{a_1}[\overline{u'w'}]^{b_1}$$

$$\Pi_2 = \left[\frac{g}{\overline{\theta}_v}\overline{w'\theta_v'}\right][z]^{a_2}[\overline{u'w'}]^{b_2}$$

To find out what the values for a_1, a_2, b_1 and b_2 should be, we need to analyse the fundamental dimensions:

$$[-] = [T^{-1}][L]^{a_1}[L^2T^{-2}]^{b_1}$$
$$[-] = [L^2T^{-3}][L]^{a_2}[L^2T^{-2}]^{b_2}$$

For the preceding equality to be true, the exponents of each fundamental dimension, in each dimensionless group should add up to zero. In principle this requires the solution of a system of equations, but often the solution is easily found. For this example:

- b_1 equals −1/2 to cancel the time dimension of the velocity gradient; as a consequence, a_1 needs to be +1 to cancel the length dimension of the stress. Thus $\Pi_1 = \frac{\partial \bar{u}}{\partial z}\frac{z}{\left(\overline{u'w'}\right)^{1/2}}$.
- b_2 equals −3/2 to cancel the time dimension of the buoyancy term; as a consequence, a_2 should be equal to one. Thus $\Pi_2 = \frac{g}{\overline{\theta}_v}\overline{w'\theta_v'}\frac{z}{\left(\overline{u'w'}\right)^{3/2}}$.

C.3 Do an Experiment

To determine the relationship between the Π_1 and Π_2, one or more experiments are needed in which all variables occurring in Π_1 and Π_2 are measured simultaneously. For the example at hand this was done for the first time in 1968 in Kansas (Businger et al., 1971).

C.4 Find the Relationship between Dimensionless Groups

Dimensional analysis does not give a prediction of the relation between the dimensionless groups. If there is one dimensionless group it will be constant, and the purpose of the experiment will only be to find the value of that constant. If there are two or more dimensionless groups (as in our example), the experiment also serves to find the functional relationship.

Figure 3.14 (in Chapter 3) shows the results of the Kansas 1968 experiment for dimensionless groups in our example.

Appendix D
Microscopic Root Water Uptake

D.1 Mass Balance Equation

Microscopic models describe the radial flow of soil water towards individual roots. The roots are considered as linear tubes. The root system as a whole can then be described as a set of such individual tubes, assumed to be regularly spaced in the soil at definable distances (Figure D.1). The density of the tubes may vary with depth, similar to root density in a root zone.

In such a geometry, a radial flow pattern towards the roots exists. Figure D.2 depicts this flow pattern for a segment with angle $\mathrm{d}\alpha$ (rad). The inflow Q_{in} ($\mathrm{m^2\ d^{-1}}$) can be written as:

$$Q_{\mathrm{in}} = qr\,\mathrm{d}\alpha \tag{D.1}$$

and the outflow Q_{out} ($\mathrm{m^2\ d^{-1}}$) equals:

$$Q_{\mathrm{out}} = \left(q + \frac{\partial q}{\partial r}\,\mathrm{d}r\right)(r + \mathrm{d}r)\,\mathrm{d}\alpha \tag{D.2}$$

where q ($\mathrm{m\ d^{-1}}$) is the soil water flux density and r (m) is the radial distance from the root centre. Calculation of the terms of Eq. (C.2), and subsequently the difference $Q_{\mathrm{in}} - Q_{\mathrm{out}}$ yields:

$$Q_{\mathrm{in}} - Q_{\mathrm{out}} = -\mathrm{d}\alpha\left[q\,\mathrm{d}r + r\frac{\partial q}{\partial r}\mathrm{d}r + \frac{\partial q}{\partial r}(\mathrm{d}r)^2\right] \tag{D.3}$$

The segment area A ($\mathrm{m^2}$) between radial distances r and $r + \mathrm{d}r$ (m) from the root centre is equal to:

$$A = \pi(r + \mathrm{d}r)^2\frac{\mathrm{d}\alpha}{2\pi} - \pi r^2\frac{\mathrm{d}\alpha}{2\pi} = \left(rdr + \frac{\mathrm{d}r^2}{2}\right)\mathrm{d}\alpha \tag{D.4}$$

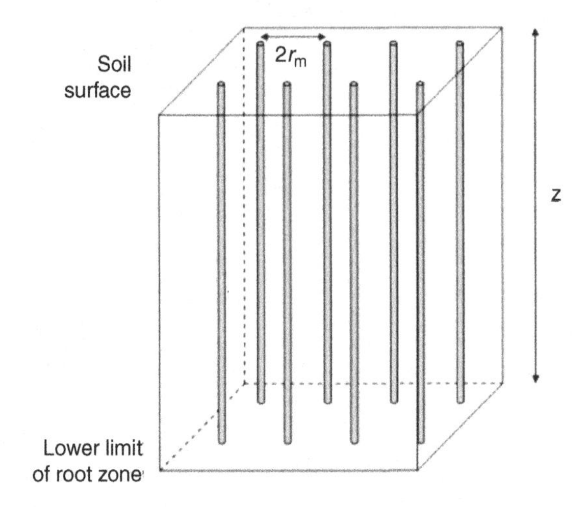

Figure D.1 Schematization of root system by equally spaced tubes. Density may differ in the vertical.

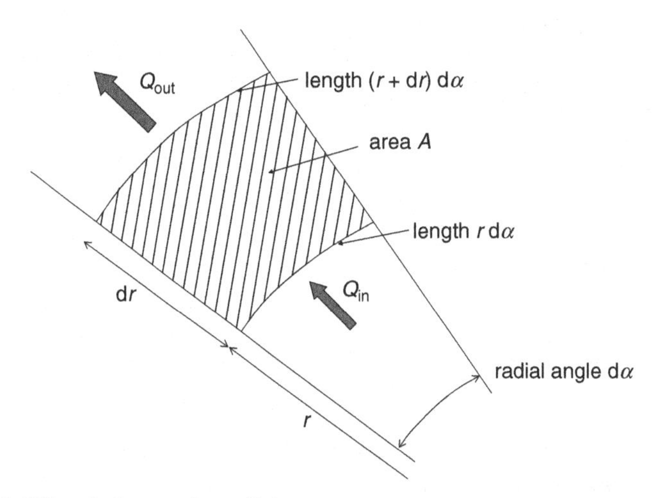

Figure D.2 Water balance of a radial segment.

As we are dealing with infinite small differences, we may omit higher-order terms with respect to first-order terms: the third term in Eq. (C.3), and the second term in Eq. (C.4). The water balance of segment A can be written with the simplified equations (C.3) and (C.4) as:

$$\frac{\partial \theta}{\partial t} A = Q_{\text{in}} - Q_{out} \quad \rightarrow \quad \frac{\partial \theta}{\partial t} r\,\mathrm{d}r\,\mathrm{d}\alpha = -\mathrm{d}\alpha \left[q\,\mathrm{d}r + r\,\mathrm{d}r \frac{\partial q}{\partial r} \right] \tag{D.5}$$

where θ is the volumetric water content (m^3 m^{-3}) and t is the time (d). Therefore we may write the water balance of the radial flow pattern towards root as:

$$\frac{\partial \theta}{\partial t} = -\frac{q}{r} - \frac{\partial q}{\partial r} \tag{D.6}$$

D.2 General Solution of Matric Flux Potential Differential Equation

Use of the matric flux potential results for axisymmetric flow towards roots into the second-order differential equation:

$$-\frac{T_p}{D_r} = -\frac{q}{r} - \frac{\partial q}{\partial r} = \frac{\partial M}{r\partial r} + \frac{\partial^2 M}{\partial r^2} \tag{D.7}$$

for which the following general solution is found:

$$M = \frac{-T_p}{4z} r^2 + C_1 \ln r + C_2 \tag{D.8}$$

where C_1 and C_2 are integration constants. We may use two boundary conditions at the root surface:

$$M = M_0 \quad ; \quad r = r_0 \tag{D.9}$$

$$\frac{\mathrm{d}M}{\mathrm{d}r} = T_p \frac{r_m^2}{2zr_0} \quad ; \quad r = r_0 \tag{D.10}$$

where M_0 (m^2 d^{-1}) is the matric flux potential at the root surface, r_0 (m) is the root radius and r_m (m) is equal to the half mean distance between roots. Applying these boundary conditions (Eqs. (C.9) and (C.10)) yields:

$$C_1 = \frac{T_p}{2z}\left(r_m^2 + r_0^2\right) \tag{D.11}$$

$$C_2 = \frac{T_p}{2z}\left[\frac{r_0^2}{2} - \left(r_m^2 + r_0^2\right)\ln r_0\right] + M_0 \tag{D.12}$$

and as general solution to Eq. (C.8):

$$M - M_0 = \frac{T_p}{2D_r}\left[\frac{r_0^2 - r^2}{2} + \left(r_m^2 + r_0^2\right)\ln\frac{r}{r_0}\right] \tag{D.13}$$

Appendix E

Crop Factors for Use with Makkink Reference Evapotranspiration

The table below contains the crop factors for use with Makkink reference evapotranspiration for use in the Netherlands.

	April			May			June			July			August			September		
	I	II	III	I	II	III	I	II	III	I	II	III	I	II	III	I	II	III
Grass	1.0	1.0	1.0	1.0	1.0	1.0	1.0	1.0	1.0	1.0	1.0	1.0	1.0	1.0	0.9	0.9	0.9	0.9
Cereals	0.7	0.8	0.9	1.0	1.0	1.0	1.2	1.2	1.2	1.0	0.9	0.8	0.6	—	—	—	—	—
Maize	—	—	—	0.5	0.7	0.8	0.9	1.0	1,2	1.3	1.3	1.2	1.2	1.2	1.2	1.2	1.2	1.2
Potatoes	—	—	—	—	0.7	0.9	1.0	1.2	1.2	1.2	1.1	1.1	1.1	1.1	1.1	0.7	—	—
Sugar beets	—	—	—	0.5	0.5	0.5	0.8	1.0	1.0	1.2	1.1	1.1	1.1	1.2	1.2	1.2	1.1	1.1
Leguminous plants	—	0.5	0.7	0.8	0.9	1.0	1.2	1.2	1.2	1.0	0.8	—	—	—	—	—	—	—
Plant onions	0.5	0.7	0.7	0.8	0.8	0.9	1.0	1.0	1.0	1.0	1.0	1.0	1.0	—	—	—	—	—
Sow onions	—	0.4	0.5	0.5	0.7	0.7	0.8	0.8	0.9	1.0	1.0	1.0	1.0	1.0	0.9	0.7	—	—
Chicory	—	—	—	—	—	—	0.5	0.5	0.5	0.8	1.0	1.1	1.1	1.1	1.1	1.1	1.1	1.1
Winter carrots	—	—	—	—	—	—	0.5	0.5	0.5	0.8	1.0	1.1	1.1	1.1	1.1	1.1	1.1	1.1
Celery	—	—	—	—	—	0.5	0.7	0.7	0.7	0.8	0.9	1.0	1.1	1.1	1.1	1.1	1.1	—
Leek	—	—	—	—	0.5	0.5	0.5	0.5	0.7	0.7	0.8	0.8	0.8	1.0	0.9	0.9	0.9	0.9
Bulb/tuber crops	—	—	—	—	0.5	0.7	0.7	0.9	1.2	1.2	1.2	1.2	1.2	1.2	1.2	1.2	1.2	1.2
Pome/stone fruit	1.0	1.0	1.0	1.4	1.4	1.4	1.6	1.6	1.6	1.7	1.7	1.7	1.3	1.3	1.2	1.2	1.2	1.2
Bare soil	0.5	0.5	0.5	0.5	0.5	0.5	0.5	0.5	0.5	0.5	0.5	0.5	0.5	0.5	0.5	0.5	0.5	0.5
Water	1.3	1.3	1.3	1.3	1.3	1.3	1.31	1.31	1.31	1.29	1.27	1.24	1.21	1.19	1.18	1.17	1.17	1.17
Deciduous forest	1.04	1.04	1.04	1.04	1.04	1.04	1.04	1.04	1.04	1.04	1.04	1.04	0.96	0.96	0.96	0.96	0.96	0.96
Coniferous forest	1.30	1.30	1.30	1.30	1.30	1.30	1.30	1.30	1.30	1.30	1.30	1.30	1.20	1.20	1.20	1.20	1.20	1.20
Mixed forest	1.17	1.17	1.17	1.17	1.17	1.17	1.17	1.17	1.17	1.17	1.17	1.17	1.08	1.08	1.08	1.08	1.08	1.08

Values are given for three decades (I, II and III) for each month.
From Feddes (1987).

Answers

Chapter 1

Question 1.1:

a) Input of solid water takes place by snow or hail, input or output by snowdrift; storage of solid water may change; exchange with other phases (see question b).

b) Solid: solid to liquid (melting) and solid to gas (sublimation); liquid: liquid to solid (freezing) and liquid to gas (evaporation); gas: gas to solid (deposition) and gas to liquid (condensation)

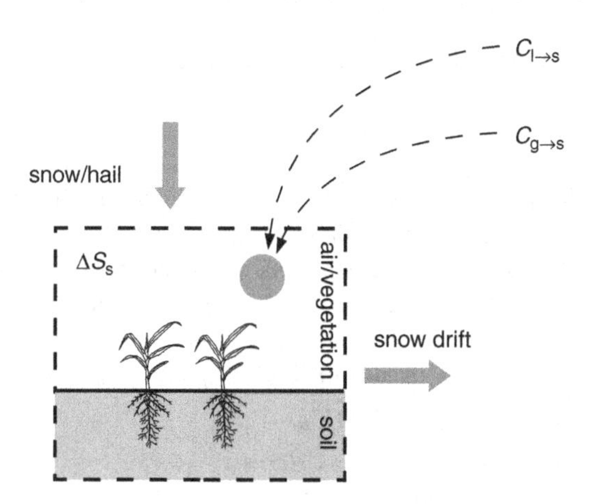

Question 1.2:

a) E_{int}: none, as only water on the vegetation is affected; T: decrease of soil moisture at all depths where roots are present; E_{soil}: only top layer.

b) E_{int}: the leaves where water is intercepted, T: leaves (location of the stomata where water changes from liquid to vapour); E_{soil}: top of the soil.

Question 1.3:

a) Units of H: W m^{-2}. Units of the right-hand side (kg m^{-3}) (J kg^{-1} K^{-1}) (m^2 s^{-1}) (K m^{-1}) = m^{-2} J s^{-1} = W m^{-2}.

b) Given: H, ρ, c_p and κ. Requested $\frac{\partial T}{\partial z} \cdot \frac{\partial T}{\partial x} = -\frac{H}{\rho c_p \kappa} = -4146\ \text{K m}^{-1}$: the temperature decreases with height.

Chapter 2

Question 2.1:

a) Solar constant = 1365 W m^{-2}. Radiation flux density decreases with the square of the distance travelled (here we take the centre of the Sun as the origin). Solar radiation originates from a sphere with radius R_{Sun} 1.34·10^6/2 = 6.7·10^5 km. The fraction of the flux density emitted by the Sun arriving at the location of Earth is $(R_{Sun}/d_{Sun})^2$ = (6.7·10^5/149.6·10^6)^2 = 2.15·10^{-5}. Black-body radiation at 5800 K is (using Stefan–Boltzmann's law) 6.42·10^7 W m^{-2}. At 1 AU from the Sun: 2.15·10^{-5} × 6.42·10^7 = 1380 W m^{-2}.

b) Black-body radiation at 293 K: 418 W m^{-2}: this is 6.5·10^{-6} times the value for 5800 K.

Question 2.2:

$\left(\frac{d_{\text{Sun},0}}{d_{\text{Sun}}}\right)^2$ varies between 1 – 0.033 and 1 + 0.033. Total difference is 0.066. Given I_0 = 1365 W m^{-2}, variation of radiation at top of atmosphere, perpendicular to the beam is 90 W m^{-2}.

Question 2.3:

a) With a gravitational acceleration g = 9.8 m s^{-2}, a vertical air mass of 10 000 kg m^{-2} implies a force of 9.8·10^4 N m^{-2} ($F = mg$). As the force is exerted on 1 m^{-2} the surface pressure is 9.8·10^4 Pa or 980 hPa, which is of the same order of magnitude as the global average sea level pressure of 1013 hPa. The data in the problem could refer to an elevated location, where the surface pressure is lower.

b) $\tau_{\lambda v} = 0.8; \tau_{\lambda v} = e^{-k_{\lambda,i} q_i m_v} \Rightarrow \ln(\tau_{\lambda,v}) = -k_{\lambda,i} q_i m_v \Rightarrow k_{\lambda,i} q_i m_v = 0.22$. Hence, $k_{\lambda,i} q_i = 2.2 \cdot 10^{-5}\,\text{kg}^{-1}$ and $k_{\lambda,i} = 0.0073$.

c) $\tau_\lambda = (\tau_{\lambda,v})^{m_r}; m_r \approx (\cos\theta_z)^{-1} = 1/0.766 = 1.305 \Rightarrow \tau_\lambda = 0.75$.

d) Transmissivity $\tau_\lambda = I_\lambda / I_{\lambda 0}$. With τ_λ = 0.75 (from c) and $I_{\lambda 0}$ =1.5 W m^{-2}μm^{-1}, this gives I_λ =1.13 W m^{-2}μm^{-1}.

Question 2.4:

a) Sunny day: approximately 0.2 (mostly direct radiation, some scattering by molecules (Rayleigh) and aerosols (Mie)).

b) Overcast day: approximately 1 (no direct radiation, all radiation scattered by cloud droplets (Mie) plus the usual scattering by molecules and aerosols).

Question 2.5:

a) $I = S/\cos\theta_z = (K^{\downarrow} - D)/\cos\theta_z$; For both days, $\cos(\theta_z) = 0.848$. May 22: $K^{\downarrow}$ =370 W m^{-2}, D = 340 W m^{-2}, hence S = 30 W m^{-2} and I = 35 W m^{-2}. May 23: $K^{\downarrow}$ = 840 W m^{-2}, D = 120 W m^{-2}, hence S = 720 W m^{-2} and I = 849 W m^{-2}.

b) I_0 = 1330 W m^{-2} (taking into account distance to the Sun); May 22: $\tau_{b\theta} = 0.026$, May 23: $\tau_{b\theta} = 0.64$.

c) With $K^{\downarrow}_0 = I\cos(\theta_z)$ = 1127 W m^{-2}; May 22: $\tau_b = 0.31$, May 23: $\tau_b = 0.75$.

d) Mainly scattering (clouds). For shortwave radiation the absorption by the atmosphere is not very variable.

Question 2.6:

Symbol	Meaning	Definition
$\tau_{\lambda\theta}$	The spectral beam transmissivity indicates the fraction of incident radiation of a given wavelength at the top of the atmosphere that is transmitted, along the direction of the solar beam.	$\tau_{\lambda\theta} \equiv \frac{I_\lambda}{I_{\lambda 0}}$
$\tau_{\lambda v}$	The spectral vertical transmissivity holds for the (for most situations) hypothetical condition that the solar radiation enters the atmosphere at an angle perpendicular to Earth's surface. It indicates the fraction of incident radiation of a given wavelength at the top of the atmosphere that would be transmitted along the vertical direction. The quantity is useful to isolate the effects of solar angle and atmosphere.	$\tau_{\lambda v} \equiv (\tau_{\lambda\theta})^{-m_r}$
$\tau_{b\theta}$	The broadband beam transmissivity indicates the fraction of total incident solar radiation at the top of the atmosphere that is transmitted, along the direction of the solar beam.	$\tau_{b\theta} \equiv \frac{I}{I_0}$
τ_b	The broadband transmissivity indicates the fraction of the solar radiation incident at the top of the atmosphere, through a plane parallel to Earth's surface, that is transmitted to the surface.	$\tau_b \equiv \frac{K^{\downarrow}}{K_0^{\downarrow}}$

Question 2.7:

a) Although the solar angle in Oslo is smaller than in Valencia (due to the more northern location), this is more than compensated by the larger day length in Oslo.

b) Estimate relative sunshine durations from the graphs: in January and June these are 0.57 and 0.61 for Valencia and 0.22 and 0.42 for Oslo. Using the daylength data from the table yields daily average sunshine durations of 5.4 and 9.0 hours for Valencia, and 1.5 and 7.6 for Oslo. For Valencia the number of sunshine hours in June is 1.6 times that in January, whereas in Oslo this factor is more than 5. The reason for this difference is twofold: the contrast in daylength between winter and summer is larger in Oslo than in Valencia. In addition, the relative sunshine duration in Oslo is longer in summer than in winter, whereas in Valencia there is little difference between the seasons.

c) Using the empirical relationship linking relative sunshine duration and global radiation, with coefficients $a = 0.25$ and $b = 0.5$ yields for Valencia $\overline{K}^{\downarrow 24}$ equal to 64 and 198 W m^{-2} in January and June, respectively. For Oslo the values are 11 and 165 W m^{-2}.

Question 2.8:

With $K^{\downarrow}_0 = I \cos(\theta_z) = 1127$ W m^{-2} for both days. Then for May 22 $\tau_b = 0.31$, for May 23: $\tau_b = 0.75$. With Eq. (2.19) the Linke turbidity can be estimated from τ_b provided one knows the relative optical mass m_r, which can be estimated as $1/\cos(\theta_z) = 1.18$. This yields $T_{L,2} = 23$ for May 22 and $T_{L,2} = 3.1$ for May 23. The value for May 22 is outside the range of value quoted in the main text because those relate to cloudless conditions. The large turbidity found here is due to the extinction by the clouds. The value for May 23 is just below the value quoted for moist, warm air.

Question 2.9:

d) If the Sun is shining parallel to rows, there is no interaction between both surface types (no shadowing). Hence the albedo is the simple mean of both surfaces: $r = 0.5\, r_{veg} + 0.5\, r_{soil} = 0.5 \cdot 0.2 + 0.5 \cdot 0.1 = 0.15$.
e) We neglect diffuse radiation; the soil is not illuminated at all and hence only the albedo of the vegetation is relevant: $r = r_{veg} = 0.2$.
f) Situation a: $r = 0.15$ so that $K^* = 0.85\, K^{\downarrow}$; situation b: $r = 0.2$ so that $K^* = 0.80\, K^{\downarrow}$; difference = $0.05\, K^{\downarrow}$ or $0.05/0.85 = 6\%$.

Question 2.10:

No difference: with purely diffuse radiation, global radiation has no directional dependence and as a result, the albedo will also be independent of the direction of the solar beam.

Question 2.11:

The time dependence of the albedo would be due to the time dependence of the function $K^{\downarrow}(\lambda, \theta_{in}, \phi_{in})$. But for purely diffuse radiation the incoming radiation has no directional dependence. Hence this function becomes – except for the time-dependent *magnitude* of the radiation – independent of θ_{in} and ϕ_{in} and could hence be approximated as $K^{\downarrow}(t) \cdot f(\lambda)$, where t is time and $f(\lambda)$ is a function of wavelength only (this function determines the distribution of radiation over different wavelengths). This will cause the resulting broadband albedo to be a weighted average of the BRDF, weighted by $f(\lambda)$:

$$r \equiv \frac{K^{\uparrow}}{K^{\downarrow}} = \frac{K^{\downarrow}(t)\int_0^{\pi/2}\int_0^{2\pi}\int_{\lambda_1}^{\lambda_2} f(\lambda)\cdot r(\lambda,\theta_{in},\varphi_{in})\cdot d\lambda \cdot d\varphi_{in} \cdot d\theta_{in}}{K^{\downarrow}(t)\int_0^{\pi/2}\int_0^{2\pi}\int_{\lambda_1}^{\lambda_2} f(\lambda)\cdot d\lambda \cdot d\varphi_{in} \cdot d\theta_{in}} = \frac{\int_0^{\pi/2}\int_0^{2\pi}\int_{\lambda_1}^{\lambda_2} f(\lambda)\cdot r(\lambda,\theta_{in},\varphi_{in})\cdot d\lambda \cdot d\varphi_{in} \cdot d\theta_{in}}{\int_0^{\pi/2}\int_0^{2\pi}\int_{\lambda_1}^{\lambda_2} f(\lambda)\cdot d\lambda \cdot d\varphi_{in} \cdot d\theta_{in}}$$

Question 2.12:
At wavelengths where a black body with a temperature of that of the atmosphere emits most. From Wien's law: at $T = 270$ K, $\lambda_{max} = 11$ μm. But for emission to occur, emission lines need to be present (the energy absorbed by a given gas is transferred to the air as a whole, and hence may be emitted by another gas than by which it was absorbed).

Question 2.13:
a) Difference in $L^{\downarrow}$ at 9 UTC between both days is 90 W m^{-2}.
b) Assume $L^{\downarrow}_{cloud} = 0.35\ \sigma T^4_{cloud} = 90$ W m^{-2}. Then $T_{cloud} = 260$ K.

Question 2.14:
a) Incoming longwave radiation at 12 UTC is 400 and 330 W m^{-2} at May 22 and 23, respectively. Determine emissivity from $\varepsilon_a = L^{\downarrow} / (\sigma T_a^4)$ with the air temperature in K. This gives values for ε_a of 0.97 and 0.79, respectively.
b) On May 22 conditions are close to overcast (judging from the graphs on shortwave radiation). Hence, $f_{cloud} = 1$ and the atmospheric emissivity is equal to one. On May 23 conditions are cloudless and we assume $f_{cloud} = 0$. To compute $\varepsilon_{a,clear}$ we need e_a. At $T_a = 19.5$ °C we find $e_{sat} = 22.7$ hPa, which gives with RH = 49%: $e_{sat} = 11.1$ hPa. With the Brunt formula this gives $\varepsilon_a = 0.74$.
c) Compared to the observed atmospheric emissivities the values are off by +3.1% and –6.4% for May 22 and 23, respectively. The relative errors in estimated down-welling radiation are of the same order: +3.0% and –7.4%. The absolute errors are +12 and –23 W m^{-2}. For L^* (–30 and –120 W m^{-2}, respectively) the relative errors incurred by the error in $L^{\downarrow}$ are much larger: +40% and –19%, respectively.

Question 2.15:
To obtain the emitted longwave radiation, we need to subtract the reflected long-wave radiation from the upwelling longwave radiation: $L_e^{\uparrow} = L^{\uparrow} - (1-\varepsilon_s)L^{\downarrow}$. Then $T_s = \left(L_e^{\downarrow} / (\varepsilon_s \sigma)\right)^{1/4}$. Observations are: May 22: $L^{\downarrow} = 397\ \mathrm{W\,m^{-2}}, L^{\uparrow} = 433\ \mathrm{W\,m^{-2}}$; May 23: $L^{\downarrow} = 331\ \mathrm{W\,and\,m^{-2}}, L^{\uparrow} = 446\ \mathrm{W\,and\,m^{-2}}$.

	Question a)		Question b)		Question c)	
	May 22	May 23	May 22	May 23	May 22	May 23
ε_s (-)	0.96	0.96	1.0	1.0	0.96	0.96
$L_e^{\uparrow}$ (W m^{-2})	417	433	433	446	433	446
T_s (K)	295.8	298.6	295.6	297.8	298.6	300.9

For a surface with $\varepsilon_s = 0.96$, the method in question a) is correct. In b) a compensation of errors occurs: overestimation of emitted longwave radiation is compensated by an underestimation of the temperature needed to cause that emission. The error is largest for sunny conditions, because then the difference between $L^{\downarrow}$ and $L_e^{\uparrow}$ is largest. In c) temperature is overestimated, because the reflected longwave radiation is erroneously accounted for as emission.

Question 2.16:
Assume ε_s = 0.96 for both surfaces. $L^{\downarrow} = 410\ \mathrm{W\,m^{-2}}$. Use $L_e^{\uparrow} = L^{\uparrow} - (1-\varepsilon_s)L^{\downarrow}$ and $T_s = \left(L_e^{\downarrow} / (\varepsilon_s \sigma)\right)^{1/4}$.

Plants: $L^{\uparrow} = 500\ \mathrm{W\,m^{-2}} \Rightarrow L_e^{\uparrow} = 484\ \mathrm{W\,m^{-2}} \Rightarrow T_s = 307.0\ \mathrm{K}$

Bare soil: $L^{\uparrow} = 590\ \mathrm{W\,m^{-2}} \Rightarrow L_e^{\uparrow} = 574\ \mathrm{W\,m^{-2}} \Rightarrow T_s = 320.4\ \mathrm{K}$

Difference in surface temperature between both surfaces is 13.4 K.

Question 2.17:
λ_s (conductivity): relationship between flux and gradient
κ_s (diffusivity): ability to diffuse thermal disturbance
c_s (specific heat capacity): energy needed for temperature change (per mass)
C_s (heat capacity, volumetric = ρc_s): energy needed for temperature change (per volume)

Question 2.18:
a) Temperature increase is from 9 to 30 °C. Total heat storage is (21 K) · (0.10 m) · ($3.0 \cdot 10^6\ \mathrm{J\,K^{-1}m^{-3}}$) = $6.3 \cdot 10^6\ \mathrm{J\,m^{-2}}$.
b) Heat storage = soil heat flux divergence (Eq. (2.30), in integrated form): $C_s\left(\frac{\Delta T}{\Delta t}\right)\Delta z = -\Delta G$). 11 hours = 39 600 s. Hence $\Delta G = -159\ \mathrm{W\,m^{-2}}$ over a depth of 10 cm.

Question 2.19:
a) Values for the total soil that can be calculated: ρ_s, c_s, C_s.
b)

Quantity	Vol. fract. x_i	ρ_i	$C_i = \rho_i c_i$	$x_i\rho_i$	x_iC_i
Unit	(m^3 component i)/ (m^3 soil)	10^3 kg m^{-3}	10^6J m^{-3} K^{-1}	10^3 kg m^{-3}	10^6 J m^{-3} K^{-1}
Quarts	0.6 · 0.2 = 0.12	2.66	2.13	0.319	0.255
Clay	0.6 · 0.5 = 0.30	2.65	2.39	0.795	0.716
Org. mat.	0.6 · 0.3 = 0.18	1.30	2.47	0.234	0.445
Water	0.4 · 0.75 = 0.3	1.00	4.18	0.300	1.254
Air	0.4 · 0.25 = 0.1	$1.2\ 10^{-3}$	$1.2\ 10^{-3}$	$1.2\ 10^{-4}$	$1.2\ 10^{-4}$
Sum	1			1.65	2.67

$c_s = C_s/\rho_s$, which gives $1.62 \cdot 10^3\ \mathrm{J\,kg^{-1}\,K^{-1}}$.

Question 2.20:
a) For sinusoidal forcing in homogeneous soil, the phase shift equals z/D. Phase shift in time: 6.5 hours = 23 400 seconds. For diurnal cycle (86 400 seconds) this phase shift corresponds to 2π 23 400/86 400 = 1.70 rad. Hence $D = z/1.7 = 0.2/1.7$ = 0.118 m. For the diurnal cycle, $\omega = 2\pi/(86\ 400\ \mathrm{s}) = 7.272 \cdot 10^{-5}\ \mathrm{s^{-1}}$, so that, with $D = \sqrt{2\kappa_s / \omega}$, $\kappa_s = 0.5 \cdot 10^{-6}\ \mathrm{m^2\,s^{-1}}$.

b) Amplitude decreases exponentially with depth: $A(z) = A(0)e^{-\frac{z}{D}}$. At 20 cm depth, with a damping depth of 0.118 m, the amplitude is 0.18 of the surface amplitude.

Question 2.21:

$G = \Lambda_{veg}(T_{veg} - T_{top})$; Given G = 50 W m^{-2} and T_{top} = 20 ºC. From the text: Λ_{veg} = 5 W m^{-2} K^{-1}. Hence, $T_{veg} = \frac{G}{\Lambda_{veg}} + T_{top}$ which gives T_{veg} = 30 ºC.

Question 2.22:

a) The soil thermal conductivity can be obtained from Table 2.2: λ_s = 2.5 W m^{-2} K^{-1}. With thickness 0.005 m and area 0.0079 m^2, the correction factor (with Eq. (2.43)) becomes 1.14. Thus the real soil heat flux is 1.14 · 55 = 63 W m^{-2}.
b) For a thickness of 0.01 m, the correction factor becomes 1.29 and hence the observed soil heat flux is 63/1.29 = 49 W m^{-2}.

Question 2.23:

a) Instantaneous rate of increase: determine slope of tangent line at 9 UTC: 6 K in 2 hours, or 3 K h^{-1}, or 8.33 · 10^{-4} K s^{-1}.
b) Storage = (8.33 · 10^{-4} K s^{-1}) · (0.05 m) · (3.0 10^6 J K^{-1} m^{-3}) = 125 W m^{-2}.
c) Decrease of soil heat flux over 5 cm = 125 W m^{-2}. Observed is 150 W m^{-2}, hence surface value is 275 W m^{-2}.

Question 2.24:

a) The solution strategy is to construct a spreadsheet that computes the soil temperature at 10 cm depth for the requested times (converted to seconds) using Eq. (2.44), with an assumed (first guess) damping depth. The unkowns in Eq. (2.44) are the A_n and ϕ_n at 10 cm for both harmonics. But these can, with an assumed D, be constructed from the amplitude and phase shift at 5 cm: $A_n(0.1) = A_n(0.05)\exp(-\sqrt{n}(0.1-0.05)/D)$ and $\phi_n(0.1) = \phi_n(0.05) - \sqrt{n}(0.1-0.05)/D$ (first convert the given phase shifts to radians). Then iteratively find the D that gives the closest correspondence between the estimated and observed temperature at 10 cm depth: 0.090 m. This corresponds to $\kappa_s = 2.95 \cdot 10^{-7}$ m^2 s^{-1}.
b) The expression for the soil heat flux using all harmonics is $G(z_d, t) = \sum_{n=1}^{M} A(0)e^{-\sqrt{n}z_d/D}\sqrt{\frac{\omega}{\kappa_s}}\lambda_s \sin\left(n\omega t + \phi_n(0) - \sqrt{n}\frac{z_d}{D} + \frac{\pi}{4}\right)$ (based on vertical derivative to Eq. (2.44), equivalent to Eq. (2.38)). In combination with observed soil heat flux this gives λ_s = 0.78 W m^{-1} K^{-1} (the temperature gradient is –45.1 K m^{-1}).
c) With the above expression for the soil heat flux, the value at the surface can be computed to be 110 W m^{-2} (the temperature amplitudes and phase shifts at the surface are: A_1 = 12.0 °C, A_2 = 3.1 °C, ϕ_1 = –8.0 hours and ϕ_2 = –7.0 hours).

Question 2.25:

a) Take density of old snow from Table 2.2. Mass of 10 cm of snow on 1 m^2 is (0.10 m) · (800 kg m^{-3}) = 80 kg m^{-2}. Energy needed to melt mass of snow times latent heat of fusion: (80 kg m^{-2}) · (0.33 · 10^6 J kg^{-1}) = 26.4 · 10^6 J m^{-2}.

b) Net energy input is 40 W m^{-2}. Time required to melt is energy needed divided by rate of energy supply: (26.4 $\cdot 10^6$ J m^{-2})/(40 W m^{-2}) = 6.60 $\cdot 10^5$ s = 7.6 days.

Question 2.26:

a) For this calculation, Eq. (2.46) can be used, if we assume that there is no exchange between the soil and the snow layer. For fresh snow we assume λ_{snow} 0.1 W m^{-2} K^{-1}. From the graph we take $L^* = -50$ W m^{-2} and $G = -40$ W m^{-2}. For $z_d = d_{snow} = 0.02$ m we find a temperature difference $T(z_d) - T(0)$ of 9 K. From the graph we find a temperature difference of 16 K between top of snow and 5 cm depth in the soil. The difference may partly be due to uncertainty in snow thickness, and partly because the temperature at 5 cm in the soil will be lower than that at the soil (or grass) surface.

b) For $d_{snow} = 1$ cm, temperature difference ia 4.5 K, for $d_{snow} = 3$ cm, the temperature difference is 13.5 K. The variability in the surface temperature is directly proportional to the variability in the snow layer thickness.

Question 2.27:

a) Maximum penetration is at the date that the frost index reaches its highest value. This happens on the last day with negative temperatures: December 22.

b) Sum up all daily mean temperatures until December 22 inclusive, starting on the first day with a negative temperature (December 16): –16.2 °C day.

c) $z_f = a\sqrt{-I_n}$. With $I_n = -16.2$ °C day and $a = 0.05$ m $K^{-1/2}$ $day^{-1/2}$ this yields z_f = 0.2 m.

d) Continue summation of temperatures. Frost has disappeared when the frost index becomes positive: at December 28.

Chapter 3

Question 3.1:

a)

Height (m)	0.3	1.3	6	15
K_h (night) (m^2 s^{-1})	0.0020	0.050	0.25	0.37
K_h (day) (m^2 s^{-1})	0.0094	0.30	2.2	14

b)

Height interval (m)	0.3–1.3	1.3–6	6–15
$\frac{\partial K_h}{\partial z}$ (night) (m^2 s^{-1} m^{-1})	0.048	0.043	0.013
$\frac{\partial K_h}{\partial z}$ (day) (m^2 s^{-1} m^{-1})	0.29	0.40	1.3

For nighttime conditions K_h increases with height less than linear; during daytime more than linear. One could also look at one height interval: the increase of K_h with height is larger during daytime than during nighttime.

Question 3.2:

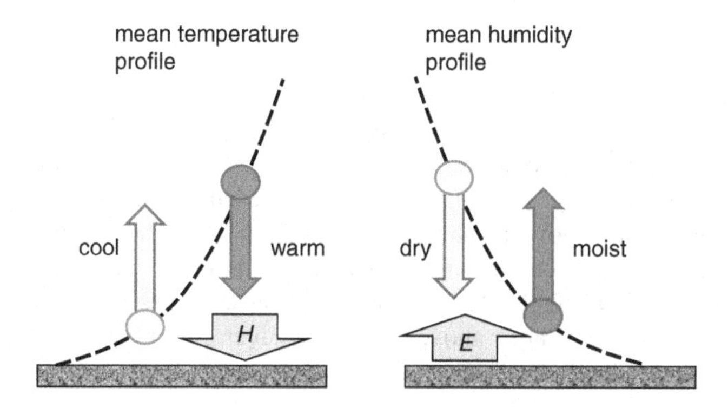

Question 3.3:

a) $c_p/c_v - 1$ is larger than 0. In order to keep $p^{c_p/c_v-1}T^{-c_p/c_v}$ constant, an increase in pressure should be paired to an increase in temperature (the exponent of temperature in the above relationship is negative).

b) $p^{c_p/c_v-1}T^{-c_p/c_v} = C_1 \Leftrightarrow p = C_2 T^{\gamma/(\gamma-1)}$; with $\rho = \dfrac{p}{RT}$ this results in $\rho = C_3 T^{\gamma/(\gamma-1)-1}$. Thus density decreases with temperature (with $\gamma = c_p/c_v$ being larger than 1, $\gamma/(\gamma-1)-1$ will always be less than 0).

c) Parallel to b): $\rho_v = C_4 T^{\gamma/(\gamma-1)-1}$.

d) Dependence of specific humidity on temperature for an adiabatic process: $\dfrac{\rho_v}{\rho} = \dfrac{C_4 T^{\gamma/(\gamma-1)-1}}{C_3 T^{\gamma/(\gamma-1)-1}} = \dfrac{C_4}{C_3} = C_5$, hence constant.

Question 3.4:

a) $\overline{X} = \dfrac{1}{N}\sum_{i=1}^{N} X_i = \dfrac{1}{5}(4+1+5+2+3) = 3$ and $\overline{Y} = \dfrac{1}{N}\sum_{i=1}^{N} Y_i = \dfrac{1}{5}(10+2+5+7+4) = 5.6$.

b) $\overline{X'} = \dfrac{1}{N}\sum_{i=1}^{N} X_i{'} = \dfrac{1}{N}\sum_{i=1}^{N}(X_i - \overline{X}) = \dfrac{1}{5}(1-2+2-1+0) = 0$ and

$\overline{Y'} = \dfrac{1}{N}\sum_{i=1}^{N} Y_i{'} = \dfrac{1}{N}\sum_{i=1}^{N}(Y_i - \overline{Y}) = \dfrac{1}{5}(4.4-3.6-0.6+1.4-1.6) = 0$. The mean of the deviations must be zero by definition.

c) $\overline{X'X'} = \dfrac{1}{N}\sum_{i=1}^{N} X_i{'}X_i{'} = \dfrac{1}{5}\left(1^2+(-2)^2+2^2+(-1)^2+0^2\right) = 2$ and

$\overline{Y'Y'} = \dfrac{1}{N}\sum_{i=1}^{N} Y_i{'}Y_i{'} = \dfrac{1}{5}\left(4.4^2+(-3.6)^2+(-0.6)^2+1.4^2+(-1.6)^2\right) = 7.44$. As this is a quadratic quantity, it is positive by definition.

d) $\overline{X'Y'} = \frac{1}{N}\sum_{i=1}^{N} X_i'Y_i' = \frac{1}{5}(1\cdot 4.4 + (-2)\cdot(-3.6) + 2\cdot(-0.6) + (-1)\cdot 1.4 + 0\cdot(-1.6)) = 1.8$.
The covariance could be positive or negative.

e) $R_{XY} = \frac{\overline{X'Y'}}{\sqrt{\overline{X'X'}\,\overline{Y'Y'}}} = 0.467$. Note that we used $\overline{X'X'} = \sigma_X^2$.

Question 3.5:

The temperature profile shows a strong decrease with height in the lower part of the surface layer, but higher up in the surface layer the gradient decreases. In the boundary layer above the surface layer potential temperature is even uniform with height.

Parcels coming from below can have temperatures up to θ_s. The actual temperature of the rising thermal depends on the height from which it comes (somewhere between the surface and observation height). Parcels from above have as lower limit a temperature θ representative of the mixed layer above the surface layer. From Figure 3.4 this temperature can be estimated to be 31.8 °C.

Question 3.6:

a) Increase of q yields an increase of virtual temperature.
b) Addition of humidity lowers the density of the air (density of water vapour is lower than the density of dry air): the dry air will have the highest density.

Question 3.7:

a) Mean kinetic energy: $\text{MKE} = \frac{1}{2}(\overline{u}^2 + \overline{v}^2 + \overline{w}^2) = \frac{1}{2}(4^2 + 2^2 + 0^2) = 10\ \text{m}^2\ \text{s}^{-2}$.

b) Turbulent kinetic energy: $\text{TKE} = \frac{1}{2}({\sigma_u}^2 + {\sigma_v}^2 + {\sigma_w}^2) = \frac{1}{2}(0.3^2 + 0.2^2 + 0.2^2) =$ $0.085\ \text{m}^2\ \text{s}^{-2}$.

c) Units of velocity squared: $\text{m}^2\ \text{s}^{-2}$.

d) Quantities given under a) and b) are energies per unit mass. As it is impossible to identify a physical body of which one could determine the mass, energies per unit volume are used instead. To obtain that we have to multiply with density to obtain: $(\text{kg m}^{-3})(\text{m}^2\ \text{s}^{-2}) = \text{kg m}^{-1}\ \text{s}^{-2} = (\text{kg m s}^{-2})\ \text{m}^{-2} = \text{N m}^{-2} = \text{N m m}^{-3} = \text{J m}^{-3}$.

Question 3.8:

The vertical gradient of mean wind speed is always positive, whereas the momentum transport $\overline{u'w'}$ is negative. With the inclusion of the minus sign the total term becomes positive.

Question 3.9:

a) $\Delta t_1 = \frac{\Delta x}{c+u} \Leftrightarrow u = \frac{\Delta x}{\Delta t_1} - c$. With $\Delta t_1 = 0.310$ ms, $\Delta x = 0.10$ m, $c = 330\ \text{m s}^{-1}$, we obtain $u = -7.4\ \text{m s}^{-1}$. Minus indicates opposite to the direction of the pulse, that is, downward in this case.

b) $\Delta t_{\mathrm{up}} = \dfrac{\Delta x}{c+u}$ and $\Delta t_{\mathrm{down}} = \dfrac{\Delta x}{c-u} \Leftrightarrow \dfrac{\Delta x}{\Delta t_{\mathrm{up}}} = c+u$ and $\dfrac{\Delta x}{\Delta t_{\mathrm{down}}} = c-u$. Hence $c + u =$ 339.0 m s^{-1} and $c - u = 331.1$ m s^{-1}. This gives $u = 3.9$ m s^{-1}, $c = 335$ m s^{-1}.

c) Assume dry air, so that $q = 0$. Then with $T = 273.15\left(\dfrac{c}{331.3}\right)^2$ this yields $T =$ 279.3 = 6.1 °C.

Question 3.10:

Webb velocity $= \dfrac{\overline{w'T'}}{\overline{T}} = \dfrac{0.1}{300} = 0.33 \text{ mm s}^{-1}$.

Question 3.11:

a) Conservation equation for water vapour: $\dfrac{\partial \overline{q}}{\partial t} = -\dfrac{\partial \overline{w'q'}}{\partial z}$. Integration from the surface to height z gives: $\overline{w'q'}_z - \overline{w'q'}_0 = -\int_0^z \dfrac{\partial \overline{q}}{\partial t}\mathrm{d}z$. Note that $\overline{w'q'}_0$ is in fact not a turbulent flux, but the surface flux of water vapour. Using the definition of the latent heat flux, we obtain: $L_v E_z - L_v E_0 = -\int_0^z L_v \overline{\rho} \dfrac{\partial \overline{q}}{\partial t}\mathrm{d}z \approx -z\overline{\left[L_v \overline{\rho} \dfrac{\partial \overline{q}}{\partial t}\right]}^z$, where the upper averaging indicates the mean over the layer from the surface to height z.

b) 0.25 g kg^{-1} per hour amounts to $6.94 \cdot 10^{-5}$ g kg^{-1} s^{-1} or $6.94 \cdot 10^{-8}$ kg kg^{-1} s^{-1}. Assume an air density of 1.2 kg m^{-3} and take for L_v $2.5 \cdot 10^6$ J kg^{-1}, the flux difference is 4.2 W m^{-2} in the lowest 20 m.

c) Relative difference is 4.2/150 is about 3%.

Question 3.12:

Definition of the Obukhov length: $L \equiv \dfrac{\overline{\theta}_v}{\kappa g} \dfrac{\rho c_p}{H} u_*^3$. Then the units are:

$$\frac{\mathrm{K}}{\mathrm{m\,s^{-2}}} \frac{\mathrm{kg\,m^{-3}\;J\,kg^{-1}K^{-1}}}{\mathrm{J\,s^{-1}\,m^{-2}}} \mathrm{m^3\,s^{-3}} = \frac{\mathrm{K}}{\mathrm{m\,s^{-2}}} \frac{\mathrm{m^{-1}\;K^{-1}}}{\mathrm{s^{-1}}} \mathrm{m^3\,s^{-3}} = \mathrm{m}$$

Question 3.13:

a) σ_T has dimension of temperature: only a temperature scale is needed. $\dfrac{\sigma_T}{|\theta_*|}$ (the absolute sign is needed because the standard deviation is always positive).

b) C_T^2 has dimensions of temperature squared times length to the power 2/3. Both a temperature and a length scale are needed: $\frac{C_T^2 z^{2/3}}{\theta_*^2}$

Question 3.14:

a) Turbulent motion moves air up and down. If the temperature *increases* with height, an upward moving parcel of air will carry relatively cool air upward. A downward moving parcel will carry relatively warm air downward. The net effect of these (and many more) turbulent motions is that heat is transported downward, that is, a negative sensible heat flux.
b) The sign of the flux Richardson number depends on the sign of the virtual heat flux (or equivalently, the sensible heat flux). Given that the term $\overline{u'w'}\frac{\partial \overline{u}}{\partial z}$ is negative, a negative sensible heat flux will give a positive Richardson number.
c) The sign of the Obukhov length depends on the sign of the sensible heat flux. With a negative sensible heat flux, the Obukhov length is positive.
d) The situation is stably stratified, which implies that turbulence is hampered by the stratification: vertical motion is suppressed.

Question 3.15:

Use the expressions given in the text and compute $\phi_m\left(\frac{z}{L}\right)$ and $\phi_h\left(\frac{z}{L}\right)$. For the given values of z/L the values for $\phi_h\left(\frac{z}{L}\right)$ are as follows. Businger–Dyer: 0.174, 0.242, 1, 3.5 and 6; Högström: 0.200, 0.277, 1, 4.9 and 4. This gives relative differences of –0.15, –0.14, 0, –0.4 and –0.47. For the given values of z/L the values for $\phi_m\left(\frac{z}{L}\right)$ are as follows. Businger–Dyer: 0.417, 0.492, 1, 3.5 and 6; Högström: 0.399, 0.471, 1, 3.4 and 5.8. This gives relative differences of –0.044, –0.043, 0, 0.029 and 0.033. Thus, the relationships for heat are very different, whereas those for momentum differ only little.

Question 3.16:

Take the wind speed gradient as an example. Under neutral conditions the only other relevant variables are the velocity scale u_* (representative of the surface friction) and the length scale z (buoyancy does not play a role and hence the Obukhov length is not relevant). As there are only two basic dimensions (length and time) and three variables, only one dimensionless group can be made (see Appendix C): $\frac{\partial \overline{u}}{\partial z}\frac{z}{u_*} = \text{constant}$. With $u_*^2 = \tau/\rho$ and the constant being equal to $\frac{1}{\kappa}$, one obtains $\tau = \rho\kappa u_* z\frac{\partial \overline{u}}{\partial z}$.

Question 3.17:

a) $r_a = \frac{\ln(8/2.5)}{0.4 \cdot 0.3} = 9.7 \text{ s m}^{-1}$

b) Resistances in series: the total resistance is the sum of resistances:

$$r_{\text{a,total}} = r_{\text{a,1}} + r_{\text{a,2}} = \frac{1}{\kappa u_*}\left(\ln\left(\frac{z_2}{z_1}\right) + \ln\left(\frac{z_3}{z_2}\right)\right) = \frac{1}{\kappa u_*}\left(\ln\left(\frac{z_2}{z_1}\cdot\frac{z_3}{z_2}\right)\right) = \frac{1}{\kappa u_*}\ln\left(\frac{z_3}{z_1}\right)$$

Question 3.18:

Latent heat flux: $L_v E = -\rho L_v \frac{\overline{q}(z_2) - \overline{q}(z_1)}{r_{\text{ae}}}$

Scalar flux: $F_x = -\rho \frac{\overline{q}_x(z_2) - \overline{q}_x(z_1)}{r_{\text{ax}}}$

Question 3.19:

a) Rewrite the log-wind profile: $u_* = \kappa \frac{\overline{u}(z_2) - \overline{u}(z_1)}{ln\left(\frac{z_2}{z_1}\right)}$. With $\overline{u}(z_1) = 2.0 \text{ m s}^{-1}$, $\overline{u}(z_2) = 2.5 \text{ ms}^{-1}$, $z_1 = 2$ m and $z_2 = 4$ m we obtain $u_* = 0.289$ m s^{-1}.

b) Rewrite the logarithmic wind profile as: $\overline{u}(z_2) = \overline{u}(z_1) + \frac{u_*}{\kappa}\ln\left(\frac{z_2}{z_1}\right)$. With $\overline{u}(z_1) = 2.0 \text{ m s}^{-1}$, z_1 = 2 m and z_2 = 10 m and u_* = 0.289 m s^{-1} we obtain $\overline{u}(z_2) = 3.16 \text{ m s}^{-1}$.

Question 3.20:

Below the steps to determine the production terms are given for question a) and b).

Quant.	$\overline{w'\theta'}$	β	$\overline{w'\theta_v'}$	$\overline{u'w'}$	z/L	ϕ_m	$\frac{\partial \overline{u}}{\partial z}$	Buoy. prod.	Shear prod.	Ratio
Unit	K m^{-1}	—	K m^{-1}	m^2 s^{-2}	—	—	s^{-1}	m^2 s^{-3}	m^2 s^{-3}	—
a	0.110	0.56	0.137	–0.09	–0.673	0.540	0.040	0.00454	0.00364	1.25
b	–0.043	–10	–0.043	–0.04	0.71	4.56	0.228	–0.0014	0.0091	–0.16

Question 3.21:

a) 0.30 m s^{-1} for both heights
b) 0.37 m s^{-1} for the interval 2–4 m, 0.33 m s^{-1} for the interval 4–6 m
c) The relative error in $(z - d)$ is smaller if z is larger.

Question 3.22:

a) As u_* occurs in the denominator of the temperature scale θ_*, which in turn is used to divide the temperature gradient, the dimensionless gradient (ϕ_h) is proportional to u_*. In the stability parameter z/L u_* occurs to the third power in the denominator.

Hence, $(z/L)^{-1/3}$ is proportional u_*. Because both sides are proportional to u_* the actual value of u_* is irrelevant.

b) The dimensionless gradient ϕ_h is proportional to height z. The same holds for z/L. Because both sides are proportional to z the actual value of z is irrelevant.

Question 3.23:

a) The recipe for the iteration is given in Section 3.6.1. Each new step in the iteration is a new line in the spreadsheet. Make columns for u_*, θ_*. From those determine L, z/L for both levels and Ψ_m and Ψ_h for both levels. And from those, in combination with the vertical temperature and wind speed differences determine new estimates for u_*, θ_* (final two columns). Those values are the starting point for the next iteration step on the next line. It may be helpful to introduce two extra columns for the help variable x in the integrated flux gradient relationships for both levels. The spreadsheet will look like shown below. The resulting values for $u_* = 0.42$ m s^{-1}, $\theta_* = -0.55$ K.

		z_1	z_2	Δ	Average							
z	(m)	2	10	8								
T	(K)	281.52	280.41	–1.11	280.965							
u	(m s^{-1})	1.7	2.9	1.2								
First estimate											Next estimate	
u_* est	T^* est	L	$z\,1/L$	$z\,2/L$	x_1	x_2	Ψ_{m1}	Ψ_{m2}	Ψ_{h1}	Ψ_{m2}	u_*	T_*
0.2982	–0.2759	–23.0860	–0.0866	–0.4332	1.2429	1.6781	0.2544	0.7340	0.4817	1.2922	0.4249	–0.5557
0.4249	–0.5557	–23.2594	–0.0860	–0.4299	1.2415	1.6754	0.2529	0.7310	0.4791	1.2874	0.4243	–0.5542
0.4243	–0.5542	–23.2587	–0.0860	–0.4299	1.2415	1.6754	0.2529	0.7310	0.4791	1.2874	0.4243	–0.5542
0.4243	–0.5542	–23.2587	–0.0860	–0.4299	1.2415	1.6754	0.2529	0.7310	0.4791	1.2874	0.4243	–0.5542

b) From the definition: $H = -\rho c_p u_* \theta_* = 274$ W m^{-2}.

Question 3.24:

Start with a logarithmic wind profile with the surface as the lowest level: $\overline{u}(z_u) = \frac{u_*}{\kappa}\ln\left(\frac{z_u}{z_0}\right)$ Rewrite to obtain an expression z_0: $z_0 = z_u \exp\left(-\frac{\kappa\overline{u}(z_u)}{u_*}\right)$. With the speed at 4 m of 2.5 m s^{-1}, and the friction velocity computed in Question 3.19 ($u_* =$ 0.289 m s^{-1}) we obtain a value for z_0 of 0.125 m. The direct way (without computing the friction velocity) is: $(\overline{u}_2 - \overline{u}_1)/(\overline{u}_2 - 0) = \ln\left(\frac{z_2}{z_1}\right)/\ln\left(\frac{z_2}{z_0}\right)$. This yields $\ln\left(\frac{z_2}{z_0}\right) = 3.74$, giving $z_0 = 0.125$ m as well.

Question 3.25:

a) $\dfrac{\kappa\left(\overline{u}(z_2) - \overline{u}(z_1)\right)}{\ln\left(\frac{z_2}{z_1}\right)} = u_* = \dfrac{0.4(10-7)}{\ln\left(\frac{10}{2}\right)} = 0.746$ m s^{-1}. To compute the roughness length, replace the velocity at 2 m by 0 m s^{-1} and 2 m by z_0. Then solve for z_0:

$$\bar{u}(z_2) - 0 = \frac{u_*}{\kappa}\ln\left(\frac{z_2}{z_0}\right) \Leftrightarrow \ln\left(\frac{z_2}{z_0}\right) = \kappa\frac{\bar{u}(z_2)}{u_*} \Leftrightarrow z_0 = z_2\exp\left(-\kappa\frac{\bar{u}(z_2)}{u_*}\right) =$$

0.047 m.

b) $r_a = \ln\left(\frac{z_2}{z_1}\right)/(\kappa u_*)$. In this case the lower level is at $z = z_0$. Hence

$$r_a = \ln\left(\frac{z_2}{z_0}\right)/(\kappa u_*) = r_a = \ln\left(\frac{2}{0.047}\right)/(0.4\cdot 0.746) = 12.6\ \mathrm{s\,m^{-1}}$$

c) The temperature decreases with height: the conditions are unstable.

d) The roughness length will not change: it is a property of the surface, not of the flow. The unstable stratification will enhance turbulence and subsequently turbulent transport as well. Because u_* reflects the turbulent transport of momentum it will increase (see Eq. (3.29)). Owing to the enhanced transport, the resistance to transport will diminish (at a given vertical difference, more transport is possible, (see Eq. (3.31)).

Question 3.26:

The expression for the temperature profile (without stability correction) is: $\bar{\theta}(z_\theta) - \theta_s = \frac{\theta_*}{\kappa}\ln\left(\frac{z_\theta}{z_{0h}}\right)$. From the sensible heat flux and friction velocity the temperature scale can be determined: $\theta_* = -\frac{H}{\rho c_p u_*}$ with the assumption ρ=1.2 kg m^{-3} and c_p = 1004 J kg^{-1} K^{-1} gives θ_*=–0.415 K. With the expression for θ_s $\left(\theta_s = \bar{\theta}(z_\theta) - \frac{\theta_*}{\kappa}\ln\left(\frac{z_\theta}{z_{0h}}\right)\right)$ and different values for z_{0h} (0.005 m, 0.0005 m and 0.00005 m), the surface temperatures are 26.2, 28.6 and 31.0 °C, respectively.

Question 3.27:

(see also question 3.23)

1. Initial conditions for u_*, θ_* based on observed vertical wind speed difference and temperature differences: $u_* = \kappa\frac{\bar{u}(z_u)}{\ln\left(\frac{z_u}{z_0}\right)}$ and $\theta_* = \kappa\frac{\bar{\theta}(z_t) - \theta_s}{\ln\left(\frac{z_t}{z_{0h}}\right)}$
2. Compute z/L.
3. Compute $\Psi_m(z_0/L)$, $\Psi_m(z_u/L)$, $\Psi_h(z_t/L)$ and $\Psi_h(z_{0h}/L)$
4. Compute new values for u_* and θ_* from $\bar{u}(z_u)$, $\bar{\theta}(z_t) - \theta_s$ and the Ψ-functions.
5. Repeat steps 2 through 4, as long as computed values of u_* and θ_* change significantly from one iteration to the next.
6. If we left the loop, compute the momentum flux and sensible heat flux from u_* and θ_* ($\tau = \rho u_*^2$ and $H = -\rho c_p \theta_* u_*$).

Question 3.28:
The graph should look like this:

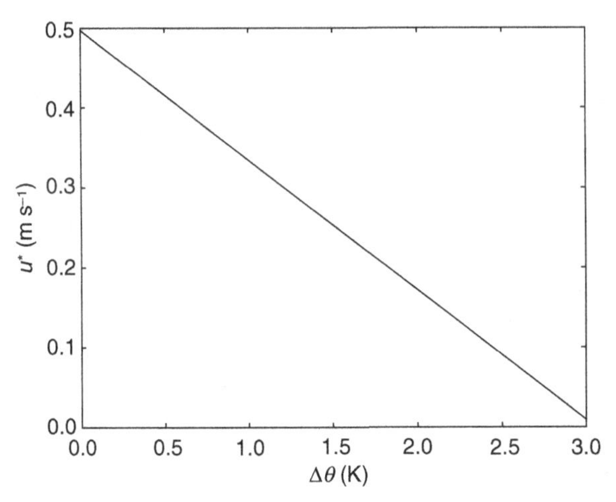

The linear dependence of u_* on $\Delta\theta$ could have been anticipated beforehand, since the bulk Richardson number depends linearly on the temperature difference.

Chapter 4

Question 4.1:
Groundwater recharge = 100 mm y^{-1} ± 100 mm y^{-1}. Thus the relative error amounts as much as 100%!

Question 4.2:
After 1 day:

Reduction factor $\beta = \dfrac{0.12-0.05}{0.15-0.05} = 0.7$;

drainage flux $D = 10\left(\dfrac{0.12-0.05}{0.30-0.05}\right)^2 = 0.784 \text{ cm d}^{-1}$;

water content $\theta = 0.12 + \dfrac{0-0.7 \times 0.5-0.784}{30} = 0.0822 \text{ cm}^3 \text{ cm}^{-3}$.

After 2 days:

Reduction factor $\beta = \dfrac{0.0822-0.05}{0.15-0.05} = 0.322$;

drainage flux $D = 10\left(\dfrac{0.0822-0.05}{0.30-0.05}\right)^2 = 0.166 \text{ cm d}^{-1}$;

water content $\theta = 0.0822 + \dfrac{2.0-0.322 \times 0.5-0.166}{30} = 0.1380 \text{ cm}^3 \text{ cm}^{-3}$.

Question 4.3:

Energy/mass = force × distance/mass = 1 × 10 (acceleration 9,81 ≈ 10 m s^{-2}) × 2 (height)/1 = 20 J kg^{-1}

Energy/volume = 1 × 10 × 2/0.001 (volume 1 kg = 0.001 m^3) = 2 · 10^4 N m^{-2}

Energy/weight = height = 2 m

Question 4.4:

a) The pressure head at the filter is equal to the water column length in the piezometer. Thus, piezometer 1: $h = -80 - (-100) = +20$ cm; piezometer 2: $h = -90 - (-200) = +110$ cm.
b) When we set $z = 0$ at the soil surface, the hydraulic head of piezometer is equal to the water level with respect to the soil surface. Thus, piezometer 1: $H = -80$ cm; piezometer 2: $H = -90$ cm.
c) Flow occurs in direction of lowest H. At $z = -100$ cm, $H = -80$ cm; at $z = -200$ cm, $H = -90$ cm. Therefore flow direction is downward.
d) Search for depth at which $h = 0$. At $z = -100$ cm, $h = 20$ cm; at $z = -200$ cm, $h = 110$ cm. Linear interpolation gives: $h = 0$ at $z = -77.8$ cm.

Question 4.5:

General for a clean glass tube:

$$z_c = \frac{2 \times 0.07 \times 1}{1000 \times 9.81 \times r} = \frac{1.43 \times 10^{-5}}{r} \text{ m}$$

For $r = 1$ mm: $z_c = 0.0143$ m = 14.3 mm

For $r = 0.1$ mm: $z_c = 0.143$ m = 143 mm

Question 4.6:

a. Hydraulic head piezometer: H_1 = level piezometer = $-x_1$ cm;
   Hydraulic head tensiometer: H_2 = height water column = $-x_2$ cm.
b. Water flows in direction of lower hydraulic head. As $H_1 > H_2$, water flows upward.
c. No problem to measure positive water pressures in a tensiometer. The only requirement is that the pressure transducer is able to handle positive pressures.

Question 4.7:

Assume $\pi = 0$; if $h = -16\,000$ cm → relative air humidity $p/p_0 = 0.988$;

If relative air humidity $p/p_0 = 0.80$ → soil water pressure $h = -2.97 \times 10^5$ cm

Question 4.8:

a) Amount water that can be extracted = 400 × (0.34 – 0.12) = 88 mm;
b) According to water balance: Depth d × (0.34 – 0.12) = 1.0 cm (rainfall amount); → d = 4.5 cm;
c) Amount = 300 (depth in mm) × (0.34 – 0.12) = 66 mm = 660 m^3 ha^{-1} (as 1 mm = 10 m^3 ha^{-1});
d) Amount = 300 × (0.45 – 0.34) = 33 mm = 330 m^3 ha^{-1}.

Question 4.9:
Assume $z = 0$ at bottom of soil column.
At top of soil column $H = h + z = 10$ (height water layer) + 50 = 60 cm.
At bottom of soil column $H = h + z = 0$ (equal to atmosphere) + 0 = 0 cm.

Apply Darcy: $q = -k_s \dfrac{\Delta H}{\Delta x} = -100 \dfrac{60-0}{50} = -120 \text{ cm d}^{-1}$; negative sign denotes downward flow.

Question 4.10:

a) Effective k from Eq. 4.15 or sum of resistances: $\dfrac{100}{k_{\text{eff}}} = \dfrac{75}{25} + \dfrac{25}{5} \rightarrow k_{\text{eff}} =$ 12.5 cm d^{-1}

b) Apply Darcy's law to the entire column: Flux density
$$q = 12.5 \frac{100+10-0}{100} = 13.75 \text{ cm d}^{-1}$$

c) The soil water pressure head at the interface of loam and sand can be calculated by applying Darcy's law either to the sand layer or to the loam layer, using the known flux density q. For instance with respect to the loam layer: $13.75 = 5 \dfrac{h+25-0}{25} \rightarrow h = 43.75 \text{ cm}$.

d) With this information we can derive the hydraulic potential diagram as depicted in the accompanying figure. Note that in the sand layer the soil water pressure head increases in the direction of flow!

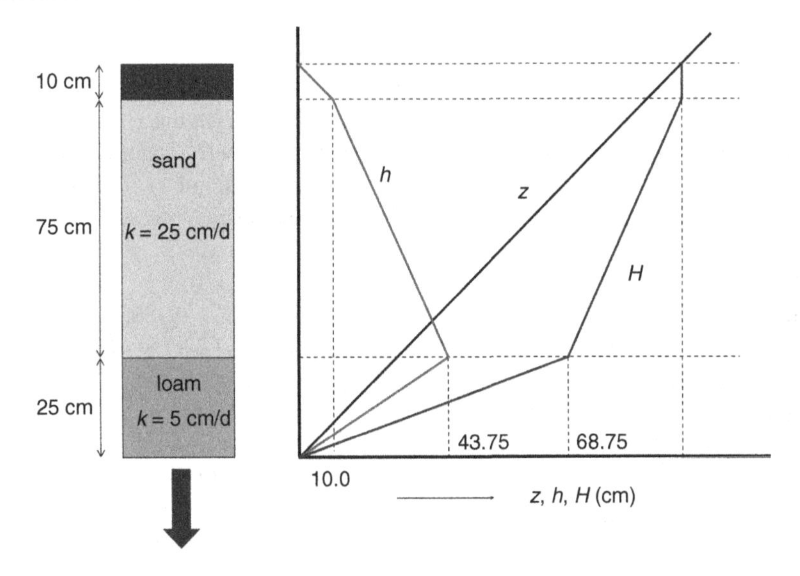

Question 4.11:

a) Use Eq. (4.26): 25 = 8 + (70 − 8) exp (−1.5 t) → exp (−1.5 t) = 0.274 → t = 0.86 hours

b) Until this time all the rain water has infiltrated. Thus $I_{cum} = 0.86 \times 25 = 21.6$ mm.
c) Infiltration between $t = 0.86$ hour and $t = 2.00$ hours is equal to the difference of cumulative infiltration (Eq. (4.27)) between these times. Thus:

$$I_{cum}(2.0) - I_{cum}(0.86) = 8 \times 2 + (70-8)/1.50 \times (1-\exp(-1.5\times 2)) \\ -8\times 0.86 + (70-8)/1.50 \times (1-\exp(-1.5\times 0.86)) = 18.44 \text{ mm}$$

d) Runoff = rainfall – infiltration = 2 × 25 – 21.6 – 18.44 = 9.96 mm

Question 4.12:
Yes, all parameters in Eq. (4.36) are assumed to be constant during the infiltration process.

Question 4.13:

a) $$\frac{s_{f,1}}{s_{f,2}} = \frac{\left(\dfrac{-2k_t h_f}{\theta_t - \theta_1}\right)^{½} t^{½}}{\left(\dfrac{-2k_t h_f}{\theta_t - \theta_2}\right)^{½} t^{½}} = \left(\frac{\theta_t - \theta_2}{\theta_t - \theta_1}\right)^{½}$$

b) Use Eq. 4.35: $$\frac{I_{cum,1}}{I_{cum,2}} = \left(\frac{\theta_t - \theta_1}{\theta_t - \theta_2}\right)^{½}$$

c) $$\left(\frac{-2k_t h_f}{\theta_t - \theta_1}\right)^{½} t^{½} = \left(\frac{-2k_t h_f}{\theta_t - \theta_2}\right)^{½} t^{½} \rightarrow \frac{t_1}{t_2} = \frac{\theta_t - \theta_1}{\theta_t - \theta_2}$$

d) Use Eq. (4.36): $$\frac{S_1}{S_2} = \left(\frac{\theta_t - \theta_1}{\theta_t - \theta_2}\right)^{½}$$

e) Combine Eqs. (4.30) and (4.34): $$\frac{I_1}{I_2} = \frac{\theta_t - \theta_1}{\theta_t - \theta_2}\left(\frac{\theta_t - \theta_2}{\theta_t - \theta_1}\right)^{½} = \left(\frac{\theta_t - \theta_1}{\theta_t - \theta_2}\right)^{½}$$

Question 4.14:

a) $$s_f = \left(\frac{-2\times 1.38 \times -40}{0.40}\right)^{½}\left(\frac{1}{48}\right)^{½} = 2.396 \text{ cm};$$

$$S = \left[-2\times 1.38\times -40 \times (0.5-0.1)\right]^{½} = 6.64 \text{ cm/d}^{½};$$

$$I_{cum} = 6.64\left(\frac{1}{48}\right)^{½} = 0.9548 \text{ cm};$$

$$I = \frac{\partial I_{cum}}{\partial t} = ½ S^{-½} = ½ \times 6.64 \times \left(\frac{1}{48}\right)^{-½} = 23.0 \text{ cm}$$

b) $s_f = 3.389$ cm; $I = 16.3$ cm d^{-1}; $I_{cum} = 1.355$ cm

Question 4.15:

a) $s_f = 3.389 \rightarrow t = 56.7$ min; $s_f = 3.43 \rightarrow t = 58.11$ min; extrapolation $\rightarrow s_f = 3.486$

b) $I = 17.21$ cm d^{-1}; $I_{cum} = 1.394$ cm

c) Initially at infiltration in dry soil the influence of gravity is small; therefore vertical infiltration gives only slightly higher values than horizontal infiltration (17.21 vs. 16.3 cm d^{-1})

Question 4.16:

The capillary flux as function of distance to groundwater level is depicted in the figure below.

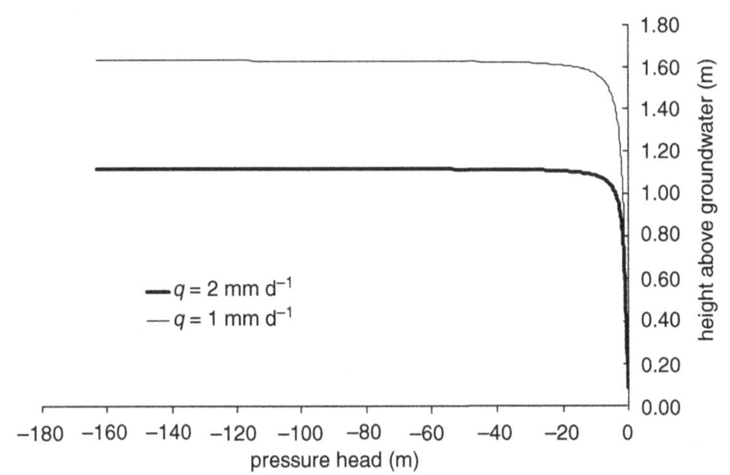

Question 4.17:

a) The hydraulic head is equal to the measured water levels. So in the case of the shallow tube $H = -80$ cm; in the case of the deep tube $H = -55$ cm.

b) The pressure head at the filter is equal to the water column inside the tube. So in case of the shallow tube $h = 170 - 80 = 90$ cm, in case of the deep tube $h = 350 - 55 = 295$ cm.

c) Apply Darcy's law: $q = 1 \times \dfrac{80-55}{50} = 0.50 \text{ cm d}^{-1}$

Question 4.18:

a)

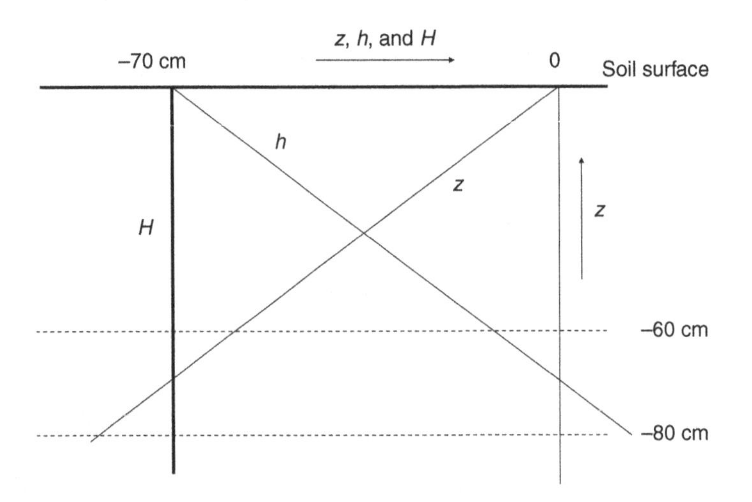

b) Hydrostatic equilibrium, thus H is constant = –70 cm. This gives $h_{gauge,2}$ = –70 + 20 = –90 cm.

Question 4.19:

a)

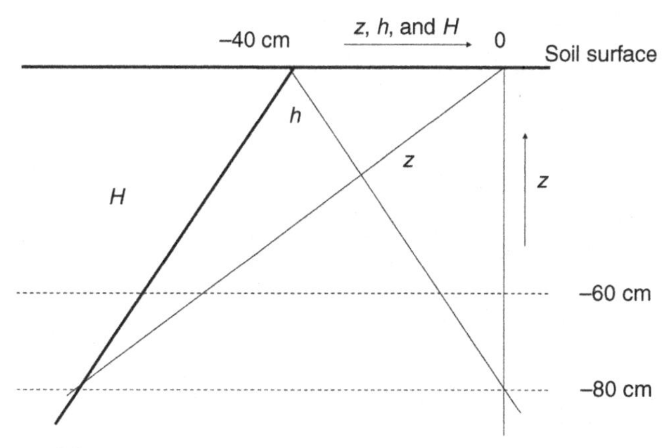

b) $h = 0 \rightarrow z = -80$ cm

c) If gauges are at the same height above soil surface: $\Delta H = h_{gauge,1} - h_{gauge,2}$

Question 4.20:

a) $H_1 = -12.6 \times 7.5 + 10.0 = -84.5$ cm; $h_1 = -84.5 + 40.0 = -44.5$ cm; $H_2 = -103.4$ cm; $h_2 = -23.4$ cm

b) Linear extrapolation with depth until $h = 0$: $\phi_{gwl} = -124.36$ cm

c) Mercury levels will be at the same height.

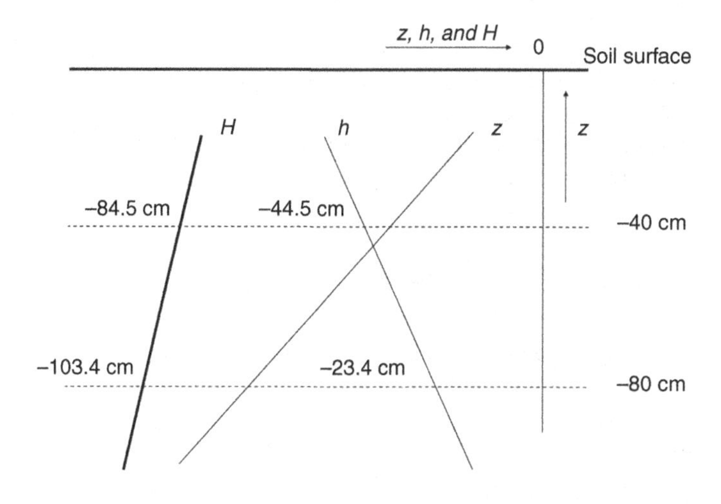

Question 4.21:

In case of capillary pores, Eq. (4.7) expresses the pressure difference Δh (m) between the air and the liquid sides of the water–air interface:

$$\Delta h = \frac{2\sigma \cos \varphi}{\rho g r}$$

where σ is the water surface tension (= 0.07 N m^{-1}), ρ_w is the water density (1000 kg m^{-3}), g is the gravitational field strength (9.81 m s^{-2}) and r is the radius of the tube (m).

$$\text{If } \Delta h = 9.0\,\text{m} : r = \frac{2\sigma \cos\varphi}{\rho_w g \Delta h} = \frac{2 \times 0.07 \times 1}{1000 \times 9.81 \times 9} = 1.58\ \mu\text{m}$$

Question 4.22:

$$\theta = \frac{V_w}{V_{total}} = \frac{\dfrac{M_w}{\rho_w}}{\dfrac{M_s}{\rho_d}} = \frac{\rho_d}{\rho_w(=1\,\text{g cm}^{-3})}\frac{M_w}{M_s} = \rho_d\, w$$

Question 4.23:

a) $w = 0.16$ g g^{-1}

b) $\theta = 0.20$ cm^3 cm^{-3}; $\rho_d = 1.25$ g cm^{-3}

Question 4.24:

Apply Eq. (4.47): in water 6×10^{-9} s; in air 0.67×10^{-9} s

Question 4.25:

a) Influx = outflux → steady state; $h_1 = -12$ cm and $h_2 = -12$ cm → unit hydraulic gradient

b) Apply Darcy: $k = \frac{4}{4} \times \frac{100}{20} = 5\ \text{cm d}^{-1}$; this unsaturated conductivity occurs at $h = -12$ cm

c) When the level of the soil column is increased, the soil water pressure head of the soil will decrease (due to the increasing gravitational head and almost constant hydraulic head). In the soil column the water content will decrease and the hydraulic resistance increase. Therefore the water flux in the system will decrease.

Chapter 5

Question 5.1:

High population density → high risks of public health

Large chemical industry → many sources of contamination during transport and chemical processing

Intensive agriculture → large diffuse sources of nutrients and pesticides

Sedimented soils → large flow domains of groundwater flow

Shallow groundwater levels → small unsaturated zone, therefore less solute decomposition and uptake

Large rainfall surplus → large groundwater fluxes

Question 5.2:

a) Use concepts for variation of soil hydraulic functions, for macropore flow or for unstable wetting fronts.
b) Determine the variation of all physical parameters (including the correlations) in a field.
c) Determine effective transport parameters by calibration, using field scale measurements.
d) Derive the variation of transport parameters from stochastic parameter distributions of comparable fields (so-called Monte–Carlo simulations).
e) Apply transfer functions which relate solutes fluxes entering and leaving a soil profile.

Question 5.3:

Calculate the dispersion coefficient: D_{dis} = 4.0 cm^2 d^{-1}. The diffusion flux equals 0.156/4.00*100% = 3.9% of the dispersion flux.

Question 5.4:

The surface below the solute profiles, multiplied with the volumetric water content, is equal to the total amount of dissolved solutes. Without adsorption and decomposition, the total amount of dissolved solutes keeps constant.

Question 5.5:

Each term in the convection–dispersion equation has the units kg m^{-3} d^{-1}.

Question 5.6:

$t = 30$ d, $C_l = 0.11$ mg L^{-1}; $t = 40$ d, $C_l = 0.29$ mg L^{-1}; $t = 50$ d, $C_l = 0.49$ mg L^{-1};
$t = 60$ d, $C_l = 0.66$ mg L^{-1}; $t = 80$ d, $C_l = 0.86$ mg L^{-1}

Question 5.7:

Veenkampen: $T_{res} = \dfrac{\text{volume}}{\text{flux}} = \dfrac{\theta L}{q} = 1.28$ years

Otterlo: T_{res} = 11.2 years

Question 5.8:

Input: $L = 100$, $v = 2$ cm d^{-1}, $D = 2$ cm^2 d^{-1}, $C_0 = 1000$ μg d L^{-1}
At $t = 40$ d: $C = 22.59$ μg L^{-1}
At $t = 50$ d: $C = 56.42$ μg L^{-1}
At $t = 60$ d: $C = 18.65$ μg L^{-1}

Question 5.9:

Residence time of mobile nitrate amounts to 6 years.

Retardation factor of adsorbing pesticide $R = 1 + \dfrac{\rho_b S_d}{\theta} = 1 + \dfrac{1.5 \times 2}{0.15} = 21.0$

Residence time adsorbing pesticide $T_{res,R} = 21.0 \times 6.0 = 126$ years

Question 5.10:

At $t = T_{50}$: $M(t) = M_0 e^{-\mu T_{50}} = 0.5 M_0$

Therefore: $e^{-\mu T_{50}} = 0.5 \quad \rightarrow \quad T_{50} = (\ln 2)/\mu$

Question 5.11:

The decomposition rate $\mu = \ln(2)/T_{50} = 0.01386\ \mathrm{d}^{-1}$
In case of piston flow: $T_{res} = 50$ d → pulse leached = $C_0 e^{-0.01386 \times 50} = 0.5 C_0$
When we include adsorption: $T_{res,R} = 200$ d → pulse leached =
$C_0 e^{-0.01386 \times 200} = 0.0625 C_0$

Question 5.12:

a) Steady state: $q(z)\ C_l$ = constant = 6.0 × 0.4 = 2.4; $q\ (0.5\ D_r)$ = 6.0 – 5.4/2 = 3.3 mm d^{-1}; thus C_l = 2.4/3.3 = 0.727 mg cm^{-3}
b) At bottom root zone q = 6.0 – 5.4 = 0.6 mm d^{-1}; → C_l = 2.4/0.6 = 4.00 mg cm^{-3}
c) In case of a triangular distribution, 75% of the rootwater uptake occurs in the upper half of the root zone. Application of Eq. (5.23) gives:

$$C_l(z) = \frac{q_0 C_0}{q(z)} = \frac{6.0 \times 0.4}{(6.0 - 0.75 \times 5.4)} = 1.231 \text{ mg cm}^{-3}$$

d) C_l = 2.4/0.6 = 4.00 mg cm^{-3}

Question 5.13:

a) Leaching fraction $L_f = 0.10$;

$$\bar{C} = C_0 \frac{1}{1 - L_f} \ln\left(\frac{1}{L_f}\right) = 0.4 \frac{1}{0.9} \ln\left(\frac{1}{0.1}\right) = 1.023 \text{ mg cm}^{-3}$$

b) The average salinity is independent of the root density profile: $\bar{C}$ = 1.023 mg cm^{-3}

Question 5.14:

The main reasons are the higher amount of soil water percolation in winter time and the lower soil temperatures (less decomposition) in winter time.

Question 5.15:

Similar to Eq. (5.35) we may derive for this case:

$$C_{out} = \frac{\ell_1}{\ell_2} C_{in} + \frac{\ell_2 - \ell_1}{\ell_2} C_{orig} = \frac{\ell_3}{\ell_4} C_{in} + \frac{\ell_4 - \ell_3}{\ell_4} C_{orig} = \frac{\frac{1}{2}L - x}{\frac{1}{2}L} C_{in} + \frac{x}{\frac{1}{2}L} C_{orig}$$

If we combine the above equation with Eq. (5.33) we can derive Eq. (5.37).

Question 5.16:

a) Mean residence time: $T_{res} = \dfrac{\theta L}{q} = \dfrac{0.2 \times 3}{0.3} = 2$ years

b) Apply Eq. (5.37): $C_{out} = 10 + (50 - 10) \exp\left(\dfrac{-0.3 \times 5}{0.25 \times 5}\right) = 22.0 \text{ mg L}^{-1}$

c) Use Eq. (5.37) inverse: $20 = 10 + (50 - 10)\ \exp\left(\frac{-0.3 \times t}{0.25 \times 5}\right) \rightarrow t = 5.78$ years

Question 5.17:

Application only during summer time → less percolation of soil water, more decomposition of pesticide due to higher soil temperatures

Increased drain depth → larger residence time in the soil, thus more decomposition

Increased ploughing depth → ploughing results in a homogeneous, well aerated soil layer without macropores, which is favourable for pesticide decomposition

Chapter 6

Question 6.1:

The parameter $M_{0,z}$ refers to atmosphere, as the pressure head at the root–soil interface is determined by the atmospheric demand. The parameters $r_{0,z}$ and $r_{m,z}$ which denote the root diameter and root radial influence as function of soil depth are plant specific. The soil hydraulic functions determine the matric flux potential in the soil matrix $M_{a,z}$.

Question 6.2:

In case of triangular root distribution: $T_p = \int_{-D_{root}}^{0} S_p\, \partial z = 0.5 \times S_p(z=0) \times D_{root}$

With values this gives: $0.8 = 0.5 \times S_p(z=0) \times 80 \rightarrow S_p(z=0) = 0.02\ \text{d}^{-1}$

Derive potential root water extraction rate at depth = 30 cm by interpolation:

depth = 0 cm → $S_p = 0.02\ \text{d}^{-1}$

depth = 80 cm → $S_p = 0.0\ \text{d}^{-1}$

depth = 30 cm → $S_p = \frac{S_p(z=0)}{80} \times (80 - 30) = 0.0125\ \text{d}^{-1}$

Question 6.3:

In case of irrigated fields, in the top soil layer soil water extraction rates are relatively high due to a high root density and soil evaporation. Also capillary rise hardly reaches this layer. Therefore water stress commonly occurs in the top soil.

The highest salinity concentrations and therefore the highest salt stress we may expect at the bottom of the root layer. This is due to the fact that solute concentrations increase when water is extracted by evaporation and plant roots.

Question 6.4:

$$\frac{\partial H_p}{\partial z} \approx 1.1 \frac{8 \eta\, v}{r^2} = 1.1 \frac{8 \times 0.001 \times 0.002 \times 10^{-3}}{(50 \times 10^{-6})^2} = 7.04\ \text{kPa m}^{-1}$$

Head loss due to gravity = 10.0 kPa m^{-1} = 0.100 bar m^{-1};

Total head loss of 20 m high tree = 20 × (0.0704 + 0.100) = 3.408 bar.

Question 6.5:

a) The expressions for the fluxes are given in the text. Do not forget to convert the specific concentrations to units of kg/kg. $E = 4.71 \cdot 10^{-5}$ kg m^{-2} s^{-1} and $A_n = 8.35 \cdot 10^{-7}$ kg m^{-2} s^{-1}.

b) WUE = A_n/E = 0.018 kg CO_2/kg H_2O.

c) The lower internal CO_2 concentration for C_4 plants implies that for a given external concentration and resistance the CO_2 uptake is higher. Because the internal CO_2 concentration does not influence the transpiration (which depends only on the resistance and leaf temperature), the WUE will be higher. In formulas: $\mathrm{WUE} = \frac{A_n}{E} = -\frac{1}{1.6}\frac{q_{\mathrm{ce}} - q_{\mathrm{ci}}}{q_{\mathrm{e}} - q_{\mathrm{i}}}$: a lower q_{ci} will increase the numerator of this ratio.

d) The air inside the leaf is saturated; hence an increase of the leaf temperature will lead to an increase in the water vapour content in the substomatal cavity. Owing to the increase of the difference in water vapour concentration inside and outside the leaf, the transpiration will increase. The CO_2 uptake is not affected (directly) by the leaf temperature. Hence the WUE will decrease.

Question 6.6:

a) E_0 can be determined by rewriting Eq. (6.33):

$E_0 = \ln\left(R_{\mathrm{s}} / R_{\mathrm{s,ref}}\right) / \left(\frac{1}{T_{\mathrm{ref}} - T_0} - \frac{1}{T - T_0}\right)$ (note: temperatures in Kelvin!). This expression is defined only for data where T is not equal to T_{ref}. Using the first night as the reference gives E_0 = 129 K.

b) For 17 °C, $R_{\mathrm{s}} = 4.3 \cdot 0^{-7}$ kg m^{-2} s^{-1}

Question 6.7:

Transpiration $T_{\mathrm{p}} = \frac{DM_{\mathrm{a}}\,\Delta e}{10\,k} = \frac{16 \times 1200}{10 \times 4.5} = 427$ mm;

Water productivity $WP_{\mathrm{T}} = \frac{DM_{\mathrm{a}} \times 10^3}{T_{\mathrm{a}} \times 10} = \frac{16 \times 10^3}{427 \times 10} = 3.75$ kg m^{-3};

In case T_{a} = 300 mm, expected yield $DM_{\mathrm{a}} = 10 \times 4.5 \times \frac{300}{1200} = 11.25$ t ha^{-1}.

Question 6.8:

c) $K^{\downarrow}(z) = K^{\downarrow} e^{-aA(z)}$. Halfway the canopy (at z = 0.5 h_{c} with h_{c} being the canopy height) the accumulated leaf area is 1.5 m^2 m^{-2}. Then the radiation at that level is $K^{\downarrow}\left(\frac{1}{2}h_{\mathrm{c}}\right) = 800 e^{-0.5 \cdot 1.5} = 378$ W m^{-2}. Likewise, the value at the bottom of the vegetation is 108 W m^{-2}.

d) Set the radiation for each of the wavelengths to 1 above the canopy. Halfway the canopy the flux densities for 0.5 and 1.0 μm are 0.35 and 0.64 respectively. Hence the ratio flux densities at the wavelengths is 0.55 (i.e., visible light has been depleted more than the near-infrared radiation). At the bottom of the canopy the flux densities are 0.061 and 0.30, respectively. This gives a ratio between of the flux densities at both wavelengths of 0.20.

Question 6.9:

a) $T_{leaf} = T_a + \frac{r_b}{\rho c_p}(Q^*_{leaf} - L_v E_{leaf})$. With the given values, the resulting leaf temperature will be $T_{leaf} = 20 + \frac{40}{1.2 \cdot 1013}\left(400 - L_v 10^{-4}\right) = 25.1\ ^{\circ}\mathrm{C}$ (assumed values for density of air: 1.2 kg m^{-3}, c_p=1013 J kg^{-1} K^{-1} and with L_v at 20 °C equal to 2.45·10^6 J kg^{-1} ; note that strictly speaking L_v should be determined at the leaf temperature).

b) Relative to a), the only change is the transpiration rate E_{leaf}:
$T_{leaf} = 20 + \frac{40}{1.2 \cdot 1013}\left(400 - L_v 0.3 \cdot 10^{-4}\right) = 30.8\ ^{\circ}\mathrm{C}$.

c) Relative to a), the only difference is that r_b has changed to 80 s m^{-1}:
$T_{leaf} = 20 + \frac{80}{1.2 \cdot 1013}\left(400 - L_v 10^{-4}\right) = 30.2\ ^{\circ}\mathrm{C}$.

Question 6.10:

- Sensible heat flux: The source of energy is mainly located at the sunlit top of the tree crowns. Hence only above that level a significant amount of sensible heat is transport upward. Within the canopy there is some net heat transport from the forest floor upward, even against the local temperature gradient. This is possible due to flow structures with a scale of the order of the canopy height: the small heat transport in the canopy is the result of some transport from the forest floor, and transport of cold air from well above the forest downward.
- Latent heat flux: Again the main source of moisture are the leaves in the forest crown, hence the main upward transport takes place above the forest. Within the forest there is some upward transport from the forest floor (and downward transport of dry air from above the forest).
- Carbon dioxide transport: Here the main *sink* of CO_2 is located at the leaves of the trees, hence a large downward transport well above trees and only a limited downward transport within the crown (at that level already part of the CO_2 has been taken up by the leaves).

Question 6.11:

a) Compute how much water vapour is contained in the canopy and convert this to a layer of liquid water. e_{sat}(10 °C) = 12.3 hPa and because RH = 100%, we have e=12.3

hPa. Using the gas law, we determine ρ_v: $\rho_v = \frac{e}{R_v T}$ which gives $\rho_v = 9.4 \cdot 10^{-3}$ kg m^{-3}. The volume of the canopy (per unit area) is 2 m^3 so that $1.9 \cdot 10^{-2}$ kg of water is contained in the canopy air. If this were liquid water (with a density of 10^3 kg m^{-3}) this would correspond to a layer of $1.9 \cdot 10^{-5}$ m = 0.019 mm. Hence it is not possible that the observed dew originates from within the canopy, even in the unlikely event that all water vapour would condense.

b) 0.25 mm of dewfall corresponds to 0.25 kg m^{-2}. The corresponding rate is 0.25/(8·3600) = $8.68 \cdot 10^{-6}$ kg m^{-2} s^{-1}. With $L_v = 2.48 \cdot 10^6$ J kg^{-1} this corresponds to a mean energy flux density of 22 W m^{-2}.

c) 0.25 mm corresponds to 0.25 kg m^{-2} of water molecules. If spread over a layer 100 meter of air this is $2.5 \cdot 10^{-3}$ kg m^{-2}. The initial $\rho_v = 9.4 \cdot 10^{-3}$ kg m^{-3} (see question a). Thus the remaining ρ_v after dewfall is $6.9 \cdot 10^{-3}$ kg m^{-2}. With $\rho = 1.24 \cdot 10^{-3}$ kg m^{-3} this corresponds to a mixing ratio of $5.6 \cdot 10^{-3}$ kg kg^{-1}.

Question 6.12:

Separation of the variables gives: $$\frac{\partial P_i}{(1-r-r_t)P_{mean} - \frac{E_{mean}}{S}P_i} = \partial t$$

For integration we might use the standard integral: $$\int \frac{\partial x}{a+bx} = \frac{1}{b}\ln(a+bx)$$

Integration gives: $$\frac{S}{E_{mean}}\ln\left[(1-r-r_t)P_{mean} - \frac{E_{mean}}{S}P_i\right] + C_1 = t$$

where C_1 is an integration constant. At $t = 0$, $P_i = 0$. Therefore the integration constant equals:

$$C_1 = \frac{-S}{E_{mean}}\ln\left[(1-r-r_t)P_{mean}\right]$$

Substitution yields: $$\frac{S}{E_{mean}}\ln\left[1 - \frac{E_{mean}}{(1-r-r_t)P_{mean}}\frac{P_i}{S}\right] = t$$

At saturation $P_i = S$ and $t = P_s/P_{mean}$. Substitution in the former equation yields Eq. (6.48).

Chapter 7

Question 7.1:

a) If $Q^* - G = 0$, then sensible and latent heat flux cancel as well: $H + L_vE = 0$; thus $H = -L_vE$ and consequently the Bowen ratio equals –1.

b) For $\beta = -1$, the fluxes are undetermined (one divides zero by zero). Any combination with $H = -L_vE$ is possible.

Question 7.2:

a) Observations of temperature and humidity available at two heights: the Bowen ratio can be computed from $\beta = \frac{c_p}{L_v}\frac{\overline{\theta}(z_2)-\overline{\theta}(z_1)}{\overline{q}(z_2)-\overline{q}(z_1)}$. First compute specific humidity from ρ ($=\rho_d+\rho_v$) and ρ_v at both heights ($8.58\cdot10^{-3}$ and $9.02\cdot10^{-3}$ kg kg^{-1}, respectively). The vertical potential temperature difference equals the temperature difference but with a correction for the dry-adiabatic lapse rate over 3 m. This reduces the temperature difference with $3 \cdot g/c_p = 0.03$ K, so that the potential temperature difference becomes 0.27 K. With $c_p = 1013$ J kg^{-1} K^{-1} and $L_v = 2.43\cdot10^6$ J kg^{-1}, this gives $\beta =$ 0.261, $H = 93$ W m^{-2} and $L_vE = 357$ W m^{-2}.

b) Use the mass Bowen ratio: $\beta_c = \frac{F_c}{E} = \frac{\overline{q}_c(z_2)-\overline{q}_c(z_1)}{\overline{q}(z_2)-\overline{q}(z_1)}$, with $q_c = \frac{\rho_c}{\rho}$ one can determine that $\beta_c = -5.44\cdot10^{-3}$. With $E = L_vE/L_v = 1.47\cdot10^{-4}$ kg s^{-1} m^{-2} this gives $-8.0\cdot10^{-7}$ kg s^{-1}m^{-2}.

Question 7.3:

Using the air temperature, aerodynamic resistance and sensible heat flux, the surface temperature can be computed: $H = -\rho c_p \frac{T_a - T_s}{r_a} \Leftrightarrow T_s = T_a + r_a \frac{H}{\rho c_p}$. The real $e_{sat}(T_s)$ can then be computed from the *real* surface temperature. The estimated $e_{sat}(T_s)$ is obtained from the linearization in Eq. (7.11).

	H (W m^{-2})	T_s (°C)	$e_{sat}(T_a)$ (hPa)	$s(T_a)$ (hPa/K)	$e_{sat}(T_s)$ estimated (hPa)	$e_{sat}(T_s)$ real (hPa)
a	0	20.0	23.4	1.45	23.4	23.4
b	100	22.5	23.4	1.45	27.0	27.2
c	300	27.4	23.4	1.45	34.1	36.6

It appears the error increases with increasing sensible heat flux. To see the consequence for the flux calculation, we should consider the effect of the error in $e_{sat}(T_s)$ on the error in $(e_a - e_{sat}(T_s))$. If we assume a relative humidity of 70%, the relative errors in $(e_a - e_{sat}(T_s))$ (due to the linearization) become 0%, 2% and 12%, respectively.

Question 7.4:

a) L_vE equals the radiation term: $L_vE = \frac{s}{s+\gamma}Q^*$

b) L_vE equals the aerodynamic term: $L_vE = \frac{\frac{\rho c_p}{r_a}\left(e_s(\overline{T}_a)-\overline{e}_a\right)}{s+\gamma}$

c) L_vE will be positive due to the aerodynamic term; hence H will be < 0: stable conditions. Hence r_a will be large.

d) L_vE increases: the first term increases (due to temperature dependence of s) and the second term increases: the aerodynamic term can be written as: $\dfrac{\frac{\rho c_p}{r_a} e_{sat}(\overline{T}_a)(1-\mathrm{RH})}{s+\gamma}$ and thus increases with temperature due to the temperature dependence of $e_{sat}(T_a)$ (which 'wins' over the temperature dependence of s (in $(s+\gamma)$).

e) An evaporation higher than the available energy implies:

$$L_vE > Q^*-G \Leftrightarrow \frac{s}{s+\gamma}(Q^*-G) + \frac{\frac{\rho c_p}{r_a}\left(e_{sat}(\overline{T}_a)-\overline{e}_a\right)}{s+\gamma} > Q^*-G$$

$$\Leftrightarrow \frac{s-s-\gamma}{s+\gamma}(Q^*-G) > -\frac{\frac{\rho c_p}{r_a}\left(e_{sat}(\overline{T}_a)-\overline{e}_a\right)}{s+\gamma}$$

$$\Leftrightarrow \gamma(Q^*-G) < \frac{\rho c_p}{r_a} e_{sat}(\overline{T}_a)(1-\mathrm{RH})$$

Thus the following conditions are favourable for high evaporation rates:

- Dry air (low RH)
- High temperatures (i.e., high $e_{sat}(T_a)$)
- Strong wind (low aerodynamic resistance)

A situation in which evaporation exceeds available energy is called the oasis effect.

Question 7.5:

a) Canopy resistance will be lower, as per unit ground area there are more stomata that can act as a pathway for water vapour transport from the plant to the atmosphere.

b) The canopy with multiple layers will have a higher canopy resistance: the first layer has intercepted part of the PAR and hence the second layer will have less PAR available and the stomata will be less open.

Question 7.6:

a) For the latent heat flux we simply fill in all given values in the Penman–Monteith equations. Variables that not have been given can be computed. $e_{sat}(T_a)$ and $s(T_a)$ can be computed based on the temperature (giving 17.1 hPa and 1.10 hPa K^{-1}, respectively). With $L_v = 2.47 \cdot 10^6$ J kg^{-1} and the given values for c_p and p, the psychrometric constant $\gamma = 0.67$ hPa K^{-1}. This gives $L_vE = 127$ W m^{-2}.

b) From the energy balance compute H: $H = Q^* - G - L_vE = 101$ W m^{-2}.

c) Given the sensible heat flux and the aerodynamic resistance, the temperature difference between air and surface can be computed: $H = -\rho c_p \dfrac{T_a - T_s}{r_a} \Leftrightarrow T_s = T_a + r_a \dfrac{H}{\rho c_p}$. This gives $T_s = 19.2$ °C.

Question 7.7:
First line in each cell is qualitative remark, second line a typical value (if appropriate) and the last line an explanation for this particular value or qualification.

	Forest (dry)	Forest (wet)	Grass (dry)	Grass (wet)	Lake
Canopy resistance, r_c	Higher	None	Low	None	None
	70 s m^{-1}	0	40 s m^{-1}	0	0
	Plant type (natural vs. agricultural), shading by higher leaf layers	No stomatal control		No stomatal control	No stomata
Aerodynamic resistance, r_a	Low	Low	Medium	Medium	High
	10 s m^{-1}	10 s m^{-1}	50 s m^{-1}	50 s m^{-1}	100 s m^{-1}
	Due to higher roughness	Due to higher roughness	Medium roughness	Medium roughness	High due to low roughness
Roughness length z_0	High	High	Medium	Medium	Low
	1 m	1 m	0.01–0.1 m	0.01–0.1 m	0.001–0.01 m
	High obstacles (note: displacement height)	High obstacles (note: displacement height)	Low obstacles	Low obstacles	No obstacles, roughness may depend on wind speed
Albedo r	Low	Low	Medium	Medium	Low
	0.1	0.1	0.25	0.25	0.05–0.1
	Absorption of PAR and trapping of radiation	Absorption of PAR and trapping of radiation	Absorption of PAR	Absorption of PAR	Note: may be higher for low solar altitudes
Evaporation of intercepted water		High		Low	
		Due to low aerodynamic resistance liquid water is easily evaporated (r_c = 0!)		Although r_c = 0, there is still a considerable r_a.	

Question 7.8:

a) In both cases the vapour pressure increases (due to surface evaporation), but in the right graph part of the water vapour added at the bottom, is lost at the top owing to exchange with the troposphere above the atmospheric boundary layer; as a result the air does not become saturated.

b) Provided that the water vapour concentration *above* the boundary layer is constant, the loss at the top will increase when traveling to the right. This is due to the fact that the contrast between the dry air above the boundary layer and the moist air inside the boundary layer increases. If the same amount of mass is exchanged (which may not be *exactly* true), the larger contrast will result in a larger exchange.

Question 7.9:
The equilibrium Bowen ratio is given in the text to be γ/s. With s equals 0.61 and 1.89 hPa K^{-1} for temperatures of 5 and 25 °C, the resulting equilibrium Bowen ratios are:

a) 1.1
b) 0.35
c) Provided that the available energy is equal for both cases, case b) will have a higher evaporation as it has a lower Bowen ratio ($L_vE = (Q^* - G)/(1 + \beta)$).

Question 7.10:
Assume that the temperature used in LE_{eq} can be taken the mean temperature in the box (for simplicity we make no distinction between temperature and potential temperature). Furthermore, the temperature and vapour pressure in the box change only due to the surface flux (no exchange at the top). Then the differential equations for temperature and vapour pressure deficit in a box of height h become: $\dfrac{\partial T}{\partial t} = \dfrac{H_{eq}}{\rho c_p h}$,

$\dfrac{\partial q}{\partial t} = \dfrac{L_v E_{eq}}{\rho L_v h} \Rightarrow \dfrac{\partial e}{\partial t} \approx \dfrac{R_v}{R_d} p \dfrac{L_v E_{eq}}{\rho L_v h}$. The differential equation for the saturated vapour pressure can be approximated as $\dfrac{\partial e_{sat}}{\partial t} \approx \dfrac{\partial e_{sat}}{\partial T}\dfrac{\partial T}{\partial t} = s\dfrac{\partial T}{\partial t}$, so that the differential equation for VPD becomes: $\dfrac{\partial VPD}{\partial t} \approx s\dfrac{\partial T}{\partial t} - \dfrac{\partial e}{\partial t}$. Next insert the expression for L_vE_{eq} and H_{eq} (the latter from the energy balance, see the text) and simplify:

$$\frac{\partial VPD}{\partial t} \approx \left[s\frac{Q^*-G}{\rho c_p h}\left(1 - \frac{s}{s+\gamma}\right) - \frac{R_v}{R_d} p \frac{Q^*-G}{\rho L_v h}\frac{s}{s+\gamma}\right] = \frac{Q^*-G}{\rho c_p h}\left[\left(s - \frac{s^2}{s+\gamma}\right) - \gamma\frac{s}{s+\gamma}\right]$$

which can be further simplified to $\dfrac{\partial VPD}{\partial t} \approx \dfrac{Q^*-G}{\rho c_p h}\left[\dfrac{s(s+\gamma)}{s+\gamma} - \dfrac{s^2}{s+\gamma} - \dfrac{s\gamma}{s+\gamma}\right] = 0$. So, under equilibrium conditions the VPD does not change in time.

Question 7.11:
According to the Penman–Monteith equation an increase in the canopy resistance (keeping all other variables the same) always results in a decrease of the evapotranspiration (physically: a higher canopy resistance hampers transport of water vapour from the stomata to the air; mathematically: the denominator of the Penman–Monteith equation increases). Because $\alpha_{PT} \equiv \dfrac{L_vE}{\dfrac{s}{s+\gamma}(Q^*-G)}$, a lower evapotranspiration with all other variables kept constant (such as s, Q^* and G) leads to a lower Priestley–Taylor coefficient.

Question 7.12:
See the text at the end of Section 7.3.3.

Question 7.13:

Method	Assumption	Data	(Dis-)advantage
Penman	Simplified energy balance (only Q^*, G, H and L_vE) is closed; wet surface; linearization to eliminate surface temperature	Q^*, G, T_a, e_a and data for r_a, p	– applicable only to wet surfaces
Penman–Monteith	Simplified energy balance is closed	Q^*, G, T_a, e_a and data for r_a, estimate/model of r_c, p	+ most realistic model of transpiration; + possibility to take into account situations where water is limited
Priestley–Taylor	Simplified energy balance is closed; aerodynamic term in Penman is fixed fraction (0.26) of energy term	Q^*, G, T_a	– only for wet surfaces or well-water surfaces + works well for those conditions + limited amount of data required
Makkink	Simplified energy balance is closed; aerodynamic term in Penman is fixed fraction (0.26) of energy term; soil heat flux negligible (daily averages) and net radiation can be estimated from global radiation	$K^{\downarrow}$, T_a	– only for wet surfaces or well-water surfaces + works well for those conditions + limited amount of data required

Question 7.14:

a) Favourable conditions are:

- Time of day: nighttime when there is no input of short wave radiation: net radiation will be negative, thus cooling the surface so that condensation can occur
- Cloudless conditions so that the downward longwave radiation is at its minimum, allowing for the largest possible radiative cooling
- High water vapour content in the air so that the aerodynamic term in the Penman equation cannot cause too much evaporation; furthermore, high water vapour contents are needed to allow for a continuous supply of water vapour once condensation has started; water vapour content should not be so large that fog occurs as that will cause extra downward longwave radiation, reducing the longwave cooling.
- Wind speed should be low to reduce the influence of the aerodynamic term but should not be zero, as some turbulence is needed to allow for downward transport of water vapour.

b) For dewfall to occur, the radiation term (negative) should be larger than the aerodynamic term (positive). The critical limit is when dew fall is just zero:

$$0 > L_v E = \frac{s(Q^* - G) + \frac{\rho c_p}{r_a}\left[e_{sat}(\bar{T}_a) - \bar{e}_a\right]}{s + \gamma} \Leftrightarrow -\frac{s(Q^* - G)}{s + \gamma} > \frac{\rho c_p}{r_a(s + \gamma)} \mathrm{VPD}$$

$$\Leftrightarrow \mathrm{VPD} < -\frac{s(Q^* - G)}{\rho c_p} r_a$$

Given that for conditions of dew fall $Q^* - G$ is negative, one can see that larger VPD is allowed when the cooling is stronger, or when the aerodynamic resistance is larger (less wind).

c) Auxiliary variables needed for the computation using the Penman equation: $e_{sat}(T_a) = 17.06$ hPa and $s(T_a) = 1.10$ hPa K^{-1}, $\gamma = 0.67$ hPa K^{-1}. This yields a latent heat flux of -16.7 W m^{-2}. To convert this into a dewfall rate in mm h^{-1}, first convert to a mass flux: $E = L_v E / L_v = -6.76 \cdot 10^{-6}$ kg m^{-2} s^{-1}. With a density of liquid water of 1000 kg m^{-3} this amounts to $-6.76 \cdot 10^{-9}$ m^{-3} m^{-2} s^{-1} or to $-6.76 \cdot 10^{-9}$ m s^{-1}. Conversion of meters to millimetres and seconds to hours leads to $E = -0.024$ mm h^{-1}.

Chapter 8

Question 8.1:
Because one is interested in the amount of energy it took to evaporate a certain amount of water, one should use the temperature at which the evaporation occurred. In general, this will be the surface temperature.

Question 8.2:
The only 'real' thing is the evapotranspiration of a given crop E. For a given set of weather data, different methods to calculate the reference evapotranspiration E_{ref} will give different answers. Hence, the crop factors using different definitions of E_{ref} will be different: $K_c \equiv \frac{E}{E_{ref}}$.

Question 8.3:
In principle the input variables for E_{ref} should be determined above similar surfaces as was done when the crop factors were determined. In practice this implies that the conditions should be similar to the reference crop (well watered short grass). But it is not sufficient that the conditions at the observation field are similar to the reference crop: the area covered by well-watered grass (or crops in general) should be so extensive that the local climate is well adapted to the well-watered conditions.

Question 8.4:
a) Bare soil evaporation factor is high when irrigation occurs (wet top soil).
b) The soil is wetted nearly instantaneously, and the top soil dries out again quite quickly (deeper soil layers remain moist).
c) As the fraction of soil covered by vegetation increases, the part of the soil exposed to high levels of energy input (direct sunlight) decreases. Hence the soil evaporation decreases reflected in a decrease K_e.

d) When the crop is left to ripe, senesce and dry out usually no irrigation takes place in late season (but this depends on the type of crop: some are irrigated also in late season towards harvest).

Question 8.5:
In the table below the optimal evapotranspiration of both grass and potatoes is determined (questions a and b).

	Crop	Quantity	Month						
			April	May	June	July	August	September	Sum
a)	Grass	K_c (-)	1.0	1.0	1.0	1.0	1.0	1.0	
		E_{opt} (mm)	54.5	82.9	86.7	91.5	80.2	48.2	444
b)	Potatoes	K_c (-)	0.0	0.8	1.1	1.1	1.1	0.2	
		E_{opt} (mm)	0.0	66.3	98.3	103.7	88.2	11.2	368

In the table below the precipitation excess for the crops under consideration is shown in mm, where for d) an initial storage of 120 mm is assumed.

		Month					
		April	May	June	July	August	September
c)	Grass	–10.0	–31.4	–46.4	–67.9	–89.9	–66.1
	Potatoes	44.5	39.7	13.1	–20.6	–50.6	10.2
d)	Grass	110.0	88.6	73.6	52.1	30.1	53.9
	Potatoes	164.5	159.7	133.1	99.4	69.4	130.2

From c) it can be seen that grass already suffers shortages in April, whereas the potatoes run into problems in July. If the initial storage in the root zone is taken into account, both crops can be grown without problems throughout the season. But note that this is based on climatological mean values (period 1971–2000). Thus in individual years things may be very different.

Question 8.6:
We start with the field water balance is presented in Chapter 4: $P+I-R=E+T+D+\Delta W$. Precipitation P, irrigation I and runoff R can be measured directly. The change in storage ΔW can be monitored using e.g. TDR (note that the change in storage in the entire column is needed, hence change in storage needs to be monitored at a number of depths). Drainage D is either forced to be zero, or monitored using a setup as shown in Figure 8.5. Assuming all other terms to be known, evapotranspiration can be determined as: $E+T=P+I-R-D-\Delta W$.

Question 8.7:

a) For the components air, water and matrix we assume the following densities: 1.2, 1000 and 2660 kg m^{-3}. The volume fractions of air, water and matrix are 10%, 30% and 60%. Finally the total volume of the lysimeter is $3.9\cdot10^{-4}$ m^3. This results in masses for air, water and matrix of $4.7\cdot10^{-5}$ kg, 0.118 kg and 0.627 kg, totalling 0.745 kg.

b) Dewfall of 0.03 mm/h corresponds to 0.0025 mm/5 min. For 1 square metre this is $2.5 \cdot 10^{-6}$ m^3/5 min. With a density of 1000 kg m^{-3}, this gives $2.5 \cdot 10^{-3}$ kg m^{-3}/5 min. The surface area of the lysimeter is $7.85 \cdot 10^{-3}$ m^3 so that the weight increase of the lysimeter is $1.9 \cdot 10^{-5}$ kg/5 min. To measure this amount of dewfall, the resolution should at least be 10^{-5} kg.

c) The relative resolution should be of the order of 10^{-5} (10^{-5} kg on a total weight of roughly 1 kg).

Chapter 9

Question 9.1:

Arithmetic average: $q = \frac{k_1 + k_2}{2} \frac{\Delta(h+z)}{\Delta z} = \frac{9.65 + 0.12}{2} \frac{100 + 10}{10} = 53.7 \text{ cm d}^{-1}$

Geometric average: $q = \sqrt{k_1 + k_2} \frac{\Delta(h+z)}{\Delta z} = \sqrt{9.65 + 0.12} \frac{100 + 10}{10} = 11.8 \text{ cm d}^{-1}$

Harmonic average: $q = \frac{2k_1 k_2}{k_1 + k_2} \frac{\Delta(h+z)}{\Delta z} = \frac{2 \times 9.65 \times 0.12}{9.65 + 0.12} \frac{100 + 10}{10} = 2.6 \text{ cm d}^{-1}$

Question 9.2:

Real water storage difference:

$\Delta W = \left(\theta_i^{j+1} - \theta_i^j\right)\Delta z_i = (0.25660 - 0.26210) \times 5.0 = -0.0275 \text{ cm}$

Approximation: $\Delta W = C_i^j \left(h_i^{j+1} - h_i^j\right)\Delta z_i = 0.00278\,(-52 - (50)) \times 5.0 = -0.0278 \text{ cm}$

Approximation: $\Delta W = C_i^{j+1} \left(h_i^{j+1} - h_i^j\right)\Delta z_i = 0.00272\,(-52 - (50)) \times 5.0 = -0.0272 \text{ cm}$.

Question 9.3:

At macroporous soils or at runoff conditions, daily rainfall rates don't suffice and actual rainfall intensities should be specified. The simulation of diurnal fluctuations of evapotranspiration rates is relevant for climate studies, for accurate root water uptake simulation, and for volatilization of pesticides.

Question 9.4:

Column 1 with free outflow condition will contain more water; at this column after 6 hours at the bottom $h_{\text{bottom}} = 0$. At column 2 with free drainage condition, the soil below applies a suction at the column. Therefore at column 2 at the bottom $h_{\text{bottom}} < 0$. This means that column 1 will contain more water than column 2.

Question 9.5:

In the saturated zone apply the principle of superposition of drainage fluxes

a) $q_{\text{drain}} = \frac{\phi_{\text{gwl}} - \phi_{\text{drain,1}}}{\gamma_1} = \frac{-2.5 - (-3.0)}{1000} = 0.0005 \text{ m d}^{-1} = 0.5 \text{ mm d}^{-1}$

b)
$$q_{\text{drain}} = \frac{\phi_{\text{gwl}} - \phi_{\text{drain,1}}}{\gamma_1} + \frac{\phi_{\text{gwl}} - \phi_{\text{drain,2}}}{\gamma_2} = \frac{-1.5-(-3.0)}{1000} + \frac{-1.5-(-2.0)}{500}$$
$$= 0.0025 \text{ m d}^{-1} = 2.5 \text{ mm d}^{-1}$$

c)
$$q_{\text{drain}} = \frac{-0.5-(-3.0)}{1000} + \frac{-0.5-(-2.0)}{500} + \frac{-0.5-(-1.0)}{250} = 0.0075 \text{ m d}^{-1} = 7.5 \text{ mm d}^{-1}$$

Question 9.6:
First make a guess of the equivalent thickness d, which should be smaller than the aquifer thickness $D = 2$ m, e.g., $d = 1.8$ m. Application of the left part of Eq. (9.8) gives:

$$L = \sqrt{\frac{4k_t m^2 + 8k_b dm}{R}} = \sqrt{\frac{4 \times 0.8 \times 1.2^2 + 8 \times 1.2 \times 1.8 \times 1.2}{0.010}} = 50.34 \text{ m}$$

Now we may calculate an update of the equivalent thickness with the right part of Eq. (9.8):

$$d = \frac{L}{8\left[\frac{\left(L - D\sqrt{2}\right)^2}{8DL} + \frac{1}{\pi}\ln\left(\frac{D}{r_d\sqrt{2}}\right)\right]} = \frac{50.34}{8\left[\frac{\left(50.34 - 2.0\sqrt{2}\right)^2}{8 \times 2.0 \times 50.34} + \frac{1}{\pi}\ln\left(\frac{2.0}{0.05\sqrt{2}}\right)\right]} = 1.627 \text{ m}$$

Next recalculate L and d until convergence: $L = 48.33$ m, $d = 1.615$ m, $L = 48.18$ m, $d = 1.614$ m, $L = 48.18$ m.

Question 9.7:

$$k_i = \frac{0.66}{1 + g_i\left(\frac{\lambda_i}{\lambda_0} - 1\right)} + \frac{0.33}{1 + (1 - 2g_i)\left(\frac{\lambda_i}{\lambda_0} - 1\right)}$$
$$= \frac{0.66}{1 + 0.144\left(\frac{7603}{492} - 1\right)} + \frac{0.33}{1 + 0.712\left(\frac{7603}{492} - 1\right)} = 0.2461$$

Question 9.8:
For wet conditions ($\theta = 0.25$):
Soil porosity $\phi = 1.0 - 0.55 - 0.08 - 0.02 = 0.35$
Air content $f_a = \phi - \theta = 0.35 - 0.25 = 0.10$

Air shape factor $g_a = 0.333 - \frac{\phi - \theta}{\phi}(0.333 - 0.035) = 0.333 - \frac{0.10}{0.35}(0.333 - 0.035) = 0.2478$

Air thermal conductivity $\lambda_a = 22 + 64 = 86$ J cm^{-1} d^{-1} K^{-1}

$$\text{Air weight factor } k_a = \frac{0.66}{1+0.2478\left(\frac{86}{492}-1\right)} + \frac{0.33}{1+0.5044\left(\frac{86}{492}-1\right)} = 1.395$$

Soil thermal conductivity:

$$\lambda_s = \frac{0.2461\times0.55\times7603 + 0.7317\times 0.08\times 2523 + 1.2602\times0.02\times216 + 0.25\times492 + 1.395\times0.10\times86}{0.2461\times0.55 + 0.7317\times0.08 + 1.2602\times0.02 + 0.25 + 1.395\times0.10}$$
$$= 2164 \text{ J cm}^{-1}\text{ d}^{-1}\text{ K}^{-1}$$

For dry conditions ($\theta = 0.02$):
Air content $f_a = \phi - \theta = 0.35 - 0.02 = 0.33$

$$\text{Air shape factor } g_a = 0.333 - \frac{\phi-\theta}{\phi}(0.333-0.035) = 0.333 - \frac{0.33}{0.35}(0.333-0.035) = 0.0520$$

$$\text{Air thermal conductivity } \lambda_a = \lambda_{da} + \lambda_v = 22 + \frac{\theta}{\theta_{wet}}64 = 22 + \frac{0.02}{0.05}64 = 48 \text{ J cm}^{-1}\text{ d}^{-1}\text{ K}^{-1}$$

Soil thermal conductivity:

$$\lambda_s = 1.25 \times \frac{0.0143\times0.55\times7603 + 0.6695\times0.08\times2523 + 0.4545\times0.02\times216 + 0.1812\times0.02\times492 + 0.33\times48}{0.0143\times0.55 + 0.6695\times0.08 + 0.4545\times0.02 + 0.1812\times0.02 + 0.33}$$
$$= 663 \text{ J cm}^{-1}\text{ d}^{-1}\text{ K}^{-1}$$

Question 9.9:

a) Wind speed: the downward transport at the top of the box is less than at the bottom: the net effect is that the air is slowed down (u and v decrease in time); temperature and humidity: the input at the bottom is higher than the output at the top, hence the net effect is that temperature and humidity increase in the grid box.
b) Integrate over the entire gridbox. Here T signifies the grid-box average temperature:

$$\Delta x\Delta y\Delta z\frac{\partial T}{\partial t} = -\Delta x\Delta y\left(\frac{H}{\rho c_p}(\Delta z) - \frac{H}{\rho c_p}(0)\right) \Leftrightarrow \frac{\partial T}{\partial t} = -\frac{1}{\rho c_p}\frac{H(\Delta z)-H(0)}{\Delta z}$$

Question 9.10:

- Each extra square metre of leaf adds extra stomata and hence extra pathways for water vapour transport. Let us consider a vegetation with an LAI of 2. If the resistance of 1 square meter is r_s then the replacement resistance of a two parallel resistances is: $r_{total} = (1/r_s + 1/r_s)^{-1} = r_s/2$. Thus in general: $r_{total} = r_s/\text{LAI}$

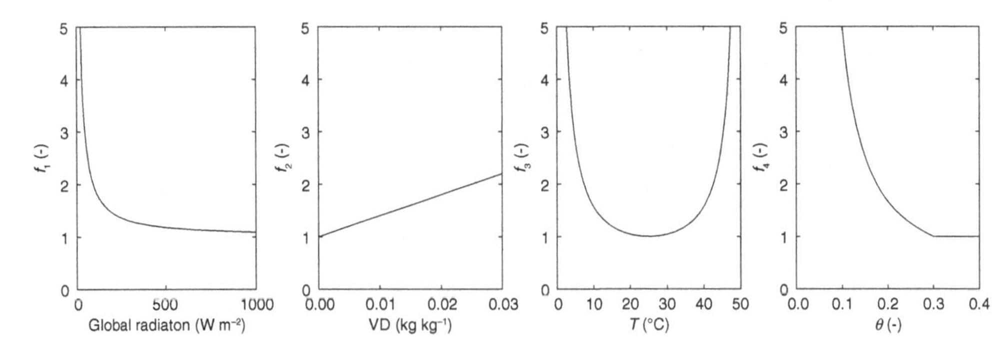

- From left to right graphical sketches of response functions f_1 through f_4,. Indeed these show similar behaviour as the corresponding subfigures in Figure 4.12 (first, third, second and fifth).

Question 9.11:

a) Derivative of A_n to q_{ci}: $\frac{\partial A_{n,c}}{\partial q_{ci}} = \rho g_m \exp\left(-\frac{\rho g_m (q_{ci} - \Gamma)}{A_{n,max}}\right)$. Close to the origin (small q_{ci}) this equals ρg_m. As long as the argument of the exponential is small, the slope is constant: a linear dependence of A_n on q_{ci}.

b) Derivative of $A_{g,cl}$ to PAR: $\frac{\partial A_{g,cl}}{\partial I_{PAR}} = \epsilon \exp\left(-\frac{\epsilon I_{PAR}}{A_{n,c} + R_d}\right)$. Close to the origin (small I_{PAR}) equals ε, the initial light use efficiency. As long as the argument of the exponential is small, the slope is constant: a linear dependence of $A_{g,cl}$ on I_{PAR}.

Question 9.12:

a) Obtain plant parameters from Table 9.5. q_{ci} depends on q_{ce} and VPD: $q_{ci} = \Gamma + f(q_{ce} - \Gamma)$ with $f = f_{min} - a_d \frac{D}{D_0}$, which yields f = 0.785 and q_{ci}=4.68·10^{-4}kg kg^{-1} (check the units of the plant parameters!).
b) All variables for calculation of $A_{n,c}$ with Eq. (9.30) are available, which yields $A_{n,c}$ = 1.69·10^{-6} kg m^{-2} s^{-1}.
c) R_d = 0.11 $A_{n,c}$ = 1.86·10^{-7} kg m^{-2} s^{-1}. Initial light use efficiency can be determined from $\epsilon = \epsilon_0 (q_{ce} - \Gamma)/(q_{ce} + 2\Gamma)$: 0.0121 mg J^{-1}. With Eq. (9.31) $A_{g,cl}$ can be obtained: = 1.61·10^{-6} kg m^{-2} s^{-1}.
d) With g_0 = 0.24 mm s^{-1}, a_1= 9.09, a_2 = 6.00 (both depending on f_{min} and f_{max}, see Section 6.4.3) $g_{s,c}$ can be obtained from Eq. (9.29): $g_{s,c}$ 0.0130 ms^{-1}. g_s = 1.6 $g_{s,c}$ which gives 0.0208 m s^{-1} for g_s.
e) Resistances are the inverse of the conductances: $r_{s,c}$ = 77.0 s m^{-1} and r_s = 48.1 s m^{-1}.

Question 9.13:

a) The method is identical to that for the previous question. R_d and $A_{n,c}$ are identical as these are not affected by radiation. $A_{g,cl}$ = 0.892·10^{-6} kg m^{-2} s^{-1}. This gives $g_{s,c}$ 0.0073 ms^{-1} hence g_s = 0.0117 m s^{-1}.
b) The canopy conductance can be considered to be the effect of two parallel conductances. Hence the conductances can be added (both refer to one unit of LAI, so they have equal weights): g_c = 0.0208 + 0.0117 = 0.0325 m s^{-1}.
c) The corresponding canopy resistance is 31 s m^{-1}.

Question 9.14:

As compared to the scheme for water vapour all canopy resistances have disappeared. The snow under vegetation is no longer connected, as most of the heat transport for that tile comes from the vegetation (the connection to the snow is important mainly for vapour transport).

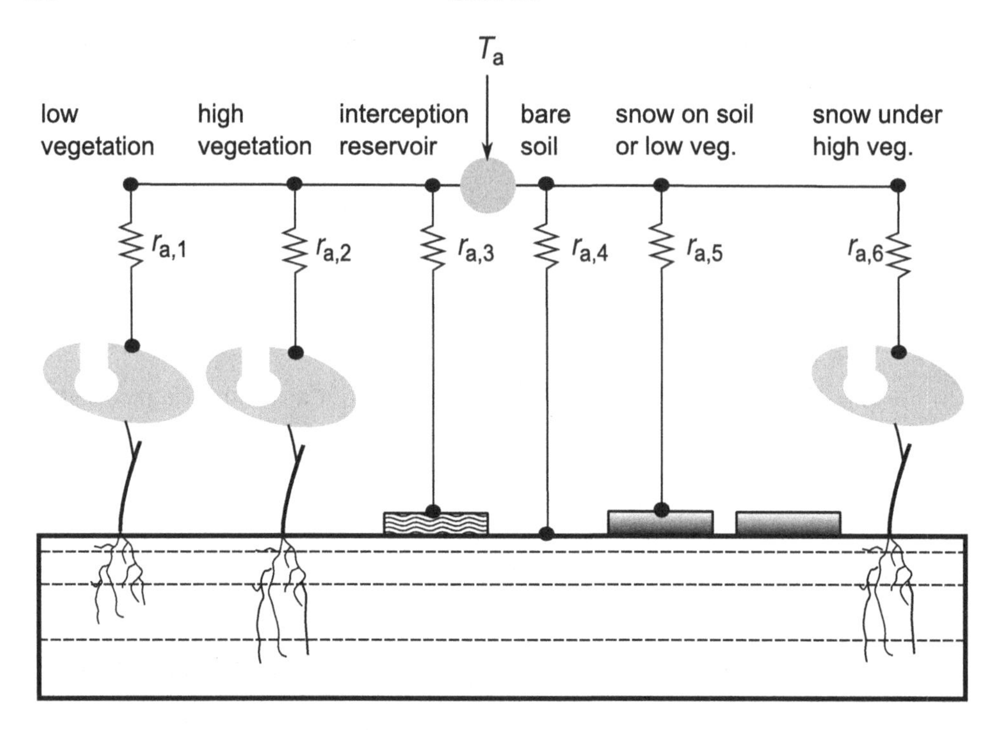

Question 9.15:

- Low vegetation: relatively high aerodynamic resistance, and low canopy resistance
- High vegetation: relatively low aerodynamic resistance, and high canopy resistance
- Interception reservoir: no canopy resistance
- Bare soil: high aerodynamic resistance and a 'canopy resistance' directly linked to soil moisture content of the upper soil layer only
- Snow on soil or low vegetation: relatively high aerodynamic resistance, high albedo and extra phase change (solid to liquid)
- Snow under high vegetation: snow dominates the exchange of water vapour, whereas the high vegetation (that is not covered by snow) dominates heat exchange.

Question 9.16:

a) Damping depth: $D = \sqrt{\frac{2\kappa}{\omega}}$. For the loamy soil we assume $\kappa = 0.205 \cdot 10^{-6}$ m^2 s^{-1}. The resulting damping depths are then 0.075 m, 0.20 m, 0.41 m and 1.43 m for periods of 1 day, 1 week, 1 month and 1 year respectively.

b) For the saturated loamy soil we assume $\kappa = 0.63 \cdot 10^{-6}$ m^2 s^{-1}, which results in the following damping depths: 0.13, 0.35, 0.72 and 2.5 m.

c) To compare the soil layer thickness to the damping depths determined above, we need to sum up the layer thicknesses for each of the periods. This yields soil depths of 0.07, 0.28, 1.0 and 2.89 m. These depths are indeed of the order of the damping depths found above (the third layer seems to be responsive to changes with a period of 2–3 months).

List of Main Symbols

Please note that some symbols are used for more than one quantity and may have different units because this book covers different research fields that all have their own nomenclature. Only those symbols are given that occur more than once in the text, or symbols that are of general importance.

Roman alphabet

Symbol	Description	Unit
A	Cross-sectional area	m^2
A_x	Advection of quantity x per unit area	[x] m^{-2} s^{-1}
C	Differential soil water capacity (d_θ/d_h)	m^{-1}
C_s	Volumetric soil heat capacity	J m^{-3} K^{-1}
C_a	Solute amount adsorbed to solid matter	kg kg^{-1}
C_l	Solute concentration in soil water	kg m^{-3}
C_T	Total solute concentration in a soil volume	kg m^{-3}
C_{dm}	Drag coefficient for momentum transport	—
C_{dh}	Drag coefficient for heat transport	—
c_s	Specific soil heat capacity	J kg^{-1} K^{-1}
c_p	Specific heat capacity of air at constant pressure	J kg^{-1} K^{-1}
c_v	Specific heat capacity of air at constant volume	J kg^{-1} K^{-1}
D_{dif}	Solute diffusion coefficient	m^2 s^{-1}
D_{dis}	Solute dispersion coefficient	m^2 s^{-1}
D_e	Effective solute dispersion coefficient	m^2 s^{-1}
DM	Dry matter amount	kg
D_r	Rooting depth	m
D	Drainage rate	kg m^{-2} s^{-1}
D	Damping depth	m

(*continued*)

Symbol	Description	Unit
D	Dewfall rate	kg m^{-2} s^{-1}
D	Vapour pressure deficit	Pa
D_q	Specific humidity deficit	kg kg^{-1}
d	Displacement height	m
E	Evapotranspiration rate (including interception)	m s^{-1} or kg m^{-2} s^{-1}
E_{int}	Evaporation of intercepted water	m s^{-1} or kg m^{-2} s^{-1}
E_p, E_{pot}	Potential evapotranspiration	m s^{-1} or kg m^{-2} s^{-1}
E_{ref}	Reference evapotranspiration	m s^{-1} or kg m^{-2} s^{-1}
E_{soil}	Soil evaporation	m s^{-1} or kg m^{-2} s^{-1}
EC_e	Electrical conductivity of saturated paste	dS m^{-1}
EC_{sw}	Electrical conductivity of soil water	dS m^{-1}
e	Water vapour pressure	N m^{-2} or Pa
e_{sat}	Saturated vapour pressure	N m^{-2} or Pa
e	Turbulent kinetic energy	m^2 s^{-2}
f_t	Factor for crop transpiration efficiency	—
G	Soil heat flux density (at the surface unless indicated otherwise)	W m^{-2}
g	Gravitational acceleration	m s^{-2}
H	Soil water hydraulic head	m
h	Soil water pressure head	m
h	Depth of the atmospheric boundary-layer	m
h_{air}	Air pressure head	m
I	Irrigation rate	m d^{-1}
I	Infiltration rate	m d^{-1}
I_n	Frost index	°C
$\bar{I}_0$	Solar constant	W m^{-2}
J	Solute flux density	kg m^{-2} s^{-1}
k	Soil hydraulic conductivity	m s^{-1}
k_s	Saturated soil hydraulic conductivity	m s^{-1}
k	Extinction coefficient for radiation in the atmosphere	m^2 kg^{-1}
K_c	Crop coefficient	—
K_{cb}	Basal crop coefficient (dual crop coefficient method)	—
K_e	Soil evaporation coefficient (dual crop coefficients method)	—
$K^{\downarrow}$	Incoming shortwave radiation flux density (global radiation)	W m^{-2}
$K_0^{\downarrow}$	Incoming shortwave radiation flux density at top of atmosphere	W m^{-2}
$K^{\uparrow}$	Upwelling shortwave radiation flux density	W m^{-2}
K^*	Net shortwave radiation flux density	W m^{-2}

Symbol	Description	Unit
K_x	Turbulent diffusivity for quantity *x*	$m^2\ s^{-1}$
L	Obukhov length	m
L_{dis}	Dispersion length	m
L_f	Leaching fraction	—
L_{root}	Root length density	$m\ m^{-3}$
L_v	Latent heat of vaporization	$J\ kg^{-1}$
$L^{\downarrow}$	Incoming longwave radiation flux density	$W\ m^{-2}$
$L^{\uparrow}$	Upwelling longwave radiation flux density	$W\ m^{-2}$
L^*	Net longwave radiation flux density	$W\ m^{-2}$
LAI	Leaf area index	$m^2\ m^{-2}$
M	Matric flux potential	$m^2\ d^{-1}$
m	Optical mass	$kg\ m^{-2}$
m_r	Relative optical mass	—
m_v	Vertical optical mass	$kg\ m^{-2}$
n	Parameter Van Genuchten soil hydraulic functions	—
P	Precipitation rate	$m\ s^{-1}$ or $kg\ m^{-2}\ s^{-1}$
P	Period of temperature forcing at soil surface	s
P_i	Rainfall interception amount	m
p	Air pressure	Pa
Q	Water discharge	$m^3\ s^{-1}$
q	Soil water flux density (positive upward)	$m\ s^{-1}$
q	Specific humidity	$kg\ kg^{-1}$
q_*	Moisture scale	$kg\ kg^{-1}$
q_c	Specific CO_2 concentration	$kg\ kg^{-1}$
R	Groundwater recharge	$m\ s^{-1}$
R	Runoff rate	$m\ s^{-1}$
R	Retardation factor	—
R	Gas constant for a gas	$J\ kg\ K^{-1}$
R_d	Gas constant for dry air	$J\ kg\ K^{-1}$
R_v	Gas constant for water vapour	$J\ kg\ K^{-1}$
R_{xy}	Correlation coefficient between quantities *x* and *y*	—
Ri_f	Flux Richardson number	—
Ri_b	Bulk Richardson number	—
Ri_g	Gradient Richardson number	—
r	Radial distance	m
r	Reflectivity (or albedo if no angular dependence)	—
r_{ax}	Aerodynamic resistance for quantity *x*	$s\ m^{-1}$
r_b	Boundary layer resistance	$s\ m^{-1}$
r_c	Canopy resistance	$s\ m^{-1}$

(*continued*)

Symbol	Description	Unit
r_s	Stomatal resistance	$s\ m^{-1}$
S	Root water extraction rate	$m^3\ m^{-3}\ s^{-1}$
S_d	Solute distribution coefficient	$m^3\ kg^{-1}$
S_x	Storage of quantity x per unit area	$[x]\ m^{-3}\ s^{-1}$
s	Slope of the saturated vapour pressure versus temperature curve	$Pa\ K^{-1}$
T	Transpiration rate	$m\ s^{-1}$ or $kg\ m^{-2}\ s^{-1}$
T_p	Potential transpiration rate	$m\ s^{-1}$ or $kg\ m^{-2}\ s^{-1}$
T_a	Actual transpiration rate	$m\ s^{-1}$ or $kg\ m^{-2}\ s^{-1}$
T	Temperature	K or °C
T_v	Virtual temperature	K or °C
T_{res}	Residence time	s
T_{50}	Half-life time	s
t	Time	s
u	Horizontal component of wind velocity	$m\ s^{-1}$
u_*	Friction velocity	$m\ s^{-1}$
v	Horizontal component of wind velocity (⊥ to u)	$m\ s^{-1}$
v	Pore water velocity	$m\ s^{-1}$
W	Soil water storage	m
WP_T	Water productivity (ratio dry matter over transpiration amount)	$kg\ m^{-3}$
w	Gravimetric soil water content	$kg\ kg^{-1}$
w	Vertical component of wind velocity	$m\ s^{-1}$
z	Vertical coordinate, positive upward	m
z_0	Roughness length for momentum	m
z_{0h}	Roughness length for heat	m

Greek alphabet

Symbol	Description	Unit
α	Parameter Van Genuchten soil hydraulic functions	m^{-1}
α_{rw}	Reduction factor root water uptake due to water stress	—
α_{rs}	Reduction factor root water uptake due to salinity stress	—
α_{PT}	Priestley-Taylor coefficient	—
β	Bowen ratio	—
γ	Psychrometric constant	$Pa\ K^{-1}$
γ_d	Drainage resistance	s
ε	Relative dielectric permittivity of a medium	—
ε	Longwave emissivity	—
κ_s	Soil thermal diffusivity	$m^2\ s^{-1}$

Symbol	Description	Unit
κ	Von Karman constant	—
Λ_s	Integrated conductivity in force-restore method	W m^{-2} K^{-1}
Λ_{veg}	Skin layer conductivity	W m^{-2} K^{-1}
λ	Parameter Van Genuchten soil hydraulic functions	—
λ	Wavelength of radiation	m
λ_s	Soil thermal conductivity	J m^{-1} s^{-1} K^{-1}
μ	First order solute transformation rate	s^{-1}
μ	Factor for crop transpiration efficiency	Pa
η	Dynamic viscosity	kg m^{-1} s^{-1}
θ	Volumetric water content	m^3 m^{-3}
θ	Potential temperature	K
θ	Zenith angle	rad
θ_*	Temperature scale	K
θ_s	Saturated volumetric water content	m^3 m^{-3}
θ_r	Residual volumetric water content	m^3 m^{-3}
θ_{fc}	Volumetric water content at field capacity	m^3 m^{-3}
θ_w	Volumetric water content at plant wilting	m^3 m^{-3}
θ_v	Virtual potential temperature	K
π	Osmotic head	m
ρ	Density	kg m^{-3}
ρ_b	Dry soil bulk density	kg m^{-3}
σ	Water surface tension	N m^{-1}
σ_x	Standard deviation of quantity x	[x]
τ	Transmission coefficient for radiation in the atmosphere	—
τ	Surface shear stress	N m^{-2}
ϕ	Soil porosity	—
ϕ_d	Drain level	m
ϕ_g	Groundwater level	m
ϕ_x	Flux-gradient relationship for quantity x	—
φ	Azimuth angle	rad
φ	Wetting angle	rad
Ψ_x	Integrated flux-gradient relationship for quantity x	—
ω	Frequency of temperature forcing at soil surface	s^{-1}

References

Aboukhaled, A., A. A. Alfaro, and M. Smith (1982). Lysimeters. Irrigation and Drainage Paper 39, FAO, Rome.

Allen, R.G., L.S. Pereira, D. Raes, and M. Smith (1998). Crop evapotranspiration. Guidelines for computing crop water requirements. Irrigation and Drainage Paper 56, FAO, Rome.

Ament, F., and C. Simmer (2008). Improved representation of land-surface heterogeneity in a non-hydrostatic numerical weather prediction model. *Boundary-Layer Meteorology*, 121, 153–174.

Andreas, E.L., and B.B. Hicks (2002). Comments on 'Critical test of the validity of Monin–Obukhov similarity during convective conditions'. *Journal of Atmospheric Sciences*, 59, 2605–2607.

Andreas, E.L, K.J. Claffey, R.E. Jordan, C.W. Fairall, P.S. Guest, P.O.G. Persson, and A.A. Grachev (2006). Evaluations of the von Kármán constant in the atmospheric surface layer. *Journal of Fluid Mechanics*, 559, 117–149.

Arain M.A., J. Michaud, W.J. Shuttleworth, and A.J. Dolman (1996). Testing of vegetation parameter aggregation rules applicable to the biosphere-atmosphere transfer scheme (BATS) and the FIFE site. *Journal of Hydrology*, 177, 1–22.

Arnfield, A.J. (2003). Two decades of urban climate research: A review of turbulence, exchanges of energy and water, and the urban heat island. *International Journal of Climatology*, 23, 1–26.

Araújo, A.C. de, B. Kruijt, A.D. Nobre, A.J. Dolman, M.J. Waterloo, E.J. Moors, and J.S. de Souza (2008). Nocturnal accumulation of CO_2 underneath a tropical forest canopy along a topographical gradient. *Ecological Applications*, 18, 1406–1419.

Ashton, G.D. (2011). River and lake ice thickening, thinning, and snow ice formation. *Cold Regions Science and Technology*, 68, 3–19.

Avissar, R. (1992). Conceptual aspects of a statistical-dynamical approach to represent landscape subgrid-scale heterogeneities in atmospheric models. *Journal of Geophysical Research*, 97D, 2729–2742.

Avissar, R., and R.A. Pielke (1989). A parameterization of heterogeneous land surfaces for atmospheric models and its impact on regional meteorology. *Monthly Weather Review*, 117, 2113–2134.

Baas P., G.J. Steeneveld, B.J.H. van de Wiel, and A.A.M. Holtslag (2006). Exploring self-correlation in flux–gradient relationships for stably stratified conditions. *Journal of the Atmospheric Sciences*, 63, 3045–3054.

Baker, C.J. (2010). Discussion of 'The macro-meteorological spectrum – A preliminary study' by R.I. Harris. *Journal of Wind Engineering and Industrial Aerodynamics*, 98, 945–947.

Baldocchi, D. (2008). 'Breathing' of the terrestrial biosphere: lessons learned from a global network of carbon dioxide flux measurement systems. *Australian Journal of Botany*, 56, 1–26.

Baldocchi, D.D., R.J. Luxmoore, and J.L. Hatfield (1991). Discerning the forest from the trees: An essay on scaling canopy stomatal conductance. *Agricultural and Forest Meteorology*, 54, 197–226.

Baldocchi, D., E. Falge, L. Gu, R. Olson, D. Hollinger, S. Running, P. Anthoni, Ch. Bernhofer, K. Davis, R. Evans, J. Fuentes, A. Goldstein, G. Katul, B. Law, X. Lee, Y. Malhi, T. Meyers, W. Munger, W. Oechel, K. T. Paw U, K. Pilegaard, H. P. Schmid, R. Valentini, S. Verma, T. Vesala, K. Wilson, and S. Wofsy (2001). FLUXNET: A new tool to study the temporal and spatial variability of ecosystem-scale carbon dioxide, water vapor, and energy flux densities. *Bulletin of the American Meteorological Society*, 82, 2415–2434.

Ball, J.T. (1987). Calculations related to gas exchange. In: Zeiger, E., G.D. Farquhar, and I.R. Cowan (eds.), *Stomatal function* (pp. 445–476). Stanford: Stanford University Press.

Balsamo, G., P. Viterbo, A. Beljaars, B. van den Hurk, M. Hirschi, A.K. Betts, and K. Scipal (2009). A revised Hydrology for the ECMWF model: Verification from field site to terrestrial water storage and impact in the integrated forecast system. *Journal of Hydrometeorology*, 10, 623–643.

Balsamo, G., F. Pappenberger, E. Dutra, P. Viterbo, and B. van den Hurk (2011). A revised land hydrology in the ECMWF model: A step towards daily water flux prediction in a fully-closed water cycle. *Hydrological Processes*, 25, 1046–1054.

Bartholomeus, R.P., J.P.M. Witte, P.M. van Bodegom, J.C. van Dam, and R. Aerts (2008). Critical soil conditions for oxygen stress to plant roots: Substituting the Feddes-function by a process-based model. *Journal of Hydrology*, 360, 147–165.

Bastiaanssen, W.G.M., Noordman, E.J.M., Pelgrum, H., Davids, G., Thoreson, B.P., and Allen, R.G. (2005). SEBAL model with remotely sensed data to improve water resources management under actual field conditions. *Journal of Irrigation and Drainage Engineering*, 131, 85–93.

Basu, S., A.A.M. Holtslag, B.J.H. van de Wiel, A.F. Moene, and G.J. Steeneveld (2008). An inconvenient 'truth' about using sensible heat flux as a surface boundary condition in models under stably stratified regimes. *Acta Geophysica*, 56, 88–99.

Bear, J. (1972). *Dynamics of fluids in porous media*. New York: Elsevier.

Beljaars, A. C. M., and A.A.M. Holtslag (1991). Flux parameterization over land surfaces for atmospheric models. *Journal of Applied Meteorology*, 30, 327–341.

Belmans, C., J.G. Wesseling, and R.A. Feddes (1983). Simulation model of the water balance of a cropped soil. *Journal of Hydrology*, 63, 271–286.

Beltman, W.H.J., J.J.T.I. Boesten, and S.E.A.T.M. van der Zee (1995). Analytical modelling of pesticide transport from the soil surface to a drinking water well. *Journal of Hydrology*, 169, 209–228.

Bessembinder, J.J.E., A.S. Dhindwal, P.A. Leffelaar, T. Ponsioen, and Sher Singh (2003). Analysis of crop growth. In J.C. van Dam and R.S. Malik (eds.), *Water productivity of irrigated crops in Sirsa district, India. Integration of remote sensing, crop and soil models and geographical information systems* (pp. 59–83). Wageningen: WATPRO Final Report.

Best, M., A. Beljaars, J. Polcher, and P. Viterbo (2004). A proposed structure for coupling tiled surfaces with the planetary boundary layer. *Journal of Hydrometeorology*, 5, 1271–1278.

Bethune, M.G., and Q.J. Wang (2004). Simulating the water balance of border-check irrigated pasture on a cracking soil. *Australian Journal of Experimental Agriculture*, 44, 163–171.

Bhumralkar, C.M. (1975). Numerical experiments on the computation of ground surface temperature in an atmospheric general circulation model. *Journal of Applied Meteorology*, 14, 1246–1258.

Bierhuizen, J.F., and R.O. Slayter (1965). Effect of atmospheric concentration of water vapor and CO_2 in determining transpiration – photosynthesis relationships of cotton leaves. *Agricultural Meteorology*, 2, 259–270.

Biggar, J.W., and D.R. Nielsen (1967). Miscible displacement and leaching phenomena. *Agronomy*, 11, 254–274.

Bird, R.E., and C. Riordan (1986). Simple solar spectral model for direct and diffuse irradiance on horizontal and tilted planes at the earth's surface for cloudless atmospheres. *Journal of Applied Meteorology*, 25, 87–97.

Black, T.A., W.R. Gardner, and G.W. Thurtell (1969). The prediction of evaporation, drainage, and soil water storage for a bare soil. *Soil Science Society of America Journal*, 33, 655–660.

Boast, C.W., and T.M. Robertson (1982). A micro-lysimeter method for determining evaporation from bare soil: Description and laboratory evaluation. *Soil Science Society of America Journal*, 46, 689–696.

Boesten, J.J.T.I., and L. Stroosnijder (1986). Simple model for daily evaporation from fallow tilled soil under spring conditions in a temperate climate. *Netherlands Journal of Agricultural Science*, 34, 75–90.

Boesten, J.J.T.I., and A.M.A. van der Linden (1991). Modeling the influence of sorption and transformation on pesticide leaching and persistence. *Journal of Environmental Quality*, 20, 425–435.

Bolt, G.H. (1979). *Soil chemistry. Part B. Physico-chemical models*. Amsterdam: Elsevier.

Bonan, G.B. (1995). Land–atmosphere CO_2 exchange simulated by a land surface process model coupled to an atmospheric general circulation model. *Journal of Geophysical Research*, 100, 2817–2831.

Bonan, G.B., D. Pollard, and S.L. Thompson (1993). Influence of subgrid-scale heterogeneity in leaf area index, stomatal resistance, and soil moisture on grid-scale land–atmosphere interactions. *Journal of Climate*, 6, 1882–1897.

Boone A., P. de Rosnay, G. Basalmo, A. Beljaars, F. Chopin, B. Decharme, C. Delire, A. Ducharne, S. Gascoin, M. Grippa, F. Guichard, Y. Gusev, P. Harris, L. Jarlan, L. Kergoat, E. Mougin, O. Nasonova, A. Norgaard, T. Orgeval, C. Ottlé, I. Poccard-Leclercq, J. Polcher, I. Sandholt, S. Saux-Picart, C. Taylor, and Y. Xue (2009). The AMMA land surface model intercomparison project. *Bulletin of the American Meteorological Society*, 90, 1865–1880.

Bosveld, F.C., and W. Bouten (2001). Evaluation of transpiration models with observations over a Douglas-fir forest. *Agricultural and Forest Meteorology*, 108, 247–264.

Bouma, J., C. Belmans, L.W. Dekker, and W.J.M. Jeurissen (1983). Assessing the suitability of soils with macropores for subsurface liquid waste disposal. *Journal of Environmental Quality*, 12, 305–311.

Braam, M., F.C. Bosveld, and A.F. Moene (2012). On Monin-Obukhov scaling in and above the atmospheric surface layer: The complexities of elevated scintillometer measurements *Boundary-Layer Meteorology*, 144, 157–177.

Braden, H. (1985). Energiehaushalts- und Verdunstungsmodell für Wasser- und Stoffhaushalts-untersuchungen landwirtschaftlich genutzter Einzugsgebiete. *Mitteilungen der Deutschen Bodenkundlichen Gesellschaft*, 42, 254–299.

Brion, J., A. Chakir, D. Daumont, J. Malicet, and C. Parisse (1993). High-resolution laboratory absorption cross section of O3. Temperature effect. *Chemical Physics Letters*, 213, 610–612.

Brunt, D. (1932). Notes on radiation in the atmosphere. I. *Quarterly Journal of the Royal Meteorological Society*, 58, 389–420.

Businger, J., J. Wyngaard, Y. Izumi, and E. Bradley (1971). Flux-profile relationships in the atmospheric surface layer. *Journal of Atmospheric Science*, 28, 181–189.

Byrd, G.T., R.F. Sage, and R.H. Brown (1992). A comparison of dark respiration between C3 and C4 plants. *Plant Physiology*, 100, 191–198.

Calvet, J.-C., J. Noilhan, J.-L. Roujean, P. Bessemoulin, M. Cabelguenne, A. Olioso, and J.-P. Wigneron (1998). An interactive vegetation SVAT model tested against data from six contrasting sites. *Agricultural and Forest Meteorology*, 92, 73–95.

Cardon, G.E., and J. Letey (1992). Plant water uptake terms evaluated for soil water and solute movement models. *Soil Science Society of America Journal*, 32, 1876–1880.

Carsel, R.F., and R.S. Parrish (1988). Developing joint probability distributions of soil water characteristics. *Water Resources Research*, 24, 755–769.

Carlson, T.N., and D.A. Ripley (1997). On the relation between NDVI, fractional vegetation cover, and leaf area index. *Remote Sensing of Environment*, 62, 241–252.

Carslaw, H.S., and J.C. Jaeger (1959). *Conduction of heat in solids*. Oxford: Oxford University Press.

Celia, M.A., E.T. Bouloutas, and R.L. Zarba (1990). A general mass-conservative numerical solution for the unsaturated flow equation. *Water Resources Research*, 26, 1483–1496.

Chehbouni A., E.G. Njoku, J.-P. Lhomme, and Y.H. Kerr (1995). Approaches for averaging surface parameters and fluxes over heterogeneous terrain. *Journal of Climate*, 8, 1386–1393.

Chen, F., and J. Dudhia (2001). Coupling an advanced land surface–hydrology model with the Penn State–NCAR MM5 modeling system. Part II: Preliminary model validation. *Monthly Weather Review*, 129, 587–604.

Collatz, G.J., J.T. Ball, C. Grivet, and J.A. Berry (1991). Physiological and environmental regulation of stomatal conductance, photosynthesis and transpiration: A model that includes a laminar boundary layer. *Agricultural and Forest Meteorology*, 54, 107–136.

Cook, F.J. (1995). One-dimensional oxygen diffusion into soil with exponential respiration: Analytical and numerical solutions. *Ecological Modeling*, 78, 277–283.

Crawford, T.M., and C.E. Duchon (1999). An improved parameterization for estimating effective atmospheric emissivity for use in calculating daytime downwelling longwave radiation. *Journal of Applied Meteorology*, 38, 473–480.

Crescimanno, G., and P. Garofalo (2005). Application and evaluation of the SWAP model for simulating water and solute transport in a cracking clay soil. *Soil Science Society of America Journal*, 69, 1943–1954.

Crescimanno, G., and P. Garofalo (2006). Management of irrigation with saline water in cracking clay soils. *Soil Science Society of America Journal*, 70, 1774–1787.

Davies, W.J., and J. Zhang (1991). Root signals and the regulation of growth and development of plants in drying soil. *Annual Review of Plant Physiology and Plant Molecular Biology*, 42, 55–76.

DeBruin, H.A.R. (1981). The determination of (reference crop) evapotranspiration from routine weather data. In *Evaporation in relation to hydrology* (pp. 25–37). The Hague: Commission of Hydrological Research TNO Proceedings and Information, Vol. 28.

DeBruin, H. A. R. (1982). The energy balance of the Earth's surface: A practical approach. Unpublished Ph.D thesis, Wageningen Agricultural.

DeBruin, H.A.R. (1983). A model for the Priestley–Taylor parameter α. *Journal of Climate and Applied Meteorology*, 22, 572–578.

DeBruin, H.A.R. (1987). From Penman to Makkink. In J. C. Hooghart (ed.), Evaporation and Weather, Committee on Hydrological Research. TNO, Den Haag, *Proceedings and Information*, 39: 5–30.

DeBruin, H.A.R. (1999). A note on Businger's derivation of nondimensional wind and temperature profiles under unstable conditions. *Journal of Applied Meteorology*, 38, 626–628.

DeBruin, H.A.R., and A.A.M. Holtslag (1982). A simple parameterization of the surface fluxes of sensible and latent heat during daytime compared with the Penman–Monteith concept. *Journal of Applied Meteorology*, 21, 1610–1621.

DeBruin, H.A.R., and C.J. Moore (1985). Zero-plane displacement and roughness length for tall vegetation derived from a simple mass conservation hypothesis. *Boundary-Layer Meteorology*, 31, 39–49.

DeBruin, H.A.R., and H.R.A. Wessels (1988). A model for the formation and melting of ice on surface waters. *Journal of Applied Meteorology*, 27, 164–173.

DeBruin, H.A.R., and W.N. Lablans (1998). Reference crop evapotranspiration determined with a modified Makkink equation. *Hydrological Processes*, 12, 1053–1062.

DeBruin, H.A.R., and J.N.M. Stricker (2000). Evaporation of grass under non-restricted soil moisture conditions. *Hydrological Sciences*, 45, 391–406.

DeBruin, H.A.R., W. Kohsiek, and B.J.J.M van den Hurk (1993). A verification of some methods to determine the fluxes of momentum, sensible heat, and water vapour using standard deviation and structure parameter of scalar meteorological quantities. *Boundary-Layer Meteorology*, 63, 231–257.

DeBruin, H.A.R., R.J. Ronda, and B.J.H van de Wiel (2000). Approximate solutions for the Obukhov length and the surface fluxes in terms of bulk Richardson numbers. *Boundary-Layer Meteorology*, 95, 145–157.

DeBruin, H.A.R., I.F. Trigo, M.A. Jitan, N. Temesgen Enku, C. van der Tol, and A.S.M. Gieske (2010). Reference crop evapotranspiration derived from geo-stationary satellite imagery: A case study for the Fogera flood plain, NW-Ethiopia and the Jordan Valley, Jordan. *Hydrology and Earth System Sciences*, 14, 2219–2228.

De Jong van Lier, Q., K. Metselaar, and J.C. van Dam (2006). Root water extraction and limiting soil hydraulic conditions estimated by numerical simulation. *Vadose Zone Journal*, 5, 1264–1277.

De Jong van Lier, Q., J.C. van Dam, K. Metselaar, R. de Jong, and W.H.M. Duijnisveld (2008). Macroscopic root water uptake distribution using a matric flux potential approach. *Vadose Zone Journal*, 7, 1065–1078.

De Vries, D.A. (1963). Thermal properties of soils. In W.R. van Wijk (ed.), *Physics of plant environment* (pp. 210–235). Amsterdam: North-Holland.

De Vries, D.A. (1975). Heat transfer in soils. In D.A. De Vries, and N.H. Afgan (eds.), *Heat and mass transfer in the biosphere. I. Transfer processes in plant environment* (pp. 5–28). Washington, DC: Scripts.

De Willigen, P., N.E. Nielsen, and N. Claassen (2000). Modelling water and nutrient uptake. In A.L. Smit, A.G. Bengough, and C. Engels (eds.), *Root methods: A handbook* (pp. 509–544). Berlin: Springer.

Della-Marta, P.M., M.R. Haylock, J. Luterbacher, and H. Wanner (2007). Doubled length of western European summer heat waves since 1880. *Journal of Geophysical Research*, 112, D15103.

Denmead, O.T., and E.F. Bradley (1987). On scalar transport in plant canopies. *Irrigation Science*, 8, 131–149.

Desborough, C.E. (1997). The impact of root weighting on the response of transpiration to moisture stress in land surface schemes. *Monthly Weather Review*, 125, 1920–1930.

Desborough, C.E. (1999). Surface energy balance complexity in GCM land surface models. *Climate Dynamics*, 15, 389–403.

Dickinson, R.E., A. Henderson-Sellers, P.J. Kennedy, and M.F. Wilson (1986). *Biosphere/atmosphere transfer scheme (BATS) for the NCAR community climate model*. NCAR Technical Note TN275. Boulder, CO: National Center for Atmospheric Research, 69 pp.

Dingman, S.L. (2002). *Physical hydrology*. Long Grove: Waveland Press.

Dirksen, C. (1979). Flux-controlled sorptivity measurements to determine soil hydraulic property functions. *Soil Science Society of America Journal*, 43, 827–834.

Dirksen, C. (1991). Unsaturated hydraulic conductivity. In K.A. Smith and C.E. Mullins (eds.), *Soil analysis, Physical methods* (pp. 209–269). New York: Marcel Dekker.

Dirksen, C. (1999). *Soil physics measurements*. Reiskirchen: Catena Verlag.

Dirksen, C., and S. Dasberg (1993). Improved calibration of time domain reflectrometry soil water content measurements. *Soil Science Society of America Journal*, 57, 660–667.

Dirksen, C., and S. Matula (1994). Automatic atomized water spray system for soil hydraulic conductivity measurements. *Soil Science Society of America Journal*, 58, 319–325.

Dirmeyer, P.A., X. Gao, M. Zhao, Z. Guo, T. Oki, and N. Hanasaki (2006). GSWP-2: Multimodel analysis and implications for our perception of the land surface. *Bulletin of the American Meteorological Society*, 87, 1381–1397.

Doorenbos, J., and W.O. Pruitt (1977). Guidelines for predicting crop water requirements. *Irrigation and Drainage Paper*, 24, 2nd ed. Rome: FAO.

Droogers, P., G. Kite, and H. Murray-Rust (2000). Use of simulation models to evaluate irrigation performance including water productivity, risk and system analysis. *Irrigation Science*, 19, 139–145.

Droogers, P., J.C. van Dam, J. Hoogeveen, and R. Loeve (2004). Adaptation strategies to climate change to sustain food security. In J.C.J.H. Aerts and P. Droogers (eds.), *Climate change in contrasting river basins* (pp. 49–74). London: CABI.

Drusch, M., and P. Viterbo (2007). Assimilation of screen-level variables in ECMWF's integrated forecast system: A study on the impact on the forecast quality and analyzed soil moisture. *Monthly Weather Review*, 135, 300–314.

Duffy, C.J., and D.H. Lee (1992). Base flow response from nonpoint source contamination: Simulated spatial variability in source, structure and initial condition. *Water Resources Research*, 28, 905–914.

Dupont, S., and E.G. Patton (2012). Influence of stability and seasonal canopy changes on micrometeorology within and above an orchard canopy: The CHATS experiment *Agricultural and Forest Meteorology*, 157, 11–29.

Dyer, A. J. (1974). A review of flux-profile-relationships. *Boundary-Layer Meteorology*, 7, 363–372.

Dyer, A.J., and B.B. Hicks (1970). Flux-gradient relationships in the constant flux layer. *Quarterly Journal of Royal Meteorological Society*, 96, 715–721.

ECMWF (2009). *IFS documentation CY31r1*. http://www.ecmwf.int/research/ifsdocs/CY31r1 (Accessed February 16, 2009).

Edwards, D. P. (1992). *GENLN2: A general line-by-line atmospheric transmittance and radiance model, Version 3.0 description and users guide*. NCAR/TN-367-STR. Boulder, CO: National Center for Atmospheric Research.

Ehlers, W., and M. Goss (2003). *Water dynamics in plant production*. Wallingford: CABI.

Ek, M.B., and A.A.M. Holtslag (2005). Evaluation of a land-surface scheme at Cabauw. *Theoretical and Applied Climatology*, 80, 213–227.

Farahani, H.J., T.A. Howell, W.J. Shuttlewort, and W.C. Bausch (2007). Evapotranspiration: Progress in measurement and modeling in agriculture. *Transactions of the ASABE*, 50, 1627–1638.

Farouki, O. T. (1986). Thermal properties of soils. Series on Rock and Soil Mechanics. *Transactions Technical*, 11.

Farquhar, G.D., and T.D. Sharkey (1982). Stomatal conductance and photosynthesis. *Annual Review of Plant Physiology*, 33, 317–345.

Feddes, R.A. (1971). *Water, heat and crop growth*. Unpublished PhD thesis, Wageningen Agricultural University.

Feddes, R.A. (1987). Crop factors in relation to Makkink's reference crop evapotranspiration. In *Evaporation and weather* (pp. 33–45). The Hague: Commission of Hydrological Research TNO Proceedings and Information, Vol. 39.

Feddes, R.A., and P.A.C. Raats (2004). Parameterizing the soil-water-plant-root system. In R.A. Feddes, G.H. de Rooij and J.C. van Dam (eds.), *Unsaturated zone modeling, Progress, Challenges and Applications* (pp. 95–144). Dordrecht: Kluwer Academic.

Feddes, R.A., P.J. Kowalik, and H. Zaradny (1978). *Simulation of field water use and crop yield*. Simulation Monographs. Wageningen: Pudoc.

Feddes, R.A., P. Kabat, P.J.T. van Bakel, J.J.B. Bronswijk, and J. Halbertsma (1988). Modelling soil water dynamics in the unsaturated zone: State of the art. *Journal of Hydrology*, 100, 69–111.

Finnigan, J. (2000). Turbulence in plant canopies. *Annual Review of Fluid Mechanics*, 32, 519–571.

Finnigan, J.J., R. Clement, Y. Mahli, R. Leuning, and H.A. Cleugh (2002). A re-evaluation of long-term flux measurement techniques. Part I: Averaging and coordinate rotation. *Boundary-Layer Meteorology*, 107, 1–48.

Flexas, J., M. Ribas-Carbó, A. Diaz-Espejo, J. Galmés, and H. Medrano (2008). Mesophyll conductance to CO_2: Current knowledge and future prospects. *Plant, Cell & Environment*, 31, 602–621.

Foken, T (2006). 50 years of Monin-Obukhov similarity theory. *Boundary-Layer Meteorology*, 119, 431–447.

Forsythe, W.E. (1954). *Smithsonian physical tables* (9th revised edition). Washington, DC: Smithsonian Institution.

Frich, P. L.V. Alexander, P. Della-Marta, B. Gleason, M. Haylock, A.M.G. Klein Tank, and T. Peterson (2002). Observed coherent changes in climatic extremes during the second half of the twentieth century. *Climate Research*, 19, 193–212.

Gardner, W.R. (1960). Dynamic aspects of water availability to plants. *Soil Science*, 89, 63–73.

Gardner, W.R. (1986). Water content. In A. Klute (ed.), *Methods of soil analysis*. Part 1 (pp. 493–544). Madison: American Society of Agronomy, Monograph 9.

Garrat, J.R. (1992). *The atmospheric boundary layer*. Cambridge Atmospheric and Space Science series. Cambridge: Cambridge University Press.

Garrat, J.R., and M. Segal (1988). On the contribution of atmospheric moisture to dew formation. *Boundary-Layer Meteorology*, 45, 209–236.

Gash, J.H.C. (1979). An analytical model of rainfall interception by forests. *Quarterly Journal of Royal Meteorological Society*, 105, 43–55.

Gash, J.H.C., C.R. Lloyd, and G. Lachaud (1995). Estimating sparse forest rainfall interception with an analytical model. *Journal of Hydrology*, 170, 79–86.

Gates D.M. (1980). *Biophysical ecology*. New York: Springer-Verlag.

Gerber, S., L.O. Hedin, M. Oppenheimer, S.W. Pacala, and E. Shevliakov (2010). Nitrogen cycling and feedbacks in a global dynamic land model. *Global Biogeochemical Cycles*, 24, GB1001.

Gerrits, M. (2010). *The role of interception in the hydrological cycle*. Unpublished PhD. thesis, Delft Technical University.

Giard, D., and E. Bazile (2000). Implementation of a new assimilation scheme for soil and surface variables in a global NWP model. *Monthly Weather Review*, 128, 997–1015.

Goudriaan, J. (1977). *Crop meteorology: A simulation study*. Simulation Monographs. Wageningen: Pudoc.

Goudriaan, J. (1986). A simple and fast numerical method for the computation of daily total of canopy photosynthesis. *Agricultural and Forest Meteorology*, 43, 251–255.

Goudriaan, J., H.H. van Laar, H. van Keulen, and W. Louwerse (1985). Photosynthesis, CO_2 and plant production. In W. Day & R.K. Atkin (eds.), *Wheat growth and modelling* (pp. 107–122). NATO ASI Series, Series A: Life Sciences Vol. 86. New York: Plenum Press.

Grachev, A. A., E.L. Andreas, C.W. Fairall, P.S. Guest, and P.O.G. Persson (2007). On the turbulent Prandtl number in the stable atmospheric boundary layer. *Boundary-Layer Meteorology*, 125, 329–341.

Graefe, J. (2004). Roughness layer corrections with emphasis on SVAT model applications. *Agricultural and Forest Meteorology*, 124, 237–251.

Graf, A., D. Schüttemeyer, H. Geiß, A. Knaps, M. Möllmann-Coers, J.H. Schween, S.

Kollet, B. Neininger, M. Herbst, and H. Vereecken. (2010). Boundedness of turbulent temperature probability distributions, and their relation to the vertical profile in the convective boundary layer. *Boundary-Layer Meteorology*, 134, 459–486.

Grimmond, C.S.B., M. Blackett, M.J. Best, J.-J. Baik, S.E. Belcher, J. Beringer, S.I. Bohnenstengel, I. Calmet, F. Chen, A. Coutts, A. Dandou, K. Fortuniak, M.L. Gouvea, R. Hamdi, M. Hendry, M. Kanda, T. Kawai, Y. Kawamoto, H. Kondo, E.S. Krayenhoff, S.-H. Lee, T. Loridan, A. Martilli, V. Masson, S. Miao, K. Oleson, R. Ooka, G. Pigeon, A. Porson, Y.-H. Ryu, F. Salamanca, G. Steeneveld, M. Tombrou, J.A. Voogt, D.T. Young, and N. Zhang (2011). Initial results from Phase 2 of the international urban energy balance model comparison. *International Journal of Climatology*, 31, 244–272.

Grinsven, J.J.M. van, C. Dirksen, and W. Bouten (1985). Evaluation of hot air method for measuring soil water diffusivity. *Soil Science Society of America Journal*, 49, 1093–1099.

Groen, K.P. (1997). *Pesticide leaching in polders. Field and model studies on cracked clays and loamy sand.* Unpublished PhD thesis, Wageningen Agricultural University.

Gryning, S-E., E. Bactchvarova, and H.A.R. DeBruin (2001). Energy balance of a sparse coniferous high-latitude forest under winter conditions. *Boundary-Layer Meteorology*, 99, 465–488.

Gueymard, C.A. (1998). Turbidity determination from broadband irradiance measurements: A detailed multicoefficient approach. *Journal of Applied Meteorology*, 37, 414–435.

Gueymard C.A. (2001). Parameterized transmittance model for direct beam and circumsolar spectral irradiance. *Solar Energy*, 71, 325–346.

Gueymard, C.A. (2004). The sun's total and spectral irradiance for solar energy applications and solar radiation models. *Solar Energy*, 76, 423–453.

Gueymard, C.A., and D.R. Myers (2008). Solar radiation measurement: Progress in radiometry for improved modeling. In V. Badescu (ed.), *Modeling solar radiation at the earth's surface: Recent advances* (pp. 1–27). Berlin: Springer-Verlag.

Hagemann, S. (2002). *An improved land* surface *parameter dataset for global and regional climate models*. Hamburg: Max-Plank Institute for Meteorologie, Report 336.

Hagemann, S, B. Machenhauer, R. Jones, O. B. Christensen, M. Déqué, D. Jacob, and P. L. Vidale (2004). Evaluation of water and energy budgets in regional climate models applied over Europe. *Climate Dynamics*, 23, 547–567.

Hainsworth, J.M., and L.A.G. Aylmore (1986). Water extraction by single plant roots. *Soil Science Society of America Journal*, 50, 841–848.

Hale, G.M., and M. R. Querry (1973). Optical constants of water in the 200 nm to 200 μm wavelength region. *Applied Optics*, 12, 555–563.

Hall, F.G, J. R. Townshend, and E.T. Engman (1995). Status of remote sensing algorithms for estimation of land surface state parameters. *Remote Sensing of Environment*, 51, 138–156.

Halldin, S., and A. Lindroth (1992). Errors in net radiometry: Comparison and evaluation of six radiometer designs. *Journal of Atmospheric and Oceanic Technology*, 9, 762–783.

Hamaker, J.W. (1972). Decomposition: quantitative aspects. In C.A.I. Goring and M. Hamaker (eds.), *Organic chemicals in the soil environment* (pp. 253–340). New York: Marcel Dekker.

Hargreaves, G.L., and Z.A. Samani (1985). Reference crop evapotranspiration from temperature. *Applied Engineering in Agriculture*, 1, 96–99.

Hartogensis, O.K., and H.A.R. DeBruin (2005). Monin–Obukhov similarity functions of the structure parameter of temperature and turbulent kinetic energy dissipation rate in the stable boundary layer. *Boundary-Layer Meteorology*, 116, 253–276.

Hayes, W.M., ed. (2011). *CRC handbook of chemistry and physics*. Boca Raton, FL: CRC Press.

Hecht, E. (1987). *Optics*. 2nd ed. Reading, MA: Addison-Wesley.

Heinen, M. (2001). FUSSIM2: Brief description of the simulation model and application to fertigation scenarios. *Agronomie*, 21, 285–296.

Henderson-Sellers, A., A.J. Pitman, P.K. Love, P. Irannejad, and T.H. Chen (1995). The Project for Intercomparison of Land Surface Parameterization Schemes (PILPS): Phases 2 and 3. *Bulletin of the American Meteorological Society*, 76, 489–503.

Hendriks, D.M.D., A. J. Dolman, M. K. van der Molen, and J. van Huissteden (2008). A compact and stable eddy covariance set-up for methane measurements using off-axis integrated cavity output spectroscopy. *Atmospheric Chemical Physics*, 8, 431–443.

Herkelrath, W.N., E.E. Miller, and W.R. Gardner (1977). Water uptake by plants: I. Divided root experiments, II. The root contact model. *Soil Science Society of America Journal*, 41, 1033–1043.

Hetherington, A.M., and F.I. Woodward (2003). The role of stomata in sensing and driving environmental change. *Nature*, 424, 901–908.

Heus, T., C.C. van Heerwaarden, H.J J. Jonker, A.P. Siebesma, S. Axelsen, K. van den Dries, O. Geoffroy, A F. Moene, D. Pino, S.R. de Roode, and J. Vila-Guerau de Arellano (2010). Formulation of the Dutch Atmospheric Large-Eddy Simulation (DALES) and overview of its applications. *Geoscientific Model Development*, 3, 415–444.

Heusinkveld, B.G. A.F.G. Jacobs, A.A.M. Holtslag, and S.M. Berkowicz (2004). Surface energy balance closure in an arid region: Role of soil heat flux. *Agricultural and Forest Meteorology*, 122, 21–37.

Heusinkveld, B.G., S.M. Berkowicz, A.F.G. Jacobs, A.A.M. Holtslag, and W.C.A.M. Hillen (2006). An automated microlysimeter to study dew formation and evaporation in arid and semiarid regions. *Journal of Hydrometeorology*, 7, 825–832.

Heusinkveld, B.G., A.F.G. Jacobs, and A.A.M. Holtslag (2008). Effect of open-path gas analyzer wetness on eddy covariance flux measurements: A proposed solution. *Agricultural and Forest Meteorology*, 148, 1563–1573.

Hill, R.J. (1989). Implications of Monin-Obukhov similarity theory for scalar quantities. *Journal of Atmospheric Sciences*, 46, 2236–2251.

Hillel, D. (1980). Uptake of soil moisture by plants. In *Applications of soil physics* (pp. 163–166). London: Academic Press.

Hillel, D. (1998). *Environmental soil physics*. London: Academic Press.

Hillel, D. (2008). *Soil in the environment: Crucible of terrestrial life*. London: Academic Press.

Hodur, R.M. (1997). The Naval Research Laboratory's coupled ocean/atmosphere mesoscale prediction system (COAMPS). *Monthly Weather Review*, 125, 1414–1430.

Högström, U. (1988). Non-dimensional wind and temperature profiles in the atmospheric surface layer: A re-evaluation. *Boundary-Layer Meteorology*, 42, 55–78.

Högström, U. (1996). Review of some basic characteristics of the atmospheric surface layer. *Boundary-Layer Meteorology*, 78, 215–246.

Holt, T.R., D. Niyogi, F. Chen, K. Manning, M.A. LeMone, and A. Qureshi (2006). Effect of land-atmosphere interactions on the IHOP 24–25 May 2002 convection case. *Monthly Weather Review*, 134, 113–133.

Hopmans, J.W., and J.N.M. Stricker (1989). Stochastic analysis of soil water regime in a watershed. *Journal of Hydrology*, 105, 57–84.

Hopmans, J.W., J. Šimůnek, N. Romano, and W. Durner (2002). Inverse methods. In J. H. Dane and G. C. Topp (eds.), *Methods of soil analysis. Part 4 – Physical methods*. Madison, WI: Soil Science Society of America Book Series 5.

Horst, T.W., and J.C. Weil (1994). How far is far enough?: The fetch requirements for micrometeorological measurements of surface fluxes. *Journal of Atmospheric and Oceanic Technology*, 11, 1018–1025.

Horton, R.E. (1933). The role of infiltration in the hydrological cycle. *Transaction of American Geophysical Union*, 14th Annual Meeting, pp. 446–460.

Horton, R.E. (1939). Analysis of runoff – plot experiments with varying infiltration capacity. *Transaction of American Geophysical Union*, 20th Annual Meeting, pp. 693–694.

Horton, R., P.J. Wierenga, and D.R. Nielsen (1983). Evaluation of methods for determining the apparent thermal diffusivity of soil near the surface. *Soil Science Society of America Journal*, 47, 25–32.

Huber, L., and T.J. Gillespie (1992). Modeling leaf wetness in relation to plant-disease epidemiology. *Annual Review of Phytopathology*, 30, 553–577.

Huete, A, K. Didan, T. Miura, E.P. Rodriguez, X. Gao, and L.G. Ferreira (2002). Overview of the radiometric and biophysical performance of the MODIS vegetation indices. *Remote Sensing of Environment*, 83, 195–213.

Hughes, R.N., and P. Brimblecombe (1994). Dew and guttation: Formation and environmental significance. *Agricultural and Forest Meteorology*, 67, 173–190.

Hupet, F., J.C. van Dam, and M. Vanclooster (2004). Impact of within-field variability in soil hydraulic properties on transpiration fluxes and crop yields: A numerical study. *Vadose Zone Journal*, 3, 1367–1379.

Hurk van den, B.J.J.M., P. Viterbo, A.C.M. Beljaars, and A.K. Betts (2000). Offline validation of the ERA40 surface scheme. *ECMWF Technical Memo*, 295.

Hurk, van den, B.J.J.M., P. Viterbo, and S.O. Los (2003). Impact of leaf area index seasonality on the annual land surface evaporation in a global circulation model. *Journal of Geophysical Research*, 108, 4191.

Iqbal, M (1983). *An introduction to solar radiation*. Toronto: Academic Press.

Ittersum, M.K. van, P.A. Leffelaar, H. van Keulen, M.J. Kropff, L. Bastiaans, and J. Goudriaan (2003). On approaches and applications of the Wageningen crop models. *European Journal of Agronomy*, 18, 201–234.

IWMI (2007). *Water for food. Water for Life. A comprehensive assessment of water management in agriculture*. Colombo: International Water Management Institute.

Ineichen, P., and R. Perez (2002). A new airmass independent formulation for the Linke turbidity coefficient. *Solar Energy*, 73, 151–157.

Intsiful, J., and H. Kunstmann (2008). Upscaling of land-surface parameters through inverse stochastic SVAT-modelling. *Boundary Layer Meteorology*, 129, 137–58.

Itier, B. (1982). Révision d'une methode simplifieé du flux de chaleur sensible. *Journal de Recherches Atmosphériques*, 16, 85–90.

Jackson, P.S. (1981). On the displacement height in the logarithmic velocity profile. *Journal of Fluid Mechanics*, 111, 15–25.

Jacobs, C.M.J. (1994). Direct impact of atmospheric CO_2 enrichment on regional transpiration. Ph.D. thesis. Agricultural University, Wageningen.

Jacobs, A.F.G., and W.A.J. van Pul (1990). Seasonal changes in the albedo of a maize crop during two seasons. *Agricultural and Forest Meteorology*, 49, 351–360.

Jacobs, C.M.J., and H.A.R. de Bruin (1992). The sensitivity of regional transpiration to land-surface characteristics: Significance of feedback. *Journal of Climate*, 5, 683–698.

Jacobs, A.F.G., W.A.J. van Pul, and A. van Dijken (1990). Similarity moisture dew profiles within a corn canopy. *Journal of Applied Meteorology*, 29, 1300–1306.

Jacobs, A.F.G., J.H. van Boxel, and R.M.M. El-Kilani (1995). Vertical and horizontal distribution of wind speed and air temperature in a dense vegetation canopy. *Journal of Hydrology*, 166, 313–326.

Jacobs, A.F.G., J.H. van Boxel, and J. Nieveen (1996). Nighttime exchange processes near the soil surface of a maize canopy. *Agricultural and Forest Meteorology*, 82, 155–169.

Jacobs, C.M.J., B.J.J.M. van den Hurk, and H.A.R. de Bruin (1996). Stomatal behaviour and photosynthetic rate of unstressed grapevines in semi-arid conditions. *Agricultural and Forest Meteorology*, 80, 111–134.

Jacobs, A.F.G, B.G. Heusinkveld, and S.M. Berkowicz (2000). Dew measurements along a longitudinal sand dune transect, Negev Desert, Israel. *International Journal of Biometeorology*, 43, 184–190.

Jacobs, A.F.G, B.G. Heusinkveld, and E.J. Klok (2005). Leaf wetness within a lily canopy. *Meteorological Applications*, 12, 193–198.

Jacobs, A.F.G., B.G. Heusinkveld, R.J. Wichink Kruit, and S.M. Berkowicz (2006). Contribution of dew to the water budget of a grassland area in the Netherlands. *Water Resources Research*, 42, W03415.

Jacovides, C.P. (1997). Model comparison for the calculation of Linke's turbidity factor. *International Journal of Climatology*, 17, 551–563.

Jacovides, C.P., F.S. Tymvios, V.D. Assimakopoulos, and N.A. Kaltsounides (2007). The dependence of global and diffuse PAR radiation components on sky conditions at Athens, Greece. *Agricultural and Forest Meteorology*, 143, 277–287

Jacquemoud, S., and F. Baret (1990). PROSPECT: A model of leaf optical properties spectra. *Remote Sensing of Environment*, 34, 75–91.

Jarlan, L., G. Balsamo, S. Lafont, A. Beljaars, J.C. Calvet, and E. Mougin (2007). Analysis of Leaf Area Index in the ECMWF land surface scheme and impact on latent heat and carbon fluxes: Applications to West Africa. *ECMWF Technical Memorandum*, 544.

Jarvis, A.J., and W.J. Davies (1998). The coupled response of stomatal conductance to photosynthesis and transpiration. *Journal of Experimental Botany*, 49, 399–406.

Jarvis, P. (1976). The interpretation of the variations in leafwater potentials and stomatal conductances found in canopies in the field. *Philosophical Transactions of Royal Society*, 273, 593–610.

Jarvis, P., and K.G. McNaughton (1986). Stomatal control of transpiration: Scaling up from leaf to region. In A. MacFadyen and E.D. Ford (eds.), *Advances in Ecological Research* 15, 1–49.

Javaux, M., Tom Schröder, Jan Vanderborght, and Harry Vereecken (2008). Use of a three-dimensional detailed modeling approach for predicting root water uptake. *Vadose Zone Journal*, 7, 1079–1088.

Jensen, M.E. (1968). Water consumption by agricultural plants. In T.T. Kozlowski (pp. 1–22). *Plant water consumption and response: Water deficits and plant growth*, Vol. II. New York: Academic Press.

Jensen, M.E., R.D. Burman, and R.G. Allen (1990). *Evapotranspiration and irrigation water requirements*. New York: ASCE Manuals and Reports on Engineering Practice, **70**.

Jhorar, R.K., A.S. Dhindwal, Ranvir Kumar, B.S. Jhorar, M.S. Bhatto, and Dharampal (2003). Water management and crop production in Sirsa Irrigation Circle. In J.C. van Dam and R.S. Malik (eds.), *Water productivity of irrigated crops in Sirsa district, India: Integration of remote sensing, crop and soil models and geographical information systems* (pp. 21–28). Wageningen: WATPRO final report.

Jhorar, R.K., J.C. van Dam, W.M.G. Bastiaanssen, and R.A. Feddes (2004). Calibration of effective soil hydraulic parameters of heterogeneous soil profiles. *Journal of Hydrology*, 285, 233–247.

Jiménez, J.I., L. Alados-Arboledas, Y. Castro-Diez, and G. Ballester (1987). On the estimation of long-wave radiation flux from clear skies. *Theoretical and Applied Climatology*, 38, 37–42.

Jiménez, P. A., and J. Dudhia (2012). Improving the representation of resolved and unresolved topographic effects on surface wind in the WRF model. *Journal of Applied Meteorology and Climatology*, 51, 300–316.

Jungk, A.O. (2002). Dynamics of nutrient movement at the soil-root interface. In Y. Waisel, A. Eshel, and U. Kafkafi (eds.), *Plant root: The hidden half*, 3rd ed. (pp. 587–616). New York: Marcel Dekker.

Jungk, A.O., and N. Claassen (1989). Availability in soil and acquisition by plants as the basis for phosphorus and potassium supply to plants. *Zeitschrift für Pflanzenernährung und Bodenkunde*, 152, 151–157.

Jury, W.A. (1982). Simulation of solute transport using a transfer function mode. *Water Resources Research*, 18, 363–368.

Jury, W.A., and G. Sposito (1985). Field calibration and validation of solute transport models for the unsaturated zone. *Soil Science Society of America Journal*, 49, 1331–1341.

Jury, W.A., W.R. Gardner, and W.H. Gardner (1991). *Soil physics*, 5th ed. New York: John Wiley & Sons.

Kasten, F. (1996). The Linke turbidity factor based on improved values of the integral Rayleigh optical thickness. *Solar Energy*, 56, 239–244.

Kasten, F., and A.T. Young (1989). Revised optical air mass tables and approximation formula. *Applied Optics*, 28, 4735–4738.

Katul, G., D. Cava, D. Poggi, J. Albertson and L. Mahrt (2005). Stationarity, homogeneity, and ergodicity in canopy turbulence. In X. Lee, W. Massman, and B. Law (eds.), *Handbook of micrometeorology: A guide for surface flux measurement and analysis* (pp. 161–180). New York: Kluwer Academic.

Katul, G.G., A.M. Sempreviva, and D. Cava (2008). The temperature–humidity covariance in the marine surface layer: A one-dimensional analytical model. *Boundary-Layer Meteorology*, 126, 263–278.

Kelliher, F.M., R. Leuning, and E.-D. Schulze (1993). Evaporation and canopy characteristics of coniferous forests and grasslands. *Oecologica*, 95, 153–163.

Kelliher, F.M., R. Leuning, M.R. Raupach, and E.-D. Schulze (1995). Maximum conductances for evaporation from global vegetation types. *Agricultural and Forest Meteorology*, 73, 1–16.

Kim, C.P. (1995). *The water budget of heterogeneous areas: Impact of soil and rainfall variability*. Unpublished PhD thesis, Wageningen University.

Kimball, B.A., and R.D. Jackson (1975). Soil heat flux determination: A null-alignment method. *Agricultural Meteorology*, 15, 1–9.

Kirkham, M.B. (2005). *Principles of soil and plant water relations*. San Diego: Elsevier Academic Press.

Klein Tank, A.M.G., et al. (2002). Daily dataset of 20th-century surface air temperature and precipitation series for the European Climate Assessment. *International Journal of Climatology*, 22, 1441–1453.

Klipp, C.L., and L. Mahrt (2004). Flux–gradient relationship, self-correlation and intermittency in the stable boundary layer. *Quarterly Journal of the Royal Meteorological Society*, 130, 2087–2103.

Klute, A. (1986). Water retention: Laboratory methods. In A. Klute (eds.), *Methods of soil analysis;* Part 1*: Physical and mineralogical methods* (pp. 635–662). Madison, WI: American Society of Agronomy, Agronomy series, **9**.

Klute, A., and C. Dirksen (1986). Hydraulic conductivity and diffusivity: Laboratory methods. In A. Klute (ed.), *Methods of soil analysis; Part 1: Physical and mineralogical methods* (pp. 687–734). Madison, WI: American Society of Agronomy, Agronomy series, **9**.

Kohsiek, W., C. Liebethal, T. Foken, R. Vogt, S.P. Oncley, Ch. Bernhofer, and H.A.R. De Bruin (2007). The Energy Balance Experiment EBEX-2000. Part III: Behaviour and quality of the radiation measurements. *Boundary Layer Meteorology*, 123, 55–75.

Kollet, S., I. Cvijanovic, D. Schüttemeyer, A.F. Moene, and P. Bayer (2009). The influence of the sensible heat of rain, subsurface heat convection and the lower temperature boundary condition on the energy balance at the land surface. *Vadose Zone Journal*, 8, 846–857

Kondo, J., O. Kanechika, and N. Yasuda (1978). Heat and momentum transfers under strong stability in the atmospheric surface layer. *Journal of the Atmospheric Sciences*, 35, 1012–1021.

Konrad W., A. Roth-Nebelsick, and M. Grein (2008). Modelling of stomatal density response to atmospheric CO_2. *Journal of Theoretical Biology*, 253, 638–658.

Koorevaar, P., G. Menelik, and C. Dirksen (1983). *Elements of soil physics*. Amsterdam: Elsevier

Kopp, G., and J. Lean (2011). A new, lower value of total solar irradiance: Evidence and climate significance. *Geophysical Research Letters*, 38, L01706.
Kormann, R., and F.X. Meixner (2001). An analytical footprint model for non-neutral stratification. *Boundary Layer Meteorology*, 99, 207–224.
Kroes, J.G., and J. Roelsma (1997). *User's Guide ANIMO 3.5; input instructions and technical programme description.* Wageningen: DLO Winand Staring Centre, Technical Document 46.
Kroes, J.G., J.C. van Dam, P. Groenendijk, R.F.A. Hendriks, and C.M.J. Jacobs (2008). *SWAP version 3.2. Theory description and user manual.* Wageningen: Alterra, Report 1649.
Lahou, F., F. Saïd, M. Lothon, P. Durand, and D. Sarça (2010). Impact of boundary-layer processes on near-surface turbulence within the West African monsoon. *Boundary-Layer Meteorology*, 136, 1–23
Lambers, H., F. Stuart Chapin III, and T.L. Pons (2008). *Plant physiological ecology*, 2nd ed. New York: Springer Science+Business Media.
Launiainen, J. (1995). Derivation of the relationship between the Obukhov stability parameter and the bulk Richardson number for flux-profile studies. *Boundary-Layer Meteorology*, 76, 165–179.
Lee, R. (1978). *Forest micrometeorology*. New York: Columbia University Press.
Lee, X., W. Massman, and B. Law, Eds. (2004). *Handbook of micrometeorology: A guide for surface flux measurement and analysis*. Dordrecht, The Netherlands: Kluwer Academic,.
Leij, F.J., W.J. Alves, M. Th. van Genuchten, and J.R. Williams (1996). *The UNSODA unsaturated soil hydraulic database. User's manual Version 1.0.* Riverside, CA: US Salinity Laboratory.
Leistra, M., A.M.A. van der Linden, J.J.T.I. Boesten, A. Tiktak, and F. van den Berg (2001). *PEARL model for pesticide behaviour and emissions in soil-plant systems. Description of processes*. Wageningen: Alterra report, 13.
Lenschow, D.H., J. Mann, and L. Kristensen (1994). How long is long enough when measuring fluxes and other turbulence statistics. *Journal of Atmospheric and Oceanic Technology*, 11, 661–673.
Lesieur M. (1993). *Turbulence in fluids*, 2nd ed. Dordrecht: Kluwer.
Lettau, H.H. (1979). Wind and temperature profile prediction surface layer including strong inversion cases. *Boundary Layer Meteorology*, 17, 443–464.
Leuning, R. (1995) A critical appraisal of a combined stomatal-photosynthesis model for C3 plants. *Plant, Cell & Environment*, 18, 339–355.
Levis, S. (2010). Modeling vegetation and land use in models of the Earth System. *Wiley Interdisciplinary Reviews: Climate Change*, 1, 840–856.
Li, D. E. Bou-Zeid, and H.A.R. DeBruin (2012). Monin–Obukhov similarity functions for the structure parameters of temperature and humidity. *Boundary-Layer Meteorology*, 145, 45–67.
Lloyd, C.R., J.H.C. Gash, W.H. Shuttleworth, and A.O. Marques (1988). The measuring and modelling of rainfall interception by Amazonian rainforests. *Agricultural and Forest Meteorology*, 43, 277–294.
Lloyd, J., and J.T. Taylor (1994). On the temperature dependence of soil respiration. *Functional Ecology*, **8**, 315–323.
Maas, E.V. (1990). Crop salt tolerance. In K.K. Tanji (ed.), *Agricultural salinity assessment and management* (pp. 262–304). New York: ASCE Manuals and Reports on Engineering Practice, no. 71.
Maas, E.V., and G.J. Hoffman (1977). Crop salt tolerance-current assessment. *Journal of the Irrigation and Drainage Divisions*, 103, 115–134.
Mahfouf J.-F. (2010). Assimilation of satellite-derived soil moisture from ASCAT in a limited-area NWP model. *Quarterly Journal of the Royal Meteorological Society*, 136, 784–798.

Mahrt, L. (1987). Grid-averaged surface fluxes. *Monthly Weather Review*, 115, 1550–1560.
Mahrt, L. (1996). The bulk aerodynamic formulation over heterogeneous surfaces. *Boundary Layer Meteorology*, 78, 87–119.
Makkink, G.F. (1957). Testing the Penman formula by means of lysimeters. *Journal of International Water Engineering*, 11, 277–288.
Malhi, Y. (1996). The behaviour of the roughness length for temperature over heterogeneous surfaces. *Quarterly Journal of the Royal Meteorological Society*, 122, 1095–1125.
Manabe, S. (1969). Climate and the ocean circulation: 1, the atmospheric circulation and the hydrology of the Earth's surface. *Monthly Weather Review*, 97, 739–805.
Marsily, G. de (1986). *Quantitative hydrogeology. Groundwater hydrology for engineers*. New York: Academic Press.
Massman, W.J. (1997). An analytical one-dimensional model of momentum transfer by vegetation of arbitrary structure. *Boundary Layer Meteorology*, 83, 407–421.
Masson, V., J.-L. Champeaux, F. Chauvin, C. Meriguet, and R. Lacaze (2003). A global database of land surface parameters at 1-km resolution in meteorological and climate models. *Journal of Climate*, 16, 1261–1282.
Mauder, M., C. Liebethal, M. Göckede, J.-P. Leps, F. Beyrich, and T. Foken (2006). Processing and quality control of flux data during LITFASS-2003. *Boundary Layer Meteorology*, 121, 67–88.
McCaughey, J.H., D.W. Mullins, and M. Publicover (1987). Comparative performance of two reversing Bowen ratio measurement systems. *Journal of Atmospheric and Oceanic Technology*, 4, 724–730.
McNaughton, K. G., R.J. Clement, and J. B. Moncrieff (2007). Scaling properties of velocity and temperature spectra above the surface friction layer in a convective atmospheric boundary layer. *Nonlinear Processes in Geophysics*, 14, 257–271.
Metselaar, K., and Q. De Jong van Lier (2007). The shape of the transpiration reduction function under plant water stress. *Vadose Zone Journal*, 6, 124–139.
Meyers, T.P., and S.E. Hollinger (2004). An assessment of storage terms in the surface energy balance of maize and soybean. *Agricultural and Forest Meteorology*, 125, 105–115.
Moene, A.F., and D. Schüttemeyer (2008). The effect of surface heterogeneity on the temperature–humidity correlation and the relative transport efficiency. *Boundary-Layer Meteorology*, 129, 99–113.
Moene A.F., D. Schüttemeyer, and O.K. Hartogensis (2006). *Scalar similarity functions: The influence of surface heterogeneity and entrainment*. Paper presented at the 17th Boundary-Layer and Turbulence Conference, 22–25 May 2006, San Diego. American Meteorological Society, Boston, p 5.1
Mogensen, V.O. (1970). The calibration factor of heat flux meters in relation to the thermal conductivity of the surrounding medium. *Agricultural Meteorology*, 7, 401–410.
Molden, D., H. Murray-Rust, R. Sakthivadivel, and I. Makin (2003). A water-productivity framework for understanding and action. In J.W. Kijne, R. Barker and D. Molden (eds.), *Water productivity in agriculture: Limits and opportunities for improvement* (pp. 1–18). Wallingford: CABI.
Molz, F.J. (1981). Models of water transport in the soil-plant system: A review. *Water Resources Research*, 17, 1245–1260.
Moncrieff, J.B., J.M. Massheder, H. de Bruin, J. Elbers, T. Friborg, B. Heusinkveld, P. Kabat, S. Scott, H. Soegaard, and A. Verhoef (1997). A system to measure surface fluxes of momentum, sensible heat, water vapour and carbon dioxide. *Journal of Hydrology*, 188–**189**, 589–611.
Monin, A.S., and A.M. Obukhov (1954). Basic laws of turbulent mixing in the surface layer of the atmosphere. *Tr. Akad. Nauk SSSR Geophiz. Inst.* 24, 163–187 (translation edited by K. G. McNaughton, available from http://www.mcnaughty.com/keith/papers/Monin_and_Obukhov_1954.pdf (Accessed February 25, 2013).

Monin, A.S., and A.S. Yaglom (1971). *Statistical fluid mechanics: Mechanics of turbulence*, Vol. I. Cambridge, MA: MIT Press.

Monteith, J.L. (1965). Evaporation and the Environment. In G.E. Fogg (ed.), *The state and movement of water in living organisms* (pp. 205–234). Cambridge: Cambridge University Press.

Monteith, J.L. (1995). A reinterpretation of stomatal responses to humidity. *Plant, Cell and Environment*, 18, 357–364.

Monteith, J.L., and M.H. Unsworth (2008). *Principles of environmental physics*, 3rd ed. Burlington: Academic Press.

Moors, E. (2002). *Hydrologische woordenlijst*. Zeist: NHV (Dutch Hydrological Society), in Dutch.

Mualem, Y. (1976). A new model for predicting the hydraulic conductivity of unsaturated porous media. *Water Resources Research*, 12, 513–522.

Mundel, G. (1992). Untersuchungen zur Evapotranspiration von Silomaisbeständen in Lysimetern. *Archiv für Acker- und Pflanzenbau und Bodenkunde*, 36, 35–44.

Muzylo, A., P. Llorens, F. Valente, J.J. Keizer, F. Domingo, and J.H.C. Gash (2009). A review of rainfall interception modelling. *Journal of Hydrology*, 370, 191–206.

Nielsen, D.R., M.Th. van Genuchten, and J.W. Biggar (1986). Water flow and solute transport in the unsaturated zone. *Water Resources Research*, 22 (Supplement), 89S–108S.

Nimmo, J.R., J. Rubin, and D.P. Hammermeister (1987). Unsaturated flow in a centrifugal field: measurement of hydraulic conductivity and testing of Darcy's law. *Water Resources Research*, 32, 124–134.

Ochsner, E.T., R. Horton, and T. Ren (2001). A new perspective on soil thermal properties. *Soil Science Society of America Journal*, 65, 1641–1647.

O'Donnell, J.A., V.E. Romanovsky, J.W. Harden, and A.D. McGuire (2009). The effect of moisture content on the thermal conductivity of moss and organic soil horizons from black spruce ecosystems in interior Alaska. *Soil Science*, 174, 646–651

Ommen, H.C. van (1988). *Transport from diffuse sources of contamination and its application to a coupled unsaturated-saturated system*. Unpublished PhD thesis, Wageningen University.

Oke, T.R. (1987). *Boundary layer climates*, 2nd ed. London: Methuen.

Papaioannou, G., G. Nikolidakis, D. Asimakopoulos, and D. Retalis (1996). Photosynthetically active radiation in Athens. *Agricultural and Forest Meteorology*, 81, 287–298.

Paulson, C.A. (1970). The mathematical representation of wind speed and temperature profiles in the unstable atmospheric-surface layer. *Journal of Applied Meteorology*, 9, 857–861.

Pauwels, V.R.N., N.E.C. Verhoest, G.J.M. De Lannoy, V. Guissard, C. Lucau, and P. Defourny (2007). Optimization of a coupled hydrology – crop growth model through the assimilation of observed soil moisture and leaf area index values using an ensemble Kalman filter. *Water Resources Research*, 43, W04421, doi:10.1029/2006WR004942.

Paw U, K.T., J. Qiu, H.B. Su, T. Watanabe, and Y. Brunet (1995). Surface renewal analysis: A new method to obtain scalar fluxes without velocity data. *Agricultural and ForestMeteorology*, 74, 119–137.

Penning de Vries, F.W.T., and H.H. van Laar (1982). *Simulation of plant growth and crop production*. Wageningen: Pudoc.

Peters-Lidard, C.D., E. Blackburn, X. Liang, and E.F. Wood (1998). The effect of soil thermal conductivity parameterization on surface energy fluxes and temperatures. *Journal of Atmosperic Science*, 55, 1209–1224.

Petty, G.W. (2004). *A first course in atmospheric radiation*. Madison, WI: Sundog Publishing.

Philippon, N., L. Jarlan, N. Martiny, P. Camberlin, and E. Mougin, (2007). Characterization of the interannual and intraseasonal variability of West African vegetation between

1982 and 2002 by means of NOAA AVHRR NDVI data. *Journal of Climate*, 20, 1202–1218.

Pitman, A.J. (2003). The evolution of, and revolution in, land surface schemes designed for climate models. *International Journal of Climatology*, 23, 479–510.

Priestley, C.H.B., and R.J. Taylor (1972). On the assessment of surface heat flux and evaporation using large scale parameters. *Monthly Weather Review*, 100, 81–92.

Prinn, R.G., R.F. Weiss, P.J. Fraser, P.G. Simmonds, D.M. Cunnold, F.N. Alyea, S. O'Doherty, P. Salameh, B.R. Miller, J. Huang, R.H.J. Wang, D.E. Hartley, C. Harth, L.P. Steele, G. Sturrock, P.M. Midgley, and A. McCulloch (2000). A history of chemically and radiatively important gases in air deduced from ALE/GAGE/AGAGE. *Journal of Geophysical Research*, 105, 17751–17792.

Raats, P.A.C. (1975). Distribution of salts in the crop root zone. *Journal of Hydrology*, 27, 237–248.

Radcliffe, E.R., and J. Šimůnek (2010). *Soil physics with HYDRUS: Modeling and applications*. Boca Raton, FL: CRC Press.

Reichstein, M., E. Falge, D. Baldocchi, D. Papale, M. Aubinet, P. Berbigier, C. Bernhofer, N. Buchmann, T. Gilmanov, A. Granier, T. Grünwald, K. Havránková, H. Ilvesniemi, D. Janous, A. Knohl, T. Laurila, A. Lohila, D. Loustau, G. Matteucci, T. Meyers, F. Miglietta, J.-M. Ourcival, J. Pumpanen, S. Rambal, E. Rotenberg, M. Sanz, J. Tenhunen, G. Seufert, F. Vaccari, T. Vesala, D. Yakir, and R. Valentini (2005). On the separation of net ecosystem exchange into assimilation and ecosystem respiration: Review and improved algorithm. *Global Change Biology*, 11, 1424–1439.

Reynolds, O. (1895). On the dynamical theory of incompressible viscous fluids and the determination of the criterion. *Philosophical Transactions of Royal Society*, 186, 123–161.

Rhoades, J.D., A. Kandiah, and A.M. Mashali (1992). *The use of saline water for crop production*. Rome: FAO, Irrigation and Drainage Paper 48.

Rijtema, P.E. (1965). *An analysis of actual evapotranspiration*. Agricultural Research Report 659. Wageningen: Pudoc.

Rijtema, P.E., P. Groenendijk, and J.G. Kroes (1997). *ANIMO, a dynamic simulation model for transport and transformation of nutrients and organic materials in soils*. Wageningen: DLO Winand Staring Centre, Report 30.

Riou, C. (1982). Une expression analytique du flux de chaleur sensible en conditions suradiabatiques à partir de mesures du vent et de la température à deux niveaux. *Journal de Recherches Atmosphériques*, 16, 15–22.

Riseborough, D., N. Shiklomanov, B. Etzelmüller, S. Gruber, and S. Marchenko (2008). Recent advances in Permafrost modelling. *Permafrost and Periglacial Processes*, 19, 137–156.

Ritsema, C.J., J.C. van Dam, L.W. Dekker, and K. Oostindie (2005). A new modeling approach to simulate preferential flow and transport in water repellent porous media: Model structure and validation. *Australian Journal of Soil Research*, 43, 361–369.

Ritzema, H.P. (1994). *Drainage principles and applications*, 2nd ed. Wageningen: ILRI, Publication 16.

Rodriguez-Iturbe, I., and A. Porporato (2004). *Ecohydrology of water-controlled ecosystems: soil moisture and plant dynamics*. Cambridge: Cambridge University Press.

Ronda, R.J., H.A.R. de Bruin, and A.A.M. Holtslag (2001). Representation of the canopy conductance in modeling the surface energy budget for low vegetation. *Journal of Applied Meteorology*, 40, 1431–1444

Roth, C.H., M.A. Malicki, and R. Plagge (1992). Empirical evaluation of the relationship between soil dielectric constant and volumetric water content as the basis for calibrating soil moisture measurements by TDR. *Journal of Soil Science*, 43, 1–13.

Rothman, L.S., et al. (2009). The HITRAN 2008 molecular spectroscopic database. *Journal of Quantitative Spectroscopy and Radiative Transfer*, 110, 533–572.

Rutter, A., A. Morton, and P. Robins (1975). A predictive model of rainfall interception in forests. II Generalization of the model and comparison with observations in some coniferous and hardwood stands. *Journal of Applied Ecology*, 12, 367–380.

Santanello, J.A., and M.A. Friedl (2003). Diurnal covariation in soil heat flux and net radiation. *Journal of Applied Meteorology*, 42, 851–862.

Sarwar, A., W.M.G. Bastiaanssen, Th.M. Boers, and J.C. van Dam (2000). Evaluating drainage design parameters for the fourth drainage project, Pakistan by using the SWAP model: Part 1: Calibration. *Irrigation and Drainage Systems*, 14, 257–280.

Sauer, T.J., D.W. Meek, T.E. Ochsner, A R. Harris, and R. Horton (2003). Errors in heat flux measurement by flux plates of contrasting design and thermal conductivity. *Vadose Zone Journal*, 2, 580–588.

Schaik, N.L.M.B., R.F.A. Hendriks, and J.C. van Dam (2010). Parameterization of macropore flow using dye-tracer infiltration patterns in the SWAP model. *Vadose Zone Journal*, 9, 95–106.

Schalkwijk, J., F. Bosveld, and A. Siebesma (2010). *Timescales and structures in vertical transport in the atmospheric boundary layer*. Technical Report WR-2010–02, KNMI.

Scharmer, K., and Greif, J., eds. (2000). *The European solar radiation atlas,* Vol. 2: *Database and exploitation software*. Paris: Les Presses de l' École des Mines.

Schmid, H.P. (1997). Experimental design for flux measurements: Matching scales of observations and fluxes. *Agricultural and Forest Meteorology*, 87, 179–200.

Schmidt, W. (1921). Wird die Luft durch Konvektion von der Erdoberfllche her erwarmt? *Meterologikal Zeitschrift*, 38, 262.

Schröder, T., M. Javaux, J. Vanderborght, B. Körfgen, and H. Vereecken (2009). Implementation of a microscopic soil-root hydraulic conductivity drop function in a three dimensional soil-root architecture water transfer model. *Vadose Zone Journal*, 8, 783–792.

Schüttemeyer, D., A.F. Moene, A.A.M. Holtslag, H.A.R. de. Bruin, and N. van de Giesen (2006). Surface fluxes and characteristics of drying semi-arid terrain in West Africa. *Boundary Layer Meteorology*, 118, 583–612.

Schüttemeyer, D., Ch. Schillings, A.F. Moene, and H.A.R. de Bruin (2007). Satellite-based actual evapotranspiration over drying semiarid terrain in West-Africa. *Journal of Applied Meteorology and Climatology*, 46, 97–111.

Schuurmans, J., P.A. Troch, A. Veldhuizen, W.G.M. Bastiaanssen, and M. Bierkens, (2003). Assimilation of remotely sensed latent heat flux in a distributed hydrological model. *Advanced Water Resources*, 26, 151–159.

Scott, H.D. (2000). *Soil physics: Agricultural and environmental applications*. Ames: Iowa State University Press.

Sellers, P.J. (1985). Canopy reflectance, photosynthesis and transpiration. *International Journal of Remote Sensing*, 6, 1335–1372.

Sellers, P.J., D.A. Randall, G.J. Collatz, J.A. Berry, C.B. Field, D.A. Dazlich, C. Zhang, G.D. Collelo, and L. Bounoua (1996). A revised land surface parameterization (SiB2) for atmospheric GCMS. Part I: Model formulation. *Journal of Climate*, 9, 676–705.

Sellers, P.J., R.E. Dickinson, D.A. Randall, A.K. Betts, F.G. Hall, J.A. Berry, G.J. Collatz, A.S. Denning, H.A. Mooney, C.A. Nobre, N. Sato, C.B. Field, and A. Henderson-Sellers (1997). Modelling the exchanges of energy, water and carbon between continents and the atmosphere. *Science*, 275, 502–509.

Seneviratne, S.I., D. Lüthi, M. Litschi, and C. Schär (2006). Land–atmosphere coupling and climate change in Europe, *Nature*, 443, 205–209.

Seth A., F. Giorgi, and R.E. Dickinson (1994). Simulating fluxes from heterogeneous land surface: Explicit subgrid method employing the biosphere-atmosphere transfer scheme (BATS). *Journal of Geophysical Research*, 99, 18651–18667.

Shuttleworth, W.J., and J.S. Wallace (1985). Evaporation from sparse crops – an energy combination theory. *Quarterly Journal of the Royal Meteorological Society*, 111, 839–855.

Šimůnek, J., M. Sejna, and M.Th. van Genuchten (1998a). *The HYDRUS-1D software package for simulating one-dimensional water, heat, and multiple solutes in variably saturated media*, Version 2.0. Riverside, CA: US Salinity Laboratory.

Šimůnek, J., R. Angula-Jaramillo, M.G. Schaap, J.-P Vabdervaere, and M.Th. van Genuchten (1998b). Using an inverse method to estimate the hydraulic properties of crusted soils from tension disc infiltrometer data. *Geoderma*, 86, 61–81.

Šimůnek, J., M.Th. van Genuchten, M. Sejna, N. Toride, and F.J. Leij (1999). *The STANMOD computer software for evaluating solute transport in porous media using analytical solutions of convection-dispersion equation*. Riverside, CA: US Salinity Laboratory.

Singh, R. (2005). *Water productivity from field to regional scale: Integration of crop and soil modelling, remote sensing and geographical information*. Unpublished PhD thesis, Wageningen University.

Singh, R., J.G. Kroes, J.C. van Dam, and R.A. Feddes (2006a). Distributed ecohydrological modelling to evaluate the performance of irrigation system in Sirsa district, India. I. Current water management and its productivity. *Journal of Hydrology*, 329, 692–713.

Singh, R., R.K. Jhorar, J.C. van Dam, and R.A. Feddes (2006b). Distributed ecohydrological modelling to evaluate the performance of irrigation system in Sirsa district, India. II. Impact of alternative water management scenarios. *Journal of Hydrology*, 329, 714–723.

Skaggs, T.H., M.T. van Genuchten, P.J. Shouse, and J.A. Poss (2006). Macroscopic approaches to root water uptake as a function of water and salinity stress. *Agricultural Water Management*, 86, 140–149.

Smith, M. (1992). *CROPWAT: A computer program for irrigation planning and management*. Rome, FAO, Irrigation and Drainage Paper 46.

Smith, R.E. (2002). *Infiltration theory for hydrologic applications*. Washington, DC: American Geophysical Union, Water Resources Monograph 15.

Smits, K.M., T. Sakaki, A. Limsuwat, and T.H. Illangasekare (2010). Thermal conductivity of sands under varying moisture and porosity in drainage–wetting cycles. *Vadose Zone Journal*, 9, 1–9.

Spitters, C.J.T., H. van Keulen, and D.G.W. van Kraalingen (1989). A simple and universal crop growth simulator: SUCROS 87. In R. Rabbinge, S.A. Ward and H.H. van Laar (eds.), *Simulation and systems management in crop protection* (pp. 147–181). Wageningen: Pudoc.

Stanhill, G., J.T.H. Cox, and S. Moreshet (1968). The effect of crop and climatic factors on the radiation balance of an irrigated maize crop. *Journal of Applied Ecology*, 5, 707.

Stefan, J., (1889). Über die Theorien des Eisbildung insbesondere über die Eisbildung in Polarmeere. *Annalen der Physik*, 278, 269–286.

Stewart, J. B. (1988). Modeling surface conductance of pine forest. *Agricultural and Forest Meteorology*, 43, 19–35.

Stewart, J.B., W.P. Kustas, K.S. Humes, W.D. Nichols, M.S. Moran, and H.A.R. de Bruin (1994). Sensible heat flux-radiometric surface temperature relationship for eight semiarid areas. *Journal of Applied Meteorology*, 33, 1110–1117.

Stull, R.B. (1988). *An introduction to boundary-layer meteorology*. Dordrecht: Kluwer Academic.

Supit, I., A.A. Hooyer, and C.A. van Diepen (1994). *System description of the WOFOST 6.0 crop simulation model implemented in CGMS.* Vol. 1: *Theory and algorithms.* Luxembourg: Agricultural series, EUR publication 15956.

Tappeiner, U., and A. Cernusca (1989). Canopy structure and light climate of different alpine plant communities: Analysis by means of a model. *Theoretical and Applied Climatology*, 40, 81–92.

Ten Berge, H.F.M. (1986). *Heat and water transfer at the bare soil surface: Aspects affecting thermal images.* Unpublished PhD thesis, Wageningen University.

Tennekes, H., and Lumley J.L. (1972). *A first course in turbulence.* Cambridge, MA: MIT Press.

Teuling, A.J., S.I. Seneviratne, C. Williams, and P.A. Troch (2006). Observed timescales of evapotranspiration response to soil moisture. *Geophysical Research Letters*, 33, L23403.

Teuling, A.J., M. Hirschi, A. Ohmura, M. Wild, M. Reichstein, P. Ciais, N. Buchmann, C. Ammann, L. Montagnani, A. D. Richardson, G. Wohlfahrt, and S. I. Seneviratne (2009). A regional perspective on trends in continental evaporation. *Geophysical Research Letters*, 36, L02404.

Teuling, A.J., S.I. Seneviratne, R. Stöckli, M. Reichstein, E. Moors, P. Ciais, S. Luyssaert, B. van den Hurk, C. Ammann, C. Bernhofer, E. Dellwik, D. Gianelle, B. Gielen, T. Grünwald, K. Klumpp, L. Montagnani, C. Moureaux, M. Sottocornola, and G. Wohlfahrt (2010). Contrasting response of European forest and grassland energy exchange to heatwaves. *Nature Geosciences*, 3, 722–727.

Thom, A.S. (1971). Momentum absorption by vegetation. *Quarterly Journal of the Royal Meteorological Society*, 97, 414–428.

Tiktak, A., F. van den Berg, J.J.T.I. Boesten, M. Leistra, A.M.A. van der Linden, and D. van Kraalingen (2000). *Pesticide Emission at Regional and Local scales: Pearl version 1.1 User Manual.* Zeist: RIVM report 711401008.

Tol, C. van der (2000). *Soil evaporation and plant transpiration as simulated with the FAO 56-method and the agrohydrological model SWAP.* Unpublished MSc thesis, Wageningen University.

Topp, G.C., J.L. Davis, and A.P. Annan (1980). Electromagnetic determination of soil water content: Measurement in coaxial transmission lines. *Water Resources Research*, 16, 574–582.

Trier, S.B., F. Chen, K.W. Manning, M.A. LeMone, and C.A. Davis (2008). Sensitivity of the PBL and precipitation in 12-day simulations of warm-season convection using different land surface models and soil wetness conditions. *Monthly Weather Review*, 136, 2321–2343.

Utset, A., F. Imma, A. Martinez-Cob, and J. Cavero (2004). Comparing Penman-Monteith and Priestley-Taylor as reference evapotranspiration inputs for modeling maize water use under Mediterranean conditions. *Agricultural Water Management*, 66, 205–219.

Valente, F., J.S. Daid, and J.H.C. Gash (1997). Modelling interception loss for two sparse eucalypt and pine forests in central Portugal using reformulated Rutter and Gash analytical models. *Journal of Hydrology*, 190, 141–162.

Van Dam, J.C., and R.A. Feddes (2000). Numerical simulation of infiltration, evaporation and shallow groundwater levels with the Richards' equation. *Journal of Hydrology*, 233, 72–85.

Van Dam, J.C., and R.S. Malik (2003). *Water productivity of irrigated crops in Sirsa district, India. Integration of remote sensing, crop and soil models and geographical information systems.* Wageningen: WATPRO final report.

Van Dam, J.C., J.N.M. Stricker, and P. Droogers (1994). Inverse method to determine soil hydraulic functions from multi-step outflow experiments. *Soil Science Society of America Journal*, 58, 647–652.

Van Dam, J.C., P. Groenendijk, R.F.A. Hendriks, and J.G. Kroes (2008). Advances of modeling water flow in variably saturated soils with SWAP. *Vadose Zone Journal*, 7, 640–653.

Van de Berg, F., and J.J.T.I. Boesten (1998). *Pesticide leaching and accumulation model (PESTLA) version 3.4; description and user's guide*. Wageningen: Alterra Green World Research, Technical Document 43.

Van de Pol, R.M., P.J. Wierenga, and D.R. Nielsen (1977). Solute movement in a field soil. *Soil Science Society of America Journal*, 41, 10–13.

Van de Wiel, B.J.H., A.F. Moene, G.J. Steeneveld, O.K. Hartogensis, and A.A.M. Holtslag (2007). Predicting the collapse of turbulence in stably stratified boundary layers. *Flow, Turbulence and Combustion*, 79, 251 – 274.

Van de Wiel, B.J.H., A.F. Moene, W.H. de Ronde, and H.J.J. Jonker (2008). Local similarity in the stable boundary layer and mixing-length approaches: Consistency of concepts. *Boundary-Layer Meteorology*, 128, 103–116.

Van de Wiel, B.J.H., A.F. Moene, and H.J.J. Jonker (2012a). The cessation of continuous turbulence as precursor of the very stable nocturnal boundary layer. *Journal of the Atmospheric Science*, 69, 3097–3115.

Van de Wiel, B.J.H., A.F. Moene and H.J.J. Jonker, P. Baas, S. Basu, J.M.M Donda, J. Sun, and A.A.M. Holtslag (2012b). The minimum wind speed for sustainable turbulence in the nocturnal boundary layer. *Journal of Atmospheric Sciences*, 69, 3116–3127.

Van der Zee, S.E.A.T.M., and W.H. van Riemsdijk (1987). Transport of reactive solute in spatially variable soil systems. *Water Resources Research*, 23, 2059–2069.

Van Dijk, A., A.F. Moene, and H.A.R. DeBruin (2004). *The principles of surface flux physics: theory, practice and description of the ECPACK library*. Internal Report 2004/1, Meteorology and Air Quality Group, Wageningen University, Wageningen, the Netherlands, 99 pp.

Van Genuchten, M.Th. (1980). A closed form equation for predicting the hydraulic conductivity of unsaturated soils. *Soil Science Society of America Journal*, 44, 892–898.

Van Genuchten, M.Th., and P.J. Wieringa (1974). *Simulation of one-dimensional solute transfer in porous media*. New Mexico: New Mexico State University Agricultural Experimental Station Bulletin 628.

Van Genuchten, M.Th., and R.W. Cleary (1979). Movement of solutes in soil: Computer simulated and laboratory results. In G.H. Bolt (ed.), *Soil chemistry B: Physico-chemical models* (pp. 349–386). Amsterdam: Elsevier.

Van Heerwaarden, C.C., J. Vilà-Guerau de Arellano, A.F. Moene, and A.A.M. Holtslag (2009). Interactions between dry-air entrainment, surface evaporation and convective boundary layer development. *Quarterly Journal of the Royal Meteorological Society*, 135, 1277–1291.

Van Wijk, W.R., and D.A. de Vries (1954). Evapotranspiration. *Netherlands Journal of Agricultural Science*, 2, 105–118.

Van Wijk, W.R., and D.A. de Vries (1963). Periodic temperature variations in a homogeneous soil. In W.R. van Wijk (ed.), *Physics of plant environment* (pp. 102–143). Amsterdam: North-Holland.

Van Wijk, A.L.M., R.A. Feddes, J.G. Wesseling, and J. Buitendijk (1988). *Effecten van grondsoort en ontwatering op opbrengst van akkerbouwgewassen. Een evaluatie over 30 jaren van de opbrengst van aardappelen en zomergraan op acht bodemprofielen bij vijftien combinaties van ontwateringsdiepte en -intensiteit*. Wageningen: Instituut voor Cultuurtechniek en Waterhuishouding (ICW), Rapport 31.

Vazifedoust, M., J.C. van Dam, W.G.M. Bastiaanssen, and R.A. Feddes (2009). Assimilation of satellite data into agrohydrological models to improve crop yield forecasts. *International Journal of Remote Sensing*, 30, 2523–2545.

Verhoef, A., B.J.J.M van den Hurk, A.F.G Jacobs, and B.G. Heusinkveld (1996). Thermal soil properties for vineyard (EFEDA-I) and savanna (HAPEX-Sahel) sites. *Agricultural and Forest Meteorology*, 78, 1–18.

Vickers, D., C. Thomas, and B.E. Law (2009). Random and systematic CO_2 flux sampling errors for tower measurements over forests in the convective boundary layer. *Agricultural and Forest Meteorology*, 149, 73–83.

Vilà-Guerau de Arellano, J., and C.C. van Heerwaarden (forthcoming). *Atmospheric boundary layer: Integrating air chemistry and land interactions*. Cambridge: Cambridge University Press.

Viterbo P., and A. C. M. Beljaars (1995). An improved land surface parameterization scheme in the ECMWF model and its validation. *Journal of Climate*, 8, 2716–2748.

Von Caemmerer, S., and G.D. Farquhar (1981). Some relationships between the biochemistry of photosynthesis and the gas exchange of leaves. *Planta*, 153, 376–387.

Von Hoyningen-Hüne, J. (1983). Die Interception des Niederschlags in landwirtschaftlichen Beständen. *Schriftenreihe des DVWK*, 57, 1–53.

Voogt, M., B.J.J.M. van den Hurk, and C. Jacobs (2006). *The ECMWF land surface scheme extended with a photosynthesis and LAI module tested for a coniferous site*. De Bilt, Royal Dutch Meteorological Institute, report WR-06–02.

Voronovich, V., and G. Kiely (2007). On the gap in the spectra of surface-layer atmospheric turbulence. *Boundary-Layer Meteorology*, 122, 67–83.

Walker, J., and P. Houser (2001). A methodology for initializing soil moisture in a global climate model: Assimilation of near-surface soil moisture observations. *Journal of Geophysical Research*, 106, 11,761–11,774.

Warrick, A.W. (2002). *Soil physics companion*. Boca Raton, FL: CRC Press.

Warrick, A.W. (2003). *Soil water dynamics*. New York: Oxford University Press.

Wallace, J.M., and P.V. Hobbs (2006). *Atmospheric science: An introductory survey*, 2nd ed. Amsterdam: Academic Press.

Walsum, P.E.V. van, and P. Groenendijk (2008). Quasi steady-state simulation of the unsaturated zone in groundwater modelling of lowland regions. *Vadose Zone Journal*, 7, 769–781.

Wan, F., and F. Porté-Agel (2011). A large-eddy simulation study of turbulent flow over multiscale topography. *Boundary-Layer Meteorology*, 141, 201–217.

Warrilow, D.A. (1986). *Indications of the sensitivity of European climate to land use variation using a one-dimensional model* (pp. 156–159). Rome: Proceedings ISLSCP conference.

Webb, E., Pearman, G., and Leuning, R. (1980). Correction of flux measurements for density effects due to heat and water vapour transfer. *Quarterly Journal of Royal Meteorological Society*, 106, 85–100.

Wehrli, C. (1985). *Extraterrestrial solar spectrum*. Davos: Physikalisch-Meteorologisches Observatorium + World Radiation Center (PMO/WRC), Publication no. 615.

Weiss, A. D.L. Lukens, J.M. Norman, and J.R. Steadman (1989). Leaf wetness in dry beans under semi-arid conditions. *Agricultural and Forest Meteorology*, 48, 149–162.

Wendroth, O., W. Ehlers, J.W. Hopmans, H. Kage, J. Halbertsma, and J.H.M. Wösten (1993). Reevaluation of the evaporation method for determining hydraulic functions in unsaturated soils. *Soil Science Society of America Journal*, 57, 1436–1443.

Wesseling, J.G. (1998). *Some equations and the computer program Helena for the calculation of daily evaporation*. Wageningen: Alterra, Technical document 48.

Wichink Kruit, R.J., A.F.G. Jacobs, and A.A.M. Holtslag (2008). Measurements and estimates of leaf wetness over agricultural grassland for dry deposition modeling of trace gases. *Atmospheric Environment*, 42, 5304–5316.

Wieringa, J. (1993). Representative roughness parameters for homogeneous terrain. *Boundary Layer Meteorology*, 63, 323–363.
Willmer, C., and M. Fricker (1996). *Stomata*, 2nd ed. London: Chapman and Hall.
Wilson, D.K. (2001). An alternative function for the wind and temperature gradients in unstable surface layers. *Boundary Layer Meteorology*, 99, 151–158.
Wilson, J.D. (2008). Monin-Obukhov functions for standard deviations of velocity. *Boundary-Layer Meteorology*, 129, 353–369.
WMO (2008). *WMO guide to meteorological instruments and methods of observation*, 7th ed. Geneva, Switzerland: WMO, No. 8.
Wolf, J., A.H.W. Beusen, P. Groenendijk, T. Kroon, R. Rötter, and H. van Zeijts (2003). The integrated modeling system STONE for calculating emissions from agriculture in the Netherlands. *Environmental Modeling and Software*, 18, 597–617.
Wong, S.-C., I.R. Cowan, and G.D. Farquhar (1985). Leaf conductance in relation to rate of CO_2 assimilation – I. Influence of nitrogen nutrition, phosphorus nutrition, photon flux density, and ambient partial pressure of CO_2 during ontogeny. *Plant Physiology*, 78, 821–825.
Wösten, J.H.M., A. Lilly, A. Nemes, and C. Le Bas (1998). *Using existing soil data to derive hydraulic parameters for simulation models in environmental studies and in land use planning*. Wageningen: Alterra Green World Research, Report 156.
Wösten, J.H.M., G.J. Veerman, W.J.M. de Groot, and J. Stolte (2001). *Waterretentie- en doorlatendheidskarakteristieken van boven- en ondergronden in Nederland. De Staringreeks*. Wageningen: Alterra Green World Research, Report 153.
Wyngaard, J.C. (2010). *Turbulence in the atmosphere*. Cambridge: Cambridge University Press.
Xiao, H., R. Meissner, J. Seeger, H. Rupp, and H. Borg (2009). Effect of vegetation type and growth stage on dewfall, determined with high precision weighing lysimeters at a site in northern Germany. *Journal of Hydrology*, 377, 43–49.
Xiu, A., and J.E. Pleim (2001). Development of a land surface model. I – Application in a mesoscale meteorological model. *Journal of Applied Meteorology*, 40, 192–209.
Yates, S.R., M.Th. van Genuchten, A.W. Warrick, and F.J. Leij (1992). Analysis of measured, predicted and estimated hydraulic conductivity using the RETC computer program. *Soil Science Society of America Journal*, 56, 347–354.
Zangvil, A. (1996). Six years of dew observations in the Negev Desert, Israel. *Journal of Arid Environments*, 32, 361–371.
Zelik, M., S.A. Gregory, and E.V.P. Smith (1992). *Introductory astronomy and astrophysics*, 3rd ed. Fort Worth, TX: Saunders College.
Zeng, X. (2001). Global vegetation root distribution for land modeling. *Journal of Hydrometeorology*, 2, 525–530.
Zhang, H., and P.S. Nobel (1996). Dependency of c_i/c_a and leaf transpiration efficiency on the vapour pressure deficit. *Australian Journal of Plant Physiology*, 232, 561–568.
Zilitinkevich, S.S., T. Elperin, N. Kleeorin, and I. Rogachevskii (2007). Energy-and flux-budget (EFB) turbulence closure model for stably stratified flows. Part I: steady-state, homogeneous regimes. *Boundary-Layer Meteorology*, 125, 167–191.
Zilitinkevich, S. S., T. Elperin, N. Kleeorin, I. Rogachevskii, and I. Esau (2013). A hierarchy of energy- and flux-budget (EFB) turbulence closure models for stably-stratified geophysical flows. *Boundary-Layer Meteorology*, 146, 341–373.

Index

Printed in the United States
By Bookmasters